THE
INTERNATIONAL SERIES
OF
MONOGRAPHS ON PHYSICS

GENERAL EDITORS

W. MARSHALL   D. H. WILKINSON

NUCLEAR SIZES AND STRUCTURE

BY

ROGER C. BARRETT AND DAPHNE F. JACKSON

CLARENDON PRESS · OXFORD

1977

Oxford University Press, Walton Street, Oxford OX2 6DP
OXFORD LONDON GLASGOW NEW YORK
TORONTO MELBOURNE WELLINGTON CAPE TOWN
IBADAN NAIROBI DAR ES SALAAM LUSAKA ADDIS ABABA
KUALA LUMPUR SINGAPORE JAKARTA HONG KONG TOKYO
DELHI BOMBAY CALCUTTA MADRAS KARACHI

ISBN 0 19 851272 4

Printed in Great Britain
by Thomson Litho Ltd,
East Kilbride, Scotland

PREFACE

When we first discussed the need for a new book on nuclear sizes in 1970, nine years after the publication of Elton's *Nuclear sizes*, the topic had become established as a major area of physics. There had been a number of conferences devoted entirely to such topics as 'Electromagnetic sizes of nuclei' and 'Intermediate energy physics', and a continuing series with the title 'High energy physics and nuclear structure'. A number of review articles and books had appeared on subjects such as electron scattering, optical isotope shifts, muonic atoms, nucleon-nucleus scattering, pionic atoms, and theories of the nuclear ground state.

It appeared that the main task would be to draw together established methods and to show the relation between results obtained from the various electromagnetic and strong interaction processes. By the time we had actually started work it was evident that the subject was going through a phase of major development and renewal. New experimental techniques and completely new accelerators have produced fresh data with vastly increased accuracy. This has completely changed our expectation of what can be deduced about nuclear sizes and shapes. At the same time new methods of analysis have led to a much better awareness of what properties are actually being studied in a particular process and what uncertainties are associated with the parameters deduced from the analyses. In response to these developments we have endeavoured to cover the study of nuclear sizes and shapes by means of a very wide range of electromagnetic, strong, and weak interaction processes, and have placed particular emphasis on the progress in experimental measurement and in theoretical techniques used to predict nuclear behaviour or to extract detailed information from the measurements.

A principal aim of the book is to explain precisely which properties of the nucleus can be obtained from the various experiments and to indicate the extent to which previously published parameters are really determined by the measurements. As far as possible we have tried to include sufficient formalism and background information to serve as an introduction to each topic for postgraduate students. We have taken critical discussion of results and comparison of methods to be of greater importance than the tabulation of parameters, although many results of analyses are presented in tables and figures.

The major part of this book was completed early in 1975 although a very few topics have been extended to include work reported in 1976. This means that the treatment of certain fast-developing topics will inevitably appear inadequate by the time the book is in the hands of the reader. Almost every week a new preprint arrives which could lead to an additional comment or revision. We hope that the shortcomings of the book in this respect will be taken as an indication of the vitality of the subject and act as a stimulus to further interest and work in the various fields.

Guildford, Surrey
September 1976

R.C.B.
D.F.J.

ACKNOWLEDGEMENTS

It is a pleasure to acknowledge the interest and help of many physicists who have sent information about their work and made detailed comments on parts of the book, especially C.J. Batty, J.S. Blair, M.Y. Chen, R.C. Johnson, F. Lenz, J. Matthews, J.W. Negele, D.O. Newton, E.F. Redish, I. Sick, D.W.L. Sprung, D.N. Stacey, M. Veneroni, L. Wilets, and C.S. Wu.

We should like especially to thank Sir Denys Wilkinson for his encouragement and advice, S. Murugesu for carrying out a very thorough literature survey on hadronic atoms, and Barbara Barrett for help in clarifying parts of the text and carrying out the formidable task of collating and typing the references.

We are indebted to the Institute of Physics for permission to use material contained in review articles written by us, and to the Institute of Physics, the American Institute of Physics, the French Physical Society, the Physical Society of Japan, Laval University Press, University of Toronto Press, Academic Press, Methuen, North-Holland Publishing Company, Plenum Press, and Springer-Verlag for permission to reproduce diagrams; to C.E. Dear for drawing some of the figures; and to Gay Shannon, Judith Smith, Sara Deane, and Mickey Fortuna for typing.

One of us (R.C.B.) carried out some of this work in Vancouver while on leave of absence from the University of Surrey, and acknowledges the hospitality of the University of British Columbia.

CONTENTS

1
INTRODUCTION AND DEFINITIONS

It will be seen that this theory makes the radius of the uranium nucleus very small, about 7×10^{-13} cm,.... It sounds incredible but may not be impossible.

Rutherford 1929

1.1 INTRODUCTION TO THE STUDY OF NUCLEAR SIZES

The determination of nuclear shapes and sizes is one of the traditional problems of nuclear physics. The extent to which we are able to make precise and meaningful statements about the nuclear matter distribution and the nuclear charge distribution and the variation in both quantities from one nucleus to another reveals quite clearly the state of our understanding of much more fundamental issues, such as the nature of the interactions between various types of particles and the role of these interactions in scattering phenomena, the subtle balance between various features of the nucleon-nucleon interactions in bound states, and the difference between the average properties of nuclei described by macroscopic models and the specific nuclear structure properties described by microscopic models.

The study of nuclear sizes involves both the study of the nuclear charge distribution by means of processes dominated by electromagnetic interactions and the study of the nuclear matter distribution by means of strong-interaction processes. By combining the information so obtained, comparison of the proton and neutron distributions can be made. Most of the discussion will be devoted to the determination of the radial shape of the distributions in spherical nuclei, but the angular dependence of the shape of nuclei which are not spherical will also be considered. One of our principal aims will be to try to determine and explain precisely which properties of the relevant distributions can be obtained from the various experiments and to indicate the extent to which previously published parameters are really determined by the measurements as opposed to being merely consistent with them.

The interaction between charged leptons (i.e. electrons, positrons, and muons) and nucleons consists of an electromagnetic and a weak term, but the latter has a completely negligible effect in the processes considered here. The first direct determination of nuclear charge radii came from the measurements on elastic electron scattering by Lyman, Hansen, and Scott (1951). This was many years after the suggestion (Guth 1934) that, for fast electrons, the finite size of the nuclear charge distribution would produce large deviations from the differential cross-section (Mott 1929) for elastic scattering from a point charge. Experimental techniques subsequently improved very rapidly, and electron scattering has been extensively used in the study of nuclear charge distributions and other properties of nuclei. Two years after the experiments of Lyman *et al.*, Fitch and Rainwater (1953) measured the energies of X-rays from muonic atoms, which provided another means of studying nuclear charge radii. The idea of stopping negative muons in matter to form muonic atoms and using the X-ray energies to obtain nuclear size information was originally due to Wheeler (1947). The accuracy of experimental measurements on muonic atoms improved slowly at first and then very rapidly after about 1960. A much older method exists for the determination of differences between nuclear charge radii of different isotopes of an element. This involves measurement of the isotope shift of spectral lines in electronic (i.e. ordinary) atoms. A similar shift, called the isomer shift, gives the change in the charge radius when a long-lived nuclear state is excited. The nuclear size contribution to the shift was first pointed out by Rosenthal and Breit (1932) long after the first measurement of an isotope shift in which finite nuclear size effects made a substantial contribution had been done by Merton (1919). The electromagnetic interaction with the nuclear magnetic moment is observable in elastic electron scattering at 180°, in inelastic electron scattering, and in hyperfine splitting of certain atomic levels, and it is therefore possible to determine the magnetic moment distribution of nuclei.

It is now known that the proton and neutron each have an intrinsic electromagnetic structure and that processes such as high-energy electron-nucleon scattering cannot be described adequately in terms of a point charge and a point magnetic moment. The theory of the electromagnetic structure of nucleons will not be discussed here, and the reader is referred to reviews such as that by Drell and Zachariasen (1961). It will, however, be necessary to use the information so far obtained for nucleons in order to make a connection between the predictions of theories for nuclear distributions due to point nucleons and the observed nuclear charge and magnetic moment distributions due to nucleons with finite electromagnetic size.

The effect of a strong-interaction radius for nuclei was seen in early experiments on α-particle scattering. The scattering data, together with the results from α-decay of heavy nuclei, were interpreted in terms of an attractive nuclear potential plus a repulsive Coulomb barrier (Rutherford 1929) and an estimate of the nuclear potential radius was obtained. Following the development of the cyclotron, the energy dependence of α-particle scattering from heavy nuclei was studied up to 40 MeV (Farwell and Wegner 1954) and the abrupt departure from pure Coulomb scattering beyond a critical energy was interpreted in terms of a radius parameter (Blair 1954). This radius parameter cannot be directly interpreted in terms of a nuclear matter radius, since the range of the potential must be connected with the finite range of nuclear forces and the size of the projectile.

The description of nucleon scattering from nuclei in terms of a complex one-body potential was placed on a sound theoretical foundation by Feshbach, Porter, and Weisskopf (1954), who showed that such a potential can be related to the averaged or gross-structure properties of the compound nuclear system. They succeeded in reproducing neutron scattering data up to a few MeV with a complex square-well potential, but fits to angular distributions for 20-MeV protons required a potential with a diffuse surface resembling the surface of the nuclear charge distribution (Woods and Saxon 1954). For scattering at ~100 MeV, Serber (1947) had suggested that the collision of an incident nucleon with the nucleus could be interpreted in terms of collisions with individual nucleons, and Fernbach, Serber, and Taylor (1949) analysed total neutron cross-sections effectively with a square-well potential whose imaginary (absorptive) part was related to the total cross-sections for nucleon-nucleon scattering. The long mean free paths derived in such work were explained by Weisskopf (1951) in terms of the exclusion principle which limits nucleon-nucleon collisions in the nucleus to those in which bound nucleons are raised above the Fermi surface. The scattering of nucleons and pions from nuclei at energies ~1 GeV were interpreted in terms of a potential which was related in a fairly intuitive way to the nuclear matter distribution (Coor *et al.* 1955, Williams 1955, Abashian, Cool, and Cronin 1956).

In recent years the data obtained in many experiments on scattering of medium-energy nucleons have been analysed in terms of a complex potential and potential parameters have been determined, while analyses of the scattering of strongly absorbed projectiles, such as α-particles, in terms of diffraction models have yielded values of diffraction radii. In addition there have been many developments in the theory of potential scattering for nuclei and in our understanding of the connection between the potential, the nucleon-nucleon force, and the nuclear ground state (Feshbach 1958, 1962, Goldberger and Watson 1964). Some progress has been made in understanding the relation between diffraction radii and potential radii (Blair 1966).

As in the case of the study of electromagnetic interactions, our intention in the study of nuclear scattering is to establish what size parameters can be determined from the detailed fits to data, and to examine the relation between these parameters and the nuclear matter distribution.

In addition to elastic scattering, many other processes yield information on nuclear sizes, usually by somewhat indirect means. These processes include inelastic scattering, certain direct nuclear reactions, and high-energy processes such as pion production. The measurement of Coulomb energy differences between appropriate nuclear states and of X-ray energies in π-mesonic and K-mesonic atoms also yields valuable information. All these topics have been studied with increasing accuracy in recent years.

In the earliest years of the work mentioned above it was sufficient to regard the nucleus as an object with a sharp radius R and uniform density as shown in Fig. 1.1(a). By 1954

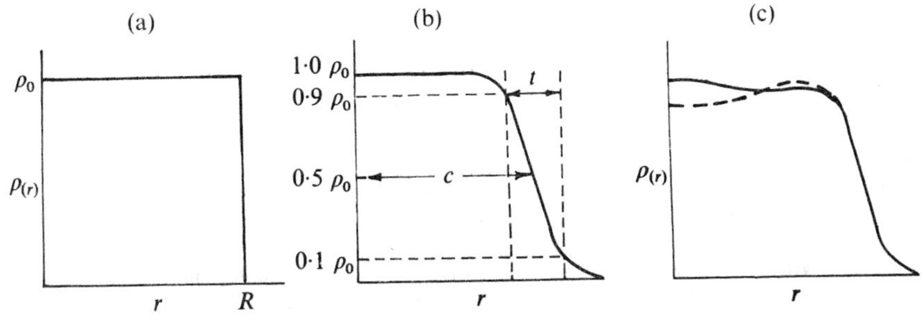

FIG. 1.1. Examples of nuclear distribution functions. (a) Billiard-ball nucleus with a well-defined radius R. (b) The Fermi distribution. (c) Shell model distributions. (From Jackson 1974.)

it was known that the nucleus had a diffuse surface which was quite well represented by the distribution shown in Fig. 1.1(b), and more recent work has suggested distributions of the form shown in Fig. 1.1(c). Thus it is no longer possible to define a single 'nuclear radius', and we are now concerned with the set of size parameters or spatial moments of certain distribution functions for nuclei.

1.2 DEFINITIONS

1.2.1 *Distribution functions and form factors*

In nuclear theory, we require density distributions or one-particle density functions for the protons, neutrons, and nucleons. There are distributions of *point* particles in the nucleus. If the ground state of a nucleus with Z protons and $N = A - Z$ neutrons is denoted by $|0\rangle$, then the proton distribution is given by

$$Z\,\rho_p(\underset{\sim}{r}) = \langle 0|\sum_{i=1}^{Z}\delta(\underset{\sim}{r}-\underset{\sim}{r}_i)|0\rangle \tag{1.1}$$

and the neutron distribution is given by

$$N\,\rho_n(\underset{\sim}{r}) = \langle 0|\sum_{i=1}^{N}\delta(\underset{\sim}{r}-\underset{\sim}{r}_i)|0\rangle \tag{1.2}$$

where the sums run only over protons or neutrons, respectively, and the normalization chosen here is

$$\int\rho_p(\underset{\sim}{r})\,d^3r = \int\rho_n(\underset{\sim}{r})\,d^3r = 1. \tag{1.3}$$

The matter or nucleon distribution is then given by

$$A\,\rho_m(\underset{\sim}{r}) = Z\,\rho_p(\underset{\sim}{r}) + N\,\rho_n(\underset{\sim}{r}). \tag{1.4}$$

In many situations involving electromagnetic interactions we actually require the nuclear charge distribution $\rho_{ch}(\underset{\sim}{r})$ of the nucleus instead of the distribution of point protons. Since the proton is not a point charge but has a finite size in the electromagnetic sense, ρ_{ch} can be obtained from ρ_p by folding in the charge distribution of the proton, i.e.

$$\rho_{ch}(\underset{\sim}{r}') = \int\rho_p(\underset{\sim}{r})\,\rho_d(|\underset{\sim}{r}-\underset{\sim}{r}'|)\,d^3r \tag{1.5}$$

where ρ_d represents the charge distribution of the proton also normalized to unity. It follows from the folding integral (1.5) that the r.m.s. radius of ρ_{ch} is increased relative to that of ρ_p, according to the formula

$$\langle r^2\rangle_{ch} = \langle r^2\rangle_p + \langle r^2\rangle_d, \tag{1.6}$$

the diffuseness is increased, and any irregularities in the shape of ρ_p are smoothed by the folding procedure.

It is also possible to define two-particle and higher density functions. These give the probability of finding a point nucleon at $\underset{\sim}{r}$ if there is another nucleon at $\underset{\sim}{r}'$, and so on. For example, the two-nucleon density function is given by

$$A(A-1)\ \rho_{\rm m}(\underset{\sim}{r},\underset{\sim}{r}') = \langle 0|\sum_{i\neq j}^{A} \delta(\underset{\sim}{r}-\underset{\sim}{r}_i)\ \delta(\underset{\sim}{r}'-\underset{\sim}{r}_j)\ |0\rangle. \qquad (1.7)$$

If $\rho_{\rm m}(\underset{\sim}{r},\underset{\sim}{r}') = \rho_{\rm m}(\underset{\sim}{r})\ \rho_{\rm m}(\underset{\sim}{r}')$, the system is said to be uncorrelated, but in general $\rho_{\rm m}(\underset{\sim}{r},\underset{\sim}{r}')$ is written as

$$\rho_{\rm m}(\underset{\sim}{r},\underset{\sim}{r}') = C(\underset{\sim}{r},\underset{\sim}{r}') + \rho_{\rm m}(\underset{\sim}{r})\ \rho_{\rm m}(\underset{\sim}{r}') \qquad (1.8)$$

where $C(\underset{\sim}{r},\underset{\sim}{r}')$ is the pair correlation function. In an extreme single-particle model only Pauli correlations arising from the exclusion principle are present, but in a more realistic model, e.g. including configuration mixing or clustering in the ground state, additional correlations of medium range arise. Short-range correlations due to the short-range behaviour of the nuclear force are also present.

The Fourier transforms of the various distributions are known as form factors. For example, the nuclear form factor is

$$F_{\rm m}(q) = \int \exp(i\underset{\sim}{q}\cdot\underset{\sim}{r})\ \rho_{\rm m}(\underset{\sim}{r})\ d^3r \qquad (1.9)$$

and the charge form factor is

$$F_{\rm ch}(q) = \int \exp(i\underset{\sim}{q}\cdot\underset{\sim}{r})\ \rho_{\rm ch}(\underset{\sim}{r})\ d^3r. \qquad (1.10)$$

Using eqn (1.5) for $\rho_{\rm ch}$ it follows that

$$F_{\rm ch}(q) = f_{\rm d}(q)\ F_{\rm p}(q) \qquad (1.11)$$

where $f_{\rm d}$, $F_{\rm p}$ are the form factors of a single proton and of the point protons in the nucleus, respectively.

1.2.2 *Some functional forms*

The distribution functions can be parametrized directly or indirectly. The direct parametrization involves the choice of a suitable functional form with parameters which may be varied to fit the experimental data. The functional form most widely used is the Fermi distribution

$$\rho(r) = \rho_0[1 + \exp\{(r-c)/a\}]^{-1} \tag{1.12}$$

shown in Fig. 1.1(b), where c is the halfway radius, a is the diffuseness parameter, and the constant ρ_0 is obtained from the normalization condition. The surface thickness t, which measures the distance over which the density falls from $0{\cdot}9\ \rho_0$ to $0{\cdot}1\ \rho_0$, is given by

$$t = 4{\cdot}39\ a \tag{1.13}$$

and the r.m.s. radius $\langle r^2\rangle^{1/2}$ is determined from the relation

$$\frac{5}{3}\langle r^2\rangle = c^2 + \frac{7}{3}\pi^2a^2. \tag{1.14}$$

The parabolic Fermi distribution has the form

$$\rho(r) = \left(1 + w\frac{r^2}{c^2}\right)\rho_0[1 + \exp\{(r-c)/a\}]^{-1} \tag{1.15}$$

where w is a constant which may be positive or negative and can cause a hump or depression near the origin. If w is negative, this density distribution must be modified so that it does not go negative but goes smoothly to zero. The modified gaussian distribution

$$\rho(r) = \rho_0[1 + \exp\{(r^2-c^2)/a^2\}]^{-1} \tag{1.16}$$

has been used, and also the trapezoidal distribution

$$\begin{aligned} \rho(r) &= \rho_0 &&, \quad r < c-b \\ &= \rho_0(c+b-r)/2b &&, \quad c-b \leqslant r \leqslant c+b \\ &= 0 &&, \quad r > c+b \end{aligned} \tag{1.17}$$

Other parametrizations have been used at various times for ρ_{ch} (Herman and Hofstadter 1960). The motivation for the introduction of various parametrizations is discussed in §3.4.1

1.2.3 *Radius parameters and moments*

The current approach to the study of nuclear distributions is to concentrate on the moments of the distribution rather than a single functional form, and a number of different procedures and notations have been adopted.

We consider a distribution

$$\rho(r) = \rho_0 \, f(r) \tag{1.18}$$

such that

$$f(0) = 1 \quad , \quad f(\infty) = 0. \tag{1.19}$$

The expectation values $\langle r^k \rangle$ are defined in the usual way as

$$\langle r^k \rangle = \frac{\int_0^\infty f(r) \, r^{k+2} \, dr}{\int_0^\infty f(r) \, r^2 \, dr}$$

so that $\langle r^2 \rangle$ is the mean square radius. The moments R_k defined by Ford and Wills (1969) are related to these expectation values by the formula

$$R_k = \left[\frac{1}{3}(k+3)\langle r^k \rangle\right]^{1/k} \tag{1.21}$$

Süssmann (1970) defines the volume moment

$$F_\nu = \int_0^\infty f(r) \, r^\nu \, dr \tag{1.22}$$

so that

$$\langle r^{\nu-2} \rangle = F_\nu / F_2. \tag{1.23}$$

This leads to the definition of the following radius parameters:

$$\text{central radius} \qquad C = F_0 \tag{1.24}$$

uniform radius $\quad U = (3F_2)^{1/3}$ (1.25)

quadratic radius $\quad Q = (5F_4/3F_2)^{1/2}$ (1.26)

For a symmetric distribution, such as the Fermi distribution, the central radius is clearly the same as the halfway radius, i.e. $C = c$. We have used the symbol U and the term uniform radius in preference to Süssmann's charge radius R in order to make the definition more general and avoid confusion with other quantities. The R_k of Ford and Wills are related to these new parameters by the relations

$$R_2 = Q = \left[\frac{5}{3}\langle r^2\rangle\right]^{1/2} \tag{1.27}$$

$$R_{-2} = (U^3/C)^{1/2} \tag{1.28}$$

The volume integral of the distribution is

$$J = 4\pi\,\rho_0\,F_2 = \frac{4\pi}{3}\,\rho_0\,U^3. \tag{1.29}$$

For the charge and matter distributions the value of the volume integral will be Z or A, respectively, or unity, depending on the normalization condition, and this determines the physical significance of the uniform radius U as the radius of the uniform sphere which contains the same amount of matter or charge as the real distribution.

Süssmann also defines the surface moments

$$G_\mu = [r^\mu] = \int_0^\infty g(r)\,r^\mu\,\mathrm{d}r \tag{1.30}$$

where

$$g(r) = -\frac{\mathrm{d}f}{\mathrm{d}r} \tag{1.31}$$

so that

$$C = [r]$$

$$U^3 = [r^3]$$

$$Q^2 = [r^5]/[r^3].$$

The surface width is defined as

$$b = (G_2 - G_1^2)^{1/2} = \{[r^2] - [r]^2\}^{1/2} \quad (1.32)$$

and determines the size of the nuclear surface. For the Fermi distribution (1.12) the surface width is

$$b = 3^{-1/2}\,\pi a \quad (1.33)$$

where a is the diffuseness parameter. General formulae expressing the parameters C and Q in terms of b and U have also been given by Süssmann.

Moments of some distributions derived from microscopic calculations are given in Table 1.1 and moments of Fermi distributions are given in Table 1.2. The values in the latter table are taken from the tabulation of Owen and Satchler (1963), for an arbitrary choice of parameters, and the corresponding distributions have not been fitted to any data. It is evident from these tables, as might be expected, that the more diffuse distributions show a wider spread in the values of the moments. This emphasizes the importance of defining clearly which radius parameter or moment is under discussion in any particular case.

Elton (1961*b*) introduced a basic length l which is related to the uniform radius by

$$l = A^{-1/3}\,U \quad (1.34)$$

and uses the symbol R to denote the radius of the uniform distribution which has the same r.m.s. radius as the true distribution so that $R = Q$. This quantity is also denoted by R_{EQ} and called the equivalent uniform radius. The usefulness of the concept of the basic length l depends on the view that all nuclei have essentially the same constant central density ρ_0 and essentially the same surface thickness. The physical basis of the role of the quadratic radius is less evident.

TABLE 1.1

Moments of distributions obtained from microscopic calculations using the Elton-Swift (ES), *Batty-Greenlees* (BG), *and Zaidi-Darmodjo* (ZD) *potentials*

	C	U	Q	$\langle r^2\rangle^{1/2}$
^{40}Ca protons (ES)	2.90	3.46	4.28	3.32
neutrons (ES)	2.82	3.38	4.20	3.25
protons (BG)	3.03	3.58	4.40	3.41
neutrons (BG)	3.05	3.63	4.45	3.44
neutrons (ZD)	2.75	3.28	4.08	3.16
^{48}Ca protons (ES)	2.93	3.51	4.25	3.29
neutrons (ES)	3.65	4.10	4.63	3.58
protons (BG)	2.97	3.57	4.38	3.39
neutrons (BG)	3.72	4.29	4.97	3.85
neutrons (ZD)	3.27	3.79	4.48	3.47
^{58}Ni protons (BG)	3.69	4.23	4.87	3.77
neutrons (BG)	4.06	4.45	5.05	3.91
neutrons (ZD)	3.63	3.98	4.58	3.55
^{60}Ni protons (BG)	3.70	4.24	4.87	3.77
neutrons (BG)	4.42	4.61	5.17	4.01
neutrons (ZD)	3.88	4.11	4.70	3.64
^{62}Ni protons (BG)	3.70	4.24	4.87	3.77
neutrons (BG)	4.61	4.76	5.26	4.08
neutrons (ZD)	4.13	4.20	4.76	3.69
^{120}Sn protons (BG)	(6.83)	(6.12)	5.93	4.59
neutrons (BG)	4.92	5.62	6.51	5.05
neutrons (ZD)	4.43	5.05	5.87	4.55
^{208}Pb protons (BG)	-	-	7.02	5.44
neutrons (BG)	-	-	7.84	6.06
neutrons (ZD)	-	-	7.01	5.43

TABLE 1.2

Moments of Fermi distributions†

A	a	C	U	Q
40	0.6	3.59	3.89	4.60
	0.5	3.59	3.81	4.32
	0.4	3.59	3.74	4.07
60	0.6	4.31	4.56	5.18
	0.5	4.31	4.48	4.93
	0.4	4.31	4.43	4.72
120	0.6	5.44	5.64	6.14
	0.5	5.44	5.58	5.94
	0.4	5.44	5.52	5.75
210	0.6	6.54	6.73	7.15
	0.5	6.54	6.67	6.98
	0.4	6.54	6.63	6.82

†From Owen and Satchler (1963).

Formulae relating C and Q to a and l are given by Elton for the Fermi distribution only and can also be deduced from the general formulae of Süssmann. It follows from this argument that U may be expected to increase as $A^{1/3}$, at least along the line of maximum β-stability, but that C and Q have a more complicated dependence on A.

The disadvantage of the view adopted by Süssmann and by Elton is that it implies a precise knowledge of the central density ρ_0, since

$$F_\nu = \rho_0^{-1} \int_0^\infty \rho(r)\, r^\nu \, dr$$

The matter density at $r=0$ is not determined at all by experiment, and the error in the charge distribution determined by electron scattering and muonic X-rays has its maximum value at the centre.† Also, if the distribution is not approxi-

†See §3.4.

mately monotonic ρ_0 must be defined as some average density in the interior of the nucleus, otherwise the results for the radii C and U are meaningless as is the case for the BG proton distribution in ^{120}Sn (see Table 1.1 and Fig. 2.2). For these reasons the generalized radial moments introduced by Ford and Rinker (1973) may be more useful. They define

$$\langle u(r)r^k \rangle = \int_0^\infty \rho(r)\, u(r)\, r^{k+2}\, dr \tag{1.35}$$

so that the equivalent radius of a uniform distribution with the same moment is given by

$$R_k^3 \langle u(r)r^k \rangle = 3 \int_0^{R_k} u(r)\, r^{k+2}\, dr \tag{1.36}$$

For the special case when $u(r) \equiv 1$ this reduces to eqn (1.21).

Freidrich and Lenz (1972) have proposed that the nuclear charge distribution can be described in terms of the average radius $R(Q)$ of the innermost fraction Q of the total charge, i.e.

$$R(Q) = \frac{4\pi}{Q} \int_0^a \rho_{ch}(r)\, r^3\, dr \tag{1.37}$$

$$Q(a) = 4\pi \int_0^a \rho_{ch}(r)\, r^2\, dr \tag{1.38}$$

where a is the radius of a sphere containing the fraction Q of the total charge. Thus the values of $R(Q)$ are the mean radii of the intervals in which the charge increases from zero to the fraction Q. If the function $R(Q)$ is known, the corresponding charge density $\rho_{ch}(r)$ can be derived from the equations

$$\frac{d}{dQ}\{Q(a)\, R(Q)\} = a \tag{1.39}$$

$$\rho_{ch}(a) = \left[4\pi a^2 \frac{da}{dQ}\right]^{-1}$$

This approach leads to a model-independent approach to charge distributions which is described in §3.4.2.

2
THEORIES OF THE NUCLEAR MATTER DISTRIBUTION

Fact is theory and fiction is experiment.
Wilets 1972

2.1 SOME SIMPLE IDEAS

The nuclear matter distribution is given, according to eqn (1.4), by

$$\rho_m(\underset{\sim}{r}) = \frac{Z}{A}\rho_p(\underset{\sim}{r}) + \frac{N}{A}\rho_n(\underset{\sim}{r}). \qquad (2.1)$$

The simplest situation would arise if the proton and neutron distributions ρ_p and ρ_n were identical. This would imply that if the proton distribution were determined from electromagnetic processes the matter distribution would be known also, and at every point in the nucleus the ratio of the neutron to proton density would be just N/Z. However, this simple picture does not take account of the effect of Coulomb repulsion between the protons. The mere presence of the Coulomb potential gives rise to the condition that $N > Z$ in stable heavy nuclei, but the shape of the potential has an important influence on the variation of the ratio of neutron to proton density in different regions of the nucleus.

Naively, we may suppose that the Coulomb repulsion between the protons forces them as far apart as possible, thus lowering the proton density in the central region of the nucleus. On the other hand, the long-range repulsive Coulomb potential forms a barrier which prevents the proton wavefunctions from penetrating far into the surface region. Johnson and Teller (1954) also drew attention to the role of stability against β-decay, which requires that the total energy of the last proton and the last neutron, i.e. those at the Fermi level, must be the same. It follows that, neglecting the neutron-proton mass difference, if the averaged potential experienced by protons in the nucleus is equal to the (algebraic) sum of the neutron potential and the Coulomb potential, i.e. if $V_p = V_n + V_c$, then the turning point for the last proton is well inside that for the last neutron, as shown in Fig. 2.1, so that the total proton density does not

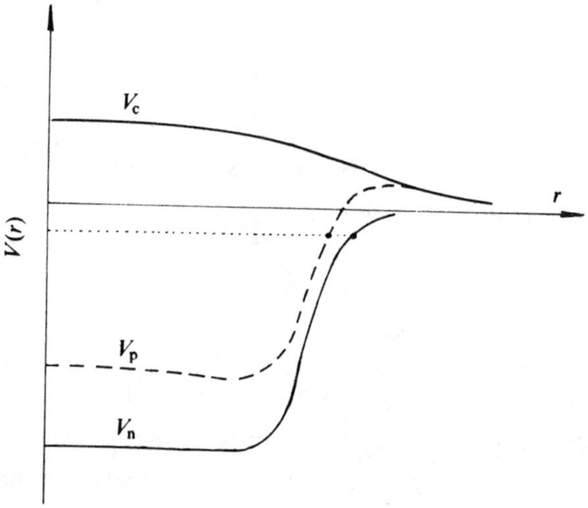

FIG. 2.1. Behaviour of neutron and proton potentials according to the assumption that $V_p = V_n + V_c$. (From Jackson 1974.)

extend so far as the neutron density. In addition, the kinetic energy of the neutrons at the top of the Fermi sea is greater than that of the corresponding protons (Moszkowski 1957) and so this too causes the neutrons to penetrate further into the nuclear surface. The Johnson-Teller effect would suggest that the proton density distribution should be concentrated inside the neutron distribution with a difference in the r.m.s. radii of ~1.0 fm, thus giving rise to a neutron-rich surface region or 'neutron skin'.

Shortly after the paper by Johnson and Teller had been published, it became clear that the averaged nuclear interaction is more strongly attractive for protons than for neutrons, i.e. $V_p \neq V_n + V_c$. Early evidence came both from the analysis of scattering data (Melkanoff *et al.* 1956, Melkanoff, Novdik, and Saxon 1957) and from studies of single-particle energies in nuclei (Green 1956, Ross, Mark, and Lawson 1956); a complete review of the early and more recent evidence is given by Satchler (1969). There are two physical effects which give rise to the difference. Firstly, owing to the nature of the nucleon-nucleon force and the operation of the Pauli exclusion principle, the interaction between unlike particles is much stronger than between like particles, which implies that protons and neutrons do not like to be separated. In a nucleus with $N > Z$, each proton has more unlike particles (neutrons) with which to interact than each neutron does; this leads to an enhancement of the attractive proton potential by an amount proportional to $\varepsilon = (N-Z)/A$ and a corresponding reduction in the neutron potential. This is known as the symmetry term, or more correctly as the asymmetry term, in the nuclear potential.

The second effect arises because the nucleon-nucleus interaction is energy dependent and becomes less attractive with increasing energy, but for the proton the bottom of the potential is pushed up by the Coulomb contribution so that the kinetic energy of the proton is reduced (Lane 1957, Satchler 1969). If the dependence of the potential on the kinetic energy K is assumed to be linear, we may write

$$U(K) = -U_0 + \alpha K. \tag{2.2}$$

The total energy of a nucleon is given by

$$E = U(K) + K + \Delta_c \tag{2.3a}$$

$$= -U_0 + (1+\alpha)K + \Delta_c \tag{2.3b}$$

where Δ_c is the Coulomb energy which is zero for a neutron and positive for a proton and U_0 is the same for protons and neutrons. It follows from eqn (2.3*b*) that $(1+\alpha)$ is equal to the ratio m/m^* of the nucleon mass to the effective mass. Using eqns (2.2) and (2.3) to rewrite the potential as a function of E we obtain

$$U(E) = -\frac{(U_0 - \alpha E + \alpha\Delta_c)}{1+\alpha} \tag{2.4}$$

so that, at a given energy E, we have

$$U_n - U_p = \alpha\Delta_c/(1+\alpha). \tag{2.5}$$

Following the usual practice for scattering problems in which E is the bombarding energy, we take the Coulomb potential to be that felt by a proton within a uniformly charged sphere of radius $1.3A^{1/3}$ fm and take $m^* \simeq 0.7$ m (Preston 1962), which yields

$$\alpha\Delta_c/(1+\alpha) \simeq 0.4Z/A^{1/3}\ \text{MeV}. \tag{2.6}$$

It is evident that there is a subtle balance between Coulomb effects and asymmetry effects, which may permit slight differences between proton and neutron distributions and could lead, in nuclei with $N > Z$, to the proton r.m.s. radius being somewhat greater or somewhat less than the neutron radius, but it is unlikely that very large differences will arise.

In the rest of this chapter an outline is given of various nuclear models used in the analysis of scattering and reactions. The success or otherwise of these models is discussed in subsequent chapters.

2.2. MICROSCOPIC MODELS FOR SPHERICAL NUCLEI

2.2.1. *Phenomenological single-particle potentials*

In a microscopic model a description is given of the properties of individual nucleons in the nucleus. Within the framework of the model it is then possible to build up a prediction of the total nuclear properties such as total energy, density distribution, electromagnetic moments, etc. The simplest model of this kind is of course the single-particle shell model in which it is assumed that each nucleon moves in an averaged nuclear potential, consisting of a central, a spin-orbit and a Coulomb part. Each single-particle energy ε_i and the corresponding single-particle wavefunctions are then determined by solving a Schrödinger equation of the form

$$(T_i + V_i)\,\psi_i = \varepsilon_i\,\psi_i. \tag{2.7}$$

The ground state wavefunction for the nucleus is constructed from the ψ_i using the appropriate angular momentum coupling scheme, and for a spherical closed-shell nucleus insertion of the wavefunction into eqns (1.1) and (1.2) yields a spherically symmetric distribution, while inclusion of quadrupole deformation for a nucleus of non-zero spin yields a distribution of the form

$$\rho(\underset{\sim}{r}) = \rho^{(0)}(r) + \rho^{(2)}(r)\,P_2(\cos\theta) \tag{2.8}$$

(Meyer-Berkhout, Ford, and Green 1959). In early studies of electron scattering the approach was very popular for light nuclei for which single-particle wavefunctions of oscillator form could be used with reasonable success (Meyer-Berkhout *et al.* 1959, Elton 1961*b*), but it is now clear that the corresponding distributions will fit electron scattering data for ^{12}C and ^{16}O only for momentum transfer of $q < 2.5\ \text{fm}^{-1}$ while for heavier nuclei they yield agreement only within the first diffraction minimum.

A very much improved single-particle model can be obtained by using a phenomenological single-particle potential where parameters may be varied to fit other experimental data. For nuclei in the 1p and 2s1d shells Elton and Swift (1967) chose a potential of the form

$$\begin{aligned} V(r) &= -V_0\,f(r) + V_{so}\left(\frac{\hbar}{m_\pi c}\right)^2 \frac{1}{r}\frac{d}{dr} f(r)\,\underset{\sim}{\ell}\cdot\underset{\sim}{\sigma} + V_C \\ f(r) &= \{1 + \exp(r-R_0)/a\}^{-1} \end{aligned} \tag{2.9}$$

and fitted the proton separation energies and spin-orbit splittings, as determined by poor-resolution (p,2p) and (e,ep) experiments,† and the electron-scattering data up to 250 MeV.

†See §8.44 and 3.41.

This procedure determines the potential parameters, the individual wavefunctions, and hence the proton density distribution. The potential radii are not proportional to $A^{1/3}$ and for a given nucleus the parameters V_0 and V_{so} are state dependent. Assuming that the symmetry term also has a Saxon-Woods shape, the state dependence of the central term can be converted to an energy dependence of the form

$$V_0 = 39 - 0.64E + 0.89Z/A^{1/3} + 25(N-Z)/A. \qquad (2.10)$$

Despite the relatively large number of parameters involved in these calculations, it appears that the strength and radius of the central potential are rather well determined by this method, while the diffuseness and spin-orbit strength are determined only to ~10 per cent. A high degree of accuracy in fitting the single-particle energies was not demanded in this calculation, and discrepancies of the order of a few hundred keV arise in some cases.

Other calculations of this type have been carried out, particularly for the calcium isotopes. Gibson and Van Oostrum (1967) found agreement using an energy-independent potential increasing in depth as $17(N-Z)/A$ and with a halfway radius proportional to $A^{1/3}$ for ^{40,48}Ca but requiring a larger radius for ^{42,44}Ca. A further study was carried out by Perey and Schiffer (1966) who used a potential radius proportional to $A^{1/3}$. The results obtained for the proton distributions in the calcium isotopes from these potentials may no longer be valid in view of a more recent examination of the electron scattering data which is discussed in §6.1.2.

In order to examine neutron distributions in the calcium isotopes, Elton (Swift and Elton 1966, Elton 1967 *a*,*b*) assumed that the symmetry potential was of the same volume shape as the central potential and simply adjusted the depth to give the observed binding energy for the least-bound neutron. This leads to almost identical proton and neutron distributions for ^{40}Ca, but the neutron r.m.s. radius increases much faster with $A^{1/3}$ than the proton radius so that the total effect is to yield a neutron-rich surface in ^{48}Ca. Batty and Greenlees

(1969) have fitted the single-particle level positions in $^{40,48}Ca$ very accurately and obtain rather similar behaviour (see Table 1.1), although they omit the symmetry term and compensate by allowing the parameters of the neutron and proton potentials to take different values.

Several analyses of level positions, in some cases together with reaction data, have led to single-particle potentials for ^{208}Pb (Dost, Hering, and Smith 1967, Zaidi and Darmodjo 1967, Rost 1968, Parkinson *et al.* 1969, Batty and Greenlees 1969) (see Table 8.5). With the exception of the work of Dost *et al.* all these studies lead to a radius parameter for the spin-orbit potential which is smaller than that of the central potential and closer to the halfway radius of the matter distribution. (This is due to the short range of the two-nucleon spin-orbit force (Bryan and Scott 1964). Calculations of the nucleon-nucleus spin-orbit potentials for spin-saturated $N = Z$ nuclei in the Hartree approximation (Blin-Stoyle 1955) and in the Hartree-Fock approximation (Scheerbaum 1969) lead to lowest-order terms proportional to $(1/r)(d\rho/dr)$.) Another very interesting feature of these calculations is that in those calculations where emphasis is placed on a good fit to the single-particle energies, e.g. those by Rost (1968) and Batty and Greenlees (1969), the radius parameters of the proton and neutron single-particle potentials are significantly different. Typical values are $R_0 \sim 1.35A^{1/3}$ for neutrons and $R_0 \sim 1.26A^{1/3}$ for protons, which would imply that the symmetry term is surface peaked but of the opposite sign to that predicted by the simple arguments in the previous section. This leads to a halfway radius and an r.m.s. radius for the neutrons which are approximately 0.6 fm larger than the corresponding proton radius. Consequently, the average† central density per particle is smaller for neutrons than for protons. In contrast, when emphasis is placed on fits to the reaction data (Zaidi and Darmodjo 1967,

†We must speak of an averaged value because the shell structure of the nucleus leads to a certain amount of undulation in the central region. This effect may be seen in Fig. 1.1(c) and Fig. 2.2.

Dost *et al.* 1967, Muehllehner *et al.* 1967, Parkinson *et al.* 1969) a smaller neutron radius is required, and it is found that the proton and neutron radii must be more nearly equal, which would imply an appreciable volume component in the symmetry potential. The neutron potential of Zaidi and Darmodjo, which has been used in a number of calculations, has $R_0 = 1.19\ A^{1/3}$ and a symmetry term of $26(N-Z)/A$. Some examples of distributions obtained from the Batty-Greenlees (BG) potential and the Zaidi-Darmodjo (ZD) neutron potential are shown in Fig. 2.2 and the moments are listed in Table 1.1.

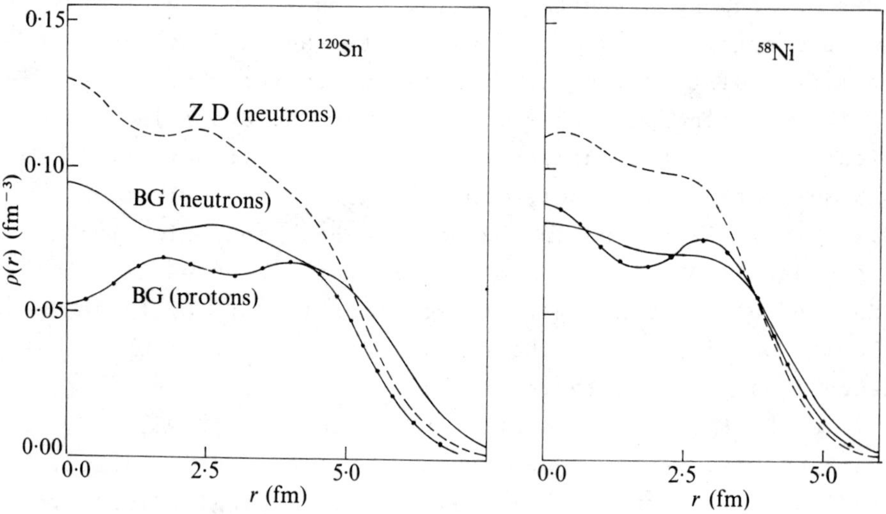

FIG. 2.2. The proton distributions $Z\rho_p(r)$ and neutron distributions $N\rho_n(r)$ predicted from the Batty-Greenlees and Zaidi-Darmodjo single-particle potentials for ^{120}Sn and ^{58}Ni. (From Jackson 1974.)

A generalized Saxon-Woods potential has been used by Beiner (1970). This has the form

$$W_{\pm}(r) = -\{\alpha_1 \pm \alpha_2 (N-Z)/A\}\, f_{\pm}(r) - \alpha_4 \frac{1}{r}\frac{\mathrm{d}f}{\mathrm{d}r} \pm \underset{\sim}{\ell}\cdot\underset{\sim}{s} \qquad (2.11)$$

$$f(r) = [1 + \exp\{(r-\tilde{R})/\alpha_5\}]^{-\alpha_6}$$

where $\tilde{R}$ is not the halfway radius but is determined by the condition $f(R) = \frac{1}{2}$ with

$$R = \alpha_3 A^{1/3} \pm \alpha_2 (C_n - C_p)/2\alpha_1$$

and the positive sign is taken for protons and the negative sign for neutrons. The parameters $\alpha_1, \alpha_2, \alpha_3$ are the same for protons and neutrons, while the constants $\alpha_4, \alpha_5, \alpha_6$ are allowed to take different values for protons and neutrons, and C_n and C_p are the radii corresponding to the maxima of $-d\rho_n/dr$ and $-d\rho_p/dr$, respectively, in the nuclear surface. This gives nine constants which are determined by fitting the observed r.m.s. radii, the mean separation energies as functions of N and Z, and the empirical sequence of single-particle levels in selected nuclei. This yields $\alpha_2 = 31.1$ MeV and $\alpha_3 = 1.26$ fm. If the term containing $C_n - C_p$ is omitted and two different values for α_3 are used, they are $\alpha_3 = 1.27$ fm for protons and 1.25 fm for neutrons. The results for individual nuclei are in reasonable agreement with experimental data.

A detailed study of single-particle states in a local potential has been carried out by Millener and Hodgson (1973) for $35 \leqslant A \leqslant 65$. They disregarded the 1s and 1p states in these nuclei but fitted carefully the centroid† values of the energies for the other single-particle states. The real central potential is taken to have isospin dependence and a linear dependence on A. Separate values of the potential parameters are listed for proton and neutron hole states and for proton and neutron particle states. This procedure gives the strength of the isospin term as 31.6 MeV. If linear energy dependence had been assumed, instead of A-dependence, the isospin strength would be 39.5 MeV and the coefficient of energy ~0.28.

Several calculations have been carried out using non-local single-particle potentials, so that the Schrödinger equation (eqn (2.7)) becomes

---

†See §§ 2.2.4 and 8.4.4.

$$(\varepsilon_i - T_i)\, \psi_i(\underset{\sim}{r}) = \int V_i(\underset{\sim}{r},\underset{\sim}{r}')\, \psi_i(\underset{\sim}{r}')\, d^3r'.$$

Grimm, McCarthy, and Storer (1971) did not succeed in finding a non-local single-particle potential which reproduced both the observed single-particle levels and the r.m.s. radii for a range of nuclei. The same result was found by Batty and Greenlees (1969) for ^{48}Ca and ^{208}Pb, although they did find that the non-locality improved the fit for ^{40}Ca. It must be noted that they did not attempt to fit the deepest single-particle levels. In constrast, Elton, Webb, and Barrett (1969) report a successful attempt to fit both electron scattering data and single-particle levels for protons in a range of nuclei from ^{12}C to ^{208}Pb using a non-local potential of the Perey-Buck type, i.e.

$$V(\underset{\sim}{r},\underset{\sim}{r}') = \pi^{-3/2}\, \beta^{-3}\, \exp\{-(\underset{\sim}{r}-\underset{\sim}{r}')^2/\beta^2\}\, U(p) \qquad (2.12)$$

where β is the non-locality parameter, $\underset{\sim}{p} = \frac{1}{2}(\underset{\sim}{r}+\underset{\sim}{r}')$, and $U(p)$ is the usual, local, single-particle potential (Perey and Buck 1962). The halfway radius of the potential so obtained decreases slightly with A while the diffuseness parameter increases from 0.35 fm to 0.8 fm. In more recent work Janiszewski and McCarthy (1972*a*) have obtained encouraging results for the single-particle energies, including the deepest proton levels, and for proton r.m.s. radii of a wide range of nuclei using a fixed set of potential parameters and a symmetry term of standard form but with a non-locality parameter which varies as a function of the nuclear density. Thus the non-locality disappears in the nuclear surface, and their results are therefore consistent with those analyses which prefer a local potential when fitting only the least-bound levels and are consistent with the observation of Brown, Gunn, and Gould (1963) that the compression of the least-bound single-particle levels suggests that $m^*/m \gtrsim 1$. The effect of the variable non-locality parameter on the single-particle levels in ^{208}Pb is shown in Fig. 2.3. In medium and heavy nuclei the level spacing and level ordering are in quite good agreement with experiment; the proton distributions are in

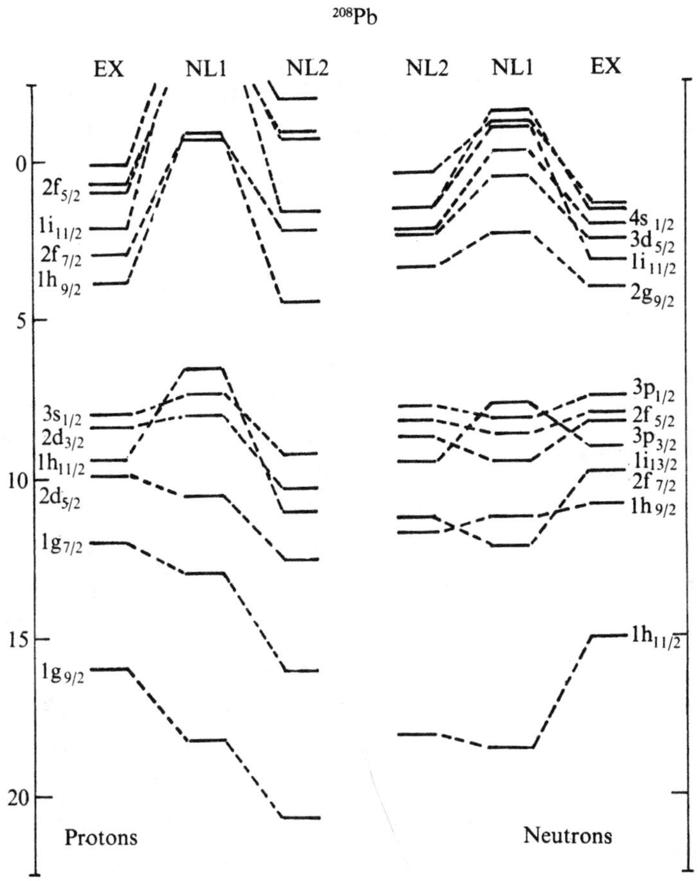

FIG. 2.3. Proton and neutron single-particle levels for a density-dependent non-local potential NL2 and a density-independent non-local potential NL1 compared with experiment. (From Janiszewski and McCarthy 1972*a*.)

good agreement in the surface region with those which fit electron scattering data, although the surface thickness and r.m.s. radius for ^{208}Pb are both slightly too small.

Meldner (1969) has constructed a phenomenological non-local interaction which has five arbitrary parameters and is density dependent. The resulting distributions have high central densities and show quite strong oscillations in the nuclear interior, but this deficiency may arise because the 1s level is taken to be at ~80 MeV in medium and heavy nuclei rather than the value of ~50 MeV obtained from recent (p,2p) measurements. For ^{88}Sr, ^{140}Ce, and ^{208}Pb he obtains $\langle r^2\rangle_n^{1/2} - \langle r^2\rangle_p^{1/2}$ ~0.05–0.07 fm. A faster method of carrying out these calculations has recently been reported by Cusson, Trivedi, and Kollo (1972).

2.2.2. *Hartree-Fock theory*

In the phenomenological single-particle model, the distribution functions are constructed from single-particle wavefunctions generated in a phenomenological potential. In principle, however, the potential derives from the interactions of the nucleons in the nucleus so that the phenomenological approach lacks self-consistency. A self-consistent calculation is carried out in the framework of Hartree-Fock theory, in which it is assumed that the ground-state wavefunction of the nucleus is a determinant of single-particle wavefunctions generated in the self-consistent potential. One common approach is to expand the single-particle functions in terms of a set of basis states which are often taken to be oscillator functions. The expansion coefficients are treated as variational parameters in order to minimize the total energy of the system. This leads to a set of eigenvalue equations which can be solved by fairly standard diagonalization procedures; for light nuclei rather few oscillator terms are needed but for heavy nuclei the dimensionality required to reproduce the correct exponential tail is much greater (Baranger 1967). An example of the oscillator approximation to the correct behaviour is shown in Fig. 2.4.

Davies *et al.* (1969) have given criteria for choice of nucleon-nucleon potentials in Hartree-Fock (HF) calculations: they require potentials sufficiently simple in form to be able to carry out systematic calculations over a wide range of nuclei, which yield a good description of the properties of nuclear matter such as total energy, density, and symmetry energy, which give rise to small second-order corrections, and which yield reasonable agreement with data for the free nucleon-nucleon system. The last requirement is not pressed

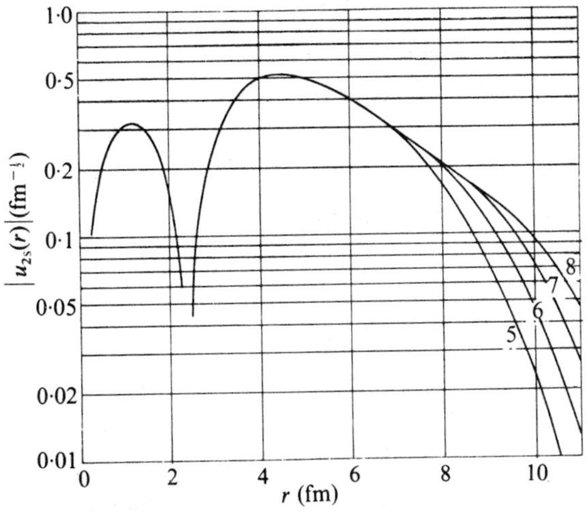

FIG. 2.4. The absolute value of the 2s wavefunction in ^{16}O, plotted for various values of the oscillator dimensionality. (From Davies, Krieger, and Baranger 1966.)

very strongly on the grounds that the interaction in the nucleus is an effective interaction and not the free interaction. The results for a range of nuclei obtained by this method (Davies, Krieger, and Baranger 1966, Krieger, Baranger, and Davies 1966, Tarbutton and Davies 1968) show a central nuclear density higher than deduced from phenomenological fits to data and a smaller r.m.s. radius. Some recent results obtained by Lee and Cusson (1971) using the oscillator basis are given in Table 2.1. Kerman, Svenne, and Villars (1966) have used Tabakin's separable potential in HF calculations for ^{16}O and ^{40}Ca, while Lande and Svenne (1971) have used the same potential to compare HF predictions for the r.m.s. radii of ^{40}Ca and ^{48}Ca.

The HF method which uses an oscillator basis and diagonalization procedures has certain disadvantages in that it is not suitable for scattering problems and the form of the self-

TABLE 2.1

Root-mean-square radii from Hartree-Fock calculations

	$\langle r^2\rangle_p^{1/2}$	$\langle r^2\rangle_n^{1/2}$	$\langle r^2\rangle_n^{1/2}-\langle r^2\rangle_p^{1/2}$	$\langle r^2\rangle_m^{1/2}$
Köhler and Lin (1969): velocity-dependent interaction				
^{16}O	2.57	2.54	-0.03	2.56
^{40}Ca	3.39	3.34	0.05	3.36
^{208}Pb	5.43	5.83	0.40	5.68
Köhler and Lin (1969): density-dependent interaction				
^{16}O	2.66	2.60	-0.06	2.63
^{40}Ca	3.46	3.34	-0.12	3.40
^{208}Pb	5.47	5.63	0.16	5.57
Vautherin (1969), Vautherin and Veneroni (1969): density-independent interaction				
^{16}O	2.56	2.65	-0.02	2.66
^{40}Ca	3.43	3.38	-0.05	3.40
^{48}Ca	3.51	3.76	0.25	3.65
^{90}Zr	4.25	4.37	0.12	4.31
^{208}Pb	5.44	5.72	0.28	5.61
Vautherin and Brink (1970): density-dependent interaction				
^{16}O	2.56	2.53	-0.03	2.55
^{40}Ca	3.31	3.27	-0.04	3.29
^{48}Ca	3.36	3.48	0.12	3.43
^{90}Zr	4.14	4.19	0.05	4.17
^{208}Pb	5.38	5.49	0.11	5.45
Negele (1970): density-dependent interaction†				
^{16}O	2.71	2.69	-0.02	2.70
^{40}Ca	3.41	3.37	-0.04	3.39
^{48}Ca	3.45	3.68	0.23	3.58
^{90}Zr	4.18	4.30	0.12	4.25
^{208}Pb	5.37 (5.45)	5.60 (5.68)	0.23	5.51

Table 2.1 continued

	$\langle r^2\rangle_p^{1/2}$	$\langle r^2\rangle_n^{1/2}$	$\langle r^2\rangle_n^{1/2}-\langle r^2\rangle_p^{1/2}$	$\langle r^2\rangle_m^{1/2}$
Nemeth and Vautherin (1970): density-dependent interaction				
^{16}O	2.66	2.63	-0.03	2.64
^{40}Ca	3.36	3.30	-0.06	3.33
^{48}Ca	3.48	3.66	0.18	3.58
^{90}Zr	4.16	4.23	0.07	4.20
^{208}Pb	5.38	5.53	0.15	5.48
Lee and Cusson (1971): density-independent interaction‡				
^{48}Ca	3.18	3.32	0.14	
^{56}Fe	3.54	3.61	0.07	
^{58}Ni	3.58	3.60	0.02	
^{62}Ni	3.61	3.72	0.11	
^{90}Zr	3.87	4.02	0.15	
^{120}Sn	4.58	4.83	0.25	
Campi and Sprung (1972): density-dependent interaction				
^{16}O	2.69	2.67	-0.02	
^{40}Ca	3.43	3.38	-0.05	
^{48}Ca	3.45	3.63	0.18	
^{90}Zr	4.21	4.28	0.07	
^{208}Pb	5.39	5.60	0.21	
Ehlers and Moszkowski (1972): density-dependent interaction				
^{16}O	2.58	2.56	-0.02	2.57
^{40}Ca	3.35	3.29	-0.06	3.32
^{48}Ca	3.42	3.61	0.19	3.53
^{90}Zr	4.19	4.28	0.08	4.24
^{208}Pb	5.45	5.64	0.19	5.57

†Figures in parentheses are revised values (see Negele and Vautherin 1972).

‡These results were derived from the published data using 0.8 fm for the r.m.s. radius of the proton.

consistent potential in co-ordinate space, which is non-local, is not readily obtainable. Solution of the HF equations in a co-ordinate representation was attempted some years ago (Brueckner, Lockett, and Rotenberg 1961, Masterson and Lockett 1963, Köhler 1965) and gave results which also showed high central densities and low r.m.s. radii, but a new technique has recently been developed (Vautherin and Veneroni 1967) in which the non-local potential is replaced by a state-dependent local potential so that the equations can be solved by an iteration method. Results for various nuclei have been obtained using a smooth nucleon-nucleon potential due to Brink and Boeker (1967), which is a sum of two Gaussian potentials, and also using Skyrme's interaction (Skyrme 1959).

Some results obtained with the local, density-independent interaction of Brink and Boeker are given in Table 2.1. The proton r.m.s. radii are in general agreement with values derived from electron scattering data. The neutron radii increase slightly, so that for ^{208}Pb the neutron r.m.s. radius exceeds that for the protons by 0.20-0.30 fm depending on the details of the interaction. A difficulty noted for ^{208}Pb is that the level density is too small and the single-particle level spacings too great (Vautherin and Veneroni 1969).

Many of the deficiencies in HF calculations have been removed by the use of density-dependent effective interactions. The effective nucleon-nucleon interaction in the nucleus differs from the free interaction owing to the presence of the other nucleons, and can be written in the form

$$V_{\text{eff}} = V + V\,\frac{Q}{e}\,V_{\text{eff}} \tag{2.13}$$

where e is the appropriate energy denominator and Q is a projection operator which excludes occupied states from the available intermediate states. It is frequently assumed (Negele 1970, Ripka 1970) that the many-body aspects of the effective interaction can be represented by a function of the density at the centre-of-mass of the pair of nucleons, i.e.

$$V_{\text{eff}}(\underset{\sim}{r},\underset{\sim}{r}',\rho) = V_1(|\underset{\sim}{r}-\underset{\sim}{r}'|) + V_2(|\underset{\sim}{r}-\underset{\sim}{r}'|)\,f(\underset{\sim}{s}) \tag{2.14}$$

where $\underset{\sim}{s} = \frac{1}{2}(\underset{\sim}{r}+\underset{\sim}{r}')$. Campi and Sprung (1972) prefer to take the local density as the average of the densities at the positions of the two interacting particles. Both forms are a statement of the local density approximation; the validity of this approximation depends on the short-range nature of the effective interaction, but this may not be a good assumption for the tensor force. A variety of density-dependent effective interactions have been used for various purposes (Ripka 1970, Lassey 1972, Nemeth and Ripka 1972). Fai and Nemeth (1973) have compared results obtained for spherical nuclei using several of these interactions.

Attempts to construct the effective interaction using eqn (2.13) meet with difficulty if V contains a repulsive core. However, it is known from many-body theory (Eden 1959) that a suitable approximation is obtained by replacing the effective interaction by the Brueckner reaction matrix which is defined by the equation†

$$G = V + V\frac{Q}{e}G \tag{2.15}$$

where

$$-e = E(p) + E(q) - E(i) - E(j),$$

i, j represent the initial states of the two nucleons, and p, q represent intermediate states, and methods have been developed to evaluate the matrix elements of G for finite nuclei (see, for example Kuo and Brown 1966). A diagrammatic representation of the equation for the reaction matrix is given in Fig. 2.5. The Yale-Shakin force, which has a hard core, has been used for such calculations (Shakin, Waghmare, and Hull 1967, Pal and Stamp 1967, Faessler *et al.* 1972*a*), but is found to give inadequate binding, small r.m.s. radii,

---

†We use the sign convention consistent with eqn (2.13) and customary in scattering theory. Articles on nuclear matter theory usually take $G = V - V(Q/e)G$ with e opposite in sign to the definition given here.

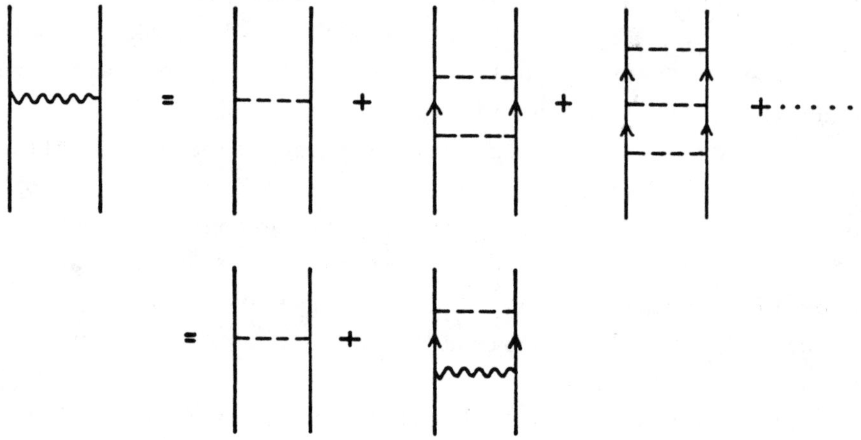

FIG. 2.5. Diagrammatic representation of the reaction matrix G. The broken line represents the single interaction V.

and high central densities. Tripathi, Faessler, and MacKellar (1973) have pointed out that if the reaction matrix is calculated in an oscillator basis the treatment of the Pauli operator Q is not self-consistent. They also use the Yale potential but treat the Pauli operator in an HF basis and include the Pauli rearrangement energy arising from the variation of $G(Q)$ with respect to the density (see § 2.2.4). The effect is to increase the r.m.s. radius of ^{16}O by ~7 per cent and to reduce the peak value of the nuclear density by ~20 per cent. This suggests that a correct microscopic treatment of the Pauli operator may remove some of the deficiencies of the density-dependent interactions.

Negele (1970) has constructed a local density-dependent, energy-dependent effective interaction from the G-matrix using the local density approximation and a nuclear matter calculation with the Reid soft-core potential but allowing slight adjustment of the parameters to give the nuclear matter binding energy as 15.68 MeV at $k_F = 1.31\ \text{fm}^{-1}$. He then performed a Hartree-Fock calculation for selected spherical nuclei from ^{16}O to ^{208}Pb. Much better results are obtained for binding energies and single-particle energies, and the proton distributions and r.m.s. radii are in accord with electron scattering data (see Chapter 6, particularly Tables 6.2 and 6.3). In ^{208}Pb the neutron r.m.s. radius exceeds the proton r.m.s. radius by 0.23 fm. Some results for r.m.s. radii are listed in Table 2.1. The proton and neutron distributions plotted by Negele appear not dissimilar, with essentially the same surface thickness. Bethe (1971) has given a more detailed examination of the behaviour of these distributions and in particular those for ^{208}Pb. He plots the difference $\Delta = N\rho_n - Z\rho_p$ for ^{208}Pb and shows that it goes through zero at 6.3 fm, which is not far from the halfway radius of 6.7 fm for the matter distribution. Inside the distance Δ is negative and oscillatory, and beyond it is positive so that there are relatively more neutrons in the exterior region.

Zofka and Ripka (1971) have compared HF calculations for light nuclei† from ^{12}C to ^{40}Ca using Negele's interaction and the Brink-Boeker interaction, and find that the interactions give the same r.m.s. radii but the density-dependent interaction gives better results for the binding energies and single-particle level densities. Lassey and Volkov (1971) have used Moszkowski's modified δ-function interaction (MDI) which is similar to that of Skyrme (Moszkowski 1970), and Krieger and Moszkowski (1972) have used a modified form of the Skyrme interaction to study s-d shell nuclei. Comparison of the latter with the density-independent work of Krieger (1970) shows that the new result for the r.m.s. radii are 5-7 per cent larger and in better agreement with experiment. In general, the r.m.s. radii calculated by Lassey and Volkov are smaller than those of Krieger and Moszkowski, and those of Zofka and Ripka are slightly higher.

Campi and Sprung (1972) have used the density and energy-dependent effective interaction designed by Sprung and Banerjee (1971) to reproduce the G-matrix elements of the Reid potential, and modified to give a nuclear matter binding energy of 16.5 MeV at $k_F = 1.35$ fm^{-1}. Some of their HF

---

†See also §2.4.1.

results for r.m.s. radii are included in Table 2.1. The r.m.s. radii of the charge distributions are in reasonable agreement with results from electron scattering data except for ^{208}Pb where the calculated proton radius is too small. For ^{208}Pb they obtain $\langle r^2\rangle_n^{1/2} - \langle r^2\rangle_p^{1/2} = 0.21$ fm. Nemeth and Vautherin (1970) have also derived a density-dependent interaction from Reid's soft-core potential, and multiply the potential energy density of Sprung and Banerjee (1971) by a factor of 1.22 to obtain the correct binding energy in nuclear matter. Otherwise their calculation is essentially the same as that of Negele (1970). The resulting binding energies are in quite good agreement with experiment and the charge distributions lead to reasonable agreement with electron scattering data. The r.m.s. radii are given in Table 2.1, from which it can be seen that the values are generally slightly smaller than those of Negele and the differences between the proton and neutron r.m.s. radii in medium and heavy nuclei are reduced.

The MDI developed by Moszkowski (1970) has only three parameters compared with the five parameters for Skyrme's interaction. These three parameters were initially determined by requiring a nuclear matter binding energy per particle of 15.75 MeV at $k_F = 1.36$ fm^{-1} and a binding energy per particle in ^{16}O of 7.98 MeV. A more recent set of parameters has been derived by Ehlers and Moszkowski (1972) by fitting the binding energy in ^{16}O as before together with the binding energy per particle and the r.m.s. radius of the charge distribution in ^{208}Pb. This leads to a nuclear matter binding energy of 16.46 MeV at $k_F = 1.33$ fm^{-1}. Compared with previous HF results, there is a reduction in the central density for ^{208}Pb of 6.5 per cent and a 15 per cent increase in the surface thickness. The results obtained by Ehlers and Moszkowki for r.m.s. radii are given in Table 2.1. The central depth of the nuclear potential can be written as

$$U_0 = 46 - 0.5E + 14(N-Z)/A + 0.7\ Z/A^{1/3}. \qquad (2.16)$$

This is not dissimilar to the behaviour of the phenomenological bound-state potential given by eqn (2.10) and the scattering potential (eqn (7.70)) derived from an HF treatment of proton scattering, but differs from the phenomenological result (eqn (7.60)) for medium-energy scattering. However, the sum of the Coulomb and symmetry terms is roughly consistent with the sum of the terms in eqn (7.60). The shapes of the isoscalar and isovector (symmetry) potentials are not the same. The halfway radius of the isoscalar part extends approximately 0.5 fm beyond the halfway radius of the density, and the isovector potential extends a further 0.7 fm. In ^{208}Pb the neutron excess distribution and the isovector potential have approximately a volume shape,† whereas in ^{48}Ca both are surface peaked owing to the role of the $f_{7/2}$ neutrons.

Negele (1970) has plotted the self-consistent equivalent local potentials which arise from his calculation. They are state dependent and in the interior region show the same oscillatory behaviour as the densities. (The same similarity between the densities and the equivalent local potentials has been noted by Vautherin and Veneroni (1969) using the Brink-Boeker force.) In the lighter nuclei the neutron and proton potentials differ by the full Coulomb energy, but in heavier nuclei the additional attraction for the protons due to the excess neutrons counteracts this so that the potentials differ by the full Coulomb effect only in the extreme tail. This makes the surface of the proton potential sharper than for neutrons. The potential extends outside the matter distribution by ~0.9 fm.

The interaction devised by Skyrme (1959) consists of a two-body term and a three-body contact interaction which, when averaged over one of the three particles, is equivalent to an effective two-body interaction which is linearly dependent on the density. Extensive HF calculations using this interaction have been carried out (Brink and Vautherin 1969, Vautherin and Brink 1970, 1972). Some results are given in Table 2.1. The proton radii and binding energies are in good

---

†See §10.3 and Fig. 10.2.

agreement with experiment, and the results for level ordering, level spacing, and the single-particle energy of the least-bound nucleon are much better than for density-independent interactions, although the binding energy of the 1s state in ^{40}Ca is smaller than the experimental value. The neutron r.m.s. radius in ^{208}Pb exceeds that of the protons by 0.11 fm. The proton and neutron distributions are very flat in the nuclear interior and the shell oscillations are much suppressed. The agreement with the electron scattering data is fair but not as good as that obtained by Negele. Vautherin and Brink (1972) also plot their equivalent local potentials and the variation of the ratio of the effective mass to the real nucleon mass as a function of distance from the centre of the nucleus. The ratio is substantially less than unity in the interior and tends fairly smoothly to unity in the surface. This result is consistent with the variation of the non-locality parameter observed by Janiszewski and McCarthy (1972a) in their calculations with a single-particle potential.

Some authors have used velocity-dependent forces in HF calculations (Köhler 1965, 1969, 1970, Krieger 1970). Köhler and Lin (1969) have compared results obtained for ^{16}O, ^{40}Ca, and ^{208}Pb with a velocity-dependent interaction, whose parameters are chosen to fit terms in the semi-empirical mass formula, and with a density-dependent interaction proportional to $\rho^{2/3}$. For ^{16}O and ^{40}Ca the central density is higher for the velocity-dependent interaction and the r.m.s. radii are smaller than the corresponding values for the density-dependent interaction. The velocity dependent interaction gives a ratio of ρ_n/ρ_p which is lower inside the nucleus and higher outside the nucleus than the ratio for the density-dependent interaction. In ^{208}Pb the ratio ρ_n/ρ_p differs rather slightly inside the nucleus, but the velocity-dependent interaction gives a large increase in the ratio in the extreme surface and hence a larger difference between the neutron and proton r.m.s. radii. However, the Coulomb interaction was omitted in the calculations for ^{208}Pb and this may affect the ratio. Some results are given in Table 2.1.

2.2.3. *Brueckner-Hartree-Fock calculations*

In the HF theory the energy of a nucleus with A nucleons is given by

$$E^{\mathrm{o}} = \sum_i \langle i|T|i\rangle + \frac{1}{2}\sum_{ij} \langle ij|\tilde{V}|ij\rangle \tag{2.17a}$$

$$= \sum \varepsilon_i - \frac{1}{2}\sum \langle ij|\tilde{V}|ij\rangle \tag{2.17b}$$

where the tilde indicates antisymmetrized matrix elements and the sum runs over occupied states. The single-particle wavefunctions and energies are given by

$$T\,\psi_i(\underset{\sim}{r}) + U(\underset{\sim}{r})\,\psi_i(\underset{\sim}{r}) - \int U(\underset{\sim}{r},\underset{\sim}{r}')\,\psi_i(\underset{\sim}{r}')\,\mathrm{d}^3r' = \varepsilon_i\psi_i(\underset{\sim}{r}) \tag{2.18}$$

where

$$U(\underset{\sim}{r}) = \sum_j \int \psi_j^*(\underset{\sim}{r}')\,V(\underset{\sim}{r},\underset{\sim}{r}')\,\psi_j(\underset{\sim}{r}')\,\mathrm{d}^3r' \tag{2.19a}$$

$$U(\underset{\sim}{r},\underset{\sim}{r}') = \sum_j \psi_j^*(\underset{\sim}{r}')\,V(\underset{\sim}{r},\underset{\sim}{r}')\,\psi_j(\underset{\sim}{r}). \tag{2.19b}$$

Self-consistency in the HF sense is obtained by taking the single-particle potential to be the HF field, so that the HF equations determine both the potential and wavefunctions.

The HF theory represents the first term in the Bethe-Goldstone equation (Eden 1959) and is valid if the second- and higher-order effects are small. Some higher-order contributions can be included in a first-order calculation by means of the Brueckner-Hartree-Fock (BHF) theory, in which the energy is given by

$$E^1 = \sum_i \langle i|T|i\rangle + \frac{1}{2}\sum_{ij} \langle ij|\tilde{G}|ij\rangle \tag{2.20}$$

Self-consistency in the Brueckner sense is satisfied by requiring that the single-particle energies in the energy denominator of the G-matrix are given by

$$E(i) = \langle i|T|i\rangle + \sum_j \langle ij|\tilde{G}(E_i+E_j)|ij\rangle \tag{2.21}$$

(McCarthy 1969). The single-particle states and potentials must satisfy HF consistency, so that

$$T\psi_i(\mathbf{r}) + \sum \int \psi_j^*(\mathbf{r}')\; G(\mathbf{r},\mathbf{r}';\mathbf{r}_1\mathbf{r}_1')\{\psi_i(\mathbf{r}_1)\psi_j(\mathbf{r}_1') - \psi_i(\mathbf{r}_1')\psi_j(\mathbf{r}_1)\}\mathrm{d}^3r'\;\mathrm{d}^3r_1\;\mathrm{d}^3r_1'$$
$$= \varepsilon_i\;\psi_i(\mathbf{r}). \qquad (2.22)$$

A variety of methods have been developed to carry out BHF calculations, since the double self-consistency requirement makes the full calculation very complicated (Davies *et al.* 1969). McCarthy (1969) ignores the HF self-consistency and uses single-particle wavefunctions of oscillator form, while Irvine (1967, 1968) expands in an oscillator basis. In general these calculations lead to incorrect saturation properties with underbinding, high central densities, and small r.m.s. radii (Köhler and McCarthy 1967, Wong 1967, Irvine 1968, Davies *et al.* 1969), but a modification of the BHF method which takes account of the depletion of the occupation of the deeply bound states (Brandow 1966, 1967) leads to improved agreement with experimental data (Becker 1970, Davies and Baranger 1970, McCarthy and Davies 1970, Davies and McCarthy 1971). This modified method is known as renormalized Brueckner-Hartree-Fock (RBHF) theory. The effect of renormalization is to raise the single-particle levels, increase the level density, increase the r.m.s. radii, and increase the binding energy. An example of the effect of renormalization of density distributions is shown in Fig. 2.6.

Unlike the procedure in most of the density-dependent HF calculations, the nucleon-nucleon potential used in RBHF calculations, usually the Reid potential, is not adjusted to give the correction saturation properties, but there are parameters which enter into the description of the two-particle intermediate states.

2.2.4. *Some fundamental problems in the microscopic model*

Both the phenomenological single-particle calculations and the HF theory suffer from the deficiency that the single-particle potentials are referred to an arbitrary origin which

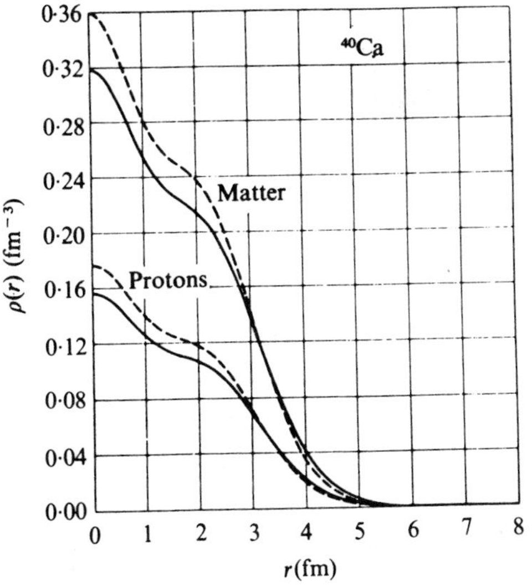

FIG. 2.6. Proton distributions and mass distributions for ^{40}Ca calculated with the BHF theory (broken lines) and with the RBHF method (solid lines). (After Davies and McCarthy 1971.)

is not the centre-of-mass of the system, so that the centre-of-mass is not stationary in the rest frame of the potential. In the simple case of the oscillator potential it is possible to show (Elliott and Skyrme 1955) that the shell-model wavefunctions of the nucleus can be factorized into the product of a translationally invariant wavefunction which is a function of relative co-ordinates and a wavefunction for the centre-of-mass. The lowest state of c.m. motion is the 1s state with kinetic energy $T_{cm} = \frac{3}{4}\hbar\omega$ and total energy $E_{cm} = \frac{1}{2}\hbar\omega$. The nuclear wavefunction is then given by

$$\Psi^{sm} = A^{1/4}\,\pi^{-3/4}\,a^{-3/2}\,\exp(-AR^2/2a^2)\Phi^{rel} \quad (2.23)$$

where $a^2 = \hbar/m\omega$, so that the form factor (eqn (1.19)) becomes

$$F^{sm}(q) = \exp(-q^2/4Aa^2)\,F^{rel}(q) \quad (2.24)$$

and the mean square radius becomes

$$\langle r^2 \rangle^{\text{sm}} = \langle r^2 \rangle^{\text{rel}} + 3a^2/2A. \tag{2.25}$$

In calculations using pure oscillator functions for light nuclei it is usual to take account of the c.m. correction in calculations of form factors for electron scattering and of total energies. In most calculations of density distributions using phenomenological potentials which are not of oscillator form (e.g. the Elton-Swift and Batty-Greenlees potentials) the correction is not included on the grounds that the density distributions are essentially phenomenological. In this case the correction should not be inserted into other calculations which make use of these distributions. In the HF calculations of Negele (1970) and Campi and Sprung (1972) the correction derived for oscillator functions has been subtracted from the calculated values of the nuclear binding energies and r.m.s. radii, and has been used to correct the density distributions even though the single-particle functions are not of oscillator form. According to Negele (1970) $-T_{\text{cm}}/A$ varies from -0.67 MeV for ^{16}O to -0.03 for ^{208}Pb, while Campi and Sprung (1972) find -0.61 MeV for ^{16}O and -0.02 MeV for ^{208}Pb.

Eqns (2.16)-(2.19) yield the HF single-particle energy as

$$\varepsilon_i^{\text{HF}} = \langle i|T|i\rangle + \sum_j \langle ij|\tilde{V}|ij\rangle \tag{2.26}$$

so that the total energy with the c.m. correction becomes

$$E^0 = \frac{1}{2}\sum_i \varepsilon_i^{\text{HF}} + \frac{1}{2}\sum_i T_i - T_{\text{cm}}. \tag{2.27a}$$

Similarly, the energy in BHF theory is given by

$$E_A^1 = \frac{1}{2}\sum_i \varepsilon_i^{\text{BHF}} + \frac{1}{2}\sum_i T_i - T_{\text{cm}}. \tag{2.27b}$$

When T_i and T_{cm} are calculated with oscillator or realistic single-particle wavefunctions and E_A, ε_i are taken from experiment, a large discrepancy in the two sides of these equations is obtained. Several such calculations have been summarized by Becker (1970) who calculates the quantity

$$\frac{1}{A}\left\{\sum_i T_i - \sum_i E_s\ (i) - 2\left(E_A^{\exp} + T_{\mathrm{cm}}\right)\right\} = \Delta^{\mathrm{B}} \tag{2.28}$$

where $E_s(i)$ is a mean separation energy obtained from experiment and it is assumed from Koopmans's theorem (Koopmans 1934) that $\varepsilon_i^{\mathrm{HF}} = -E_s(i)$. He finds that this quantity, usually called the Brueckner rearrangement energy, lies in the range 6-13 MeV for light- and medium-mass nuclei. Thus the HF and BHF theories cannot simultaneously reproduce the separation energies and the total energy, although the reasons are different in the two cases (Becker and Patterson 1971) and, in the case of BHF theory, Koopmans's theorem does not hold. A further consequence of the rearrangement energy is that the asymptotic behaviour of wavefunctions and density distributions is different in different models. In the phenomenological single-particle models, the single-particle energy is always taken to be equal to the mean separation energy, while in the HF or BHF calculations the single-particle energy can be represented as

$$\varepsilon_i^{\mathrm{HF}} = -E_s(i) - E_{\mathrm{Re}}(i), \tag{2.29}$$

where $\Sigma_i E_{\mathrm{Re}}(i) = A\Delta^{\mathrm{B}}$, and consequently the asymptotic behaviour is changed.

These difficulties can be overcome by a generalization of the self-consistent field in such a way that the rearrangement energies are reduced. This can be done in the RBHF theory, although the c.m. correction must be handled with care in this formalism (Davies and Becker 1971) as it can substantially modify the single-particle energy of the 1s state in light nuclei (Becker and Davies 1969). According to Becker and Patterson (1971), Koopmans's theorem is recovered for the RBHF theory.

In the HF theory with density-dependent forces the single-particle energies are defined as

$$\varepsilon_i^{\mathrm{DDHF}} = \langle i|T|i\rangle + \sum_j \langle ij|V|ij\rangle + \frac{1}{2}\sum_{jl} \langle jl|\delta\rho_i \frac{\mathrm{d}V}{\mathrm{d}\rho}|jl\rangle \tag{2.30}$$

where V is an effective force as defined in eqn (2.14) and

$$\delta\rho_i = |\psi_i(\underset{\sim}{s})|^2.$$

The third term in eqn (2.30) would vanish if the force were density independent. When a nucleon is put into orbit i the density changes from ρ to $\rho + \delta\rho_i$ and the interactions change from V to $V + \delta\rho_i(\mathrm{d}V/\mathrm{d}\rho)$ so there is a change in the energy (Ripka 1970). The separation energy for removal of a nucleon from state i is given by

$$-E_{\mathrm{s}}(i) = \langle i|T|i\rangle + \sum_j \langle ij|V|ij\rangle + \frac{1}{2}\sum_{jl} \langle jl|\delta\rho_i \frac{\mathrm{d}V}{\mathrm{d}\rho}|jl\rangle + \sum_j \langle ij|\delta\rho_i \frac{\mathrm{d}V}{\mathrm{d}\rho}|ij\rangle. \tag{2.31}$$

The additional term in eqn (2.31) is spurious since it represents the change in the interaction of the nucleon i due to its own density. An important feature of Skyrme's three-body interaction (Skyrme 1959) is that it gives similar effects to the two-body density-dependent interaction but does not lead to the spurious term. If this term can be neglected, it follows from eqns (2.30) and (2.31) that $-E_{\mathrm{s}}(i) = \varepsilon_i^{\mathrm{DDHF}}$, and the total energy is given by

$$E_A = \frac{1}{2}\sum_i \left(\varepsilon_i^{\mathrm{DDHF}} + T_i\right) - \frac{A}{2}e_{\mathrm{R}} - T_{\mathrm{cm}} \tag{2.32}$$

where

$$Ae_{\mathrm{R}} = \sum_i \frac{1}{2}\sum_{jl} \langle jl|\delta\rho_i \frac{\mathrm{d}V}{\mathrm{d}\rho}|jl\rangle = \sum_i \langle\Phi|\delta\rho_i \frac{\mathrm{d}V}{\mathrm{d}\rho}|\Phi\rangle.$$

Comparison with eqn (2.28) shows e_{R} is the rearrangement energy in the DDHF theory. Campi and Sprung (1972) find that $-Ae_{\mathrm{R}}$ is comparable in magnitude and of the same sign as

$\frac{1}{2}\ \Sigma(\varepsilon_i + T_i)$ in light nuclei and is about twice as large in heavy nuclei. Negele (1970) also finds this term to be important.

There remains the problem of the precise definition of the separation energy and its relation to experimental data. A definition which involves the difference between ground-state binding energies is usually called *the* separation energy, and can be deduced from atomic mass tables. The experimental definition for removal of one nucleon in a nuclear reaction is

$$E_{\mathrm{s}}^{\alpha\gamma}(i) = -(E_A^{\alpha} - E_{A-1}^{\gamma}) \tag{2.33}$$

where α is the initial state, usually the ground state of the system of A nucleons, and γ is the final state of the A-1 system. If there is only one state γ this definition is identical with the previous one and with the quantity which Brueckner, Meldner, and Perez (1972) call the adiabatic separation energy. In general the removal of a single nucleon from state α can lead to several states in the residual nucleus. If all of these final states are excited to their maximum strength, the mean separation energy is defined as

$$E_{\mathrm{s}}^{\alpha}(i) = \sum_{p} E_{\mathrm{s}}^{\alpha p}(i) = -(E_A^{\alpha} - \sum_{p} \mathscr{F}_{\mathrm{p}}^{2}(i)\ E_{A-1}^{p}) \tag{2.34}$$

where $\mathscr{F}_{\mathrm{p}}$ is the fractional parentage coefficient and the sum runs over all states which can be reached by removal of a single nucleon in state i from the initial state α. This definition corresponds to the sudden separation energy defined by Brueckner *et al.* The second term clearly corresponds to the centroid of the single-particle strength. In a high-resolution transfer or knock-out reaction, specific transitions of the least-bound nucleons leading to definite states in the residual nucleus are observed. It is usually assumed that in a poor-resolution experiment at high energies the separation energy $E_{\mathrm{s}}^{\alpha}(i)$ is measured, although it may not be true that the whole of the single-particle strength is observed (see §8.4.4). The definition given by eqn (2.34) is the natural one for use in HF or BHF calculations (Eden 1959, Pinkston and Satchler 1965, Brueckner *et al.* 1972), but this definition leads to large values of the rearrangement energy Δ^B, as noted above.

There is a further rearrangement which may be discussed, namely the rearrangement of the self-consistent field. This arises from the difference between the single-particle wavefunctions in the nuclei with A and A-1 nucleons, respectively, and leads to an orbital rearrangement energy (Koopmans 1934)

$$\Delta_i^{SCF} = \tilde{E}_{A-1} - E_{A-1} \tag{2.35}$$

where $\tilde{E}_{A-1}$ is the ground state energy of the A-1 nucleus calculated with single-particle wavefunctions in the A nucleus but with state i unoccupied. This orbital rearrangement energy is usually neglected but may be significant in certain cases (Faessler and Wolter 1969, Becker and Patterson 1971, Dieperink, Brussard, and Cusson 1972).

Alternative definitions of the single-particle energies and potentials have been given (Brueckner and Goldman 1959, Baranger 1970, Engelbrecht and Weidenmüller 1972) and involve comparisons with different experimental quantities.

2.3. MACROSCOPIC MODELS

2.3.1. *The Thomas-Fermi approximation*

The properties of finite nuclei can also be examined from a macroscopic point of view. Many recent studies of the bulk properties of nuclei have been based on the energy-density formalism in which the total energy of the system is expressed as a function of the local density. The energy is then minimized by variations in the density subject to the correct normalization conditions.

The total energy can be written as

$$E = \int V(|\underset{\sim}{r}-\underset{\sim}{r}'|)\, \rho(\underset{\sim}{r})\, \rho(\underset{\sim}{r}')\, d^3r\, d^3r' + \text{exchange term}$$
$$+ \int T\, \rho(\underset{\sim}{r})\, d^3r + \text{Coulomb term} \tag{2.36}$$

where T is the kinetic energy operator and numerical coefficients have been omitted. If the interaction is written as the sum of a long-range part v and a short-range part, the total energy becomes

$$E = \int \{T\,\rho(\underset{\sim}{r}) + F(\rho)\}\,d^3r + \int v(|\underset{\sim}{r}-\underset{\sim}{r}'|)\,\rho(\underset{\sim}{r})\,\rho(\underset{\sim}{r}')\,d^3r\,d^3r'$$

$$+ \text{ Coulomb term} \qquad (2.37)$$

where $F(\rho)$ represents the effect of short-range repulsion and exchange. The kinetic energy term is normally evaluated using the Thomas-Fermi approximation, i.e. by constructing the density from plane wavefunctions so that $\int T\,\rho(\underset{\sim}{r})\,d^3r \sim \rho^{5/3}$, and some similar approximation is made for the two-particle density function which enters the exchange term.

The direct part of the potential energy term may be expanded in powers of the derivatives of the density. Berg and Wilets (1956) and Wilets (1956) took the second term in the expansion as $\zeta(\hbar^2/8M)(\nabla\rho)^2/\rho$ which is analogous to the inhomogeneity correction to the kinetic energy introduced by Weizsäcker (1935). Other expansion procedures are also used for this term (Lombard 1970, Brueckner *et al.* 1968, Stocker 1971), and that part of the energy density which corresponds to an infinite homogeneous medium is also expanded in terms of the density (Wilets 1956, Lombard 1973) with phenomenological parameters which are determined by the binding energy E/A per particle in nuclear matter, the Fermi momentum k_F, and the compressibility K at equilibrium density. The Euler-Lagrange formalism, together with Poisson's equation, then leads to a set of differential equations which can be solved to determine the proton and neutron densities. A more fundamental procedure (Bethe 1968, Nemeth and Bethe 1968) is to use the density-dependent interaction derived from nuclear matter theory and to evaluate the potential energy terms within the Thomas-Fermi approximation.

The calculation by Wilets (1956) gave proton and neutron distributions with essentially the same halfway radii. For a heavy nucleus the neutron distribution has a larger surface thickness t = 2.8 fm, compared with $t \simeq 2.15$ fm for the proton distribution, and a longer tail. An initial calculation by Bethe (1968) led to a density distribution of the form

$$\rho(r) = \rho_0[1 - \exp\{(r-R)/a\}]^2 \tag{2.38}$$

with a = 1.20 fm, in the interior region. Since the Thomas-Fermi approximation breaks down when $\rho(r)/\rho_0 \lesssim 0.15$-$0.18$ (Bethe 1968, Nemeth and Gadioli Erba 1971), the form (2.38) is not valid at large distances and was therefore matched to an exponential function of the form

$$\rho(r) = \text{const.} \times \exp\{-\gamma(r-R)\} \tag{2.39}$$

where γ is determined from the binding energy. Improvement of this calculation leads to proton and neutron distributions with a central dip, and this type of distribution has been shown to be in good agreement with data for electron scattering and muonic X-rays (Bethe and Elton 1968). Further calculations by Lin (1970) using an improved density-dependent interaction gave a surface thickness $t \simeq 2.50$ fm for the matter distribution. He also obtained a surface thickness of ~3.3 fm for the charge distribution, which is much larger than the value deduced from precise fits to the electron scattering data although the fits given by his distribution for Au are not too bad. Some further modifications to the theory lead to a value of 2.18 fm for the surface thickness of the matter distribution, although the predicted surface energy is too high which suggests that there is still too much matter in the surface region.

Dahll and Warke (1970) have carried out Thomas-Fermi calculations for ^{90}Zr, ^{120}Sn, ^{140}Ce, and ^{208}Pb. They choose the Fermi shape as a parametrization of the proton and neutron distributions and determine the halfway radii and diffuseness parameters from the minimization procedure. Some of their results are listed in Tables 2.2 and 2.3. In general they find values for the proton r.m.s. radii in agreement with those deduced from experiment and neutron r.m.s. radii which

TABLE 2.2

Root-mean-square radii from Thomas-Fermi calculations

	$\langle r^2\rangle_p^{1/2}$	$\langle r^2\rangle_n^{1/2}$	$\langle r^2\rangle_n^{1/2}-\langle r^2\rangle_p^{1/2}$
Dahll and Warke (1970)			
^{90}Zr	4.30	4.35	0.05
^{120}Sn	4.69	4.80	0.11
^{140}Ce	4.91	5.02	0.11
^{208}Pb	5.49	5.62	0.13
Nemeth (1970)			
^{40}Ca	3.77	3.78	0.01
^{48}Ca	3.79	3.86	0.07
^{90}Zr	4.38	4.45	0.07
^{208}Pb	5.53	5.64	0.11
Lombard (1973) set II			
^{40}Ca	3.38	3.33	-0.05
^{48}Ca	3.44	3.56	0.12
^{90}Zr	4.20	4.28	0.08
^{116}Sn	4.52	4.63	0.11
^{140}Ce	4.77	4.90	0.13
^{208}Pb	5.44	5.60	0.16

TABLE 2.3

Parameters of Thomas-Fermi distributions†

		^{90}Zr	^{120}Sn	^{140}Ce	^{208}Pb
Halfway radius	protons	5.24	5.80	6.05	6.80
	neutrons	5.23	5.87	6.10	6.82
	matter	5.23	5.84	6.09	6.82
Surface thickness	protons	1.7	1.6	1.7	1.8
	neutrons	1.9	1.8	2.0	2.3
	matter	1.8	1.7	1.9	2.1

†From Dahll and Warke (1970).

are slightly larger than the proton values. For ^{208}Pb the difference is 0.13 fm. They find rather small values for the surface thickness. Results for a range of nuclei are also given by Nemeth (1970) and by Lombard (1973); some of their results are included in Table 2.2. The distributions obtained by Lombard have a rather large asymmetry in the surface and a short tail; for ^{208}Pb the charge radius is in agreement with experiment but the surface is not quite correct, and the neutron r.m.s. radius exceeds the proton radius by ~0.16 fm. Lombard also discussed some methods for improving the description of the surface by incorporating some single-particle features.

Local single-particle potentials can be determined from the derivative of the potential energy with respect to the density (Wilets 1956, Dahll and Warke 1970, Lin 1970). These potentials generally extend ~0.8 fm beyond the corresponding distribution and fall less rapidly.† Dahll and Warke find that the halfway radii are (1.23-1.26)$A^{1/3}$ fm for neutrons and (1.25-1.28)$A^{1/3}$ fm for protons. A comparison of their potentials and distributions is shown in Fig. 2.7.

Several studies of the validity of the Thomas-Fermi approximation have been made (Siemens 1970, Nemeth and Gadioli Erba 1971) and comparisons of TF and HF calculations have been reported (Nemeth 1970, Lombard 1970). The TF approximation leads to distributions with a flat central part of the density and a much lower value for central density than is given by HF methods. The conclusions reached from a comparison of the r.m.s. radii seem to depend on the choice of effective interaction. A comparison of matter distributions for ^{40}Ca and ^{208}Pb obtained by Lombard using Skyrme's interaction is shown in Fig. 2.8.

2.3.2. *The hydrodynamical model*

A further type of variational procedure is used in the hydrodynamical model (Swiatecki 1951, Damgaard, Scott, and Osnes

---

†This is a general result due to nuclear saturation (Berg and Wilets 1956).

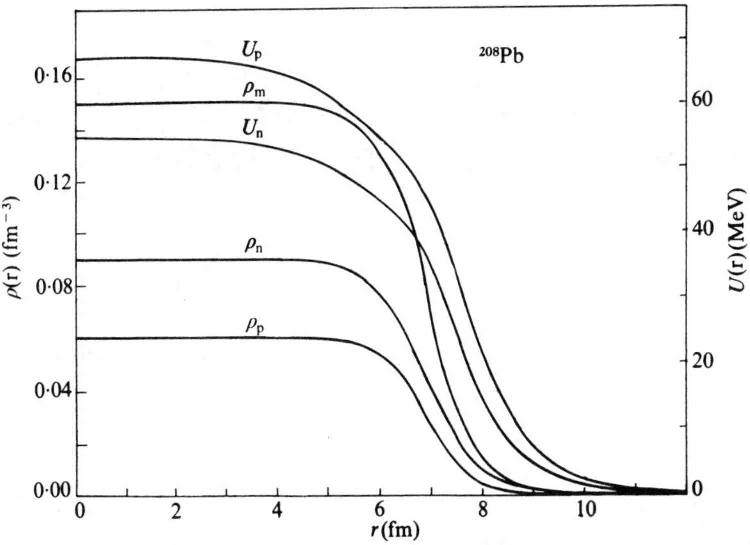

FIG. 2.7. Comparison of potentials calculated by the Thomas-Fermi method and the corresponding distributions. (From Dahll and Warke 1971.)

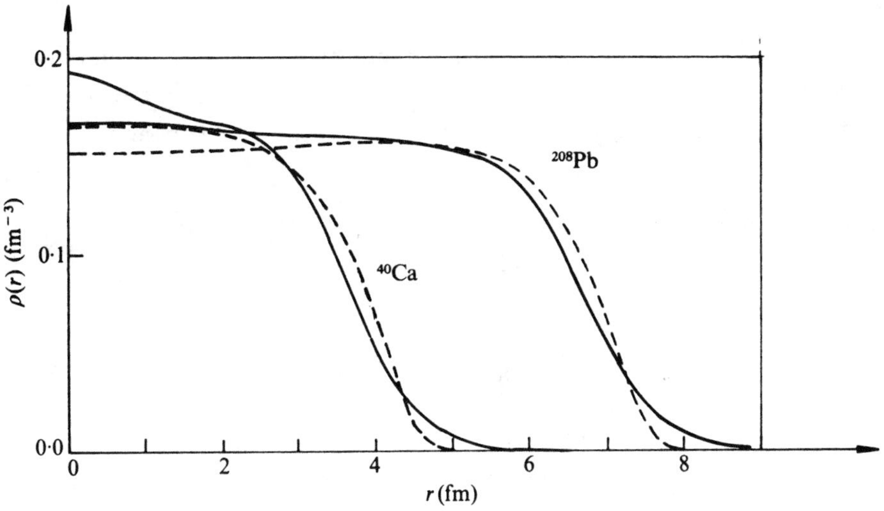

FIG. 2.8. Comparison of distributions calculated by the Hartree-Fock method (solid line) and the Thomas-Fermi method (broken line). (From Lombard 1970.)

1970, Friedman 1971a). We have already seen, in §2.1, that differences between the neutron and proton distributions arise from the interplay between the effect of the Coulomb interaction and the symmetry term. In a local-density approximation the sum of the Coulomb and the symmetry energies can be written as

$$E = \frac{Z}{2}\int\tilde{\rho}_{\rm p}(\underset{\sim}{r})\;\; V_{\rm c}(\underset{\sim}{r})\;{\rm d}^3r + \frac{1}{2}\,b_{\rm sym}\int\frac{\{Z\tilde{\rho}_{\rm p}(\underset{\sim}{r}) - N\tilde{\rho}_{\rm n}(\underset{\sim}{r})\}^2}{\rho_0(\underset{\sim}{r})}\,{\rm d}^3r \tag{2.40}$$

where $\tilde{\rho}_{\rm p}$ and $\tilde{\rho}_{\rm n}$ include the finite size of the proton. Also

$$\tilde{\rho}_0(\underset{\sim}{r}) = Z\,\tilde{\rho}_{\rm p}(\underset{\sim}{r}) + N\tilde{\rho}_{\rm n}(\underset{\sim}{r}), \tag{2.41}$$

$b_{\rm sym} \simeq 50$ MeV is the coefficient of the symmetry term in the semi-empirical mass formula, and $V_{\rm c}$ is the Coulomb potential. The procedure of the hydrodynamical model is then to minimize this energy subject to the constraint

$$Z\int\delta\tilde{\rho}_{\rm p}(\underset{\sim}{r})\;{\rm d}^3r = -\,N\int\delta\tilde{\rho}_{\rm n}(\underset{\sim}{r})\;{\rm d}^3r = 0. \tag{2.42}$$

Friedman (1971*a*) has studied neutron distributions using the model and the assumption that the charge distributions are known. The variational procedure gives the difference

$$Z\tilde{\rho}_{\rm p}(\underset{\sim}{r}) - N\tilde{\rho}_{\rm n}(\underset{\sim}{r}) = \tilde{\rho}_0(\underset{\sim}{r})\,\frac{Z-N}{A} + \frac{1}{2b_{\rm sym}}\,\tilde{\rho}_0(\underset{\sim}{r})\,\{\bar{V}_{\rm c} - V_{\rm c}(\underset{\sim}{r})\} \tag{2.43}$$

where

$$\bar{V}_{\rm c} = \frac{1}{A}\int\tilde{\rho}_0(\underset{\sim}{r})\;V_{\rm c}(\underset{\sim}{r})\;{\rm d}^3r \tag{2.44}$$

which, since $\tilde{\rho}_{\rm p} = \tilde{\rho}_{\rm ch}$ may be regarded as known, may be iterated to obtain $\tilde{\rho}_{\rm n}$. His results for a range of nuclei are given in Table 2.4. The values of the r.m.s. radii for neutrons are slightly smaller than those for protons, even for heavy nuclei, and for ^{208}Pb the halfway radius of the neutron distribution is ~0.1 fm smaller than the value for protons. This model also gives predictions for the distribution of the excess neutrons which will be discussed in connection with

Coulomb energy differences in Chapter 10.

TABLE 2.4

Root-mean-square radii predicted from the hyrdodynamical model†

	$\langle r^2\rangle_p^{1/2}$	$\langle r^2\rangle_n^{1/2}$	$\langle r^2\rangle_n^{1/2}-\langle r^2\rangle_p^{1/2}$	$\langle r^2\rangle_m^{1/2}$
^{48}Ca	3.39	3.39	0	3.39
^{60}Ni	3.77	3.73	-0.04	3.76
^{120}Sn	4.57	4.53	-0.04	4.55
^{208}Pb	5.44	5.39	-0.05	5.41

†From Friedman 1971*a*.

2.3.3. *The droplet model*

A macroscopic model based on the liquid-drop model has been developed by Myers and Swiatecki (Myers 1968, Myers and Swiatecki 1969). In this droplet model, the nuclear properties are expanded about the values from the liquid-drop model in terms of two quantities,

$$\varepsilon = -\frac{1}{3}(\rho-\rho_0)/\rho_0 \ , \ \delta = (\rho_n-\rho_p)/\rho_0$$

where $\rho = \rho_n + \rho_p$, $\rho_0 = \left(\frac{4}{3}\pi r_0^3\right)^{-1}$ and $r_0A^{1/3}$ is the radius in liquid-drop model. The potential energy in the droplet model consists of a volume energy, a surface energy, and the Coulomb energy, and is a functional of the neutron and proton distributions. This is minimized with respect to density variations, which leads to a droplet model formula with nine parameters. Four of these parameters come from the liquid-drop model and the remaining parameters are estimated from Thomas-Fermi calculations for semi-infinite nuclear matter.

The average values of ε and δ are determined as functions of N,Z, and A in terms of the model parameters, and the separate uniform radii of the neutron and proton distributions and the neutron skin thickness $U_n - U_p$ are then defined in terms of these average values and $(N-Z)/A$. The model predicts that the neutron skin thickness reaches a value of +0.3 fm at

$A \simeq 200$ (Myers 1969). The parameters of a single-particle potential can also be deduced within this model (Myers 1970).

2.4. DEFORMED NUCLEI

2.4.1. *Hartree-Fock theory for deformed nuclei*

The characteristic features of deformed nuclei, such as ground state quadrupole moments, strong E2 transitions, and excited states forming rotational bands, have been recognized in many regions of the periodic table. The Hartree-Fock method has been used to study the shapes of deformed nuclei, using modifications appropriate to particular mass regions.

The methods used for light nuclei and the early results obtained have been reviewed by Ripka (1968). The intrinsic Hartree-Fock state is taken to be a Slater determinant of single-particle states generated in a deformed field. In restricted Hartree-Fock calculations the core is regarded as inert and only the nucleons in the last unfilled shell participate in defining the HF field, while in unrestricted Hartree-Fock calculations self-consistency is sought between the single-particle potential and the single-particle states. The HF states are expanded in a suitable basis, usually deformed oscillator functions, and an initial guess is made for the expansion coefficients and for which states are filled; the usual diagonalization procedures are then carried out to reach self-consistency. For large deformations the HF states have expansion coefficients very similar to those of the Nilsson model, but there is a significant difference in that the HF calculations produce an energy gap between the filled and unfilled states. This gap arises from the non-local exchange part of the HF potential. Owing to the deformation the intrinsic HF wavefunction does not have good angular momentum in general, but must be used to generate a rotational band with suitable projection techniques.

The HF energy can be determined as a function of the deformation by minimizing the quantity $H + f(\mu, \hat{Q})$, where μ is a Lagrange multiplier and $\hat{Q}$ is the quadrupole moment operator (Flocard *et al.* 1973), and allowing μ to take a range of values. Bassichis, Kerman and Svenne (1967) used a linear constraint, $f(\mu,\hat{Q}) = \mu\hat{Q}$, while Giraud, le Tourneux, and Wong (1970) and Flocard *et al.* (1973) used a quadratic form.

Bassichis *et al.* (1967) used Tabakin's separable, non-local potential and surveyed nuclei in the 1p and 2s1d shells. For ^{12}C they found that the minimum-energy solution corresponds to oblate deformation, while the spherical solution corresponds to a local energy maximum. Similar results were found for ^{28}Si and ^{32}S. The first-order energy is not in good agreement with experiment but there is a large second-order correction which improved the agreement.

Zofka and Ripka (1971) have studied the double-even $N = Z$ nuclei from ^{12}C to ^{40}Ca using Negele's interaction and the density-independent Brink-Boeker interaction, and have compared their results with those of Krieger (1970), who used a velocity-dependent interaction which contains spin-orbit terms, with those of Muthukrishnan (1967) who used a Yamaghuchi separable, non-local potential, and with those of Brink *et al.* (1970), who used the Brink-Boeker interaction within the framework of an α-particle model. The behaviour of Negele's interaction was modified to remove the non-linear dependence on the density which becomes difficult to handle when the nucleus is deformed. The main effect of the density dependence of the interaction was found to be a compression of the single-particle energies which raised the deeply bound states and lowered the empty states so that the energy gap was reduced roughly by a factor of 2. The calculated values of the r.m.s. radii of the charge distributions are compared in Table 2.5. Compared with experiment (see Table 6.2) the the calculations of Zofka and Ripka give satisfactory agreement for ^{16}O and ^{40}Ca, but the calculated radii for the deformed nuclei are 5-8 per cent larger than the experimental values, which suggests that the deformation is too great. In every case the r.m.s. radius of the proton distribution is found to exceed that of the neutron distribution. The values of $\langle r^2 \rangle_{ch}$ obtained by Krieger are too small, except for ^{12}C. The electric quadrupole moments for odd-A nuclei are given in Table 2.6 and have been calculated in the rotational model by adding or subtracting one nucleon from the HF state for the

TABLE 2.5

Root-mean-square radii of charge distributions from deformed HF *calculations*

Nucleus	$\langle r^2 \rangle^{1/2}_{ch}$ A	B	C	D	E
^{12}C	2.68	2.69	2.73	2.45	-
^{16}O	2.72	2.72	2.73	2.53	3.06
^{20}Ne	3.05	3.04	3.05	2.77	3.37
^{24}Mg	3.24	3.24	3.22	2.82	-
$^{28}Si_{PR}$	3.38	3.40	3.38	-	3.56
$^{28}Si_{OB}$	3.35	3.35	3.33	3.00	3.56
^{32}S	3.42	3.44	-	3.12	-
^{36}Ar	3.46	3.48	-	3.16	3.60
^{40}Ca	3.50	3.50	-	3.20	3.62

A, Zofka and Ripka (1971) — Negele interaction.

B, Zofka and Ripka (1971) — Brink-Boeker interaction.

C, Brink *et al.* (1970).

D, Krieger (1970).

E, Muthukrishnan (1967).

even-even system. The good agreement for ^{21}Ne, ^{23}Na, ^{25}Mg, and ^{35}Cl suggests that the nuclei ^{20}Ne, ^{24}Mg, and ^{36}Ar are quite well described by deformed HF solutions, but the disagreement in sign for some of the other nuclei suggests that ^{12}C, ^{28}Si, and ^{32}S are not satisfactorily represented by this means.

Friar and Negele (1975*a*) have pointed out that the earlier constrained HF calculations for ^{12}C did not include the two-body spin-orbit potential; their results indicate that, when the spin-orbit potential is included, the deep energy minimum corresponding to an oblate deformation disappears and is replaced by a rather flat behaviour for the deformation-energy curve. They infer that it is preferable to use a spherical HF basis for ^{12}C and show that the spheri-

TABLE 2.6

Electric quadrupole moments of the ground states of odd-A nuclei†

Nucleus	Core	Quadrupole moment (fm^2)		
		Negele interaction	Brink-Boeker interaction	Experiment (Fuller and Cohen 1969)
^{11}C	^{12}C	-4.1	-4.2	±3.1
^{11}B	^{12}C	-3.2	-3.3	+4
^{21}Ne	^{20}Ne	9.9	9.5	+9
^{23}Na	^{24}Mg	12.3	11.8	+14
^{25}Mg	^{24}Mg	24.5	23.8	+22
^{27}Al	$^{28}Si_{OB}$	-22.4	-22.0	+15
^{33}S	^{32}S	-12.7	-12.9	-5.5
^{35}Cl	^{36}Ar	-10.7	-10.7	-7.9

†From Zofka and Ripka (1971).

cal HF calculations with a variety of forces yield values for $\langle r^2 \rangle_{ch}^{1/2}$ closest to the experimental value and about 0.2 fm smaller than those obtained with a deformed basis. It is necessary to use finite-range DDHF to reproduce the small central dip in the charge distribution indicated by the electron scattering data.

Studies of heavy nuclei have been carried out by Flocard *et al.* (1973) and Flocard, Quentin, and Vautherin (1973). These calculations include the effect of pairing correlations in an approximate but self-consistent way (Vautherin 1973) by minimizing the quantity $H - \Delta\Sigma_i u_i v_i$, where u_i and v_i are the usual quantities of BCS theory which are subject to the constraints

$$u_i^2 + v_i^2 = 1 \;, \; \sum_i v_i^2 = N.$$

The variational procedure leads to the standard HF equations, with the HF fields depending on the occupation probabilities v_i^2 and the energy given by

$$E_A = \frac{1}{2}\sum_i (\varepsilon_i + T_i)\, v_i^2 - \frac{1}{2}\Delta \sum_i u_i v_i$$

coupled with the BCS equations

$$v_i^2 = \frac{1}{2}\left\{1 - \frac{\varepsilon_i - \lambda}{\{(\varepsilon_i - \lambda)^2 + \Delta^2\}^{1/2}}\right\}$$

where the chemical potential λ is determined from the condition for conservation of particle number. Vautherin (1973) has derived these equations for the Skyrme force; for density-dependent and energy-dependent forces, such as that used by Campi, Sprung, and Martorell (1974), the coupled HF and BCS equations are not based on a variational principle.

Flocard *et al.* (1973) use the Skyrme force and take a constant value for the pairing gap Δ. The pairing force $G = 2\Delta/\Sigma u_i v_i$ then varies with the deformation. Flocard *et al.* (1973) have studied the cerium isotopes and Flocard, Quentin, and Vautherin (1973) have examined rare-earth nuclei. In the latter case the r.m.s. radii of the charge distributions are consistently larger than the experimental values, but only by ~1 per cent, and the quadrupole moments are in satisfactory agreement except for ^{190}Os. Campi *et al.* (1974) use a fixed pairing force G and the force of Sprung and Banerjee (1971) to study the isotopes of tin. The r.m.s. radii of the charge distributions are about 0.02 fm too small but show the correct trend with A.

These calculations represent an approximation to the Hartree-Fock-Bogolyubov (HFB) method which provides a self-consistent description of the HF field and the pairing correlations. The Hamiltonian expressed in terms of particle operators is transformed and expressed in terms of quasi-particle operators; the requirement that the new Hamiltonian is diagonal with respect to the number of quasi-particles yields the HFB equations (Eisenberg and Greiner 1972). A comparison of HF and HFB results for 2s1d and 2p1f shell nuclei has been made by Wolter, Faessler, and Sauer (1971) using G-matrix elements calculated from the Yale potential. Both calculations give underbinding and small radii, but in most cases the change in the r.m.s. radii is negligible.

2.4.2. *Bubble nuclei*

The possible existence of nuclei with exceptionally low density in the central region arising from depletion of s-states was first examined by Wong (1972*b*, 1973) using a phenomenological model. Subsequently, Davies, Wong, and Krieger (1972) carried out HF calculations using a velocity-dependent potential, and they proposed ^{36}Ar, ^{84}Se, ^{138}Ce, ^{174}Yb, ^{184}Hg, ^{200}Hg, and ^{250}104 as possible bubble nuclei.

Further calculations have been made by Campi and Sprung (1973) using a density-dependent interaction which gives a good description of spherical nuclei. They argue that the occupation probabilities of the single-particle states, which are so crucial in the formation of the bubble situation, depend rather critically on the pairing effect; consequently they have taken pairing into account in the approximate manner described in the preceding section, and they also used an external constraint to enforce a bubble solution. Without pairing they found a significant depletion of the central density for both ^{36}Ar and ^{200}Hg, but the bubble effect is much less pronounced when pairing is included. Similar calculations have been carried out by Beiner and Lombard (1973), who consider the HF + BCS solutions and also the bubble configurations proposed by Wong. They note that the maximum fluctuations in $\rho_m(0)$ resulting from filling of s-shells are less than 10 per cent when pairing is taken into account, except for the region around A = 32. From detailed calculations they conclude that the bubble configurations cannot compete energetically with the shell model and BCS configurations and that none of the candidates proposed by Davies *et al.* show any sign of significant central depletion. Nevertheless, the remarkable behaviour of the r.m.s. radii of the Hg isotopes shown in Fig. 6.3 does imply rather unusual changes in the proton distributions (see §6.1.4).

3
ELECTRON SCATTERING

The analysis of electron-scattering experiments gives more information about charge distributions than any other technique. Since the scattering measurements of Lyman *et al.* (1951) the analysis has progressed from the one- or two-parameter fits of the 1950s to the eleven-parameter 'model-independent' fits of the present. The separation of the charge and magnetic contributions to scattering using the technique of Rosenbluth (1950) or by measuring 180° scattering (Peterson and Barber 1962) has given information about the magnetic-moment distribution. Electron-scattering techniques and the extraction of information about the nuclear charge and current distributions have been reviewed by Hofstadter (1956, 1957), Elton (1961*b*), Bishop (1965), deForest and Walecka (1966), deForest (1967), Überall (1971), Barrett (1974), and Donnelly and Walecka (1975). A review of model-independent analyses has been given by Friar and Negele (1975*b*).

3.1. MOTION OF ELECTRONS AND MUONS IN THE NUCLEAR ELECTROMAGNETIC FIELD

We first consider a lepton of charge $-e$ ($e > 0$) in an external electromagnetic field $A_\mu \equiv (\underset{\sim}{A}, iA_0)$. The steady-state motion is described by the time-independent Dirac equation

$$(H-E)\,\psi = 0 \tag{3.1}$$

where ψ is a four-component spinor and the Dirac Hamiltonian H is given by[†]

$$H = \underset{\sim}{\alpha}.(c\underset{\sim}{p} + e\underset{\sim}{A}) + \beta mc^2 - eA_0. \tag{3.2}$$

[†]See for example Dirac (1958), Bjorken and Drell (1964), Sakurai (1967).

The most convenient choice of representations for the 4 × 4 matrices α and β depends on the physical situation. For A = 0 and A_0 *spherically symmetric* we can find simultaneous eigenstates of H, the angular-momentum operators J^2 and J_z, and parity.† A representation of α and β which is frequently used is

$$\alpha = \begin{pmatrix} 0 & \sigma \\ \sigma & 0 \end{pmatrix} \qquad \beta = \begin{pmatrix} I & 0 \\ 0 & -I \end{pmatrix} \tag{3.3}$$

where the components of σ are the usual Pauli matrices and I is the 2 × 2 unit matrix. For a given angular momentum and parity we can write ψ in this representation as

$$\psi(\mathbf{r}) \equiv \begin{pmatrix} \phi_A \\ \phi_B \end{pmatrix} = \frac{1}{r}\begin{pmatrix} G(r)\,Y_{ljm} \\ iF(r)Y_{l'jm} \end{pmatrix} \tag{3.4}$$

where ϕ_A and ϕ_B are two-component spinors called respectively the 'large' and 'small' components of ψ (ϕ_B vanishes in the non-relativistic limit). The two-component spinor Y_{ljm} is obtained by coupling spin and orbital angular momentum states (of magnitude l) to form the state of total angular momentum j, z-component m. The parity of ψ is $(-1)^l$ and the value of l' is $2j - l$.

Another representation, sometimes used for electron-scattering calculations (Yennie, Ravenhall, and Wilson 1954), can be obtained from eqn (3.3) by the unitary transformation

$$u = u^{-1} = \frac{1}{\sqrt{2}}\begin{pmatrix} I & I \\ -I & I \end{pmatrix} \tag{3.5}$$

which results in the following matrices

$$\alpha = \begin{pmatrix} \sigma & 0 \\ 0 & -\sigma \end{pmatrix} \qquad \beta = \begin{pmatrix} 0 & I \\ I & 0 \end{pmatrix} \tag{3.6}$$

The eigenstates of angular momentum and parity become

---

†The parity operator is βP where P is the non-relativistic parity operator defined by $P\,U(\mathbf{r}) \equiv U(-\mathbf{r})$.

$$\psi(r) = \frac{1}{r}\begin{pmatrix} G(r)\, Y_{ljm} + iF(r)\, Y_{l'jm} \\ G(r)\, Y_{ljm} - iF(r)\, Y_{l'jm} \end{pmatrix} \tag{3.7}$$

For both representations the spin operator $\frac{1}{2}\underset{\sim}{\Sigma}$ is given by

$$\underset{\sim}{\Sigma} \equiv \frac{1}{2}\, i\, \underset{\sim}{\alpha} \times \underset{\sim}{\alpha} = \begin{pmatrix} \underset{\sim}{\sigma} & 0 \\ 0 & \underset{\sim}{\sigma} \end{pmatrix} \tag{3.8}$$

In the field-free case, if the energy is so high that we can neglect the lepton rest mass, we can find eigenstates of angular momentum and the helicity operator $\underset{\sim}{\Sigma}.\underset{\sim}{p}/|\underset{\sim}{p}|$ instead of the parity:

$$\psi_R \equiv \begin{pmatrix} \phi_R \\ 0 \end{pmatrix} = \frac{1}{r}\begin{pmatrix} G(r)\, Y_{ljm} + iF(r)\, Y_{l'jm} \\ 0 \end{pmatrix} \tag{3.9a}$$

$$\psi_L \equiv \begin{pmatrix} 0 \\ \phi_L \end{pmatrix} = \frac{1}{r}\begin{pmatrix} 0 \\ G(r)\, Y_{ljm} - iF(r)\, Y_{l'jm} \end{pmatrix} \tag{3.9b}$$

where the subscripts R and L indicated right-handed (spin parallel to the momentum) and left-handed helicity respectively.

For the spherically symmetric potential $V(r) = -e A_0(r)$ the radial functions $G(r)$ and $F(r)$ satisfy the equations

$$\hbar c \left(-\frac{dF}{dr} + \frac{\kappa}{r} F\right) = (E - mc^2 - V)G \tag{3.10a}$$

$$\hbar c \left(\frac{dG}{dr} + \frac{\kappa}{r} G\right) = (E + mc^2 - V)F \tag{3.10b}$$

where κ is an integer given in terms of the angular momentum j and parity π by

$$\kappa = (j + \tfrac{1}{2})\, (-1)^{j+1/2}\, \pi. \tag{3.11}$$

For atomic states the non-relativistic spectroscopic notation is frequently used (even though orbital angular momentum is not conserved), for example the symbols $s_{1/2}$, $p_{1/2}$, $p_{3/2}$, $d_{3/2}$, and $d_{5/2}$ correspond to the κ = -1, 1, -2, 2, and -3 states respectively.

The motion of the nucleus and its constituent nucleons can be taken into account if we generalize eqn (3.1) to a combined Dirac-Schrödinger equation for the system (i.e. we are treating the nucleus non-relativistically)

$$[H_{\text{free}} + H_{\text{N}} + H_{\text{int}} - E]\Psi = 0 \tag{3.12}$$

where

$$H_{\text{free}} = -i\hbar c\, \underset{\sim}{\alpha}.\nabla + \beta mc^2 \tag{3.13}$$

is the Dirac Hamiltonian for a free electron, H_{N} is the nuclear Hamiltonian and H_{int} the lepton-nucleus interaction. Defining the eigenstates of H_{N} by the equation

$$[H_{\text{N}} - \varepsilon_I]\phi_I(\xi) = 0 \tag{3.14}$$

and using the expansion

$$\Psi = \sum_I \psi_I(\underset{\sim}{r})\ \phi_I(\underset{\sim}{\xi}) \tag{3.15}$$

we obtain coupled Dirac equations for the spinors ψ_I

$$\begin{aligned}[H_{\text{free}} + \langle I|H_{\text{int}}|I\rangle + \varepsilon_I - E]\psi_I \\ = -\sum_{I\neq I'} \langle I|H_{\text{int}}|I'\rangle\psi_{I'}\end{aligned} \tag{3.16}$$

These can be reduced to coupled radial equations for the large and small components of the ψ_I. In practice we do not need to solve for all functions ψ_I but only for certain linear combinations of them chosen to satisfy a certain asymptotic behaviour or certain angular momentum conditions.

The interaction term is given by

$$H_{\text{int}} = e\ \underset{\sim}{\alpha}.\underset{\sim}{A} - eA_0 \tag{3.17}$$

where $A_\mu = (\underset{\sim}{A}, iA_0)$ is the potential derived from the nuclear total current (J_μ)

$$\Box A_\mu = -eJ_\mu \tag{3.18}$$

where

$$J_\mu = (\underset{\sim}{J}, i\rho_{ch})$$

and

$$\underset{\sim}{J} = \underset{\sim}{j} + c\,\nabla \times \underset{\sim}{\mu}. \tag{3.19}$$

Here $\underset{\sim}{j}$ is the convection-current operator and $\underset{\sim}{\mu}$ the magnetic-moment operator.

3.2 THE BORN APPROXIMATION

The scattering amplitude which comes from the solution of eqn (3.12) is readily obtained in the Born approximation, i.e. to first order in H_{int}. This approximation may be represented by the diagram in Fig. 3.1 which shows an electron of initial

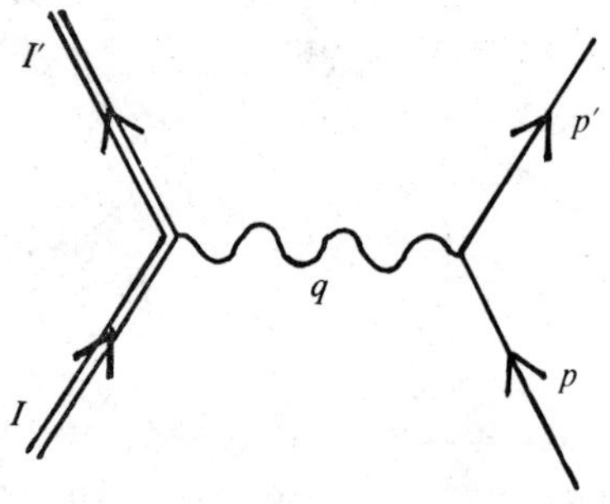

FIG. 3.1. Scattering of an electron by a nucleus.

four-momentum $(\underset{\sim}{p}, iE)$ being scattered into a state of final momentum $(\underset{\sim}{p}', iE')$ transferring momentum $(\underset{\sim}{q}, i\omega)$ to the nucleus. For scattering by a fixed nucleus (i.e. elastic scattering neglecting nuclear recoil) there is no transfer of energy and we have

$$q^2 = 4p^2 \sin^2 \frac{\theta}{2} \tag{3.20}$$

where θ is the angle between $\mathbf{p}$ and $\mathbf{p}'$.

The cross-section for scattering by a fixed-point nucleus is given to lowest order in the interaction by the 'Mott scattering formula' (Mott 1929)†

$$\left(\frac{d\sigma}{d\Omega}\right)_M = \left(\frac{2Ze^2E}{q^2c^2}\right)^2 \left(1 - \frac{q^2c^2}{4E^2}\right) . \tag{3.21}$$

If we assume that the motion of the nucleus is non-relativistic, the effect of nuclear recoil may be taken into account by dividing the cross-section by a factor‡

$$1 + \frac{2E}{Mc^2} \sin^2 \frac{\theta}{2} \tag{3.22}$$

where M is the nuclear mass. For high energies we can neglect the electron mass so that eqn (3.21) becomes

$$\left(\frac{d\sigma}{d\Omega}\right)_M = \left(\frac{Ze^2}{2E}\right)^2 \frac{\cos^2 \frac{1}{2}\theta}{\sin^4 \frac{1}{2}\theta} . \tag{3.23}$$

In the non-relativistic limit the expression (3.21) reduces to the Rutherford scattering formula. For an extended nucleus the Born approximation for the non-relativistic scattering amplitude is proportional to

$$Ze^2 \int \rho_{ch}(\mathbf{r}') \exp(i\mathbf{q}.\mathbf{r}')\, d^3r' \int \frac{\exp\{i\mathbf{q}.(\mathbf{r}-\mathbf{r}')\}}{|\mathbf{r}-\mathbf{r}'|}\, d^3r = \frac{Ze^2}{q^2} F_{ch}(\mathbf{q}) \tag{3.24}$$

where $F_{ch}(\mathbf{q})$ is the charge form factor defined in eqn (1.11).

---

†The term 'Mott scattering' is sometimes used to refer simply to the scattering of relativistic electrons by a fixed point charge, but the term 'Mott scattering formula' almost invariably means the approximate expression (3.21).

‡See for example Hofstadter (1957).

The effect of the extended charge is taken into account if the Rutherford cross-section is multiplied by the factor $|F_{ch}(\mathbf{q})|^2$. The same factor gives the relativistic cross-section in Born approximation in terms of the lowest-order Mott scattering cross-section

$$\left(\frac{d\sigma}{d\Omega}\right) = \left(\frac{d\sigma}{d\Omega}\right)_M |F_{ch}(\mathbf{q})|^2 \tag{3.25}$$

If $\rho_{ch}(\mathbf{r})$ is spherically symmetric then $F_{ch}(\mathbf{q})$ is a function of $\mathbf{q}^2$ only

$$F_{ch}(\mathbf{q}) = \frac{4\pi}{q}\int_0^\infty r \sin qr\, \rho_{ch}(r)\, dr \tag{3.26}$$

$$= 1 - \frac{q^2\langle r^2\rangle}{3!} + \frac{q^4\langle r^4\rangle}{5!} - \dots \tag{3.27}$$

where the symbol q is being used to denote the scalar quantity $|\mathbf{q}|$. The condition for validity of the Born approximation is

$$\frac{Ze^2}{\hbar v} \ll 1 \tag{3.28}$$

which in the relativistic limit becomes

$$\frac{Ze^2}{\hbar c} = Z\alpha \ll 1 \tag{3.29}$$

where $\alpha \approx 1/137$ is the fine-structure constant. Part of the effect of higher-order contributions to the cross-section may be taken into account by using an effective momentum transfer q_{eff} given by (Ravenhall and Yennie 1957)

$$q_{eff} = q(1 - V(0)/E) \tag{3.30}$$

$$= q(1 + \frac{3Z\hbar c\alpha}{2ER}) \tag{3.31}$$

where we obtain the second expression if we approximate the nuclear charge distribution by a uniform sphere of radius R. The expression for the second Born approximation has been given for point-charge scattering by McKinley and Feshbach

(1948) and Dalitz (1951) and for an extended charge by Budini and Furlan (1959). The second Born approximation amplitude is actually infinite but the contribution to the cross-section is finite. It is instructive to write down the integral which gives the second-order contribution to the amplitude:

$$f_2(\underset{\sim}{p},\underset{\sim}{p}') \propto \int d^3k \, \frac{F_{ch}(\underset{\sim}{p}'-\underset{\sim}{k})}{|\underset{\sim}{p}'-\underset{\sim}{k}|^2} \langle \underset{\sim}{k}|G_0|\underset{\sim}{k}\rangle \frac{F_{ch}(\underset{\sim}{k}-\underset{\sim}{p})}{|\underset{\sim}{k}-\underset{\sim}{p}|^2} \tag{3.32}$$

where $\langle \underset{\sim}{k}|G_0|\underset{\sim}{k}\rangle$ is the free-electron propagator. This integral shows how much more difficult it is to obtain information about the form factor and hence the charge when the lowest-order Born approximation is not valid. When this is so, the ratio at fixed $\underset{\sim}{q}^2$ of the experimental to the Mott cross-section (for a given nucleus) is no longer independent of energy and this provides a means of checking the validity of the Born approximation.

In the case of magnetic scattering and scattering by non-spherical nuclei the Born approximation again gives the scattering amplitude readily. For unpolarized electrons and target nuclei of initial energy, momentum, and angular-momentum states ε, $\underset{\sim}{p}$, and I_i respectively and final states ε', $\underset{\sim}{p}'$ and I_f the cross-section can be written (deForest and Walecka 1966)

$$\frac{d\sigma}{d\Omega} = \frac{8\pi\hbar^2\alpha^2 Z^2|\underset{\sim}{p}'|}{(\underset{\sim}{q}^2-\omega^2/c^2)^2|\underset{\sim}{p}|(2I_i+1)} \Big[V_L(\theta) \sum_{\ell=0}^{\infty} |\langle I_f \| \hat{M}_\ell^{Coul}(q) \| I_i\rangle|^2$$

$$+ V_T(\theta) \sum_{J=1}^{\infty} \{|\langle I_f \| \hat{T}_J^{el}(q) \| I_i\rangle|^2 + |\langle I_f \| \hat{T}_J^{mag}(q) \| I_i\rangle|^2\}\Big] \tag{3.33}$$

where

$$\omega \equiv \varepsilon - \varepsilon' \tag{3.34}$$

$$V_L(\theta) \equiv \frac{1}{2}\left(1 - \frac{\omega^2}{c^2\underset{\sim}{q}^2}\right)^2 \left\{\frac{(\varepsilon+\varepsilon')^2}{c^2} - \underset{\sim}{q}^2\right\} \tag{3.35}$$

$$V_T(\theta) \equiv \frac{1}{4}(\underset{\sim}{p}+\underset{\sim}{p}')^2 - \frac{\{(\underset{\sim}{p}+\underset{\sim}{p}')\cdot(\underset{\sim}{p}-\underset{\sim}{p}')\}^2}{4(\underset{\sim}{p}-\underset{\sim}{p}')^2} + \frac{1}{2}\left(\underset{\sim}{q}^2 - \frac{\omega^2}{c^2}\right). \tag{3.36}$$

The double vertical lines indicate the reduced matrix elements defined by (for example Edmonds 1957)

$$\langle I_f M_f | \hat{M}^{\text{Coul}}_{lm}(q) | I_i M_i \rangle = \frac{(I_i M_i lm | I_f M_f)\langle I_f \| \hat{M}^{\text{Coul}}_{l}(q) \| I_i \rangle}{(2I_f+1)^{1/2}}. \quad (3.37)$$

With this definition the reduced matrix element is symmetric in I_i and I_f. The operators for the electric and magnetic multipole interactions whose matrix elements give the charge scattering, the orbital part of the electron interaction with the nuclear magnetic moment, and the electron-spin interaction with the nuclear moment respectively are given by

$$M^{\text{Coul}}_{lm}(q) \equiv \int d^3r\, j_l(qr)\, Y_{lm}(\theta,\phi)\, \rho_{\text{ch}}(\mathbf{r}) \quad (3.38)$$

$$T^{\text{el}}_{JM}(q) \equiv \frac{1}{q}\int d^3r\, \nabla \times \{j_{\text{T}}(qr)\, \mathbf{Y}^{M}_{JJ'}(\theta,\phi)\}.\mathbf{J}(\mathbf{r}) \quad (3.39)$$

$$T^{\text{mag}}_{JM}(q) \equiv \int d^3r\, j_{\text{T}}(qr)\, \mathbf{Y}^{M}_{JJ'}(\theta,\phi).\mathbf{J}(r) \quad (3.40)$$

For elastic scattering the matrix element of the transverse electric multipole operator $\hat{T}^{\text{el}}_{J}$ vanish due to parity conservation and time-reversal invariance. Eqn (3.38) will be used for the scattering from deformed nuclei considered in §3.5, and the operators (3.39) and (3.40) contribute to the magnetic scattering discussed in §3.6.

3.3. CHARGE SCATTERING BY SPHERICAL NUCLEI

To calculate the elastic-scattering cross-section for large momentum transfers or for heavy nuclei it is necessary to solve the Dirac equation numerically. The earliest calculation in which this was done for an extended charge distribution was carried out by Elton (1950) who calculated the cross-section for the scattering of 20 MeV electrons on mercury. A detailed description of the partial-wave phase-shift analysis was given by Yennie *et al.* (1954). The equations (3.10) can be solved by standard techniques for each value of κ which contributes to the scattering. This involves calculating the functions F and G near the origin using a power-series expansion and then integrating step by step to a radius outside of the nuclear region. The functions F and G must then be matched on to a linear combination of the regular and irregular relativistic point-charge Coulomb functions (obtained using an asymptotic series) and the particular linear combination needed gives the phase shift for the particular κ value under consideration. In practice there are a number of difficulties in doing the numerical calculations, especially for high-energy scattering where the number of partial waves is large and there is almost complete cancellation between the different partial waves. This means that the phase shifts must be calculated to an accuracy many orders of magnitude greater than that required for the final cross-section. Another difficulty occurs in calculating the asymptotic series for the relativistic Coulomb functions. Here the problem is that the asymptotic series may not converge fast enough at the matching radius and it becomes necessary to evaluate them at some larger radius (the 'asymptotic radius') and integrate inwards to the matching point. The appropriate value of this asymptotic radius may be different for the different partial waves. In order to improve convergence of the partial-wave expansion of the scattering amplitude Yennie *et al.* (1954) developed a method of obtaining 'reduced series', i.e. series for the function $(1 - \cos\theta)^m$ times the scattering amplitude. These series converge much more rapidly for suitably chosen values of m (e.g. $m = 3$ for 125 MeV electrons on gold).

3.4. MODEL-DEPENDENT AND MODEL-INDEPENDENT ANALYSES OF SCATTERING

There are a number of corrections to the simple theory of the scattering of an electron by a single structureless fixed nucleus and they are considered in §§3.7 and 3.8. Once these corrections have been calculated and the numerical difficulties of accurately solving the Dirac equation overcome, it is easy to obtain the scattering cross-section for any given shape of the charge distribution. It would be convenient for nuclear charge distribution studies if we could perform the inverse process of obtaining the charge distribution from electron-scattering cross-sections, but this is not possible. Until recently the analysis of scattering experiments was done by taking a simple function $\rho_{ch}(\underset{\sim}{r}, x_1, x_2, \ldots)$ of the position vector $\underset{\sim}{r}$ and a small number of size and shape parameters x_i, and adjusting the parameters to give a least-squares deviation (minimum χ^2) between theoretical and experimental cross-sections. The resulting density is 'model dependent' in the sense that it depends on the assumed form of the function ρ_{ch} and we shall refer to it as a 'model density'. Sometimes several densities have been found by carrying out a least-squares fit for several functions ρ_{ch} and the range of the resulting model densities (which gave good fits) has been taken to be an indication of the error in the density. A much more rigorous estimate of the errors has been obtained in recent 'model-independent' analyses in which the inverse process was carried out after 'physically reasonable assumptions' had been made about the minimum wavelength of oscillations in the radial shape and about the radius beyond which the charge density could be assumed to be zero. Some of these model-independent analyses are described below.†

3.4.1. *Model-dependent analyses*

The earliest electron-scattering experiments were at low momentum transfer so that only the first two terms of the expansion (3.27) were important. This meant that the scattering could be reproduced using any distribution with the right mean-square radius (Bodmer 1953), and there was nothing to distinguish between extremes such as an exponential shape and a uniform charge distribution, or even a spherical shell. Since the nuclear binding energy per nucleon was known to be very nearly constant, nuclear matter and nuclear charge densities were expected to be approximately constant. Further

---

†A recent compilation of nuclear charge distribution parameters from elastic electron scattering contains a brief review of these analyses (de Jager, de Vries, and de Vries 1974). A detailed discussion has been given by Friar and Negele (1975*b*).

evidence for this picture came when higher momentum-transfer experiments for heavy nuclei showed a diffraction pattern which ruled out exponential or gaussian shapes and indicated some sort of edge to the nucleus (Ravenhall and Yennie 1954). A number of different two-parameter radial distributions were used successfully by various authors (for example, Yennie *et al.* 1954, Brown and Elton 1955), and these gave a density which was more or less constant in the interior and had a diffuse edge. Gradually the Fermi distribution†

$$\rho_{\mathrm{ch}}(r) = \rho_0\left[\exp\{(r-c)/a\} + 1\right]^{-1} \tag{3.41}$$

became more popular than other shapes and remained so until very recently. A common modification has been a 'parabolic Fermi' shape in which a factor $(1 + wr^2/c^2)$ multiplied the Fermi distribution. (For $\omega > 0$ this is sometimes called a 'wine-bottle' shape.) The most recent experiments have indicated that a Fermi distribution does not give quite the right shape to the tail of the density, for example the analysis of high-energy scattering from ^{208}Pb by Heisenberg *et al.* (1969) has shown that a better fit is obtained using a 'modified gaussian' distribution of the form

$$\rho_{\mathrm{ch}}(r) = \rho_0(1 + w^2r^2/c^2)\left[\exp\{(r^2-c^2)/a^2\} + 1\right]^{-1}. \tag{3.42}$$

When experiments were done on Ca isotopes at very high momentum transfer it was found that the usual two- or three-parameter functions gave cross-sections which did not fit the third diffraction minimum (Bellicard *et al.* 1967), and the approach used in the analysis was to choose an additional three-parameter distribution $\Delta\rho_{\mathrm{ch}}$ of zero total charge and of such a shape that its form factor was a gaussian peaked at large momentum transfer (3 fm^{-1}).

In a number of recent analyses the muonic-atom transition energies have been fitted simultaneously with the data

---

†See §1.2.2.

from electron scattering (for example Frosch *et al.* 1968, Heisenberg *et al.* 1969). We shall see in Chapter 4 that the information about charge distributions which is provided by muonic-atom experiments is to a large extent independent of that provided by electron scattering, except in the case of 2p-1s X-rays and low-momentum-transfer scattering on light nuclei in which both types of experiments give the mean-square radius of the charge.† The analysis of experiments in terms of phenomenological densities is useful because it checks the consistency of the data to some extent and provides a starting point for some model-independent analyses, but there has been a tendency to attach more significance to these densities than was justified, based on the belief that, when they gave really good fits to the data, they were themselves very nearly model independent. Calculations carried out very recently have shown that this belief was too optimistic (for example Friar and Negele 1973). This means that parameters of the charge distribution such as the half-density radius, 90-10 per cent skin thickness, and mean square radius are in most cases not determined to the accuracies quoted in the literature. In fact the first two of these can be meaningfully defined only for very simple forms of ρ_{ch}.

Another approach which has been used in electron scattering calculations is a semi-phenomenological approach in which constraints on ρ_{ch} based on nuclear theory are introduced.‡ The earliest examples were densities obtained by

---

†For electron scattering on light nuclei this is apparent from the expansion in q^2 of the form factor. For heavy nuclei the result remains true even though the Born approximation is not valid, i.e. the cross-section is not given by the product of the Mott formula and the square of the form factor (van Niftrik 1969).

‡Such constraints are always present in the sense that any choice of the functional form of ρ_{ch} is affected by knowledge of our prejudice regarding the charge density. We are considering here very explicit constraints.

assuming that the motion of the protons was described by a single-particle shell model with the protons moving in a potential (sometimes state or energy dependent) and adjusting the parameters of the nuclear *potential* to fit the electron scattering. Such *single-particle model densities* were first obtained for light nuclei using state-dependent harmonic-oscillator wavefunctions (for example, Amaldi, Fidecaro, and Mariani (1950), Hofstadter (1957), and many references therein). This method was extended to medium and heavy nuclei by Elton and collaborators (Shaw, Swift, and Elton, 1965, Swift and Elton 1966, Elton and Swift 1967) who obtained wavefunctions for protons moving in a Woods-Saxon potential:

$$V(r) = V_{\mathrm{C}} - V_{\epsilon} f(r) + V_{\mathrm{s}\epsilon}\left(\frac{\hbar}{m_{\pi}c}\right)^2 \frac{1}{r}\frac{\mathrm{d}f}{\mathrm{d}r}\,\underset{\sim}{\ell}\cdot\underset{\sim}{\sigma} \qquad (3.43)$$

where V_{C} is the Coulomb potential and $f(r)$ has the same shape as a Fermi distribution,

$$f(r) = \left[\exp\left(\frac{r-R}{a}\right) + 1\right]^{-1}. \qquad (3.44)$$

At first the strengths V_{ϵ} and $V_{\mathrm{s}\epsilon}$ for the central and spin-orbit parts of the potential were taken to be the same for all levels. The parameters of the potential were varied, subject to the constraint that the single-particle energies agreed fairly well with the proton separation energies, until the resulting charge distribution gave the best fit to the electron-scattering cross-sections. It was subsequently found that if V_{ϵ} and $V_{\mathrm{s}\epsilon}$ were made energy dependent to simulate the non-locality of a Hartree-Fock potential, it was possible to obtain better fits to the separation energies. This method does not allow much freedom in varying the shape of the charge distribution (Elton 1967*a*) and the success of the method in fitting medium-energy experiments was therefore impressive. To obtain good fits, however, it was necessary to adjust the parameters individually for each nucleus and the resulting parameters sometimes showed unexpectedly large variations between nuclei (for example Elton and Swift 1967). When the momentum transfer of the experiments reached 3 fm^{-1}, the single-particle model failed to give satisfactory fits and it became necessary to take account of correlations. Elton and Webb (1970) did this by introducing additional arbitrary parameters which made this type of analysis much less convincing.

Although it is possible to feel a certain amount of confidence in the accuracy of single-particle-model densities which give good fits to electron scattering, it is extremely difficult to calculate the errors due to uncertainties and simplifying assumptions in the theory. Several authors have recently been carrying out 'model-independent' analyses of scattering which differ from previous calculations in two important respects: firstly, a much more general function (with many more parameters) is used, and secondly a serious effort is made to estimate the errors. The errors previously quoted after fitting Fermi-type distributions were simply a (very low) lower limit to the errors.

3.4.2. *Model-independent analyses*

The success of an analysis in which $\rho(r)$ is expanded in a complete set of functions $\{f_i(r)\}$ depends critically on the choice of the functions. It is convenient first to consider the analysis of a low-Z nucleus for which the Born approximation is valid so that the scattering determines the form factor $F(q)$ over a range of momentum transfer q up to a certain measured maximum value $q_{\max}$. For complete knowledge of the charge distribution (without any *a priori* assumptions) it would be necessary to know $F(q)$ for further values of q ranging from $q_{\max}$ to infinity. One criterion for the choice of $\{f_i\}$ is that the functions whose coefficients are determined by $F(q)$ for $q < q_{\max}$ should be orthogonal to those whose coefficients depend on $F(q)$ for $q > q_{\max}$. A simple expansion which satisfies this criterion very well is a Fourier sine series suggested by eqn (3.26) which defines $F(q)$:

$$\rho_{\text{ch}}(r) = \frac{1}{r} \sum_{n=1}^{\infty} C_n \sin(n\pi r/R) \quad r \leqslant R \tag{3.45}$$

R is chosen so that $\rho(r)$ is negligible and the density is assumed to be zero for $r > R$. The coefficients are given by

$$C_n = q_n F(q_n)/2\pi R \tag{3.46}$$

where

$$q_n = n\pi/R. \tag{3.47}$$

Such an expansion was used by Meyer-Berkhout *et al.* (1959) in the analysis of scattering by several light nuclei. They terminated the series for a maximum value of $n(n_{max})$ of 5 or 6. The results for ^{16}O are shown in Fig. 3.2.

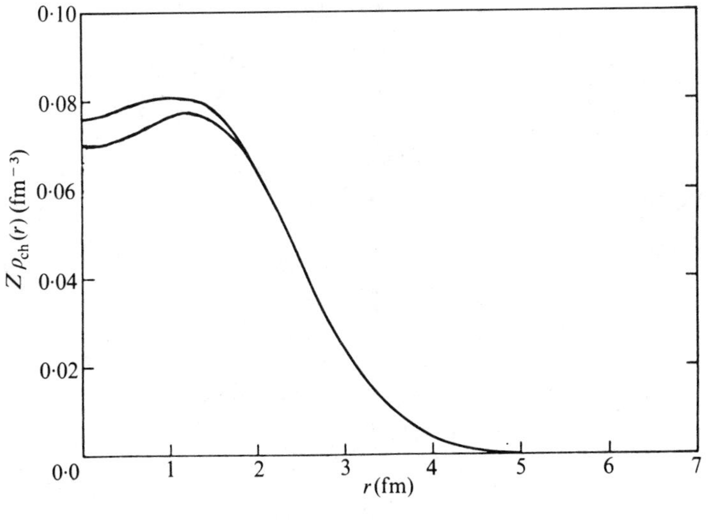

FIG. 3.2. Charge densities for ^{16}O obtained using a Fourier sine series for $r\rho(r)$. The upper and lower curves result from taking a cut-off at R = 6 fm and R = 5 fm respectively (From Meyer-Berkhout *et al.* 1959).

This approach has been modified by Friar and Negele (1973) who have developed a method (applicable also to heavy nuclei) of combining as much model independence as possible with theoretically motivated model dependence. The tail of $\rho_{ch}(r)$, which is poorly determined by experiment, can be taken with reasonable certainty to be that given by a Hartree-

Fock calculation. The method of Friar and Negele is to assume that the density is of the form $\rho_0 + \delta\rho$ where ρ_0 is the Hartree-Fock density (or other reasonable density), and $r\delta\rho$ is assumed to be zero outside a maximum radius R and is expanded for $r < R$ as a Fourier sine series. If the Dirac equation is solved exactly for the density ρ_0 to obtain cross-sections $\sigma_0(q_i)$ and muonic atom X-ray energies $E_0^{\,i}$, then, for a change $\delta\rho$ in the density, first-order perturbation theory gives integrals for the changes in these quantities which are linear in $\delta\rho$ so that the coefficients of the series may be obtained for the best fit to the data. The error due to using perturbation theory may be removed by solving the Dirac equation exactly for $\rho_0 + \delta\rho$ and using perturbation theory to find the coefficients corresponding to an additional change in the density. For ^{208}Pb Friar and Negele found that with 11 terms in the sine series the method converged with three iterations.

The error $\Delta\rho_{\text{ch}}$ on the charge density is given immediately by the errors on the coefficients, from which the uncertainty in integral quantities of the form

$$\langle G \rangle = \int_0^\infty g(r)\rho_{\text{ch}}(r)r^2 \mathrm{d}r \tag{3.48}$$

is easily obtainable. These errors do not have any significance unless all contributions to the cross-section have been calculated, and the experimental errors are purely statistical. The density and errors obtained by Friar and Negele for ^{208}Pb are shown in Fig. 3.3. They used 11 terms in the expansion and a value of R of 11 fm. The fact that the errors are smaller than the distance between the curves obtained with and without the inclusion of muonic X-ray data indicates that there are systematic experimental errors or that the theory is not sufficiently accurate. Another quantity which has been given by Friar and Negele is the correlation between the errors at different radii; only a small fraction of all possible densities in the band $\rho_{\text{ch}}(r) \pm \Delta\rho_{\text{ch}}(r)$ would give a good fit to the scattering, and this feature is described by the correlation function

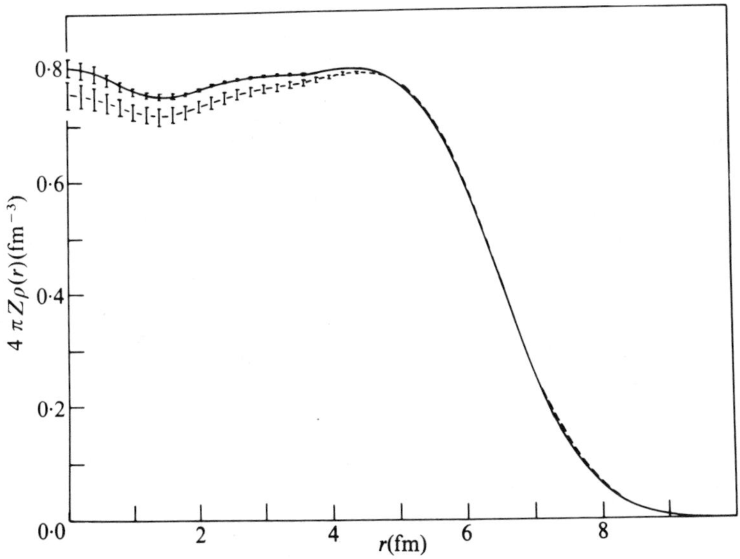

FIG. 3.3. The charge density of ^{208}Pb from the model-independent analysis of Friar and Negele (1973). The solid curve comes from the simultaneous analysis of muonic X-ray and electron-scattering data and the broken curve from electron scattering alone. The vertical bars represent the statistical errors in the density.

$$f(r,r') = \frac{\overline{\{\rho(r) - \overline{\rho(r)}\}\{\rho(r') - \overline{\rho(r')}\}}}{\Delta\rho(r)\ \Delta\rho(r')} \tag{3.49}$$

Fig. 3.4 shows the function $f(r,r')$ for ^{208}Pb with $r' = 0.25$ fm. The significance of the function $f(r, 0.25)$ is as follows: if the densities $\rho(r)$ and $\rho(r) + \delta\rho(r)$ both give statistically probable fits to the data and $\Delta\rho(0.25)$ is positive (negative), then $\rho(r)$ tends to have the same (opposite) sign as (to) $f(r, 0.25)$, and certainly has the same (opposite) sign if $|f(r, 0.25)| = 1$.

It is obviously impossible to rule out high-frequency oscillations corresponding to values of n greater than $q_{max}R/\pi$ on the basis of scattering experiments alone. Two arguments are used to suggest that the maximum frequency of such oscillation is limited: firstly, by the finite size of

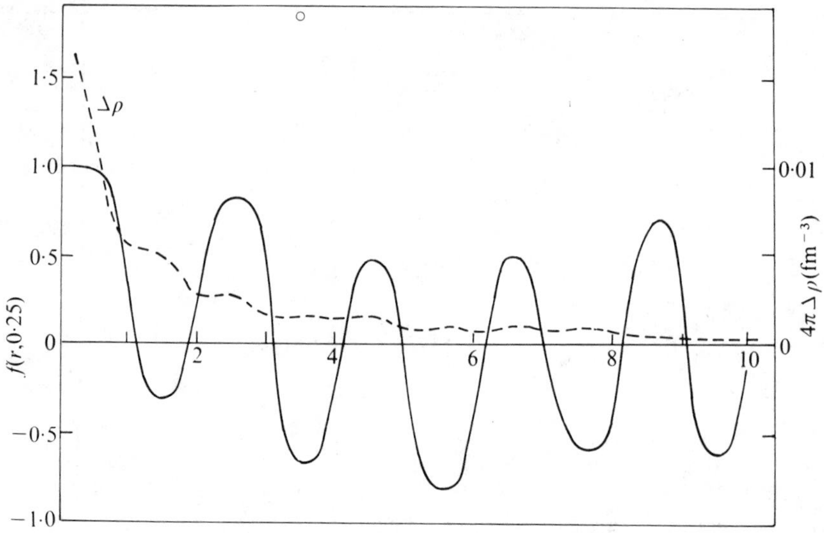

FIG. 3.4. The correlation function $f(r, 0.25)$ defined by eqn (3.49) and the error Δρ (broken curve) obtained in the analysis of ^{208}Pb by Friar and Negele (1973).

the proton whose r.m.s. radius is 0.83 fm (see Table 6.1) and secondly by the minimum wavelength of the interior oscillations of the wavefunctions of the protons, which is of the order of 1 or 2 fm. For ^{208}Pb the wavelength of the first neglected term (n = 12) is 2 fm, which suggests that the experimental q_{max} of 2.7 fm^{-1} is not quite large enough for the uncertainties to be eliminated by the proton size and wavefunction arguments. The *completeness* error, i.e. the error due to the fact that an arbitrary shape, even if it is quite smooth, cannot be represented by a finite Fourier series, was found by Friar and Negele to be small compared with the errors in the density due to statistical experimental errors.

A number of other complete sets of functions have been used to represent the charge distribution, for example Hermite polynomials (Meyer-Berkhout *et al.* 1959), Laguerre polynomials (Friar and Negele 1973), and cosine, spline, and unspline functions (Borysowicz and Hetherington 1973). The first two of these do not come close to satisfying the desired orthogonality between functions determined by values of q below and above q_{max}. The last three were all used in the analysis of ^{3}He and ^{4}He experiments to check the dependence of the derived charge distributions on the series expansion. The dependence was very slight (except in the case of the discontinuous slab function). These functions do not seem to be as suitable as the sine series expansion for $r\rho_{ch}(r)$ which has the advantage that each coefficient is determined by the measurements over a very small range of momentum transfer.

Another method for obtaining information about the charge density has been developed by Hetherington and Borysowicz (1974). They have an elegant technique for making the analysis of electron-scattering experiments independent of the basis used in the expansion of $\rho_{ch}(r)$: the necessary model dependent assumptions about the asymptotic behaviour of $\rho(r)$ and the form of $F(q)$ beyond the measured region are incorporated into the analysis by adding the terms $\int X(r)\{\rho(r)\}^2\,dr$ and $\int Y(q)\{F(q)\}^2\,dr$ to χ^2 before minimizing,† and choosing $X(r)$ and $Y(q)$ appropriately. For example they choose exponentially increasing functions of r and q which constrain $\rho(r)$ and $F(q)$ to be within exponentially decreasing envelopes. Minimizing the modified χ^2 gives an operator equation for the best density which can be solved by inverting the 'error operator' H. The last step is achieved by calculating the matrix elements of H in a convenient basis such as harmonic-oscillator functions. The coefficients (and errors in the coefficients) for the expansion of the density in any set of basis functions can be obtained immediately, and the most appropriate basis is the set of eigenstates of the error operator H.

A very different approach to the problem of obtaining information about the charge distribution was introduced by

---

†The form of χ^2 used is the weighted sum of the squared deviations between the square root of the cross-sections and the form factor, instead of the usual form which is defined in terms of the squared deviations between the cross-sections and the square of the form factor.

Lenz (1968, 1969), Friedrich and Lenz (1972), and in modified form by Dreher *et al.* (1974) and Sick (1973*a*,*b*). The principal features of this method are the complete freedom to vary one region of the density without affecting any other and the fact that calculations can be carried out so rapidly that it is possible to investigate thousands of charge distributions. (Friedrich and Lenz calculated the cross-section for scattering by ^{208}Pb for 90 000 charge distributions using a random-number generator and found several hundred which gave good fits.) The expansion used was a set of spherical shells

$$\rho_{\text{ch}}(r) = \sum_{i=1}^{N} p_i \; \delta(r-R_i)/4\pi R_i^{\,2} \tag{3.50}$$

where the radius R_i of each shell and the fraction p_i of the total charge which it contains are treated as parameters of the charge distribution. (In practice many sets of fixed values of p_i were used and the radii R_i were varied to fit the scattering.) This is obviously not intended to be a realistic representation of ρ_{ch} but the many different densities obtained can be used to calculate integral properties of the charge density, and the range of values obtained gives a measure of the error in the integral property. Two of the integral properties investigated by Friedrich and Lenz were the root mean kth moment of the density M_k and the partial average radius $R(Q)$

$$M_k \equiv (\langle r^k \rangle)^{1/k}$$

$$\equiv \{\int r^k \rho_{\text{ch}}(r)\; \mathrm{d}^3 r\}^{1/k} \tag{3.51}$$

$$R(Q) \equiv \frac{4\pi}{Q} \int_0^a r \; \rho_{\text{ch}}(r)\; r^2 \mathrm{d}r \tag{3.52}$$

where

$$Q \equiv 4\pi \int_0^a \rho_{\text{ch}}(r)\; r^2 \mathrm{d}r. \tag{3.53}$$

The quantity a is the radius of the sphere which contains a fraction Q of the charge and $R(Q)$ is the mean radius of the charge in this sphere. Figs. 3.5 and 3.6 show the functions as determined in an analysis of scattering on ^{208}Pb (with $q_{max} = 1.3\ fm^{-1}$). The errors represent the separation from

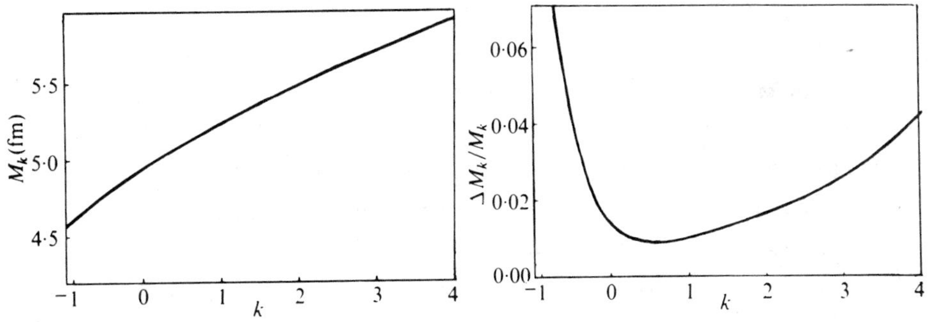

FIG. 3.5. The graph on the left shows an average (over all the δ-shell charge distributions which fit the scattering) of the moment M_k plotted against k. The graph on the right shows values of $\Delta M_k/M_k$ where ΔM_k is the spread in M_k over the different distributions. (From Friedrich and Lenz 1972.)

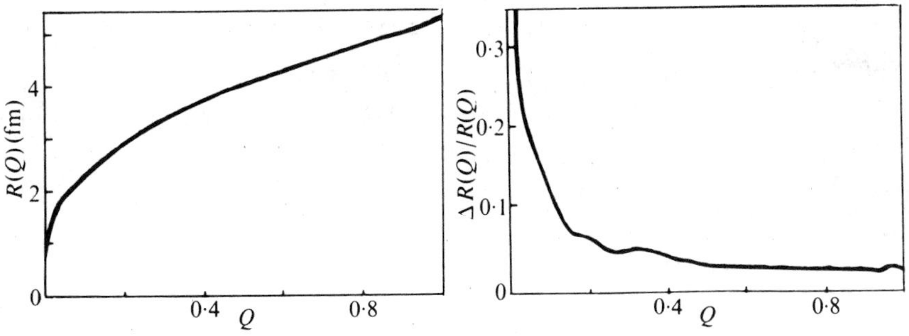

FIG. 3.6. The curve on the left shows an average (over δ-shell distributions) of the mean radius $R(Q)$ of the innermost Qth fraction of the charge density plotted against Q; the right-hand curve shows values of $\Delta R(Q)/R(Q)$ where $\Delta R(Q)$ is the spread in $R(Q)$ over the different distributions. (From Friedrich and Lenz 1972.)

the mean of the most extreme curves obtained from hundreds of spherical shell densities. They come from densities which are 'compatible' with the cross-sections and represent the uncertainty due to the lack of data for $q > q_{max}$ as well as that due to the statistical errors in the cross-sections.

If we want to obtain a charge density from a δ-shell analysis it is necessary to smooth out the delta functions. The method used by Friedrich and Lenz was to fit parametrized functions to $R(Q)$ and then calculate $\rho_{ch}(r)$ from $R(Q)$. Less arbitrary densities together with errors have been obtained using modifications of the Lenz method by Sick (1973 , 1974) who carried out calculations in which the sum of spherical shells was replaced by a sum of gaussians (SOG) centred about radii R_i

$$\rho_{ch}(r) = \sum_i \frac{p_i}{2\pi^{3/2}\,\gamma(\gamma^2+2R_i^2)}\left[\exp\{-(r-R_i)^2/\gamma^2\} + \exp\{-(r+R_i)^2/\gamma^2\}\right] \quad (3.54)$$

where p_i is the fraction of charge in the ith gaussian and the (very small) second term is added to eliminate a cusp at $r = 0$. The width γ is chosen either so that the mean-square radius of the gaussian is equal to that of the proton or so that the full width $\Gamma = 2\gamma \ln 2$ of the gaussian is equal to the minimum full width of oscillation obtained using Hartree-Fock wavefunctions. (The corresponding values are $\Gamma \approx 1.1$ and 1.8 fm respectively.) The form factor corresponding to the SOG charge density is

$$F(q) = \exp(-q^2\gamma^2/4) \sum_i \frac{p_i}{\gamma^2+2R_i^2} \left\{\gamma^2\cos(qR_i) + 2R_i^2 \frac{\sin(qR_i)}{qR_i}\right\} \quad (3.55)$$

$$\underset{\gamma\to 0}{\to} \sum p_i \frac{\sin(qR_i)}{qR_i} \quad (3.56)$$

which in the $\gamma = 0$ limit is the form factor for the δ-shell density. The main effect on the form factor of the smearing of the δ-shells is to dampen the oscillations at large a by the rapidly decreasing factor $\exp(-q^2\gamma^2/4)$. In his analyses Sick randomly generates many sets of radii R_i evenly distributed (on average) in the range $0 < R_i < R_{max}$ and adjusts the p_i to give the best fit to the data for each set. Sick found that in the case of ^{12}C and ^{32}S the variation in the densities obtained from the different sets of R_i was much greater than the error due to the statistical errors in the measurements, i.e. the errors which he quotes are due to the lack of knowledge of higher q data. (For these nuclei the value of q_{max} is ~4 fm^{-1}.) Examples of densities and errors obtained by Sick (1974) are shown in Figs. 3.7 (for ^{12}C) and and 3.8 (for ^{32}S) and by Sick *et al.* (1975) in Fig. 3.9 (for ^{58}Ni). The cross-sections used for the ^{58}Ni analysis are shown in Fig. 3.10 (Sick *et al.* 1975).

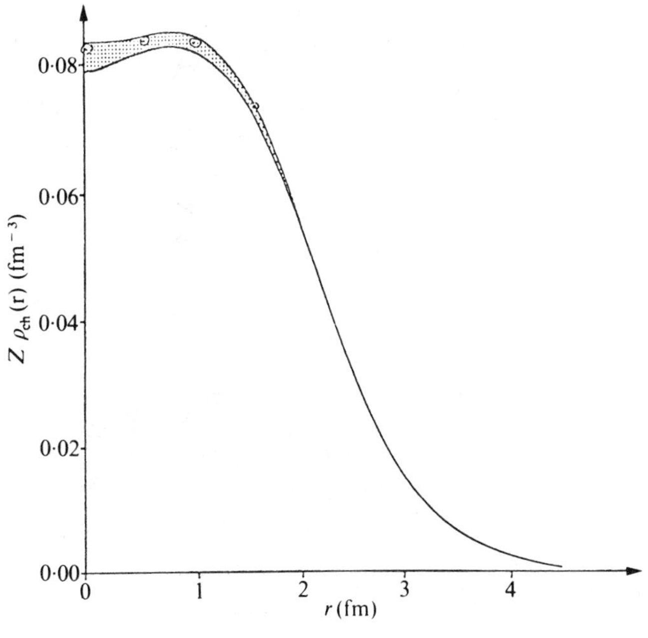

FIG. 3.7. Charge density of ^{12}C. The circled points correspond to a modified harmonic-oscillator model based on the experiment and analysis of Sick and McCarthy (1970). The shaded area gives the range of allowed densities obtained from a model-independent analysis of the same data. (From Sick 1974.)

In order to reduce the amount of computation necessary, Sick linearizes the calculation by replacing the experimental

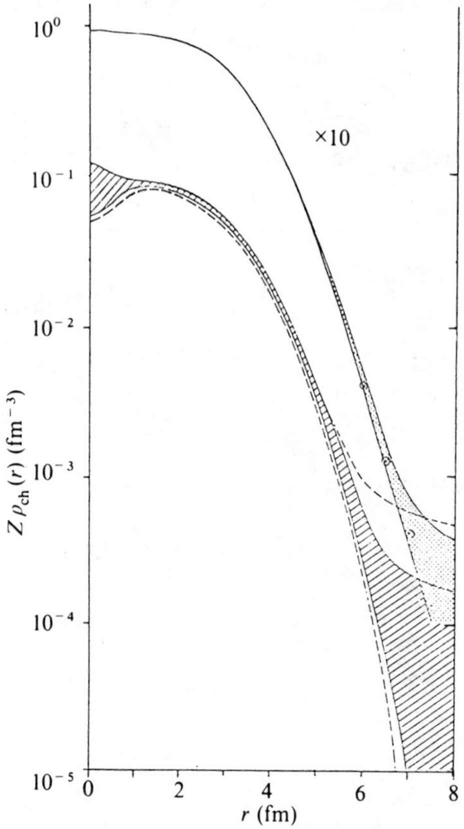

FIG. 3.8. A semi-log plot of the charge density of ^{32}S. The circled points show 10 times the model density obtained from the high-q data of Li, Sick, and Yearian (1974). The dotted shaded area shows the range of densities obtained by Sick (1974) from a model-independent (SOG) analysis of the high-q data and muonic X-ray energies. The lined shaded area comes from his analysis of the low-q data of Hultzsch (1970) and muonic energies (Backenstoss *et al.* 1967*b*; Suzuki 1967). The broken curves show the result of omitting the muonic data from the low-q analysis.

cross-sections by Born approximation pseudo-data (Negele 1970). These are obtained using a phenomenological density $\rho_0(r)$ which gives cross-sections $(d\sigma/d\Omega)^0$ close to the experimental values. The pseudo-data are then taken to be the quantities

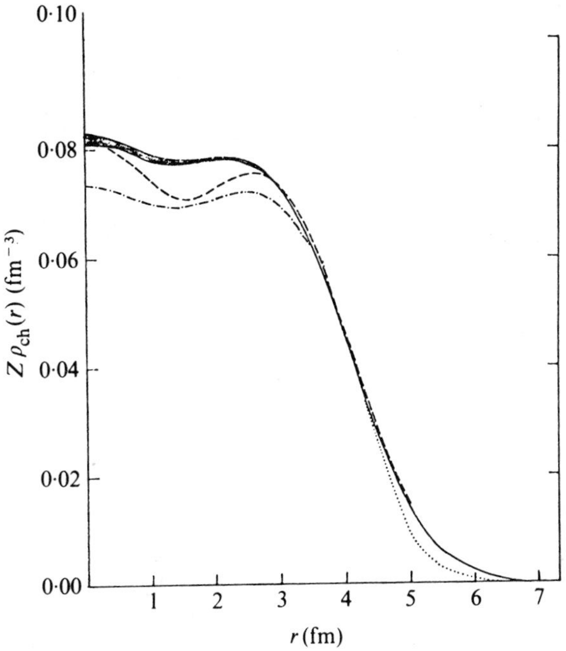

FIG. 3.9. Charge density of ^{58}Ni. The full curve gives the range of densities which fit the Saclay and Stanford electron scattering experiments (see Fig. 3.10) and the muonic X-ray data of Ehrlich (1968). The broken curve is from a density-dependent Hartree-Fock calculation of Campi (1975) using an effective interaction derived from a Reid soft-core potential. The dash-dot curve is from a similar calculation of Flocard (1975) using a Skyrme force. (From Sick *et al.* 1975.)

$$\{F_{ps}(q_{eff})\}^2 \equiv \frac{(d\sigma/d\Omega)^{exp}}{(d\sigma/d\Omega)^0}\{F_0(q_{eff})\}^2 \tag{3.57}$$

where $F_0(q_{eff})$ is the form factor corresponding to $\rho_0(r)$ at the appropriate effective momentum transfer (as defined by eqn (3.30)). The SOG densities can then be obtained very rapidly using the expression (3.55) and varying the coefficients p_i to give a best fit. The validity of this method has been checked by calculating the exact cross-section with a SOG density obtained from the pseudo-data.

The method used by Dreher *et al.* (1974) to obtain a continuous density from the δ-shell model also depends on the use

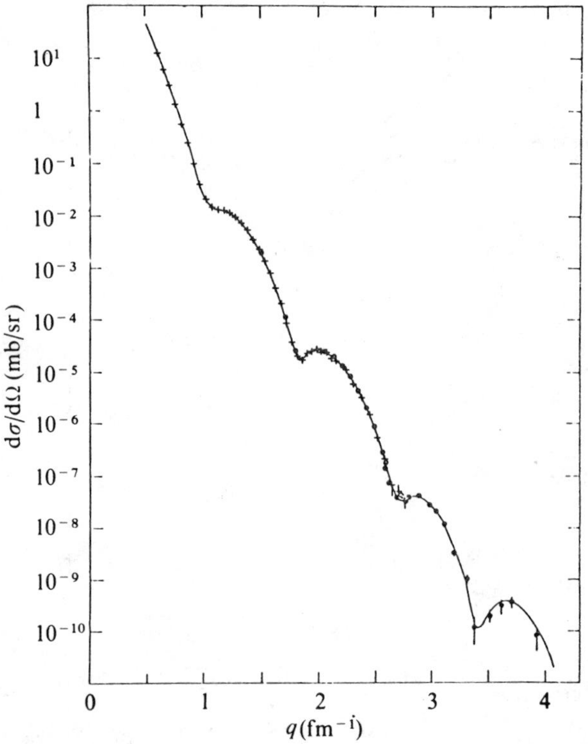

FIG. 3.10. Experimental cross-sections for elastic scattering on ^{58}Ni as a function of momentum transfer. The circles are from the 449.5 MeV Saclay experiment of Sick *et al.* (1975). The crosses are from the Stanford data of Ficenec *et al.* (1970) transformed to 449.5 MeV. The solid curve is the theoretical cross-section from the fitted density (see Fig. 3.9).

of form factors for a situation in which the Born approximation is not valid, namely the analysis of the scattering on ^{208}Pb. Dreher *et al.* obtain the form factors $F_{\delta}(q)$ for all the δ-distributions which give good fits to the scattering and find that the spread in the region of measured q values is very small, especially outside of the minima and that (in the measured range of q) the maxima of the function $qF(q)$ are very nearly equally spaced and fall on a curve proportional to $\exp(-aq/h)$ as can be seen in Fig. 3.11 (Dreher *et al.* 1974). They then extrapolate $qF(q)$ into the region for $q > q_{max}$ and

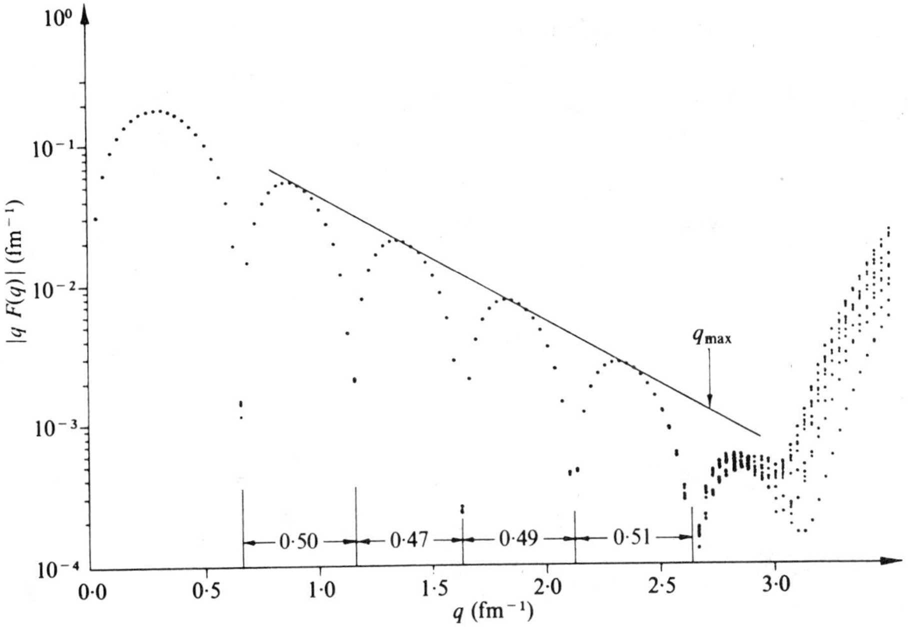

FIG. 3.11. Plot of $|qF(q)|$ for ^{208}Pb for all δ-shell densities obtained by Dreher *et al.* (1974) which fit the high-energy data of Heisenberg *et al.* (1969).

take the Fourier transform to obtain a charge density. Dreher *et al.* calculated errors on the charge density from the statistical errors in the cross-sections and found that they were approximately proportional to $1/r$.

The arguments given by Sick suggest that an exponential gives too slow a fall off in $qF(q)$, in which case the corresponding coefficients $q_n F(q_n)/2\pi R$ of the Fourier sine expansion for $r\rho_{ch}(r)$ could be used to estimate an upper limit to the error in the density due to lack of knowledge of $F(q)$ for $q > q_{max}$. A simple argument gives the result $\Delta(r\rho_{ch}(r)) \lesssim 0.00004\ \text{fm}^{-2}$. This is about the same order of magnitude as the uncertainty due to the statistical experimental errors. Dreher *et al.* have also analysed the ^{208}Pb data using the same expansion as Friar and Negele (eqn 3.45) together with the assumption that the form factor beyond the measured region is proportional to $q^{-4}\exp(-bq^2)$. Here the first factor is deduced from the single-particle-model wavefunctions and the second is the functional form assumed to take account of the proton form factor. The densities obtained by Dreher *et al.* using the two methods are shown together with the Hartree-Fock calculation of Faessler, Galonska, and Goeke (1972*b*) in Fig. 3.12. Apart from the region $r < 1$ fm the errors are almost entirely due to the statistical errors on the experiments. This is achieved by making fairly restrictive assumptions about $F(q)$ beyond the measured region.

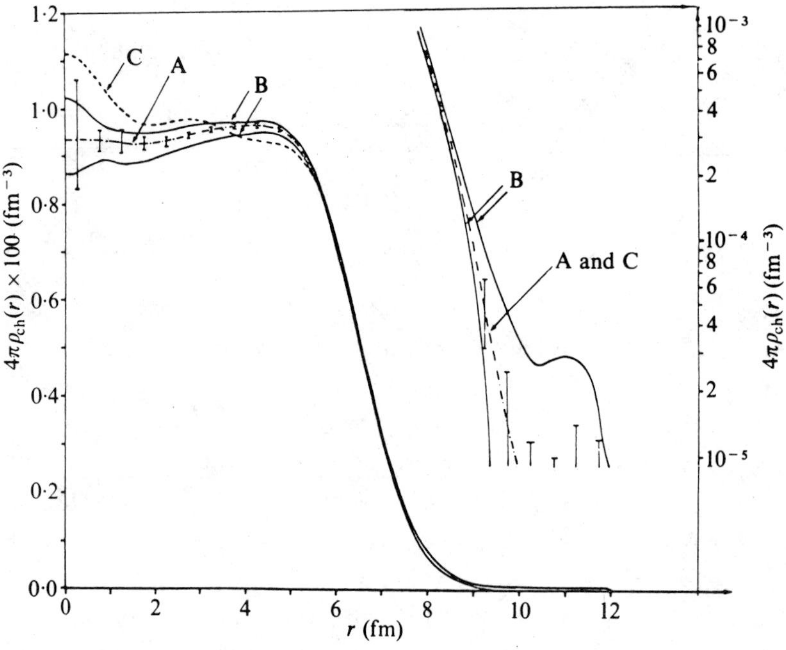

FIG. 3.12. The charge density of ^{208}Pb. The broken curve is from a Hartree-Fock calculation of Faessler *et al.* (1972*b*) and the remaining curves come from two analyses of the data of Heisenberg *et al.* (1969) by Dreher *et al.* (1974); the solid curves represent the range of densities (with 90% confidence limits) obtained using a Fourier sine series for $r\rho_{ch}(r)$; the dash-dot curve and the error bars come from a Fourier transform of the form factor (obtained from a δ-shell analysis) extrapolated beyond the measured region.

The analysis of data in terms of the very much more general functions for $\rho_{ch}(r)$ described above, which began with the method of Lenz (1968), has demonstrated that earlier analyses in terms of simple functional forms with only a few parameters were very misleading and resulted in a misplaced confidence in the accuracy to which it was possible to determine charge densities. The new methods have produced densities with error bars which are realistic (subject to the physically very reasonable assumptions about the density which have to be made before any analysis is possible). In certain cases the inclusion of additional data, such as muonic-atom energy levels, shifts the densities outside the limits obtainable without the additional data (Friar and Negele 1973). This indicates that there are significant systematic errors in addition to statistical errors in the data, or, alternatively, that there are errors in the theory. The reliability of the present methods of analysis means that, as more accurate measurements are completed for different nuclei, particularly those involving measurements up to large momentum transfer, the actual charge densities in configuration space will become one of the most accurately known properties of nuclei.

3.5. CHARGE SCATTERING BY DEFORMED NUCLEI

In order to consider the contribution of charge scattering by a deformed nucleus to the Born-approximation cross-section given in eqn (3.33) we need to obtain the matrix elements of $M^{\text{Coul}}_{lm}(q)$ defined in eqn (3.38)

$$\hat{M}^{\text{Coul}}_{lm}(q) \equiv \int d^3r \; j_l(qr) \; Y_{lm}(\theta,\phi) \; \rho(\underset{\sim}{r}) \qquad (3.38)$$

To do this we take the deformed nucleus[†] to have an intrinsic

---

[†]The reader is referred to the paper of Bohr and Mottelson (1953) in which the collective model for a deformed nucleus was introduced, and to expositions in many books (for example de Shalit and Feshbach 1974) which emphasize that a spherical nucleus of spin zero may nevertheless possess a non-zero intrinsic quadrupole moment. Strictly speaking we should use

charge density $\rho_{ch}^{int}(\underset{\sim}{r}')$ about its (body-fixed) axis of symmetry, and expand this in multipoles of charge density referred to general (space-fixed) axes:

$$\rho_{ch}^{int}(\underset{\sim}{r}') = \sum_{l} \rho_l(r') \; Y_{l0}(\theta') \tag{3.58}$$

$$= \sum_{lm} \rho_l(r) \left(\frac{4\pi}{2l+1}\right)^{1/2} Y^*_{lm}(\theta,\phi) \; Y_{lm}(\theta_N,\phi_N). \tag{3.59}$$

where the collective co-ordinates (θ_N,ϕ_N) give the orientation of the symmetry axis. It is convenient to define the following factors

$$U_l F_l(q) = (2l+1)!! \int j_l(qr) \; \rho_l r^2 \; \mathrm{d}r \tag{3.60}$$

and

$$U_l = \int r^{l+2} \; \rho_l(r) \; \mathrm{d}r \tag{3.61}$$

so that $F_l(q) \to q^l$ as $q \to 0$. The U_l are closely related to the multipole moments defined in terms of

$$Q_l = 2Z\left(\frac{4}{2l+1}\right)^{1/2} \int r^l \; Y_{l0}(r) \; \rho_{ch}^{int}(r) \; \mathrm{d}^3r. \tag{3.62}$$

We shall use the definitions

$$Q = Q_2 = Z\left(\frac{16\pi}{5}\right)^{1/2} U_2$$

$$= Z \int (3z^2-r^2) \; \rho_{ch}^{int}(r) \; \mathrm{d}^3r \tag{3.63}$$

for the intrinsic quadrupole moment, and

---

deformed nuclear wavefunctions, instead of a deformed intrinsic charge density, and take the appropriate linear combinations to give angular-momentum eigenstates. For a nucleus with a quadrupole deformation the difference which this makes in the amplitudes for elastic scattering, or for inelastic scattering from an $I = 0$ to an $I = 2$ state, is of second order in the deformation parameter.

$$\Pi = \frac{1}{2} Q_4 = Z\left(\frac{4\pi}{9}\right)^{1/2} U_4$$

$$= Z\left(\frac{4\pi}{9}\right)^{1/2} \int r^4 \, Y_{40} \, \rho_{\mathrm{ch}} \, (\underset{\sim}{r}) \, \mathrm{d}^3 r \qquad (3.64)$$

for the hexadecapole moment. We now obtain

$$\hat{M}^{\mathrm{Coul}}_{lm}(q) = \left(\frac{4}{2l+1}\right)^{1/2} \frac{1}{(2l+1)!!} U_l F_l(q) \; Y_{lm}(\theta_N, \phi_N) \qquad (3.65)$$

so that the reduced matrix elements required in eqn (3.33) are

$$\langle I_{\mathrm{f}} \| \hat{M}^{\mathrm{Coul}}_{l} \| I_{\mathrm{i}} \rangle = \left(\frac{4\pi}{2l+1}\right)^{1/2} \frac{1}{(2l+1)!!} U_l F_l(q) \; \langle I_{\mathrm{f}} \| Y_l \| I_{\mathrm{i}} \rangle . \qquad (3.66)$$

For elastic charge scattering (which dominates except at backward angles) we have $I_{\mathrm{f}} = I_{\mathrm{i}} = I$ and the sum over l contains non-vanishing terms only for even values up to $l = 2I$. In particular for nuclei with spin $I = 0$ or $\frac{1}{2}$ only the $l = 0$ term contributes and the scattering becomes that of a spherical nucleus with charge density $\rho_0(r)/4\pi)^{1/2}$. For $I \geqslant 1$ the cross-section contains a sum of two or more terms. These terms have different q-dependence but their contributions cannot be separated in a model-independent way. More information is obtained if the target nucleus is polarized and measurements are made for different orientations (or for both unpolarized and polarized nuclei). The (plane-wave) Born approximation is not accurate enough to analyse experiments on rare-earth nuclei and analyses have been carried out using the distorted-wave Born approximation (Wright 1969) to calculate elastic and inelastic scattering. The inelastic scattering gives much more definite information about deformations, since it does not contain any contribution from the spherical part of the charge density, and can be used to obtain quadrupole and hexadecapole intrinsic deformations in spin-zero nuclei. At the present time the analysis has been based on the assumption that the transition charge densities (i.e. off-diagonal elements of the density operator $\hat{\rho}$) can be simply obtained from an intrinsic density $\rho^{\mathrm{int}}_{\mathrm{ch}}(\underset{\sim}{r})$. There is not enough experimental information to determine the functions $\rho_2(r)$ and $\rho_4(r)$ (the quadrupole and hexadecapole terms in the multipole expansion of $\rho_{ch}^{int}(\underset{\sim}{r})$) without making arbitrary assumptions about the functional form of the intrinsic density (see for example Bertozzi *et al.* 1972*a*, Kalinsky *et al.* 1973). Electric multipole moments and deformation parameters obtained from experiments are discussed in §6.1.7. and tabulated in Table 6.7.

Both the plane-wave and distorted-wave Born approximations give the cross-section to second order in the matrix elements (except for scattering from polarized targets, in which case first-order terms also appear). To obtain higher-order contributions it is necessary to solve the coupled equations (3.16) which take into account the contributions of excited nuclear states. In practice only a few nuclear states can be included explicitly, and the calculation is very much more difficult and tedious than the (non-trivial) calculation for spherical nuclei. The effect of excited nuclear states (any states, whether collective or not) is called the dispersion effect or dispersion correction and is discussed in more detail in §3.7.

3.6. MAGNETIC SCATTERING

The earliest electron-scattering experiments were carried out at low momentum transfer and the cross-sections were dominated by charge scattering. More recently it has become possible to separate the magnetic contribution to scattering on nuclei with non-zero spin. This subject has been reviewed by Donnelly and Walecka (1973, 1975). The expression for the cross-section in Born approximation (eqn (3.33)) is considerably simplified in the case of elastic scattering, and, neglecting the electron mass and nuclear recoil, is proportional to the lowest-order Mott cross-section:

$$\frac{d\sigma}{d\Omega} = (4\pi)\left(\frac{d\sigma}{d\Omega}\right)_M \{F_L{}^2 + (\tfrac{1}{2} + \tan^2 \tfrac{1}{2}\theta)F_T{}^2/Z^2\} \qquad (3.67)$$

where

$$F_L{}^2(q) \equiv \frac{1}{2I+1} \sum_{J\ \mathrm{even}} |\langle I \| M_J^{\mathrm{Coul}}(q) \| I \rangle|^2 \qquad (3.68)$$

$$\underset{q\to 0}{\to} \frac{1}{4\pi} \tag{3.69}$$

and

$$F_T^{\,2}(q) \equiv \frac{1}{2I+1} \sum_{J\ \mathrm{odd}} |\langle I \| T_J^{\,\mathrm{mag}}(q) \| I \rangle|^2 \tag{3.70}$$

$$\equiv \frac{1}{4\pi} \sum_J \{F_J^{\,\mathrm{mag}}(q)\}^2\, q^{2J} \left(\frac{M_J}{m_N}\right)^2$$

$$\times \frac{(J+1)\ (2I-J)!\ (2I+J+1)!}{4J\ (2J-1)!!\ (2J+1)!!\ (2J)!\ (2J+1)!} \tag{3.71}$$

$$\underset{q\to 0}{\to} = \frac{1}{4\pi}\left(\frac{I+1}{I}\right) \frac{q^2 M^2}{6m_N^{\,2}} \tag{3.72}$$

where m_N is the nuclear mass and M_J the nuclear magnetic 2^J moment ($M_1 \equiv \mu$). Conservation of angular momentum requires $J \leqslant 2I$; since the charge and current operators are Hermitian and invariant under time reversal the matrix elements of M_l^{Coul} vanish for l odd and those of T_J^{mag} for J even (see for example Überall 1971). The factor $(\frac{1}{2} + \tan^2 \frac{1}{2}\theta)$ is the ratio $V_{\mathrm{T}}(\theta)/V_{\mathrm{L}}(\theta)$, defined by eqns (3.35) and (3.36), for elastic scattering neglecting the electron mass. The factor $F_{\mathrm{T}}^2(q)/Z^2$ in eqn (3.67) is small (so that charge scattering dominates) except for very light nuclei or large values of q. For backward scattering $(\mathrm{d}\sigma/\mathrm{d}\Omega)_{\mathrm{M}}$ vanishes but the product $(\mathrm{d}\sigma/\mathrm{d}\Omega)_{\mathrm{M}}\tan^2 \frac{1}{2}\theta$ remains finite so that the scattering for $\theta = \pi$ becomes purely magnetic. This backward-scattering technique for probing the magnetic properties of nuclei was pioneered by Peterson and Barber (1962). An alternative method of separating the magnetic and charge scattering is to carry out the analysis using a Rosenbluth plot, i.e. to plot the ratio of the experimental to the lowest-order Mott cross-section against $\tan^2 \frac{1}{2}\theta$ for fixed q (Rosenbluth 1950). If the Born approximation is valid then this should be a straight line whose slope is $4\pi F_{\mathrm{T}}^2/Z^2$. An example of a Rosenbluth plot, for electron-proton scattering, is shown in Fig. 3.13

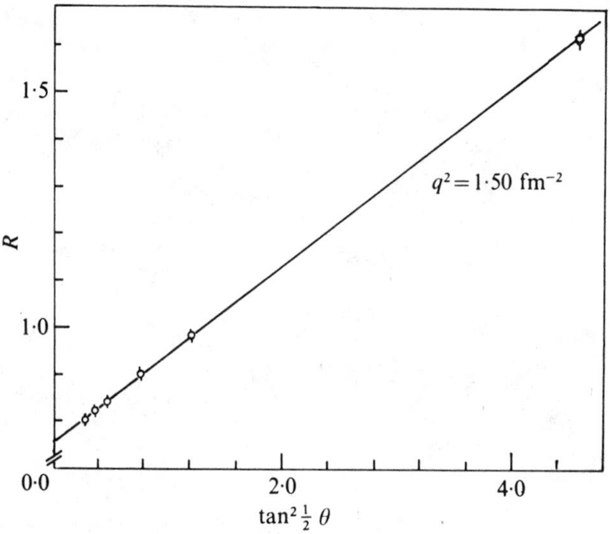

FIG. 3.13. Rosenbluth plot for electron-proton scattering obtain from the experiments of Borkowski *et al.* (1974) for a value of q^2 of 1.50 fm^{-2}.

(Borkowski *et al.* 1974). The sum over J in eqn (3.70) extends from 1 to $2I$ and it is not possible in principle to separate the contributions from different multipoles. If we take a reasonable model for the wavefunction of the odd nucleon, however, then the form factors are strongly peaked at different q values and it is possible to obtain the contribution of the different multipole moments from dipole to 2^{2I}-pole. An illustration of this is given in Fig. 3.14 (Donnelly and Walecka 1973). A selection of results for radii of dipole-moment form factors is given in §6.2.

3.7. DISPERSION CORRECTIONS

Whenever it is necessary to improve on the lowest-order Born approximation for elastic electron scattering, then one of the contributions in the next order represents the virtual excitation of the nucleus. This is represented diagramatically in Fig. 3.15. If there is a non-negligible contribution from this type of diagram, then the scattering process cannot be described as the interaction of an electron with a fixed

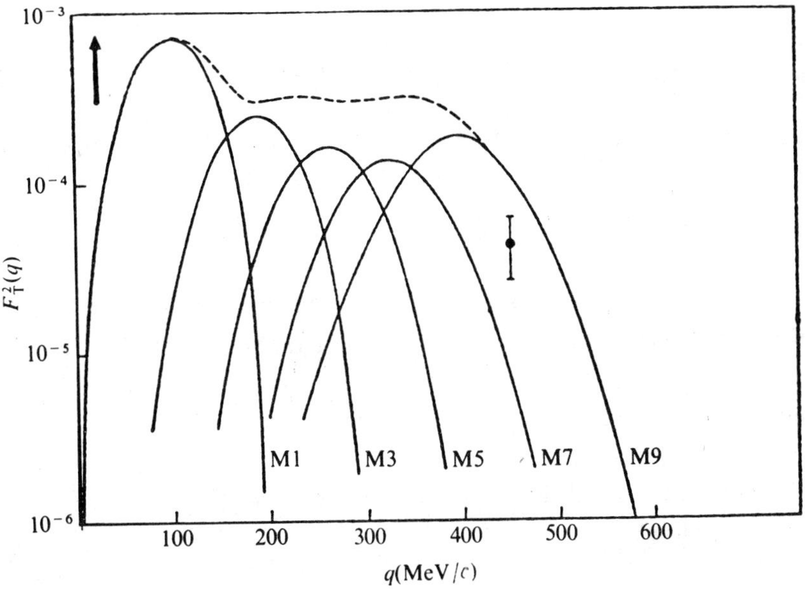

FIG. 3.14. Magnetic form factor for ^{209}Bi. The solid curves are the individual M1 to M9 form factors (whose sum is shown as the broken curve) obtained from a single-particle harmonic-oscillator calculation (Donnelly and Walecka 1973), and the experimental point is from Li *et al.* (1970).

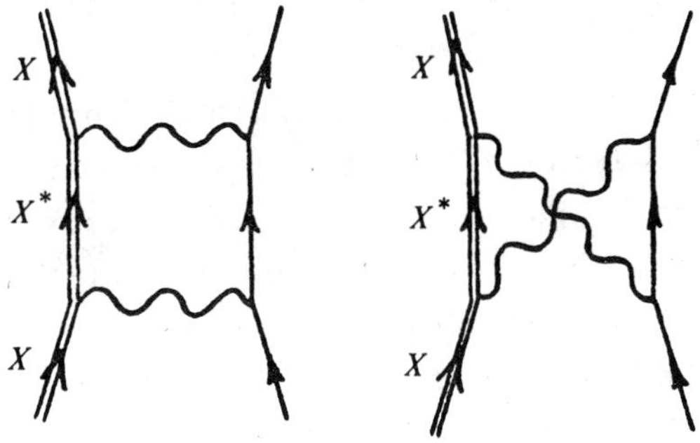

FIG. 3.15. Dispersion corrections.

charge and magnetic-moment distribution. We consider first corrections due to the electrostatic interaction between the electron and the nucleons, writing the Hamiltonian as

$$H = H_0 + \delta H \qquad (3.73)$$

where

$$H_0 = H_{\text{free}} + V(r) + H_{\text{N}}(\underset{\sim}{r}_1, \underset{\sim}{r}_2, \ldots, \underset{\sim}{r}_A) \tag{3.74}$$

and

$$\delta H = -e^2 \sum_{i=1}^{A} v(|\underset{\sim}{r}_i - \underset{\sim}{r}|) - V(r) \tag{3.75}$$

where $v(|\underset{\sim}{r}_i - \underset{\sim}{r}|)$ is the interaction between the electron and the ith nucleon and $H_{\text{N}}(\underset{\sim}{r}_1, \underset{\sim}{r}_2, \ldots, \underset{\sim}{r}_A)$ is the Hamiltonian for the nucleus in isolation. The term $V(r)$ is an arbitrary function of the electron co-ordinates and we choose it in such a way that, if the nucleus remains in its ground state, δH makes no first-order contribution to the elastic-scattering amplitude. This means that

$$\int \phi_{\text{N}}{}^* \delta H \; \phi_{\text{N}} \, \mathrm{d}^3 r_1 \, \mathrm{d}^3 r_2 \ldots \mathrm{d}^3 r_A = 0 \tag{3.76}$$

i.e.

$$V(r) = -e^2 \int \phi_{\text{N}}{}^* \sum_{i=1}^{A} v(|\underset{\sim}{r} - \underset{\sim}{r}_i|) \phi_{\text{N}} \, \mathrm{d}^3 r_1 \, \mathrm{d}^3 r_2 \ldots \mathrm{d}^3 r_A \tag{3.77}$$

where ϕ_N is the wavefunction for the ground state of the nucleus. Some of the higher-order terms in δH which affect the elastic cross-section can be obtained from inelastic cross-sections by making reasonable assumptions (deForest and Walecka 1966), but in the absence of complete knowledge of the inelastic processes they are usually estimated on the basis of a simplified model of the nucleus. Estimates have been made of the effect of the monopole part of δH (Rawitscher 1966*b*, 1967, 1970) and the dipole part (Onley 1968). Rawitscher solved the coupled Dirac equations in the adiabatic approximation (i.e. neglecting nuclear excitation energies) and also did a calculation to second order in δH but to all orders in $V(r)$ (i.e. a second-order DWBA calculation). Lin (1972) has calculated the corrections for ^{40}Ca using a single-particle-model wavefunction for the ground state and plane waves for the intermediate states. Using a similar model

Toepffer and Drechsel (1970) have obtained an imaginary part to be added to the electron-nucleus potential. Friar and Rosen (1972) have calculated the dispersion and other corrections for ^{17}C and ^{16}O using harmonic-oscillator wavefunctions for the nuclear states. Three different methods have been used by Bethe and Molinari (1971) to calculate the corrections. All of these estimates have indicated that the dispersion corrections have much less structure than the elastic cross-sections and make an important contribution at the minima but not near the peaks of the cross-sections. It turns out that in the derivation of charge densities from the cross-sections very little additional information comes from the measurements in the minima, which may thus be deleted in the determination of $\rho_{ch}(r)$, eliminating the necessity for accurate calculations of the dispersion corrections.

3.8. RADIATIVE AND STRAGGLING CORRECTIONS

The simple picture of an electron moving relativistically in the field of a nucleus must be corrected for the effect of the interaction with the radiation field and for multiple scattering by nuclei and atomic electrons. The main effect of the former comes from the emission and reabsorption of photons which can be represented by the diagrams in Fig. 3.16. If a

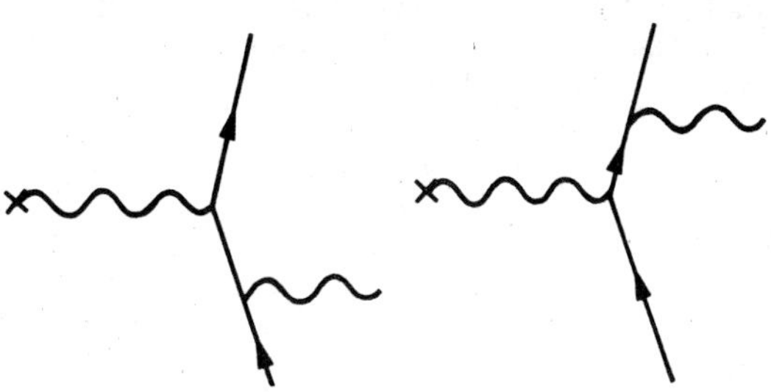

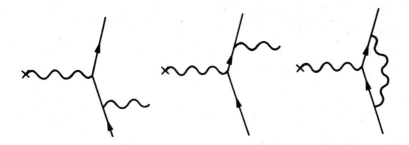

FIG. 3.16. Schwinger corrections to electron scattering.

photon is emitted and not reabsorbed (i.e. a real photon), then the process is not elastic, but photons are always emitted which are sufficiently soft (i.e. of low energy) that their total energy is less than the resolution of the electron detector (so that the electrons are counted as being elastically scattered). The contributions to the cross-section of each of the diagrams in Fig. 3.16 is infinite but the sum of the contributions is finite. The correction factor can be written in the form (Maximon 1969, Mo and Tsai 1969)

$$\left(\frac{d\sigma}{d\Omega}\right)_{\text{corr.}} = \left(\frac{d\sigma}{d\Omega}\right)_{\text{uncorr.}} (1-\delta_2)\exp(-\delta_1) \tag{3.78}$$

where δ_1 and δ_2 are functions of the electron energy, scattering angle, and the energy resolution ΔE of the detector:

$$\delta_1 = \frac{2\alpha}{\pi}\left(\ln\frac{q^2}{m^2c^2} - 1\right)\ln\frac{\Delta E}{E} \tag{3.79}$$

$$\delta_2 = -\frac{2\alpha}{\pi}\frac{13}{12}\left(\ln\frac{q^2}{m^2c^2} - 1\right) + \frac{17}{32} + \frac{\pi^2}{12} - \frac{1}{2}L_2(\cos^2\frac{\theta}{2}) \tag{3.80}$$

where

$$L_2(x) = -\int_0^x \frac{\ln(1-t)}{t}\,dt. \tag{3.81}$$

The multiple scattering from atomic electrons (ionizing collisions) has been calculated by Landau (1944) and is important for low-q scattering. Where the momentum transfer is high enough to provide nuclear structure information, it can be neglected. A much more important correction comes from multiple small-angle scattering by nuclei which leads to a correction factor (Mo and Tsai 1969)

$$1 - bt\ln\frac{E}{\Delta E} \tag{3.82}$$

where b is a function of the atomic number and t is the target thickness.

There are many other much smaller radiative corrections, including virtual photons or particle-antiparticle pairs, and these have always been neglected. The most important of these is the vacuum polarization due to the creation and annihilation of an electron-positron pair. The diagram corresponding to this is shown in Fig. 4.5(c). With present-day experimental accuracies this probably has a non-negligible effect on the cross-sections.

4

MUONIC ATOMS

When negative muons are stopped in matter they are captured by atomic nuclei to form muonic atoms. The properties of these atoms have been reviewed by Wu and Wilets (1969), Devons and Duerdoth (1969) and Scheck, Hüfner and Wu (1976). The electromagnetic and weak interactions of muons have been shown to very great accuracy to be identical to those of electrons, so that there are many similarities between the energy levels of muonic and 'electronic' atoms. For a hydrogen-like atom (with a point nucleus) the main difference comes from the scaling of the energies, which are increased by the factor m_μ/m_e (~207), and of the size of the orbits, which are reduced by the same factor. The initial part of the capture process is very complicated because of the interactions with the electrons; the muon loses energy both by direct transfer to the electrons (Auger processes) and, to a much smaller extent, by radiation. For the lower levels the latter process dominates, and by the time the muon has reached the $n = 10$ level more than 90 per cent of the charge of the 1s electrons lies outside its orbit. After this the motion of the muon can be described to great accuracy as that of a Dirac particle moving in the electrostatic field of the nucleus. The lowest angular momentum states penetrate the nucleus so that a measurement of their energies gives information about the shape and size of the charge distribution. The transitions with highest intensities are those between 'circular orbits', i.e. those with $l = n-1$.[†] The effects of electron screening, excitation of the nucleus, and radiative corrections can be calculated by perturbation theory.

[†] In nuclear physics the quantum numbers n and l are also used to label states but n has a different meaning, namely the number of radial nodes or the number of radial nodes plus 1.

4.1. ENERGY LEVELS FOR SPHERICAL NUCLEI

The bound-state energy eigenvalues of the Dirac equation (3.1) for a muon moving in the Coulomb field due to a fixed-point nucleus can be written in closed form in terms of the muon mass m, the atomic number Z, and the fine-structure constant $\alpha = e^2/\hbar c = (137.03604(11))^{-1}$ (Barasch-Schmidt *et al.* 1974):

$$E = \frac{mc^2}{\{1 + (Z\alpha/n')^2\}^{1/2}} \tag{4.1}$$

where

$$n' = n - (j+\tfrac{1}{2}) + \{(j+\tfrac{1}{2})^2 - Z^2\alpha^2\}^{1/2}. \tag{4.2}$$

Here n is the principal quantum number and $n' \to n$ in the non-relativistic limit ($Z\alpha \to 0$). For states with $j = n - \frac{1}{2}$, the formula reduces to

$$E = mc^2\left(1 - \frac{Z^2\alpha^2}{n^2}\right)^{1/2}. \tag{4.3}$$

The binding energy is defined to be

$$B = mc^2 - E \tag{4.4}$$

which reduces to the non-relativistic expression $B = Z^2\alpha^2mc^2/2n^2$ if we neglect terms of order $(Z\alpha)^4$. Most of the effect of nuclear recoil can be taken into account by using the reduced mass $m' = m/(1 + m/M)$ in the expression for B. This does not give the exact recoil correction except in the non-relativistic limit, and the small relativistic correction to the recoil is discussed in §4.7 together with radiative corrections to the energy levels.

For medium and heavy nuclei the lowest ($n \leqslant 2$) levels depend critically on the size and shape of the nuclear charge density. In practice it is necessary to assume a functional form $\rho(\underset{\sim}{r}, x_1, x_2, \ldots)$ for the charge, solve the Dirac equation numerically† for the potential due to this charge, calculate

---

†This is the standard eigenvalue problem of finding the value of E so that

various corrections to the resulting energies, and then compare with experiment. The parameters x_i are then adjusted to bring agreement, where possible, with the measured energies. The 4f-3d and higher transition energies are so nearly independent of the parameters of ρ (provided that these are constrained to fit the lower transitions) that only two or three parameters can be determined. The determination of parameters such as the half-density radius and skin thickness is very model-dependent, however, and it is much more meaningful to carry out a model-independent analysis as described in the next section.

The effect of the finite size on the different muonic energy levels is given to first order by the expectation values $\langle \Delta V \rangle$ of the quantity $\Delta V = V(r) - V_{\text{point}}(r)$ where $V(r)$ and $V_{\text{point}}(r)$ are the potential energy of the muon moving in the field of a distributed charge and a point charge respectively (neglecting radiative and other corrections). These expectation values are plotted as a function of Z in Fig. 4.1. With present experimental accuracies and uncertainties in theoretical corrections, muonic d-state and higher orbital angular momentum states provide no information about $\rho_{\text{ch}}(r)$. The term 'finite-size effect' is commonly used (for example in Table 4.2) to define a quantity which is calculated as follows: given a charge density $\rho_{\text{ch}}(r)$ which reproduces the experimental spectrum when radiative and nuclear polarization corrections are taken into account, we calculate for each level (or transition energy) the difference between the point-charge Dirac energy obtained using eqn (4.1) and the energy

---

the solutions F and G of eqns (3.10) are finite both at the origin and at infinity. For a guessed value E_0 we solve the Dirac equation for the interior region obtaining solutions F_i, G_i and for the exterior region obtaining F_e, G_e. If we scale G_e so that G is continuous at the matching radius R and normalize the wavefunctions so that $\int (F^2+G^2)\, dr = 1$, we obtain a prediction for the eigenenergy: $E = E_0 + \hbar c\, G(R)\{F_i(R) - F_e(R)\}$. Iteration of this procedure produces rapid convergence in most cases although it may be necessary to damp the shift in energy.

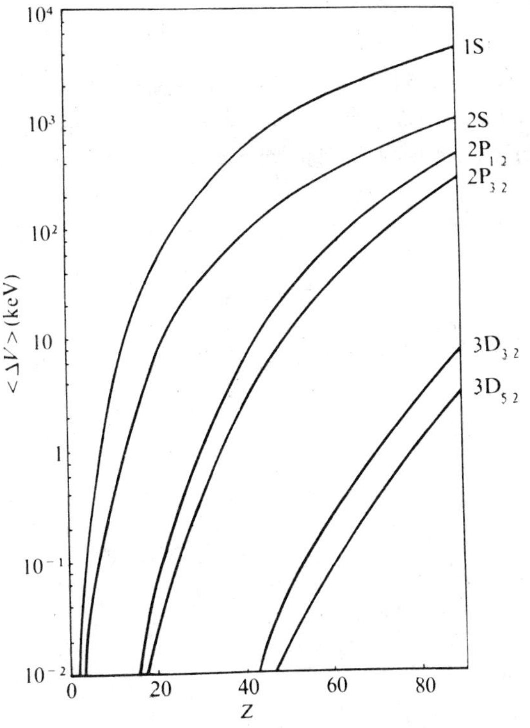

FIG. 4.1. The nuclear finite-size effect $\langle \Delta V \rangle$ in muonic atoms for different levels. (From Barrett 1974.)

obtained by solving the Dirac equation in the field of a static charge $\rho_{ch}(r)$.† This definition gives 10.5 MeV for the finite-size effect in the 1s state of ^{208}Pb, whereas the value of $\langle \Delta V \rangle$ is only 3.6 MeV. The latter quantity gives a better idea, however, of the accuracy to which we can obtain the appropriate property of the charge density by measuring the 1s energy.

†The effect of the finite size on the corrections is not usually added although it can be substantial, for example in the 1s state in muonic Bi the finite-size effect for the vacuum polarization is about 140 keV and for the Lamb shift it is about 70 keV (Barrett *et al.* 1968).

4.2. MODEL-INDEPENDENT ANALYSIS OF MUONIC X-RAY SPECTRA

Just as in the case of electron scattering, it is possible to obtain many charge densities which give identical predictions of measured quantities. An example of two shapes which fit the ^{208}Pb spectrum is shown in Fig. 4.2. It is desirable to

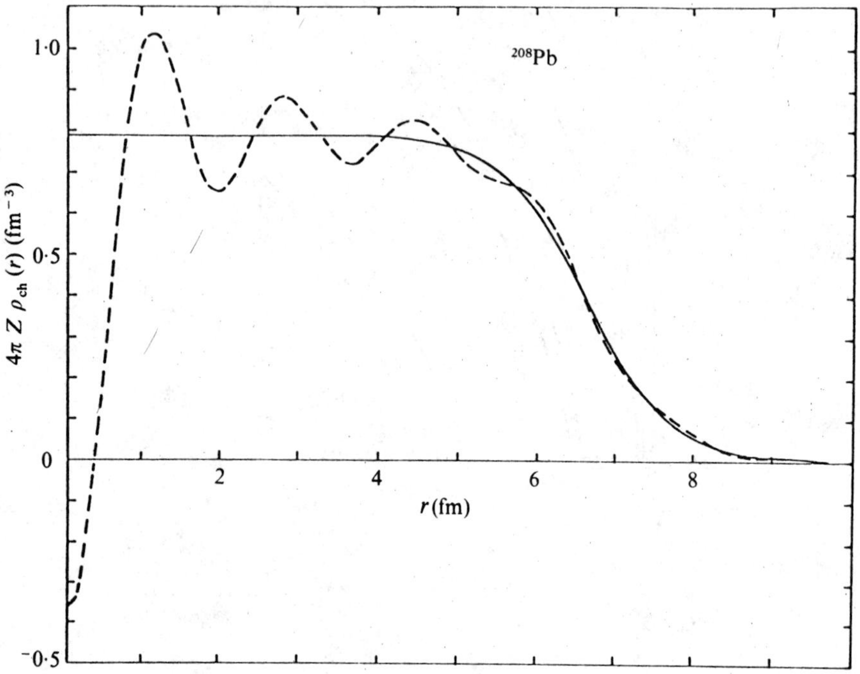

FIG. 4.2. Two ^{208}Pb distributions which are equivalent for muonic X-ray transition energies and electron-scattering cross-sections up to q_{max} = 2.73 fm^{-1}. (From Friar and Negele 1975*b*.)

describe in a simple way the common properties of these densities. In the case of electron scattering at low momentum transfer the common property is the mean-square radius of the charge distribution. The effect of the size of the nucleus on the energy of an atomic s-state is determined by the same quantity provided the wavefunction is constant over the region of the nucleus. In electronic atoms and in light muonic atoms this is a good approximation. We can understand this result and see how it must be modified for heavy muonic atoms by using first-order perturbation theory. We consider a muonic state j and the energy E_j^0 and corresponding wavefunction $\psi_j(\underset{\sim}{r})$ calculated for an assumed charge distribution $\rho_0(\underset{\sim}{r})$. For a different density $\rho(\underset{\sim}{r})$ the change in energy is given to first order by

$$\Delta E_j = -Ze^2 \iint \frac{\psi_j^\dagger(\underset{\sim}{r}')\psi_j(\underset{\sim}{r}')\{\rho(\underset{\sim}{r})-\rho_0(\underset{\sim}{r})\}}{|\underset{\sim}{r}-\underset{\sim}{r}'|} \,\mathrm{d}^3r'\,\mathrm{d}^3r \tag{4.5}$$

$$= \int f_j(r)\{\rho(r)-\rho_0(r)\}r^2\,\mathrm{d}r \tag{4.6}$$

where we have assumed ρ and ρ_0 to be spherically symmetric. The function

$$\frac{1}{Ze} f_j(r) = -e\int \frac{\psi_j(\underset{\sim}{r}')\ \psi_j(\underset{\sim}{r}')}{|\underset{\sim}{r}-\underset{\sim}{r}'|}\,\mathrm{d}^3r'\,\mathrm{d}\Omega \tag{4.7}$$

is the spherically averaged electrostatic potential due to the 'muon cloud'. If the two charge densities give the same energy then both will give the same value for the integral:

$$\langle f_j\rangle \equiv \int f_j(r)\ \rho(r)\ r^2\mathrm{d}r. \tag{4.8}$$

Thus, to the extent that first-order perturbation theory is valid, a measurement of E_j determines the quantity $\langle f_j\rangle$. If the wavefunction $\psi_j(\underset{\sim}{r})$ is constant over the region where $\rho(r)$ is non-zero (which is approximately true for 1s states of light nuclei), then f_j is of the form $a + br^2$ and different densities which give the same energy will have the same mean square radius. A transition energy between states i and j determines the quantity

$$\langle f_{ij}\rangle \equiv \langle f_i\rangle - \langle f_j\rangle \tag{4.9}$$

but in the case of most transitions, especially those from states (n,l) to $(n-1,l-1)$, the function f_{ij} is almost the same as the function f_i corresponding to the final state (apart from an unimportant constant). In practice if $\rho(r)$ and $\rho_0(r)$ give the same energies for the 2p-1s and 3d-2p X-rays, then first-order perturbation theory is very accurate indeed

(Barrett 1970).

It is not difficult to calculate the functions $f_i(r)$ numerically after choosing a density $\rho_0(r)$ to fit the transition energies (see §4.1). It would be possible to tabulate these values of $f_i(r)$ for use in calculating the quantities $\langle f_i \rangle$ for any other density $\rho(r)$, but this is very cumbersome and almost as inconvenient as solving the Dirac equation again for $\rho(r)$ itself. A more convenient method is to use a power series for $f_i(r)$ and this was first done by Bethe and Negele (1968) for lead. They used a uniform density for $\rho_0(r)$, for which the radial part of $\psi^\dagger\psi$ is an even power series whose leading term for a state of angular momentum j is proportional to r^{2j+1}, resulting in an expansion for f_i of the form

$$f_i = a_0 + \sum_{m=1}^{\infty} a_m r^{2j+2m-1} \tag{4.10}$$

It is not difficult to calculate similar power series using a diffuse charge density, but the coefficients in the expansion have alternating signs and many terms are required to specify the f_i over the whole of the region of the nucleus. A much simpler representation of the f_i was suggested by the work of Ford and Wills (1969) who found that a set of Fermi densities with different diffuseness parameters, and the half-density radius adjusted in each case to give the same value for a particular transition energy, gave the same value of the kth moment $\langle r^k \rangle$. The value of k depended on the nucleus and on the transition, and for the $2p_{3/2}$-1s transition it was found to be $k = 2 - Z/68.5$ (which gives $k = 0.8$ for Pb). The Ford and Wills result came from sets of densities whose differences were strongly peaked at the nuclear surface and they were effectively fitting the $f_i(r)$ in that region (Wu and Wilets 1969) with a functional form $a + br^k$. This does not give a good fit over a wide enough range of r for some transitions in heavy nuclei and a very much better fit can be obtained using the function (Barrett 1970)

$$a + br^k \exp(-\alpha r). \tag{4.11}$$

This form was found after calculations with a large number of other two-parameter functions such as $\exp(-\gamma r^2)r^k$ and $(1 + br^2)r^k$; the expression (4.11) gave a better fit than any other for almost every transition over a wide range of nuclei. The fit was achieved by minimizing the function

$$\int |f_i - a - br^k \exp(-\alpha r)| \rho_0(r)\, r^2 dr. \qquad (4.12)$$

Fig. 4.3 shows the resulting values of k and α for the $2p_{3/2}$-1s transition obtained by taking ρ_0 to be a Fermi distribution with parameters $c = 1.12\, A^{1/3}$ fm and $a = 0.5$ fm.

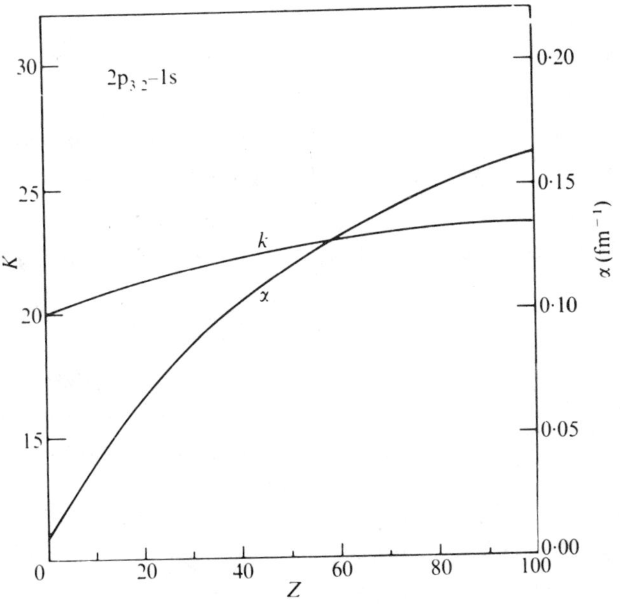

FIG. 4.3. Parameters α and k in the function $r^k \exp(-\alpha r)$ for the $2p_{3/2}$-1s transition plotted against atomic number Z. (From Barrett 1970.)

The extra parameter needed to obtain a good fit to the f_i makes the size information from muonic X-rays considerably more complicated. However, the values of k and α are strongly correlated and it is possible to obtain a good fit over a range of values of α by varying a, b, and k. Thus α can be chosen to be constant for all transitions in a particular nucleus; for example Engfer *et al.* (1974) used the value

$$\alpha = 0.03661 + 1.4194 \times 10^{-3} Z \quad (4.13)$$

(determined by fitting the $2p_{1/2}$-1s transition alone) in their compilation of data on muonic atoms. The generalized moments $\langle r^k \exp(-\alpha r) \rangle$ determined by this procedure have been found to be model-independent to a remarkable degree of accuracy (Barrett 1970, 1974, Rinker 1971).

In a comprehensive study of Pb isotopes Ford and Rinker (1973) expressed the generalized radial moments $\langle r^k \exp(-\alpha r) \rangle$ in terms of equivalent uniform radii R_k defined by

$$\int_0^{R_k} \exp(-\alpha r)\ r^{k+2}\ dr = \frac{1}{3} R_k^3 \int_0^{\infty} \exp(-\alpha r)\ \rho(r)\ r^{k+2}\ dr \quad (4.14)$$

(i.e. R_k is the radius of a uniform charge density that gives the same moment). We can thus plot values of R_k corresponding to measured energies (i.e. obtained using $\rho_0(r)$ in eqn (4.14)) against k and compare with the R_k curves obtained directly using any other trial densities $\rho(r)$ in eqn (4.14). An example of such a plot is given in Fig. 4.4. We can also use the plot of R_k versus k to check the muonic X-ray results since inconsistencies in the analysis, whether due to experimental error or to neglected theoretical corrections, are revealed by departures from a smooth curve, such as for example the point for the $3d_{3/2}$ - $3d_{5/2}$ energy difference at k = 4.6 (see Ford and Rinker 1973).

From muonic X-ray measurements alone it is impossible to make a definite statement about the charge density itself rather than derived quantities like R_k; these measurements do, however, provide a very important constraint when combined with electron-scattering measurements. Values of R_k for the $2p_{3/2}$ - 1s and $3d_{5/2}$ - $2p_{3/2}$ transitions in medium and heavy nuclei are given in Table 6.3.

4.3. NUCLEAR POLARIZATION CORRECTIONS

In this section we consider the error in the calculation of the muonic-atom energy levels which is introduced by the

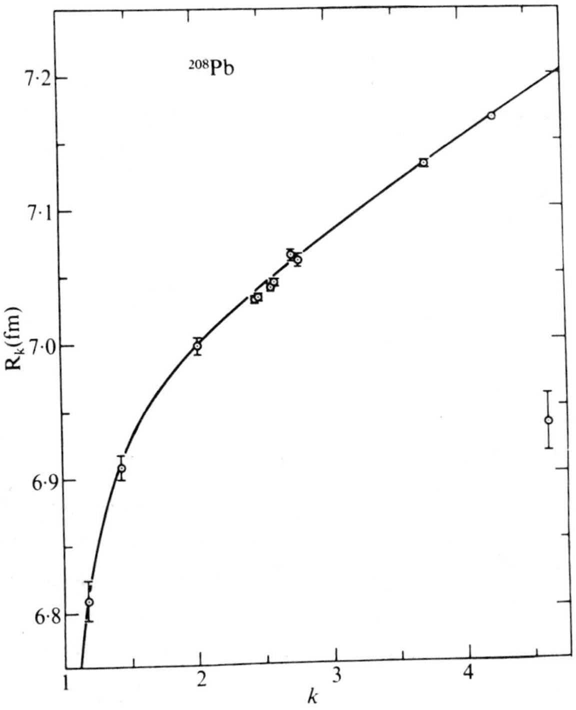

FIG. 4.4. Equivalent radii R_k for generalized moments $\langle r^k \exp(-\alpha r)\rangle$ with $\alpha = 0.17\ \mathrm{fm}^{-1}$. The circles are obtained from the measured X-ray energies (Kessler 1971) and the solid curve corresponds to a phenomenological density which fits these energies and is also very close to that obtained from the analysis of electron-scattering cross-sections. (From Ford and Rinker 1973.)

assumption that the nucleus can be represented by a static charge density; the resulting correction is called the nuclear polarization correction and is completely analogous to the dispersion corrections to electron scattering described in §3.7. Nuclear polarization corrections have been carried out by Cole (1968), Chen (1968, 1970*a*), and Skardhamar (1970). In this improvement on the static-nucleus approximation we have to include the contribution of the term

$$\delta H = -e^2 \sum_{i=1}^{A} v_i(|\underset{\sim}{r}_i - \underset{\sim}{r}|) - V(r) \qquad (4.15)$$

$$\equiv \sum_L \delta H_L \tag{4.16}$$

where the summation is over the different multipoles contributing to δH. Assuming $v_i(r) = 0$ for neutrons and $1/r$ for protons we obtain

$$\delta H_0 = -e^2 \sum_{i=1}^{Z} \frac{1}{r_>} - V(r) \tag{4.17}$$

and for $L > 0$

$$\delta H_L = -e^2 \sum_{i=1}^{Z} \frac{r_<^L}{r_>^{L+1}} P_L(\cos\theta_{i\mu}). \tag{4.18}$$

Except for the case of strongly deformed nuclei, which is discussed in the next section, we can use perturbation theory to calculate the shift in energy due to δH. We construct eigenstates $|FM\rangle$ of the unperturbed Hamiltonian in terms of eigenstates $|nljm\rangle$ and $|IKM_I\rangle$ describing the muon and the nucleus respectively. (In order to cover the case of rotational states of deformed nuclei we include the quantum number K which gives the projection of the nuclear spin on the nuclear symmetry axis.) We have

$$(H_{\text{free}} + V - E_{nlj})\,|nljm\rangle = 0 \tag{4.19}$$

$$(H_{\text{N}} - \varepsilon_I)\,|IKM_I\rangle = 0. \tag{4.20}$$

In first order the energy shift due to δH_L vanishes for $L = 0$ by construction (see eqn (3.75)) and for $L \neq 0$ is given by

$$\begin{aligned}\delta E_L(1) &= \langle nljIKFM|\,\delta H_L\,|nljIKFM\rangle \\ &= A_L(ljIKF)\langle nljI\|\,\delta H_L\,\|nljI\rangle\end{aligned} \tag{4.21}$$

where

$$\begin{aligned}A_L(ljIKF) = {}&(2L+1)(2j+1)(2I+1)^{1/2}\,(j\,{-\tfrac{1}{2}}\,j\,\tfrac{1}{2}|L0) \\ &\times(L0IK|IK)\,W(IIjj;LF)\end{aligned} \tag{4.22}$$

This vanishes for odd values of L and for nuclei of spin 0 or $\frac{1}{2}$. (Strictly speaking we should include magnetic interactions in δH but these are small owing to the large mass of the muon. They can cause a magnetic hyperfine splitting as discussed in §4.5.)

In second order the shift in a state $|njlIFM\rangle$ is given by

$$\delta E_L^F(2) = \sum_{n'l'j'I'}{}' \frac{|\langle n'l'j'I'KFM'|\delta H_L|nljIKFM\rangle|^2}{E_{nlj} + \varepsilon_I - E_{n'l'j'} - \varepsilon_{I'}} \tag{4.23}$$

where the prime on the summation excludes the term with the diagonal matrix element. If the energies of the state $|nljIKFM\rangle$ are independent of F or the hyperfine splitting is not resolved, the weighted average of the $\delta E_L^F(2)$ is given by

$$\delta E_L(2) = \sum_{n'l'j'I'm'M_{J'}} \frac{|[\langle n'l'j'm'|\langle I'KM_{I'}|\delta H_L|IKM_I\rangle|nljm\rangle]|^2}{E_{nlj} + \varepsilon_I - E_{n'l'j'} - \varepsilon_{I'}}. \tag{4.24}$$

Although we need complete knowledge of the nuclear and muon spectra to evaluate the corrections, it turns out that for $L \geqslant 2$ the results are insensitive to the nuclear energies, while for $L = 1$ they depend on the known energy of the giant dipole state. The greatest uncertainty comes from the $L = 0$ contribution which depends on the excitation energy of experimentally unknown monopole excitations ('breathing modes') of the nucleus. These points are described in detail by Chen (1970*a*) who obtained the contribution of the muon excited states from the perturbation of the muon wavefunction using the reference spectrum method.

4.4. ENERGY LEVELS FOR DEFORMED NUCLEI

The nuclear polarization correction to some levels of muonic atoms with deformed nuclei can be so large that second-order perturbation theory does not give an accurate correction. This is due to strong mixing of the two eigenstates corresponding to a fine-structure doublet, particularly the $2p_{1/2}$ and $2p_{3/2}$ states. Calculations of this 'dynamic hyperfine splitting' were first carried out by Wilets (1954) and Jacobsohn (1954). A recent description of the method, together with experimental results for a number of nuclei, has been given by Hitlin *et al.* (1970).

The problem can be solved by constructing a 'model space' of the states which are strongly mixed and diagonalizing the Hamiltonian in this space. We restrict our discussion to the quadrupole interaction (δH_2) since the higher terms are much less important and generalization of the theory to include them is straightforward. We write the basis functions for the model space as $|nljIKFM\rangle$, so that the matrix elements of δH_2 become

$$\langle n'l'j'I'KFM| -e^2 \sum_{i=1}^{Z} \frac{r_<^2}{r_>^3} P_2(\cos\theta_{i\mu}) |nljIKFM\rangle$$

$$= -\frac{e^2}{2}\left(\frac{4\pi}{5}\right)^{1/2} (-1)^{I+j'-F}\, W(I'j'Ij;F2)$$

$$\times \langle I'K\| M_2 \| IK\rangle \langle n'l'j'\| \frac{1}{r^3} f_{I'I}(r) \| nlj\rangle \qquad (4.25)$$

where

$$\langle I'K\| M_2 \| IK\rangle \equiv \langle I'K\| \sum_{i=1}^{Z} r_i^2 Y_2(\theta_i) \| IK\rangle \qquad (4.26)$$

and $f_{I'I}(r)$ defined by

$$\frac{1}{r^3} f_{I'I}(r) \langle I'K\| M_2 \| IK\rangle = \langle I'K\| \sum_{i=1}^{Z} \frac{r_<^2}{r_>^3} Y_2(\theta_i) \| IK\rangle \qquad (4.27)$$

is a penetration factor which is equal to unity when r is outside the nucleus.

As an example we consider the case of ^{182}W and construct a model space from the muon 2p doublet and the nuclear $I = 0$ and 2 states. (In this case $K = 0$.) Table 4.1 shows the unperturbed energies, the first-order corrections, and the perturbed energies obtained by diagonalizing the submatrix. We have chosen only the 2p muon states for our model space since matrix elements with the 2s states vanish and all other muon states are so far removed in energy that they are only weakly mixed. There is no splitting of the $n = 1$ levels ($I = 0$, $F = \frac{1}{2}$ and $I = 2$, $F = \frac{3}{2}$ or $\frac{5}{2}$) since the matrix element of the quadrupole operator vanishes between $j = \frac{1}{2}$ states.

TABLE 4.1

Dynamic hyperfine splitting in ^{182}W

State			Unperturbed energy (keV)	First-order correction (keV)	Perturbed energy (keV)
J	I	F			
$\frac{1}{2}$	0	$\frac{1}{2}$	-3951.1	0	-3970.1
$\frac{3}{2}$	2	$\frac{1}{2}$	-3707.5	-45.2	-3733.8
$\frac{1}{2}$	2	$\frac{3}{2}$	-3851.0	0	-3900.3
$\frac{3}{2}$	0	$\frac{3}{2}$	-3807.6	0	-3780.7
$\frac{3}{2}$	2	$\frac{3}{2}$	-3707.5	0	-3685.4
$\frac{1}{2}$	2	$\frac{5}{2}$	-3851.0	0	-3854.3
$\frac{3}{2}$	2	$\frac{5}{2}$	-3707.5	+32.3	-3672.0
$\frac{3}{2}$	2	$\frac{7}{2}$	-3707.5	-12.9	-3720.5

For the n = 3 levels there will be mixing between the $3d_{5/2}$, $3d_{3/2}$ and 3s levels, but here the off-diagonal matrix elements are very much smaller and the mixing is not strong. To the extent that the mixing of the nuclear I = 2 levels can be ignored when the muon is in the n = 3 states, we can assume that the $F = \frac{5}{2}$ and $F = \frac{7}{2}$ levels in the 2p multiplet will not be populated. The reason for this is that electric dipole (E1) radiation dominates and this does not occur between an I = 2 and an I = 0 state. We would therefore expect to see five levels in the 2p multiplet, each giving rise to transitions to the 1s, I = 0 and 1s, I = 2 levels. So far only three of the five levels have been observed and the transitions determine (on the assumption of the rotational model) three parameters of the charge density, for example the values of c, a, and β_2 of a deformed Fermi density:

$$\rho_{ch}(\underset{\sim}{r}) = \frac{\rho_0}{[e^x+1]} \tag{4.28}$$

where

$$x = \frac{r-c\{1+\beta_2 Y_{20}(\theta)\}}{a}. \tag{4.29}$$

Observation of the remaining levels as well as the hyperfine splitting for $n = 3$ should make it possible to obtain much more information, including the ratio $\langle 2\| M_2 \| 2 \rangle / \langle 2 \| M_2 \| 0 \rangle$ which is predicted to be unity by the rotational model. Measurement of the relative intensities of the transitions also contributes to our knowledge of the quadrupole matrix elements.

For deformed nuclei the correction due to the admixture of states outside the model space is particularly important (Chen 1970*b*). Just as in the case of spherical nuclei the second-order correction due to the admixture of muon states can be taken into account by solving for the perturbed muon wavefunctions. The resulting correction to the energies can be taken into account by defining an effective interaction which operates in the model space and to a very good approximation can be written in the form (Chen 1970*b*)

$$\delta H_2^{\text{eff}} = H_2(1+\eta) + \gamma \qquad (4.3(4.30)$$

where γ is the polarization correction which would be obtained for a spherical nucleus and provides the same correction to each hyperfine state. The parameter η simply renormalizes the interaction by 3 to 5 per cent. The accuracy of this approximation has been checked by comparison with the coupled-channels calculation of McKinley (1969) and found to be very good (Chen 1970*b*).

4.5. MAGNETIC HYPERFINE STRUCTURE

The magnetic interaction between a muon and a nucleus is relatively unimportant since, unlike the electrostatic interaction, it is inversely proportional to the muon mass. This is in contrast to the situation in electronic (ordinary) atoms where the magnetic and electric hyperfine effects are of the same order of magnitude. In muonic atoms, therefore, magnetic splitting in most states is buried in electric quadrupole structure even for spherical nuclei, but the situation is simpler in the muonic $j = \frac{1}{2}$ states where the quadrupole interaction vanishes in first order. This topic has been reviewed recently by Scheck, Hüfner, and Wu (1976). The magnetic dipole interaction can be written in the form

$$\langle nljIFM|\hat{T}_1^{\text{mag}}|nljFM\rangle = \frac{A_1(nljI)\{F(F+1) - I(I+1) - j(j+1)\}}{2jI} \tag{4.31}$$

In the point-nucleus approximation A_1 is proportional to the nuclear magnetic moment μ_N

$$A_1(nljI)_{\text{point}} = (-1)^{j+l+1/2}\frac{e^2\hbar}{2m_N c}\frac{2j+1}{j+1}\left\{\mu_N\int_0^\infty \frac{1}{r^2}F_{nlj}(r)G_{nlj}(r)\,\mathrm{d}r\right\} \tag{4.32}$$

where m_N is the nucleon mass. For a finite nucleus using the operator for the magnetic moment

$$\mu_N = \sum_{i=1}^{A}\left(g_\ell^{(i)}\underset{\sim}{\ell}_i + \frac{1}{2}g_s^{(i)}\underset{\sim}{\sigma}_i\right) \tag{4.33}$$

where $g_\ell^{(i)} = 1$ for protons and 0 for neutrons and $g_s^{(i)} = 5.58$ for protons and -3.82 for neutrons, we obtain

$$A_1(nljI) = A_1(nljI)_{\text{point}} - (-1)^{j+l+1/2}\frac{e^2}{2m_N c}\frac{2j+1}{j+1}$$

$$\times\langle II|\sum_i\int_0^{r_i}\left[g_\ell^{(i)}l_z^i\left\{1-\left(\frac{r}{r_i}\right)^3\right\} + g_s^{(i)}\{\sigma_z^i + K_z^i\}\right]$$

$$\times\frac{1}{r^2}F_{nlj}\,G_{nlj}\,\mathrm{d}r|II\rangle \tag{4.34}$$

where K_z^i is defined by

$$K_z^i = \left(\frac{\pi}{2}\right)^{1/2}\sum_\lambda (2\lambda 1\mu{-\lambda}|10)Y_{2\lambda}(\theta_i\phi_i)\sigma_i^{-\lambda}. \tag{4.35}$$

The reduction of the interaction energy due to the finite distribution of magnetization is known as the Bohr-Weisskopf effect (Bohr and Weisskopf 1950) or hyperfine anomaly and is well known in atomic spectra. In muonic atoms the splitting depends on the radial distribution of the magnetization, but it is not possible to obtain accurate or detailed information about this distribution since the effects are small and the quantities A_1 obtainable only for the 1s and $2p_{1/2}$ states. In practice the experimental values are used to check models of nuclear wavefunctions (which in general must contain configuration mixing otherwise they would not give the correct values for the nuclear magnetic moment μ_N).

4.6. ISOTOPE, ISOMER, AND ISOTONE SHIFTS

For different atoms with almost identical nuclei measurements of the differences in atomic transition energies can be used to obtain information about the differences between the nuclei. For ordinary atoms this is a long-established technique which is discussed in detail in Chapter 5. For muonic atoms such difference measurements are much more recent. The subject has been reviewed by Wu and Wilets (1969). If the nuclei differ only in neutron number, the difference in a particular transition energy is called an isotope shift. The isomer shift is the change which occurs when the nucleus is excited (into a state which has a lifetime much longer than the atomic excited states). If the nuclei differ in atomic but not neutron number then the change in energy is very much larger and is referred to as the isotone shift.

In the following sections we describe how these shifts may be used to obtain information about changes in charge distributions. Results of analyses for isotope pairs are given in §6.1.4, and for isotone and isomer shifts in §6.1.5 and §6.1.6 respectively.

4.6.1. *Isotope shifts*

The study of isotope shifts in electronic atoms is a complicated subject involving knowledge of the many-body electron wavefunction and it is discussed in Chapter 5. In muonic atoms the shifts are much easier to interpret, and the change in the energy E of a particular state can be separated immediately into a mass term (which is easily calculated) and a field term[†]

[†]We are for the time being ignoring the shifts in radiative corrections,

$$\delta E = \delta E_{\text{field}} + \delta E_{\text{mass}}. \tag{4.36}$$

The field term represents the shift due to the change in the nuclear region, i.e. it depends on the size and shape of $\rho_{ch}(\underset{\sim}{r})$, while the mass term depends on the change in the mass M of the nucleus. The situation differs from the more familiar case of optical isotope shifts because there is no specific mass shift and the normal shift δE_{mass} *may have a strong dependence on the size of the nucleus*. If we represent the latter by the generalized equivalent radius R_k of §4.2, then we can define

$$\delta E_{\text{field}} = \delta R_k \left.\frac{\partial E}{\partial R_k}\right|_M \tag{4.37}$$

$$\delta E_{\text{mass}} = \delta M \left.\frac{\partial E}{\partial M}\right|_{R_k} \tag{4.38}$$

For a point nucleus the binding energy is proportional to the reduced mass, but for a distributed charge there is a contribution of the opposite sign since the length scale for an atomic wavefunction is inversely proportional to the reduced mass. We thus obtain

$$\left.\frac{\partial E}{\partial M}\right|_{R_k} = \frac{mE}{M^2}\left(1 + \frac{R_k}{E}\,\frac{E}{R_k}\right) \tag{4.39}$$

(neglecting terms of higher order in m/M). For the 1s state in lead the finite-size contribution reduces the factor in parentheses in eqn (4.39) to 0.47. Provided that we have tables of R_k and $\partial E/\partial R_k$ we can obtain the field shift, and hence the change δR_k in the moments of the charge distribution, without solving the Dirac equation, but in the past analyses were usually carried out in terms of a model density;

---

which are easy to take into account, and in nuclear polarization, which are expected to be small for spherical nuclei. The case of deformed nuclei is more complicated as discussed below.

the Dirac equation was solved separately for each isotope and the parameters adjusted to fit the spectrum. The size effect could thus be obtained directly (without explicitly calculating δE_{mass} and δE_{field}), for example for a Fermi shape the isotope shift was frequently analysed in terms of changes δc and δa in the parameters. This had an unfortunate effect: these changes were frequently compared with those obtained from other experiments, particularly electron-scattering measurements, and if there was a difference it was often ascribed to a discrepancy in the measurements or shortcomings in the theory, rather than to the assumption that $\delta\rho$ is the difference between two Fermi densities. A more useful procedure is to obtain an estimate for the shift in the mean-square radius and to calculate a realistic error (i.e. based on minimal assumptions about the shape of ρ, as described below). Alternatively the shifts in the model-independent quantities R_k can be calculated as has been done for Pb isotopes by Ford and Rinker (1973, 1974) and by Engfer *et al.* (1974) in their compilation of data from muonic atoms.

In order to compare muonic-atom isotope shifts with optical and X-ray shifts it is desirable to calculate the muonic $\delta\langle r^2\rangle$ values as well as possible, i.e. without making unnecessary assumptions. One way of doing so is to carry out 'model-independent' analyses as described in §§3.4 and 4.2. If there are not enough data for such extensive analyses it may still be possible to calculate a model-independent value of $\delta\langle r^2\rangle$.

To do this we consider the field shifts $\delta E^{(i)}_{\text{field}}$ in the energy of the ith transition and use eqn (4.6) to obtain an integral expression for the change in charge density $\delta\rho(r)$:

$$\delta E^{(i)}_{\text{field}} = \int f_i(r)\delta\rho(r)\; r^2 \mathrm{d}r. \tag{4.40}$$

We make the assumption that $\delta\rho(r)$ vanishes outside a certain radius $R_{\max}$ and define the scalar product $\langle f_i|f_j\rangle$ by

$$\langle f_i|f_j\rangle = \int_0^{R_{\max}} f_i(r)f_j(r)\; r^2 \mathrm{d}r. \tag{4.41}$$

Eqn (4.40) may then be written

$$\delta E^{(i)}_{\text{field}} = \langle f_i | \delta\rho \rangle. \tag{4.42}$$

Let $\{h_i\}$ be the set of dual vectors spanning the space $\{f_i\}$ and having the property

$$\langle f_i | h_j \rangle = \delta_{ij}. \tag{4.43}$$

We expand the unknown vector $|\delta\rho\rangle$ in terms of the $|h_i\rangle$ and the vector $|r^2\rangle$ in terms of the $|f_i\rangle$:

$$|\delta\rho\rangle = \sum_i a_i |b_i\rangle + |D\rangle \tag{4.44}$$

$$|r^2\rangle = \sum_i b_i |f_i\rangle + |R\rangle \tag{4.45}$$

where the remainder terms $|D\rangle$ and $|R\rangle$ are orthogonal to the space $\{f_i\}$. Then

$$\delta\langle r^2 \rangle = \langle r^2 | \delta\rho \rangle \tag{4.46}$$

$$= \sum_i a_i b_i + \langle R | D \rangle. \tag{4.47}$$

The coefficients b_i are easily calculated, and the unknown coefficients a_i are given approximately by the experimental quantities $\delta E^{(i)}_{\text{field}}$ with experimental errors σ_i. Assuming that the errors are uncorrelated, we obtain

$$\text{statistical error in } \delta\langle r^2 \rangle = \left(\sum b_i^2 \sigma_i^2\right)^{1/2}. \tag{4.48}$$

We can guess at the 'completeness error' $\langle R|D\rangle$ in $\delta\langle r^2\rangle$ by using Schwarz's inequality:

$$|\langle R|D\rangle|^2 \leqslant \langle R|R\rangle \langle D|D\rangle. \tag{4.49}$$

The vector $|R\rangle$ is known but $|D\rangle$ is completely unknown and some physical model must be used to guess at an upper limit for the quantity $\langle D|D\rangle$. The model used by Barrett (1975*a*) was to take the value

$$\langle D|D\rangle \lesssim \langle \rho_F|\rho_F\rangle \int_0^{R_{max}} [\rho_F(r)]^2 r^2 dr \tag{4.50}$$

where $\rho_F(r)$ is a Fermi density. In practice the estimate of the completeness error may be much smaller than the statistical error and it is then more meaningful to give more weight to the more accurate measurements. This can be done by making a different choice of the coefficients in the expansion of $|r^2\rangle$:

$$|r^2\rangle = \sum c_i|f_i\rangle + |R'\rangle \tag{4.51}$$

where $|R'\rangle$ is no longer orthogonal to $\{f_i\}$. Taking the square of the total error to be the sum of squares of the statistical and completeness errors we obtain

$$\delta\langle r^2\rangle \simeq \sum c_i \delta E_i \tag{4.52}$$

$$[\text{error in } \delta\langle r^2\rangle]^2 = \sum c_i^2\sigma_i^2 + \langle R'|R'\rangle\langle D|D\rangle \tag{4.53}$$

$$= \sum c_i^2\sigma_i^2 + \langle D|D\rangle[\langle r^2|r^2\rangle + \sum_{ij} c_i c_j\langle f_i|f_j\rangle - 2\sum_i c_i\langle r^2|f_i\rangle].$$

Minimizing this with respect to the coefficients c_i gives

$$\sum_i c_i\langle f_i|f_j\rangle\langle D|D\rangle + c_j\sigma_j^2 = \langle r^2|f_j\rangle\langle D|D\rangle. \tag{4.54}$$

In the limit of vanishing errors the coefficients c_i become identical to b_i defined in eqn (4.45). This method for estimating $\delta\langle r^2\rangle$ has the advantage that the total error is always reduced by the addition of an experimental quantity. Its disadvantages are the arbitrariness of the estimate of $\langle D|D\rangle$ and the strong increase of the completeness error as the value of R_{max} increases. Nevertheless small values of the completeness error can be obtained in many cases using very conservative estimates for these quantities, so that the method is useful in practice.

The shift δE_{field} is sometimes expressed in terms of a quantity called the 'standard shift' which is a function of the mass number A. For muonic atoms the standard shift is not well defined but is often taken to be the field shift calculated for a change in $\rho_{ch}(r)$ corresponding to a Fermi distribution whose diffuseness is constant and whose r.m.s. radius is $\sqrt{(0.6)}\ 1.2\ A^{1/3}$ fm. The standard shift has no fundamental significance but is merely used as a convenient unit. For spherical nuclei most field shifts are, not unexpectedly, considerably less than the standard shift, since the shifts are due to a change in the *charge* distribution caused by the addition of *uncharged* neutrons.

4.6.2. *Isomer shifts*

The excited states of nuclei in general have much longer lifetimes than excited muonic-atom states so that muonic transitions can occur while the nucleus is excited. The excitation of the nucleus causes a shift in the energy of the muonic X-rays which has been observed indirectly in transitions to the 1s state. It is customary to define the 'isomer shift' ΔE_{isomer} to be *the field shift for the muonic 1s state*, i.e. the change in energy of the 1s state due to the change in the nuclear charge distribution caused by the nuclear excitation. The actual shift in a transition energy is also due to changes in the hyperfine structure and nuclear polarization as well as the field shift in the energy of the upper level of a transition. The last effect is negligible for $l \neq 0$ states and the nuclear polarization change has been shown to be small in some cases (Chen 1968, Gal, Grodzins, and Hüfner 1968) and is usually neglected. Calculations of isomer shifts were first carried out by Wilets and Chinn (1963) and Hüfner (1964). In practice the contribution of the isomer shift to a muonic transition is too small to be measured, but the shift for the muon 1s state can also be obtained from the change in energy of the nuclear de-excitation γ-ray caused by the presence of the muon. This can be seen in Fig. 4.5 where ε_2' and $E'_{2p_{1/2}-1s}$ are the 'isomer-shifted' quantities. The nuclear excited state must be one which can be excited by the muon during its cascade, and as we noted in §4.4 rotational levels are

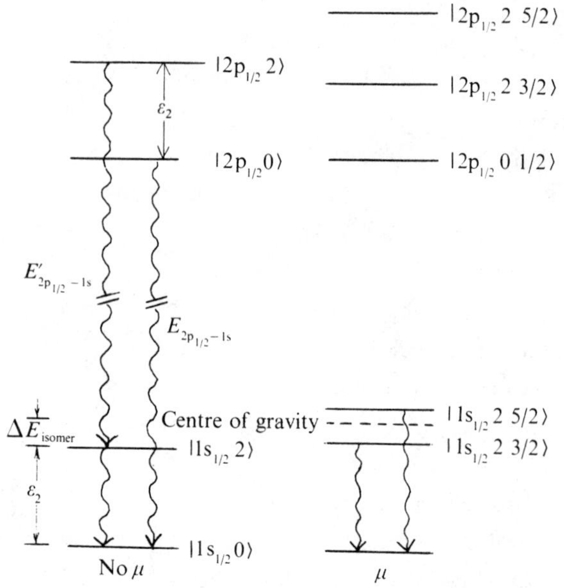

FIG. 4.5. Level diagram for a muonic atom and a nucleus with a ground-state spin 0 and excited-state spin 2. The muon states $|nlj\rangle$ and nuclear states $|I\rangle$ are coupled to form atomic states $|nljIF\rangle$ of total angular momentum F. These atomic states are split by the magnetic hyperfine interaction for $l = 0$ and also by the electric quadrupole interaction for $l = 1$.

particularly strongly excited in the 3d-2p transition in deformed nuclei. The first measurements of muonic isomer shifts were carried out by Bernow *et al.* (1967) for the 2^+ nuclear rotational state in ^{152}Sm.

As indicated in Fig. 4.5 it is necessary to find the centre of gravity of the magnetic hyperfine doublet in order to determine the isomer shift. This requires the calculation of the magnitude of the splitting which depends on the magnitude and distribution of the nuclear-excited-state magnetic moment. If the splitting is not experimentally resolved then we also need a calculation of the intensities of transitions to the $F = \frac{3}{2}$ and $F = \frac{5}{2}$ states which are not statistically populated (Gal *et al.* 1968), and we need the calculation of the depopulation of the $\frac{5}{2}$ state by an $F = \frac{5}{2} \to F = \frac{3}{2}$ M1 transition which takes place by Auger emission of electrons (Gal *et al.* 1968). The first effect tends to raise the apparent centre of gravity while the second, which is much larger, tends to lower it. Despite these difficulties a large number of isomer shifts have been studied and many of them are reviewed by Wu and Wilets (1969). The corresponding shifts in radius may also be obtained from the shifts in electronic X-rays as described in Chapter 5. Results are presented in §6.1.6.

4.6.3. *Isotone shifts*

When measurements are made on atoms with neighbouring nuclei which have the same number of neutrons, the differences in energies are known as *isotone shifts*. The change in energy is typically a few per cent, which is several orders of magnitude larger than isotope and isomer shifts. Since the muon wavefunctions differ considerably, the study of a pair of isotones is to some extent the individual study of each nucleus. If the isotones differ by a single proton, then information about the change in charge distribution is of particular interest because we can compare this with the wavefunction of a single proton and draw conclusions about the extent of the validity of the single-particle model for protons and perhaps about the accuracy of the single-particle wavefunctions.

The change in the energy of a particular transition may be written as the sum of separate contributions:

$$\delta E = \delta E_{\text{charge}} + \delta E_{\text{volume}} + \delta E_{\text{nuc. pol.}} \tag{4.55}$$

where the first term is simply due to the change in atomic number and the second to the change in the shape and size of $\rho_{\text{ch}}(\underset{\sim}{r})$. Unfortunately δE_{volume} is sometimes referred to as the 'field shift', a particularly misleading misnomer, since most of the change in the electric field due to the nucleus is caused by the change in the total charge. We define δE_{charge} to be the isotone shift which would occur if the size and shape of $\rho_{\text{ch}}(\underset{\sim}{r})$ for the heavier isotone were equal to that of the lighter isotone and if there were no change in the nuclear polarization. The quantity δE_{charge} contains a small contribution from the change in radiative corrections as well as a very small contribution from the change in reduced mass. Its value depends on the assumed shape and size of $\rho_{\text{ch}}(\mathbf{r})$. If the latter is constrained to give a fixed transition ener energy, however, then δE_{charge} varies very slowly with $\rho_{\text{ch}}(\mathbf{r})$ (Kast *et al.* 1971).

For comparison with theoretical nuclear densities the quantity of interest is $\delta\rho_{\text{ch}}(\mathbf{r})$, and the muonic X-ray measurements are of particular value when combined with electron-scattering calculations to determine this quantity. When it cannot be determined in a model-independent way, a test of the theoretical charge densities can be made by using the theoretical $\delta\rho_{\text{ch}}(\mathbf{r})$ to calculate δE_{volume}.

4.7. RADIATIVE CORRECTIONS IN MUONIC ATOMS

If we assume the nucleus to be a fixed charge then the muon-nucleus potential obtained using the Coulomb law of force is

$$V(\mathbf{r}) = -eA_0(\mathbf{r}) = -\hbar c Z\alpha \int \frac{\rho_{\text{ch}}(\mathbf{r}')\mathrm{d}^3\mathbf{r}'}{|\mathbf{r}-\mathbf{r}'|}. \tag{4.56}$$

Apart from the nuclear polarization effects there are a number of corrections to the energies obtained using $V(r)$ in the Dirac equation, namely the recoil correction and the vacuum-polarization, self-energy, and anomalous-magnetic-moment corrections. The dominant radiative correction in muonic atoms is the second-order photon self-energy or vacuum-polarization correction. The situation is reversed from that in ordinary atoms where the electron self-energy correction is much more important. If we construct an *effective potential* to take account of the corrections to the energy due to these radiative processes, then the (lowest-order) effective self-energy potential is smaller for muons by the ratio $(m_e/m_\mu)^2$. There is no such reduction factor in the effective vacuum-polarization potential, since it is due to the presence of *electron-positron* pairs both in ordinary and in muonic atoms.

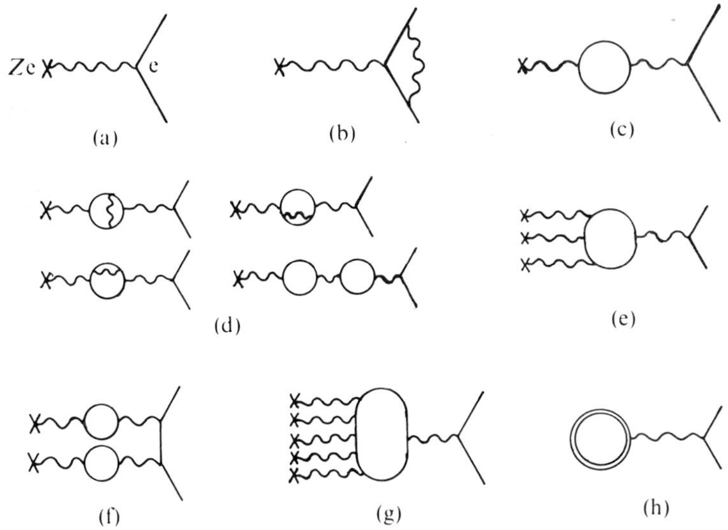

FIG. 4.6. Some of the Feynman diagrams for the muon-nucleus interaction: (a) Lowest-order interaction (order $Z\alpha$). (b) Second-order vertex correction (total order $\alpha(Z\alpha)$). The remaining diagrams are vacuum-polarization corrections: (c) second-order (total order $\alpha(Z\alpha)$); (d) fourth order ($\alpha^2(Z\alpha)$); (e) sixth order ($\alpha(Z\alpha)^3$); (f) sixth-order ladder diagram ($\alpha^2(Z\alpha)^2$); (g) tenth-order ($\alpha(Z\alpha)^5$); (h) diagram with e^+ and e^- moving in an external field.

4.7.1. *Vacuum polarization*

The lowest-order diagram in Fig. 4.6(c) results in a second-order effective potential $V_{\mathrm{vp}}^{(2)}(r)$. We call this a second-order correction because it is of second order in e *relative* to the classical Coulomb interaction. For low momentum transfer $(q/m_{\mathrm{e}}c)^2 \ll 1$ we may approximate $V_{\mathrm{vp}}^{(2)}(r)$ by the expression (Uehling 1935)

$$V_{\mathrm{vp}}^{(2)}(r) \approx -\frac{\alpha}{15\pi}\left(\frac{\hbar c}{m_{\mathrm{e}}}\right)^2 \nabla^2 V(r) \tag{4.57}$$

which is used to calculate the vacuum-polarization correction in ordinary atoms. The condition of validity is that the Fourier transform of the wavefunction should be negligible where the inequality is not satisfied. Alternatively the approximation is valid when the probability of the electron being inside a radius of the order of $\lambda_e = \hbar c/m_e = 386$ fm is negligible.

In muonic atoms eqn (4.57) is adequate to take account of the vacuum polarization due to $\mu^+\mu^-$ pairs, but the orbits for most states of interest lie well inside λ_e, and for the e^+e^- vacuum polarization we must use the exact expression (Akhiezer and Berestetskii 1965)

$$V_{\mathrm{vp}}^{(2)}(r) = -\frac{2}{3}\frac{Z\alpha^2}{\pi}\hbar c\int\frac{d^3r'\rho(\underset{\sim}{r}')}{|\underset{\sim}{r}-\underset{\sim}{r}'|}\int_1^\infty \exp\left(\frac{2y|\underset{\sim}{r}-\underset{\sim}{r}'|}{\lambda e}\right)\left(1+\frac{1}{2y^2}\right)\left(y^2-1\right)^{1/2}\frac{1}{y^2}\,dy. \tag{4.58}$$

For $\rho_{\mathrm{ch}}(\underset{\sim}{r})$ spherically symmetric this becomes

$$V_{\mathrm{vp}}^{(2)}(r) = -\frac{4}{3}Z\alpha^2\hbar c\frac{1}{r}\int dr'\rho(r')r'\{H(|r-r'|) - H(r+r')\} \tag{4.59}$$

where

$$H(r) = \frac{1}{2}\lambda e\int_1^\infty \exp\left(\frac{2yr}{\lambda e}\right)\left(1+\frac{1}{2y^2}\right)(y^2-1)^{1/2}\frac{1}{y^3}\,dy. \tag{4.60}$$

For values of r up to about $\frac{1}{2}\lambda c$ we may use the expansion

$$H(r) \approx H(0) + r\{\ln\left(\frac{r\gamma}{\lambda c}\right) - \frac{1}{6} + \frac{3\pi}{8}\left(\frac{r}{\lambda c}\right) + \frac{1}{2}\left(\frac{r}{\lambda c}\right)^2 - \frac{\pi}{12}\left(\frac{r}{\lambda c}\right)^3\} \tag{4.61}$$

where $\ln\gamma \approx 0.557$ is Euler's constant.

In light nuclei the corrections due to higher-order processes are very much smaller than experimental errors, but in heavy nuclei they become non-negligible. The fourth- and sixth-order diagrams are equally important because of the extra Z^2 factor in the latter. The fourth-order calculations have been done for point nuclei (Sundaresan and Watson 1972) and the results may be taken as upper limits on the corrections. The results are small but not negligible. The lowest-order diagram in Fig. 4.6(c), the light-by-light scattering diagram of Fig. 4.6(e), and the tenth-order diagram of Fig. 4.6(g) are the first three members of an infinite set of diagrams whose sum is represented by Fig. 4.6(h). Here the virtual electron-positron pair is not free but moving in the field of the nucleus. The summed diagram has been calculated by Wichmann and Kroll (1956) for point nuclei and by Rinker and Wilets (1975) for a finite nucleus. Rinker and Wilets find that the finite size causes a considerable reduction even for the 4f state in lead. Effective potentials for the higher-order diagrams have been given by Blomqvist (1972).

4.7.2. *Self-energy and anomalous-magnetic-moment corrections* The vertex diagram in Fig. 4.6(b) gives rise to an anomalous-magnetic-moment correction which may be represented by the effective potential

$$\frac{\alpha}{4\pi m_\mu c}\,\beta\underset{\sim}{\alpha}.\underset{\sim}{p}V. \tag{4.62}$$

Since this is a small correction it is sufficiently accurate to calculate it using only the large components of the wave-function. For a spherically symmetric potential this leads to a non-relativistic effective potential.

$$\frac{\alpha}{8\pi}\left(\frac{\hbar}{m_\mu c}\right)^2\left(\nabla^2 V + \frac{2}{r}\frac{\partial V}{\partial r}\,\underset{\sim}{\sigma}.\underset{\sim}{L}\right) \tag{4.63}$$

where the components of $\underset{\sim}{\sigma}$ are the Pauli spin matrices and $\underset{\sim}{L}$ is the orbital angular momentum operator.

The remainder of the vertex diagram is more difficult to calculate. Its expectation value for a state $|k\rangle$ may be written in the form

$$\frac{\alpha}{3\pi}\left(\frac{\hbar}{m_\mu c}\right)^2\left[\ln\frac{m_\mu c^2}{2\Delta\varepsilon_k} + \frac{11}{24}\right]\langle k|\nabla^2 V|k\rangle \tag{4.64}$$

where the 'Bethe logarithm' is defined by

$$\ln\frac{m_\mu c^2}{2\Delta\varepsilon_k} = \sum_{k'\neq k}\frac{\langle k|\underset{\sim}{p}|k'\rangle.\langle k'|[V,\underset{\sim}{p}]|k\rangle}{\frac{1}{2}\langle k|\nabla^2 V|k\rangle}\ln\frac{m_\mu c^2}{2|E_k - E_{k'}|} \tag{4.65}$$

and $\Delta\varepsilon_k$ is approximately given by the binding energy for the state $|k\rangle$ (Barrett *et al.* 1968, Barrett 1968).

Adding together the expressions (4.63) and (4.64) and the vacuum polarization correction for $\mu^+\mu^-$ pairs (eqn (4.57)) we obtain a shift in energy

$$\Delta E_k = \frac{\alpha}{3\pi}\left(\frac{\hbar}{m_\mu c}\right)^2 \left[\langle k|\nabla^2 V|k\rangle \left\{\ln\frac{m_\mu c^2}{2\Delta\varepsilon_k} + \frac{11}{24} + \frac{3}{8} - \frac{1}{5}\right\} + \frac{3}{8}\langle k|\frac{2}{r}\frac{\partial V}{\partial r}\,\underset{\sim}{\sigma}.\underset{\sim}{L}|k\rangle\right]. \tag{4.66}$$

With the electron mass and hydrogen wavefunctions this expression gives almost the whole of the Lamb shift for hydrogen, i.e. the $2s_{1/2} - 2p_{1/2}$ splitting. (The remainder of the splitting, which is less than 0.1 per cent, is due to the finite size of the proton, the relativistic recoil correction, and higher-order radiative corrections.) For this reason the total correction in eqn (4.66), evaluated for muonic atoms, is usually called the 'Lamb shift' in the muonic-atom level. It does not give an approximation to the $2s_{1/2} - 2p_{1/2}$ splitting in muonic atoms in which the finite-size effect and electron-positron pair vacuum polarization are far more important.

4.7.3. *Recoil correction*

In non-relativistic theory the recoil of the nucleus (mass M) can be obtained exactly by replacing the muon mass m_μ by the reduced mass $m = m_\mu/(1+m_\mu/M)$ and using the relative co-ordinate in the equations of motion. If the nucleus moves a non-negligible distance during the time that a light signal takes to cross the atom, then its motion must be treated relativistically. This can be done by using the Bethe-Salpeter equation to describe the muon and nucleus as a two-body system. Grotch and Yennie (1969) have developed a modified Dirac equation which corrects for the motion of the nucleus. Almost all of the correction is obtained by using the reduced mass in the Dirac equation since the nuclear recoil motion is almost completely non-relativistic. If we ignore terms of second order in m_μ/M then the remainder of the correction is given in terms of the binding energy B by Barrett *et al.* (1973)

$$\Delta E = -\frac{B^2}{2Mc^2} - \frac{B}{Mc^2}R\frac{\partial B}{\partial R} - \langle V_G\rangle \tag{4.67}$$

where $\langle V_G \rangle$ is the expectation value of the quantity

$$V_G = \{V^2 + \frac{2}{r^2}\frac{dV}{dr}\int_0^r r'^2 V(r')dr'\}/2Mc^2 \tag{4.68}$$

and R is the equivalent radius (see §4.2). The first two terms in eqn (4.68) are obtained by using an effective reduced mass

$$m' = \frac{m_\mu(1 + B/M)}{(1 + m_\mu/M)} \tag{4.69}$$

and the second is due to the change of the length scale in the atomic wavefunctions caused by the change in the reduced mass. These relativistic recoil corrections are very small indeed, except for the lowest muon states where they are of the order of the experimental error. In these lowest states, however, the uncertainty in the nuclear polarization correction is considerably larger (see Chen 1970*a* and Kessler *et al.* 1975).

4.8. ELECTRON SCREENING

The penetration of the muon orbits by the electron cloud produces a change in energy which is negligible for $n \leqslant 2$ but is an important correction for $n \geqslant 5$. The first quantitative calculations of this effect were carried out by Barrett *et al.* (1968) for ^{209}Bi by scaling relativistic Hartree atomic wavefunctions for Hg calculated by Cohen (1960). The screening is produced almost entirely by the 1s electrons. The most accurate calculation of screening requires a relativistic Hartree-Fock calculation of the atomic electrons moving in the field of the nucleus and the muon. This has been done by Vogel (1973*a*, 1974).† In principle such calculations assuming

†A very simple calculation which gives almost the same results (Tauscher 1973) uses point-nucleus relativistic wavefunctions with an effective atomic number. For the 1s electrons this is taken to be Z-1 (i.e. assuming complete screening of the nucleus by the muon) and for the n=2 electrons the value Z-3 is used (complete screening by the muon and 1s electrons).

full occupation of the lowest electron states give only an upper limit to the screening correction since depopulation of these states by Auger transitions may occur during the muon cascade; however, Vogel (1973*b*) has calculated the probability of these Auger transitions and the rate of refilling of vacancies, and concludes that there is only a very low probability that the 1s level is vacant. The screening corrections for ^{208}Pb are shown in Table 4.2.

4.9. THE COMPARISON OF THEORY AND EXPERIMENT FOR HIGH TRANSITIONS

When muonic X-ray measurements in medium and heavy nuclei are analysed to give information on nuclear charge distributions it is usually assumed that the spectra can be explained in terms of a Dirac particle moving in the field of the nucleus, and the corrections to the static nucleus approximation and the Coulomb interaction described in the previous sections come from well-established theories of atomic and nuclear physics and quantum electrodynamics. For the lowest transitions the nuclear size effect is so large that errors or neglected terms in these corrections might be masked; a change in the charge density could compensate for the errors without changing the fits to electron scattering. The situation is different in higher transitions such as the ^{208}Pb $5g_{9/2}$ - $4f_{7/2}$ for which the finite-size correction is about 1 part in 10^5. In this case it has been possible to carry out a stringent test of the theory (in particular the quantum electrodynamics) by calculating all the non-negligible corrections to the transition energy.

The calculation of these corrections was spurred on by persistent discrepancies between theory and experiment. They have been reviewed by Rafelski *et al.* (1974) and by Watson and Sundaresan (1975) who also compiled tables of different corrections. The Feynman diagrams for the relevant radiative corrections are shown in Figs. 4.6 and 4.7. The original discrepancy was very much reduced when a mistake was found in the calculation of one of the vacuum-polarization diagrams, and there was then a period of several years during which the

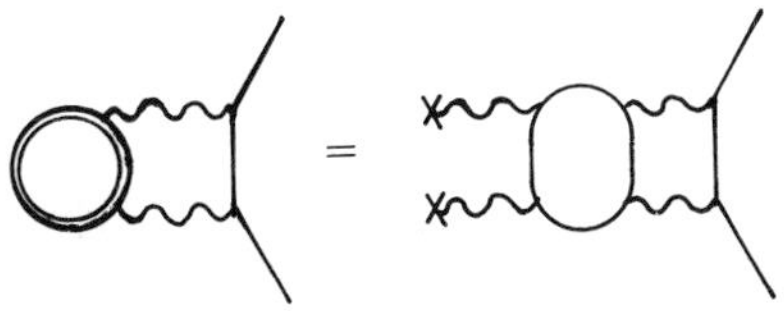

FIG. 4.7. Higher-order correction to the muon self-energy. The diagram on the right is of order $\alpha^2(Z\alpha)^2$.

vacuum-polarization diagrams, nuclear polarization effect, and the electron screening were evaluated more and more carefully. Apart from the possibility of errors in calculations or the neglect of non-negligible diagrams there were suggestions of completely new effects. A coupling between the muon and the nucleus via a scalar meson ϕ suggested by Weinberg (1967, 1971) would have removed the shift if the mass of the ϕ were about 14 MeV (Watson and Sundaresan 1975). The effect of a scalar coupling or of a correction due to non-linear terms in the coupling of the muon to the electromagnetic field was calculated by Adler (1974). This conjecture was tested by looking at the 2p-1s energy in muonic ^{4}He which was measured using laser techniques by Bertin *et al.* (1975) and calculated by Barbieri (1975) who found no evidence for an anomalous additional interaction. A suggestion that the discrepancy was due to the unjustified neglect of a Z^2 self-energy diagram was contradicted by a calculation of Wilets and Rinker (1975).

The discrepancy was finally removed by measurements of the 5g-4f transitions in μ^- Pb and the 5g-4f and 4f-3d transitions in μ^- Ba by Tauscher *et al.* (1975) and also by Dixit *et al.* (1975). The corrections for ^{208}Pb are shown in Table 4.2.

TABLE 4.2

^{208}Pb *muonic-atom binding energies*† *and corrections (in eV)*

State		$1s_{1/2}$	$2p_{1/2}$	$2p_{3/2}$	$3d_{3/2}$	$3d_{5/2}$	$4f_{5/2}$	$4f_{7/2}$	$5g_{7/2}$	$5g_{9/2}$	Notes
Point nucleus		20992568	5385466	4837295	2166549	2121988	1197386	1188312	761721	758968	1
Finite-size correction		-10471316	-604009	-237803	-4129	-1637	-10	-4	-0	-0	2
Vacuum	$\alpha(Z\alpha)$	67514	32346	29891	10560	9897	3791	3682	1600	1575	3
Polarization	$\alpha^2(Z\alpha)$	550	250	230	73	68	26	26	11	11	4
Corrections	$\alpha^2(Z\alpha)^n\ n \geqslant 3$	-492	-348	-335	-186	-180	-95	-95	-53	-53	5
Lamb shift		-2950	-331	-664	49	-52	13	-9	4	-3	6
Electron screening		0	7	9	32	35	85	88	169	172	7
Relativistic recoil		377	109	94	15	14	4	4	2	2	8
Nuclear polarization		6800	1900	1800	30	0	5	5	1	1	9
Total theory		10593051	4815390	4630517	2172992	2130132	1201205	1192009	763455	760673	10
Experiment		10593170	4815180	4630290	2173050	2129940	—	—	763461	760656	11
Experimental error		140	60	50	50	50	—	—	16	14	11
Discrepancy		119	210	-233	58	-192	—	—	6	-17	12

†The 'experimental binding energies' were obtained from the transition energies by fixing the 4f levels at the theoretical values.

Notes

1. Calculated using m = 105.6595 MeV and α^{-1} = 137.03602.
2. Based on the usual definition described in §4.1 using a three-parameter Fermi distribution with c = 6.712, a = 2.271/ℓn 81, w = -0.064 obtained from fitting the energies (Kessler *et al.* 1975).

3. Calculated using a two-parameter Fermi distribution with $c = 6.641$, $a = 2.344/\ln 81$ (Barrett 1975).
4. Obtained using the expression given by Blomqvist (1972) derived from the Kallen-Sabry calculation.
5. Rinker and Wilets (1975).
6. The values include the anomalous-magnetic-moment correction and the vacuum polarization due to $\mu^+ \mu^-$ pairs and are calculated using the expression given by Barrett *et al.* (1968) using the Bethe and Negele method for the Bethe logarithm (see Barrett 1968).
7. Values relative to the screening for a 1s muon from calculations by Engfer *et al.* using a computer program of Vogel (1973*a*).
8. Values from Barrett *et al.* (1973).
9. Values from Skardhamar (1970).
10. For the 5g-4f transition there is a theoretical uncertainty of about 7 eV which comes from the uncertainty in the muon mass (contributing 1 eV) and the fine-structure constant (1 eV) as well as the electron screening (5 eV) and radiative corrections (5 eV).
11. Values from Kessler *et al.* (1975) (lower transitions) and Tauscher *et al.* (1975) (5g - 4f transitions).
12. For the lowest levels the values of the discrepancy have little significance since they could be almost entirely removed by using a different form of the charge density.

5

ISOTOPE AND ISOMER SHIFTS IN OPTICAL AND X-RAY SPECTRA

The term 'isotope shift' is used to denote the change in the energy of a given atomic transition when one or more neutrons are added to the nucleus; the 'isomer shift' is the change which occurs when the nucleus is excited. Isotope shifts in which there is a significant contribution from the change in the charge distribution were measured (Merton 1919) long before a theoretical analysis of the size of this effect was carried out (Rosenthal and Breit 1932). The measurements of very large isotope shifts in the rare-earth region led Brix and Kopfermann (1959) to predict the occurrence of large intrinsic deformations in this region. Work on isotope shifts has been reviewed by Brix and Kopfermann (1958), Breit (1958), Wilets (1958), Kuhn (1969), Stacey (1966), and Heilig and Steudel (1974). In §4.6 we discussed the shifts which occur in muonic transitions and separated them into two parts: the *mass shift* which is caused by the change in the mass of the nucleus, and the *field shift* which is due to the change in the electrostatic field, i.e. to the change in the size and shape of the nuclear charge distribution. In muonic atoms we can easily calculate the mass shift and hence obtain the field shift from experiment (assuming that the change in the nuclear polarization correction is sufficiently well known). In many-electron atoms the situation is quite different and it is not easy to extract nuclear size information from the measurements for two reasons: firstly, the mass shift is not easily obtained, and secondly it is difficult to calculate the magnitude of the electron wavefunctions at the nucleus. In the case of isomer shifts there is no change in mass and most of the discussion of isotope field shifts can be applied directly to isomer shifts. Results of analyses for isotope pairs are given in §6.1.4 and for isomer shifts in §6.1.6.

5.1. THE FIELD SHIFT

The field shift in a transition between atomic states 1 and 2 due to a change $\delta\rho_{\text{ch}}(r)$ in the nuclear charge density is (to first order)

$$\delta E_{\text{field}} = -Ze^2 \iint \frac{\{\rho_{\text{e}}^{(1)}(\underset{\sim}{r}) - \rho_{\text{e}}^{(2)}(\underset{\sim}{r})\}\ \delta\rho_{\text{ch}}(\underset{\sim}{r}')}{|\underset{\sim}{r}-\underset{\sim}{r}'|}\ \mathrm{d}^3r\ \mathrm{d}^3r' \quad (5.1)$$

$$= -e \int \Delta\rho_{\text{e}}(\underset{\sim}{r})\ \delta V(\underset{\sim}{r})\ \mathrm{d}^3r \quad (5.2)$$

$$= Ze \int \Delta V_{\text{e}}(\underset{\sim}{r}')\ \delta\rho_{\text{ch}}(\underset{\sim}{r}')\ \mathrm{d}^3r' \quad (5.3)$$

where the electron charge density is given in terms of single-particle states ψ_{n_i} by

$$\rho_{\text{e}}(\underset{\sim}{r}) = \sum \psi_{n_i}^{\dagger}(\underset{\sim}{r})\ \psi_{n_i}(\underset{\sim}{r}). \quad (5.4)$$

Assuming that the difference in states 1 and 2 is simply due to the change in the wavefunction of a single electron from ψ_{n_1} to ψ_{n_2} then

$$\Delta\rho_{\text{e}}(\underset{\sim}{r}) \equiv \rho_{\text{e}}^{(1)}(\underset{\sim}{r}) - \rho_{\text{e}}^{(2)}(\underset{\sim}{r}) = \psi_{n_1}^{\dagger}(\underset{\sim}{r})\ \psi_{n_1}(\underset{\sim}{r}) - \psi_{n_2}^{\dagger}(\underset{\sim}{r})\ \psi_{n_2}(\underset{\sim}{r}). \quad (5.5)$$

If there is a hyperfine splitting in one (or both) of the isotopes, then the shift is defined with respect to the centre of gravity of the splitting. If we expand $\delta V(\underset{\sim}{r})$ in multipoles, then only the monopole part $\delta V^{(0)}(r)$ (the angular average of $\delta V(\underset{\sim}{r})$) contributes to the shift of the centre of gravity. This monopole part vanishes outside the nucleus so that we need to consider only the first few terms in the expansion of $\Delta\rho_{\text{e}}(\underset{\sim}{r})$. Assuming that the potential near the origin can be expanded in even powers of r the radial solutions of the Dirac equation can be expanded in a series of even or odd powers of r so that the monopole part of $\Delta\rho_{\text{e}}(\underset{\sim}{r})$ is given by an even power series. If we expand in the ratio r/R where R is the approximate nuclear radius, then the ratio of successive terms is of the order of $-Zr/a_{\text{H}}$ where a_{H} is the Bohr radius of hydrogen. We thus obtain for the monopole part of $\Delta\rho_{\text{e}}(\underset{\sim}{r})$

$$\Delta\rho_e^{(0)}(r) = r^{2j-1}(a_1 + a_2r^2 + a_3r^4 + \ldots) \tag{5.6}$$

where j is the smaller of the total angular momentum quantum numbers of states n_1 and n_2. This gives for the monopole part of $\Delta V_e(\underset{\sim}{r})$

$$\Delta V_e^{(0)}(r) = b_0 + r^{2j+1}(b_1 + b_2r^2 + b_3r^4 + \ldots) \tag{5.7}$$

where, for $n > 0$, $b_n = -e\ a_n/(2j+2n-1)(2j+2n)$. Substituting in eqn (5.3) we obtain

$$\delta E_{\text{field}} = C_1\delta\langle r^{2j+1}\rangle + C_2\delta\langle r^{2j+3}\rangle + C_3\delta\langle r^{2j+5}\rangle + \ldots \tag{5.8}$$

where

$$\delta\langle r^k\rangle \equiv \int r^k\ \delta\rho_{\text{ch}}(\underset{\sim}{r})\ d^3r \tag{5.9}$$

and $C_n = Zeb_n$.

In an electronic atom the field shift in a particular level is very small unless that level has spin $j = \frac{1}{2}$; for $p_{1/2}$ states, however, it is the *small components* of the wavefunction which is non-zero at the origin so that the shift is small for heavy (high-Z) atoms and negligible for light atoms. Thus in most cases the field shift is due to an $s_{1/2}$ initial or final state of the electron undergoing the transition,† and to a good approximation all but lowest term in eqn (5.6) can be neglected; if this is so, the magnitude of the shift is proportional to the *square of the electron wavefunction at* $r = 0$ and the *change in the mean-square radius of the nuclear charge density*.

For heavy nuclei the terms b_2r^4 and b_3r^6 in $\Delta V_e^{(0)}(r)$ (for the case $j = \frac{1}{2}$) give a non-negligible contribution to the

---

†A substantial field effect can also be produced in a transition which does *not* involve an s-electron jump if the screening of inner closed s-electron shells is changed (Heilig and Steudel 1974).

field shifts, and one way of taking this into account is to approximate the function $\Delta V_e^{(0)}(r)$ by the parametrization (Wu and Wilets 1969)

$$a + br^k \tag{5.10}$$

where k is not an integer. The additional factor $\exp(-\alpha r)$ needed for the muonic-atom case (eqn 4.11) is not necessary here because the higher terms are much less important. The field shift is given by

$$\delta E_{\text{field}} = Ze\delta\langle \Delta V_e^{(0)} \rangle = Zeb\delta\langle r^k \rangle. \tag{5.11}$$

If we choose k so that $a + br^k$ has the same slope and second derivative as $\Delta V_e^{(0)}(r)$ at $r = R$ we obtain the expression

$$k = 2 + \frac{4C_2R^2 + 12C_3R^4}{C_1 + 2C_2R^2 + 3C_3R^4}. \tag{5.12}$$

If we choose R to be the half-density radius of a Fermi distribution then we can expect the approximate form to have the desired property of fitting $\Delta V_e^{(0)}(r)$ accurately where $r^2\delta\rho_{\text{ch}}(r)$ is largest. The coefficients C_n are very nearly independent of the assumed form of $\rho_{\text{ch}}(\underset{\sim}{r})$ used in obtaining the electron wavefunctions and have been calculated by using a Fermi distribution for K X-ray transitions by Seltzer (1969). Some of his results, together with values of k obtained from eqn (5.12), are shown in Table 5.1. The ratios of successive coefficients in the expansion of the electron wavefunction are determined almost completely by the magnitude at the origin of the potential due to the nuclear charge, so that the ratios C_2/C_1 and C_3/C_1 are almost the same for 2p-1s and for higher transitions such as 6s-6p.

Assuming that we know the value of δE_{field} it is difficult to extract the quantity $\delta\langle r^2 \rangle$ (or $\delta\langle r^k \rangle$) for optical transitions because of uncertainties in the magnitude of the electron wavefunctions. These may be known for the ground state of the atom from Hartree-Fock calculations but not for excited states. The latter may be obtained approximately from

TABLE 5.1

Values of C_n and k for $1s_{1/2}$ states†

Z	C_1	C_2	C_3	k
30	6.83	0.00232	0.95×10^{-5}	1.98
50	84.5	-0.0523	1.61×10^{-4}	1.95
60	238	-0.185	5.34×10^{-4}	1.93
70	623	-0.579	1.58×10^{-3}	1.91
82	1880	-2.11	5.59×10^{-3}	1.88
92	4700	-6.01	1.56×10^{-2}	1.85

†C_n values from Seltzer (1969). The units for C_n are meV fm^{-2}.

the magnitude of magnetic hyperfine splittings, since the coefficient A_1 (eqn 4.32) depends strongly on the magnitude of the wavefunction at the origin. If measurements are made on more than two isotopes then the ratios of successive shifts are equal to the corresponding ratios of $\delta\langle r^2\rangle$. Since optical measurements are available for far more isotopes than are other measurements, they make an important contribution to our knowledge of nuclear charge radii, especially for highly radioactive or rare isotopes.

5.2. THE MASS SHIFT

The correction to the atomic energy levels due to the motion of the nucleus is given in the non-relativistic limit by the kinetic energy of the nucleus relative to the centre of mass of the atom. This may be written as the sum of two terms†

$$T_{\text{nucleus}} = \frac{1}{2M}\Big(\sum_i \mathbf{p}_i\Big)^2$$

$$= \frac{1}{2M}\Big(\sum_i \mathbf{p}_i^{\,2} + \sum_{i \neq j} \mathbf{p}_i \cdot \mathbf{p}_j\Big) \qquad (5.13)$$

†The presence of non-trivial mass terms was pointed out by Ehrenfest (1922) and a quantitative treatment was given by Hughes and Eckart (1930).

where M is the nuclear mass and $\underset{\sim}{p}_i$ the momentum of the ith electron. Since the atomic wavefunctions are almost identical for different isotopes the expectation value of the expression within the brackets in eqn (5.1) is independent of M so that the mass shift is proportional to the change in $1/M$. The first term in eqn (5.13) leads to the well-known reduced mass correction which results in the *normal mass shift*. For isotopes of mass numbers A and A' we denote this by the symbol $n^i_{AA'}$, where the superscript i refers to a particular transition (of energy E^i) between atomic states $|1\rangle$ and $|2\rangle$. The second term in eqn (5.13) gives a contribution resulting from correlations between electron pairs and is called the *specific mass shift* $s^i_{AA'}$.

Denoting the field shift by $f^i_{AA'}$, we obtain for the isotope shift

$$\delta E^i_{AA'} = f^i_{AA'} + n^i_{AA'} + s^i_{AA'} \tag{5.14}$$

with

$$f^i_{AA'} = Ze\int \Delta V^i_{\text{e}}(\underset{\sim}{r})\,\delta\rho_{AA'}(\underset{\sim}{r})\,\mathrm{d}^3r$$

$$\equiv Ze\delta\langle\Delta V^i_{\text{e}}\rangle_{AA'} \tag{5.15}$$

$$n^i_{AA'} = \frac{A'-A}{AA'}\,\frac{m_{\text{e}}}{m_{\text{p}}}\,E^i \tag{5.16}$$

$$s^i_{AA'} = \frac{A'-A}{AA'}\,\frac{m_{\text{e}}}{m_{\text{p}}}\,[\langle 1|\sum_{i\neq j}\underset{\sim}{p}_i\cdot\underset{\sim}{p}_j|1\rangle - \langle 2|\sum_{i\,j}\underset{\sim}{p}_i\cdot\underset{\sim}{p}_j|2\rangle] \tag{5.17}$$

$$= \frac{A'-A}{AA'}\,\frac{2m_{\text{e}}}{m_{\text{p}}}\sum_{k=1}^{N}\int\{\psi^{\dagger}_{k_1}(\underset{\sim}{r}_1)\psi^{\dagger}_{k}(\underset{\sim}{r}_2)\,\underset{\sim}{p}_1\cdot\underset{\sim}{p}_2\,\psi_k(\underset{\sim}{r}_1)\psi_{k_1}(\underset{\sim}{r}_2) - \psi^{\dagger}_{k_2}(\underset{\sim}{r}_1)\psi^{\dagger}_{k}(\underset{\sim}{r}_2)\,\underset{\sim}{p}_1\cdot\underset{\sim}{p}_2\,\psi_k(\underset{\sim}{r}_1)\psi_{k_2}(\underset{\sim}{r}_2)\}\,\mathrm{d}^3r_1\,\mathrm{d}^3r_2 \tag{5.18}$$

where N is the number of electrons and we have assumed that the electron wavefunctions for states $|1\rangle$ and $|2\rangle$ are Slater determinants whose difference is due solely to the replacement of the single-particle state k_1 by k_2. The specific mass shift is thus due to the Pauli correlations (i.e. the antisymmetry) in the electron wavefunction between states of opposite parity, and it may be positive or negative depending on the nature of these correlations.

In very light elements the field shift is negligible and the relative isotope shifts (for the same line in different isotopes of the same element) are proportional to $(A'-A)/AA'$; calcium is at present the lightest element in which the field shift is clearly observed through the breakdown of this proportionality. In medium-heavy elements the mass and field shifts are comparable; even in heavy elements ($Z \gtrsim 58$) the specific mass shifts cannot be neglected because although $s^i_{AA'}$ is proportional to $1/AA'$ it may fall off more slowly because of the increase in the number of electrons which may contribute to the sum in eqn (5.18). The fact that the specific mass shifts cannot be neglected for medium and heavy atoms was established by King (1963) using a procedure which may be described as follows. For each transition i and isotope pair A, A' we define the quantity

$$d^i_{AA'} = \frac{AA'}{A'-A}\,(\delta E^i_{AA'} - n^i_{AA'}) \tag{5.19}$$

and rewrite this as

$$d^i_{AA'} = S^i + \left\{\frac{AA'}{A'-A}\, Ze\delta\langle \Delta V_e{}^i\rangle_{AA'}\right\} \tag{5.20}$$

where S^i (which is proportional to the specific mass shift) is independent of A and A'. After obtaining the corresponding quantities $d^j_{AA'}$ for another transition j we plot a graph in which each point has co-ordinates $(d^j_{AA'}, d^i_{AA'})$ and the different points represent different isotope pairs. We now assume that the ratio $f^i_{AA'}/f^j_{AA'}$ is independent of AA'. This is justified since the quantities $\Delta V_e{}^i$ and $\Delta V_e{}^j$ correspond to changes in the electron s-wave density and their functional forms are almost identical. If the experiments are accurate, then the points lie on a straight line which would pass near the origin if the specific mass shifts were small (or almost equal). King constructed such plots for a number of pairs of lines i, j and found that they did not pass near the origin. His method does not give the values of the specific mass shifts unless one of them can be calculated.

The calculation of specific mass shifts for optical transitions has been carried out by Bauche (1969, 1974), but the errors in such theoretical values are very difficult to estimate. A variation of the King plot suggested by Stacey (1966) is to use values of $d^{i}_{AA'}$ from muonic X-ray isotope shifts and plot them against the optical values. In this case $s^{i}_{AA'}$ is zero so that the intercepts on the axis $d_{AA'} = 0$ give the specific mass shifts $s^{j}_{AA'}$. This method has been used to estimate the specific mass in some Nd lines (Macagno 1968, Macagno *et al.* 1970), but in this case we cannot make the assumption that the ratio $\delta\langle \Delta V_{\mathrm{e}}(r)\rangle_{AA'}/\delta\langle \Delta V_{\mu}(r)\rangle_{AA'}$ is independent of A, A' so that the resulting King plot may not be a straight line. Since the method requires extrapolation over a considerable distance from the closely spaced experimentally derived points, it is not sufficiently accurate to give useful estimates of the specific mass shifts.

A more successful method of deriving optical specific mass shifts relies on the calculation of a shift for (electronic) K X-rays and a plot of optical shifts versus X-ray shifts. The calculations for X-rays have been done for selected nuclei by Chesler (1967) and they contain fewer uncertainties than optical calculations. A King plot using X-rays can be carried out as long as there are measurements on two pairs of isotopes which have different field shifts. There is good reason to do this, since optical measurements can be carried out on extremely small samples so that rare earth radioactive isotopes can be studied. Once an optical specific mass shift has been determined, the field shifts for such isotopes can be obtained. In the case of Hg this has been done for 20 different isotopes and the relative mean square radii are plotted in Fig. 6.3 on p. 167.

The problems involved in estimating specific mass shifts have been reviewed by Heilig and Steudel (1974), and a comparison of five different ways of estimating the shifts in Nd has been given by King, Steudel, and Wilson (1963).

5.3. ODD-EVEN STAGGERING

For the majority of even-*N* nuclei the field shift caused by the addition of one neutron is considerably less than half that due to the addition of two neutrons, and this phenomenon is called odd-even staggering. A striking example is the case of Hf isotope shifts (Tomlinson and Stroke 1964). On a microscopic level this is partly explained in terms of the pairing correlations between neutrons. There is a tendency for the unpaired neutron to occupy a lower angular momentum state than the pair of neutrons. Since the higher angular momentum states have a larger average radius, a neutron in such a state will tend to pull the protons out further. The situation is complicated by the density dependence of effective forces and the 'blocking' of a pair of states by the odd nucleon (Barrett 1966, Lande, Molinari, and Brown 1968). Another effect of blocking which contributes to staggering is the lowering of the mean square deformation β^2 in the odd isotope (Reehal and Sorensen 1971).

5.4. MÖSSBAUER MEASUREMENTS OF ISOMER SHIFTS

Just as in the case of muonic atoms there is a field shift in optical transitions when the nucleus is excited, and if the lifetime of the nuclear state is of the order of several hours then the change in energy of an optical transition may be measured directly (see for example, Tomlinson and Stroke 1964). In this case the theoretical analysis is the same as that of isotope shifts without the complication of the mass shift. Many lifetimes of nuclear states are much too short for this method and yet much longer than atomic-transition lifetimes, so that the shift in the 1s level can be measured indirectly by observing the shift in the energy of the nuclear de-excitation γ-ray. In principle the effect is the same as that described in §4.6.2 for muonic atoms, but the shifts are so much smaller that the only way to observe them is by using Mössbauer techniques. The method has been reviewed by Kienle (1968). It depends on choosing an emitter and an absorber of different compounds containing the element being studied such that the electron configurations, and hence the field shifts (due to the change in the size of the nucleus when it is excited), are different and result in a difference in the energy of the nuclear γ-ray in the two compounds. This is given by

$$\Delta E_\gamma = Ze \int [V_e^{\mathrm{A}}(r) - V_e^{\mathrm{E}}(r)]\ \delta\rho_{\mathrm{ch}}(r)\ \mathrm{d}^3r \qquad (5.21)$$

where $V_e^{\mathrm{A}}(r)$ and $V_e^{\mathrm{E}}(r)$ are the electrostatic potentials due to the electron cloud in the absorber and emitter respectively. The problem of extracting nuclear size information from such a measurement is similar to that in optical-isotope-shift measurements, namely to obtain the density of the electrons in the region of the nucleus. The function $V_e^{\mathrm{A}}(r) - V_e^{\mathrm{E}}(r)$ may be expanded in a power series in r^2 or approximated by the formula $a + br^k$ as in eqns (5.7) and (5.10), and the ratios of the coefficients or the value of k for $j = \frac{1}{2}$ states will to a considerable accuracy depend only on the atomic number. The magnitude of b or of the coefficient of r^2 depends on the quantity $\rho_e^{\mathrm{A}}(0) - \rho_e^{\mathrm{E}}(0)$ which can be obtained only by lengthy relativistic Hartree-Fock calculations. The electron states which contribute to the density in the region of the nucleus are the s-states and, for heavy nuclei (to a lesser extent), the $p_{1/2}$-states. In many cases the absorber and emitter have configurations which differ only in states other than s- and p-states, and the change in density comes from the rearrangement of the electrons in the latter, especially those in the highest shells (Tucker *et al.* 1969). Values of differences in electron densities have been given by Fricke and Waber (1972) who carried out relativistic Hartree-Fock calculations for several nuclei. For heavier nuclei there have been attempts to express the isomer shift in terms of changes in the parameters c and t of a Fermi distribution (eqn (1.12)), but such a model-dependent procedure is even less appropriate here than it is in the case of muonic atoms.

One way of eliminating the need to calculate the function $V_e^{\mathrm{A}}(r) - V_e^{\mathrm{E}}(r)$ is to make measurements on two or more γ-rays. The ratios of the shifts then give the ratios of the generalized moments of $\delta\rho_{\mathrm{ch}}$.

6
EXPERIMENTALLY DERIVED CHARGE AND CURRENT DISTRIBUTIONS

6.1. CHARGE DISTRIBUTIONS

In this chapter we give the results of analyses of experiments using the theories and techniques described in the previous three chapters. Some of the resulting charge distributions are almost purely phenomenological in the sense that functional forms have been used which satisfy certain vague intuitive ideas about the expected shape of nuclei and in many cases have been chosen mainly for their simplicity or becuase they have been used frequently in the past. There is no clear dividing line between these phenomenological analyses and those in which an attempt has been made to constrain the densities using results from other branches of nuclear theory. The 'model-independent analyses' described in §§3.4 and 4.2 are the most convincing way of obtaining densities, but they necessarily depend on assumptions which go beyond the electron-scattering and muonic X-ray experiments. The main aim of these model-independent analyses has been to remove constraints on the density which have been implicitly contained in the earlier phenomenological analyses and which are *not* justified on any physical grounds. These analyses are much more difficult, however, and have been done for only a few nuclei. The results given here are selected from the vast body of experimental and theoretical literature. More comprehensive tabulations and reviews of the experiments and their analysis have been carried out by de Jager *et al.* (1974) for electron scattering, Engfer *et al.* (1974) for muonic atom data, Boehm (1974) for electronic X-ray measurements, and Heilig and Steudel (1974) for optical isotope shifts.

6.1.1. *Protons and neutrons*

The electromagnetic structure of protons and neutrons is of particular importance not only for its intrinsic interest but also because theoretical calculations of nuclear densities usually give the 'point' densities of protons and neutrons into which the intrinsic structure of the latter must be folded before the charge density is obtained. The effect of folding the neutron density is small but not always negligible (Bertozzi *et al.* 1972*b*). The mean-square radius is obtained from the proton density $\rho_p(r)$ (normalized to unity) by the equation

$$\langle r^2\rangle_{\text{charge}} = \int r^2 \rho_p(r)\mathrm{d}^3r + \langle r^2\rangle_{\text{proton}} - \frac{N}{Z}\langle r^2\rangle_{\text{neutron}}. \quad (6.1)$$

The proton form factor is assumed to be of 'dipole' form, corresponding to a Yukawa shape for the density, or a sum of dipoles:

$$F(q) = \sum_i \frac{A_i M_i^2}{q^2+M_i^2} \quad (6.2)$$

where

$$\sum A_i = 1. \quad (6.3)$$

As can be seen from eqn (3.27) the value of $\langle r^2\rangle$ is given by

$$\langle r^2\rangle = -6\left.\frac{\mathrm{d}F}{\mathrm{d}q^2}\right|_{q^2=0} \quad (6.4)$$

and the value of the slope must be obtained by interpolating between $F(0)$ (=1) and $F(q_{\text{min}})$ (where q_{min} is the lowest momentum transfer of the measurements). Thus we would expect the most reliable values of the mean-square radius to come from very low q measurements. The radii of the proton and neutron distributions are shown in Table 6.1.† The results for the proton are in good agreement with each other and with the r.m.s. magnetic radius which is expected to be the same as the charge radius. The much smaller magnitude of the neutron

---

†In Tables 6.1 to 6.8 the numbers in brackets are the errors in the least significant figures, for example 43.2(23) means 43.2 ± 2.3.

TABLE 6.1

Proton and neutron charge radii

	$\langle r\rangle^{1/2}$ (fm)	q_{min} (fm^{-1})	Ref.
p	0.81(2)	0.44	a
	0.85(3)	0.33	b
	0.84(2)	0.36	c
	0.81(4)	0.3	d
Mean	0.83		
n	-0.3359(36)	0.00	e

a Akimov *et al.* (1972)
b Theissen (1972)
c Borkowski *et al.* (1975)
d Murphy, Shin, and Skopik (1974)
e Krohn and Ringo (1973)

r.m.s. charge radius is due to the fact that the neutron has positive and negative charge and the minus sign is a reminder that the negative charge extends further out.

6.1.2. *Light nuclei* (A = *2 to* A = *48)*

For these nuclei elastic electron-scattering measurements give the Fourier transform of the charge distribution for a range of q-values and, with certain assumptions about the very-low-density region of the charge density, the mean-square radius $\langle r^2\rangle$.† Muonic X-ray measurements on the other hand give a

†The experiments do not give any information about the shape of the form factor between $q = 0$ and $q = q_{min}$ (the lowest momentum transfer of the measurements) so that in principle the slope at $q = 0$ may have any value. To obtain the mean-square radius it is necessary to assume that beyond a certain radius, which we may take to be near the radius at which the density has fallen to about 3 per cent of its maximum value, the fall-off is fairly rapid, i.e. not much slower than the exponential fall-off predicted by Hartree-Fock theory.

model-independent value of $\langle r^2 \rangle$† and, at the heavier end of the range, a very much less accurate value of $\langle r^4 \rangle$. The values of $\langle r^2 \rangle$ obtained from muonic-atom measurements can also be very much more precise than those obtained from scattering, for example by a factor of 3 in the case of ^{4}He, as shown in Table 6.2. The determination of properties of the charge distribution by both methods assumes that the interactions of the lepton with the nucleus and with the radiation field are well understood. The radiative corrections are particularly important in the case of the lightest muonic atoms. Far more nuclei have been studied by electron-scattering experiments, however, and most of the results tabulated in Table 6.2 come from analyses of these experiments. The table contains r.m.s. radii with errors quoted in the literature and most of these are obtained from model densities or polynomial form factors, so that the radii and errors are merely the values which these quantities would have (assuming no systematic errors) if the model used were correct. In spite of the inevitable doubts about the meaningfulness of the numbers extracted from the data it seems worthwhile to give the results of a selected group of analyses, particularly those on the data from the lowest-momentum-transfer experiment. Ideally a compilation such as this would include graphs of all the form factors and experimental errors, together with form factors corresponding to various theoretical, model-independent, or model densities, but the magnitude of this task is beyond the scope of this book.

The doubly-closed shell nuclei are of particular interest on account of the number of calculations of the properties of these nuclei based on nucleon-nucleon forces. Examples of

---

†Here the term 'model independent' does not have an absolute meaning, for example if a fraction of the nuclear charge extended out to a radius several orders of magnitude greater than Hartree-Fock theory predicts, the value of $\langle r^2 \rangle$ deduced would be wrong. Since the restriction on the density for determining $\langle r^2 \rangle$ from muonic X-ray measurements is so weak, we feel justified in using the term 'model independent'.

TABLE 6.2

Charge densities of light nuclei

Nucleus	Particle	Momentum transfer (fm^{-1})	Shape	c (fm)	a (fm)	w	$\langle r^2 \rangle^{1/2}$ (fm) Model density	Model independent	References Experiment	Analysis	Notes
^{2}H	e	0.2-0.7	WF				2.095(6)		Berard *et al.* 1973		1
^{3}H	e	1.0-2.8						1.70(5)	Collard *et al.* 1965		2
^{3}He	e	0.5-4.5						1.88(5)	McCarthy 1970		2
^{4}He	e	0.6-2.5	F3	0.964	0.322	0.517	1.74(4)	1.674(12)	McCarthy *et al.* 1974	Sick *et al.* 1976	3
	μ							1.674(4)	Bertin *et al.* 1975	Rinker 1976	
^{6}Li	e	0.1-0.9	$F(q)$				2.56(10)		Bumiller *et al.* 1972		4,8
^{7}Li	e	0.1-0.9	$F(q)$				2.41(10)		Bumiller *et al.* 1972		4
^{9}Be	e	0.3-0.7	SM1				2.519(12)		Jansen *et al.* 1972		5
^{10}Be	e	0.7-2.8	SM1				2.45(12)		Stovall *et al.* 1966		5
^{11}B	e	0.6-1.8	SMD				2.37		Riskalla 1971		6
^{12}C	e	0.3-2.3	MI				2.446(10)		Merle 1973		9

	e	1.1-4.0	SM1				2.460(25)		Sick & McCarthy 1970	5
	μ							2.49(5)	Dubler *et al.* 1973	
	μ,e	0.2-3.6	MI					2.447(6)	Friar & Negele 1975*a*	
^{13}C	e	0.3-3.5	SM2				2.440(25)		Heisenberg *et al.* 1970	7
^{14}C	e	1.0-2.2	SM2				2.56(5)		Kline *et al.* 1973	7
^{14}N	e	0.3-0.5	SM1				2.54(2)		Schütz 1973	5
	μ							2.55(3)	Dubler *et al.* 1973	
^{15}N	e	0.2-0.5	SM1				2.580(26)		Schütz 1973	5
^{16}O	e	0.3-0.5	SM1				2.718(21)		Schütz 1973	5
	e	1.1-4.0	F 6	2.608	0.513	-0.051	2.730(25)		Sick & McCarthy 1970	3
	μ							2.71(2)	Dubler *et al.* 1970	5
^{17}O	e	0.6-1.0	SM				2.662(26)		Singhal *et al.* 1970	5
^{18}O	e	0.5-2.5	SM1				2.789(27)		Schütz 1973	5
	μ							2.71(14)	Backenstoss *et al.* 1967*b*	
^{19}F	e	0.6-1.0	F2	2.59	0.564		2.900(15)		Hallowell *et al.* 1973	3
	μ							2.85(9)	Backenstoss *et al.* 1967*a*	

Table 6.2. continued

Nucleus	Particle	Momentum transfer (fm^{-1})	Shape	c (fm)	a (fm)	w	$\langle r^2 \rangle^{1/2}$ (fm) Model density	Model independent	References Experiment	Analysis	Notes
^{nat}Ne	μ							3.07(8)	Backenstoss *et al.* 1971		
^{20}Ne	e	0.2-0.5	F2	2.74	0.569		3.00(3)		Fey *et al.* 1973*a*		3
	e	0.2-1.0	F2	2.805	0.571		3.040(25)		Moreira *et al.* 1971		3
^{22}Ne	e	0.2-1.1	F2	2.782	0.549		2.969(21)		Moreira *et al.* 1971		3
^{23}Na	e	0.4-2.0					2.94(6)		Savitskiĭ *et al.* 1969		10
	μ							2.94(4)	Backenstoss *et al.* 1967*b*		
^{nat}Mg	μ							3.01(3)	Backenstoss *et al.* 1967*b*		
^{24}Mg	e	0.2-1.2	F2	2.99	0.548		3.08(5)		Curran *et al.* 1972	Lees 1974	3
	e	0.7-3.5	F6	3.192	0.604	-0.249	2.985(30)		Curran *et al.* 1972	Li *et al.* 1974	3
^{26}Mg	e	0.2-1.2	F2	3.05	0.524	3.06(5)			Curran *et al.* 1972	Lees 1974	3
^{27}Al	e	0.2-0.6	F2	2.84	0.569		3.05(5)		Fey *et al.* 1973		3

	μ							3.025(23)	Backenstoss 1967		3
^{nat}Si	e	0.2-0.6	F2	2.93	0.569		3.10(5)		Fey *et al.* 1973		3
	e	1.0-2.1	MI				3.078(3)		Averdung 1974		9
	μ							3.086(18)	Backenstoss *et al.* 1967*b*		
^{28}Si	e	0.7-3.7	G3	1.95	2.076	0.286	3.10(30)		Li *et al.* 1974		11
	e	0.2-1.1	F2	3.14	0.537		3.15(5)		Brain 1974		3
^{29}Si	e	0.2-1.1	F2	3.77	0.52		3.12(6)		Brain 1974		3
^{31}P	e	0.4-2.3	MI				3.187(16)		Merle 1973		9
	μ							3.188(18)	Backenstoss *et al.* 1967*b*		
^{nat}S	e	0.4-2.3	MI				3.238(15)		Merle 1973		9
	μ							3.244(18)	Backenstoss *et al.* 1967*b*		
^{40}A	e	0.8-1.8	MI				3.48(8)		Wendling & Walther 1974		9
	μ							3.423(6)	Pfeiffer *et al.* 1973		
^{nat}K	e	0.6-3.4	F6	3.743 (25)	0.585 (6)	-0.201 (22)	3.408(27)		Sinha *et al.* 1973		3
			F1	3.661 (14)	0.523			3.439(9)	Ehrlich 1968	Engfer *et al.* 1974	3
^{40}Ca	e	0.5-3.2	F6	3.766 (23)	0.586 (5)	-0.161 (23)	3.482(25)		Sinha *et al.* 1973		3
	μ,e	1.4-2.5	F3	3.697	0.587	-0.083		3.53	Frosch *et al.* 1968		3

Table 6.2. continued

Nucleus	Particle	Momentum transfer (fm^{-1})	Shape	c (fm)	a (fm)	w	$\langle r^2\rangle^{1/2}$ (fm) Model density	Model independent	References Experiment	Analysis	Notes
^{48}Ca	e	0.5-3.4	F6	3.7369	0.5245	-0.030	3.470		Bellicard *et al.* 1967		3
	μ,e	1.4-2.7	F3	3.697	0.587	-0.083		3.52	Frosch *et al.* 1968		3

Notes

1. The shapes used in the analysis were obtained from various deuteron wavefunctions.
2. Radius obtained by analysis based on assuming polynomial and exponential forms for $f(q^2)$.
3. The shapes F1, F2, F3, and F6 are based on Fermi distributions; F2 is the usual two-parameter density defined by $\rho(r) = \rho_0(e^x + 1)^{-1}$ where $x = (r-c)/a$.

   F1 is a Fermi distribution with the parameter a fixed and c variable.

   F3 is a 'parabolic Fermi distribution', namely the two-parameter density multiplied by a factor $(1 + wr^2/c^2)$.

   F6 is a sum of two densities, one of which is F3 and the other a three-parameter oscillating density containing negligible total charge which is the Fourier transform of $A \exp\{-B(q-q_0)^2\}$.

   In the case of ^4He the model independent analysis of electron scattering is based on the assumption that the extreme tail of the distribution is that given by theory.

4. $F(q)$ assumed to be of the form $1 + aq^2 + bq^4$.

5. The density was taken to be of the form predicted by the harmonic-oscillator shell model with the radius as an adjustable parameter.

6. The density was taken to be that predicted by deformed harmonic-oscillator wavefunctions.

7. The density was taken to be that given by the harmonic oscillator with an additional adjustable parameter.

8. Using corrected ^{12}C radius of Jansen *et al.* 1972.

9. Analysis using δ-shell densities.

10. Density taken to be a gaussian folded into a uniform sphere.

11. The abbreviation G3 is used for a density of the form

$$\rho(r) = \frac{\rho_0(1 + wr^2/c^2)}{\exp(x^2) + 1}$$

where $x = (r - c)/a$. In Table 6.3 G6 is used for a density which is the sum of a density of the form G3 and a three-parameter oscillating density whose Fourier transform is given in note 3.

such calculations are shown in Fig. 6.1 together with model densities obtained from electron scattering. One of the most surprising results is the *decrease* in $\langle r^2 \rangle^{1/2}$ between ^{40}Ca and ^{48}Ca which is shown by the muonic X-ray results. Bertozzi *et al.* (1972*b*) have shown that this is consistent with a slight

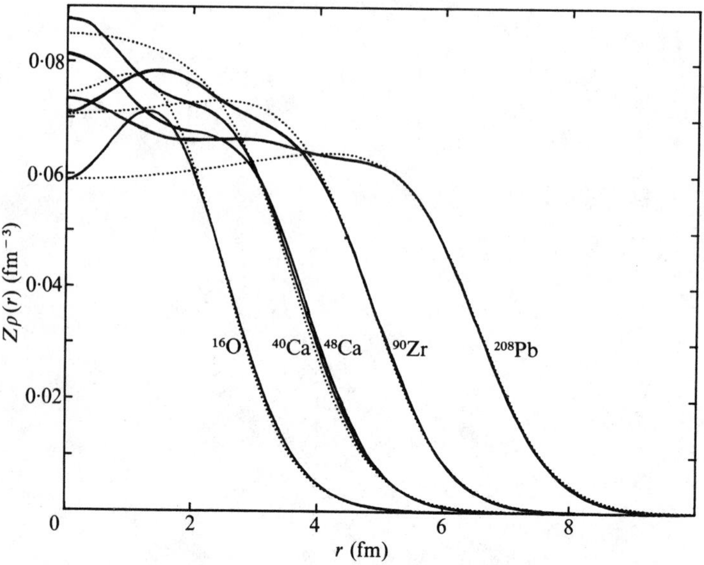

FIG. 6.1. Phenomenological and Hartree-Fock densities for closed-shell nuclei. The theoretical densities are from the Hartree-Fock calculations of Campi and Sprung (1972), while the phenomenological densities (shown as broken curves) are from the following sources: ^{16}O, Ehrenberg *et al.* (1959); ^{40}Ca, Frosch *et al.* (1968); ^{90}Zr, Fajardo *et al.* (1971); ^{208}Pb, Heisenberg *et al.* (1969).

increase in the point-proton density if the lepton-neutron interaction is taken into account (see §6.1.1). This results in a contribution to the charge distribution itself arising from the neutron charge distribution and an 'effective charge' from the neutron spin-electron orbit force. The shapes and sizes of the neutron contribution are shown in Fig. 6.2.

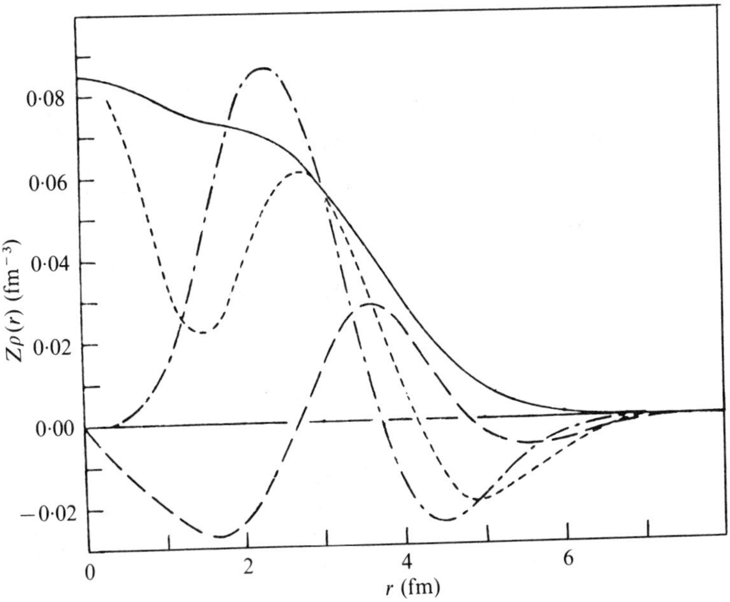

FIG. 6.2. The proton density (full curve), the charge density × 100 due to the neutron form factor from the core (short-dashed curve) and from the $f_{7/2}$ neutrons (long-dashed curve), and the effective charge density × 100 (dash-dot curve) resulting from the spin-orbit force between the muon and $f_{7/2}$ neutrons. (From Bertozzi *et al.* 1972*b*.)

6.1.3. *Medium and heavy spherical nuclei*

In Table 6.3 values of $\langle r^2 \rangle^{1/2}$ from model dependent and model independent analyses are given (see §3.4 and §4.2). The most reliable values are those in which a simultaneous fit to muonic and electron scattering has been made without using model densities. Those results which come from simultaneous fits using model densities can also be expected to be very reliable although quoted errors may be too optimistic. The accuracy of values of $\langle r^2 \rangle^{1/2}$ obtained from low-q electron scattering is almost independent of Z (van Niftrik 1969). One might expect the high-q data to give less accurate values of $\langle r^2 \rangle^{1/2}$ (since higher terms in the expansion of $F(q)$ become important), but in a model-independent analysis of ^{12}C scattering Sick (1974) has shown that the effect of the higher terms is less important than the effect of statistical errors

TABLE

Charge densities of

Nucleus	Particle	Momentum transfer (fm^{-1})	Shape	c (fm)	a (fm)	w	α	k_1
^{nat}Ti	e	0.2-0.5	F1	3.75 (4)	0.567			
	μ		F1	3.912	0.5234		0.068	2.117
^{51}V	e	0.2-0.8	F2	3.91 (4)	0.532 (29)			
^{nat}V	μ		F1	3.913	0.5234		0.069	2.116
^{nat}Cr	e	0.4-1.3	F2	3.975 (25)	0.530 (11)			
	μ		F1	3.89 (12)	0.5234		0.071	2.115
^{55}Mn	e	0.2-0.5	F1	3.89 (12)	0.567			
	μ		F1	4.08 (10)	0.5234		0.072	2.120
^{nat}Fe	e	0.2-0.6	F1	3.98 (12)	0.569			
	μ		F1	4.116 (17)	0.5234		0.074	2.121
^{59}Co	e	0.2-0.6	F1	4.08 (5)	0.569			
	μ		F1	4.16 (9)	0.5234		0.075	2.121
^{58}Ni	μ,e	0.6-2.6	F6	4.3092	0.5169	-0.1308		
^{60}Ni	μ,e	0.5-2.3	F6	4.4891	0.5369	-0.2668	0.076	2.123
^{61}Ni	μ,e	0.5-2.3	F6	4.4024	0.5401	-0.1983		
^{62}Ni	μ,e	0.5-2.3	F6	4.4425	0.5386	-0.2090		
^{64}Ni	μ,e	0.5-2.6	F6	4.5211	0.5278	-0.2284		
^{63}Cu	e	0.2-0.8	F2	4.214 (26)	0.586 (18)			
	μ		F1	4.357 (9)	0.5234		0.078	2.127

6.3

medium and heavy nuclei

R_{k_1} (fm)	k_2	R_{k_2} (fm)	$\langle r^2\rangle^{1/2}$ (fm) MD	MI	References: Experiment	References: Analysis	Notes
			3.59 (4)				1
4.619 (12)			3.60		Bardin *et al.* 1965	Engfer *et al.* 1974	
			3.615 (31)		Gompelman *et al.* 1976		
4.619 (61)			3.60		Bjorkland *et al.* 1965	Engfer *et al.* 1974	
			3.656		Bellicard *et al.* 1964		
4.60 (10)			3.58		Johnson *et al.* 1962	Engfer *et al.* 1974	
			3.68 (11)		Theissen *et al.* 1969		
4.76 (8)			3.71 (6)		Quitman *et al.* 1964	Engfer *et al.* 1974	
			3.74 (11)		Fey *et al.* 1973*a*		
4.788 (2)	4.105	5.12 (13)	3.735 (2)		Acker *et al.* 1966	Engfer *et al.* 1974	
			3.80 (5)		Fey *et al.* 1973*a*		
4.82 (8)			3.76 (6)		Quitman *et al.* 1964	Engfer *et al.* 1974	
			3.764 (10)			Ficenec *et al.* 1976	
4.886 (9)			3.796 (10)		Ficenec *et al.* 1976 (electron scattering)	Ficenec *et al.* 1976	
			3.806 (10)			Ficenec *et al.* 1976	
			3.822 (10)		Ehrlich 1968 (muonic X-ray)	Ficenec *et al.* 1976	
			3.845 (10)			Ficenec 1976	
			3.952 (22)		Gompelman *et al.* 1976		
4.993 (8)			3.896 (6)		Ehrlich 1968	Engfer *et al.* 1974	1

Table 6.3 continued

Nucleus	Particle	Momentum transfer (fm^{-1})	Shape	c (fm)	a (fm)	w	α	k_1
^{65}Cu	e	0.2-0.8	F2	4.271 (25)	0.579 (18)			
^{64}Zn	e	0.3-1.1	F2	4.285 (9)	0.584 (9)			
^{66}Zn	e	0.3-1.1	F2	4.318 (8)	0.595 (8)			
	μ		F1	4.440 (4)	0.5234		0.079	2.130
^{68}Zn	e	0.3-1.1	F2	4.378 (9)	0.569 (9)			
	μ		F1	4.466 (4)	0.5234		0.079	2.131
^{70}Zn	e	0.3-1.1	F2	4.409 (10)	0.583 (9)			
^{75}As	μ		F1	4.665 (3)	0.5234		0.083	2.139
^{88}Sr	e	0.4-1.0	F2	4.83 (1)	0.496 (11)			
^{nat}Sr	μ		F1	4.845 (2)	0.5234		0.091	2.148
^{89}Y	e	0.4-1.0	F2	4.86 (1)	0.542 (11)			
	μ		F1	4.869 (1)	0.5234		0.092	2.149
^{90}Zr	e,μ	0.5-2.8	G3	4.434 (20)	2.528 (3)	0.350 (25)		
^{91}Zr	e	0.5-2.4	G3	4.325 (20)	2.581 (3)	0.433 (25)		
^{92}Zr	e,μ	0.5-2.4	G3	4.455 (20)	2.550 (3)	0.334 (25)		
^{94}Zr	e	0.5-2.4	G3	4.494 (20)	2.585 (3)	0.296 (25)		
^{96}Zr	e	0.9-2.8	G3	4.503 (20)	2.602 (3)	0.341 (25)		
^{93}Nb	e	0.5-1.8	F2	4.87 (5)	0.573 (29)			
	μ		F1	4.9859 (3)	0.5234		0.095	2.155

R_{k_1} (fm)	k_2	R_{k_2} (fm)	$\langle r^2\rangle^{1/2}$ (fm) MD	$\langle r^2\rangle^{1/2}$ (fm) MI	References: Experiment	References: Analysis	Notes
			3.947 (22)		Gompelman *et al.* 1976		
			3.965 (17)		Neuhausen *et al.* 1972	de Jager *et al.*	
			4.009 (15)		Neuhausen *et al.* 1972	de Jager *et al.* 1974	
5.064 (5)	4.108	4.8 (8)	3.951 (6)		Jenkins *et al.* 1970*a*	Engfer *et al.* 1974	
			3.996 (16)		Neuhausen *et al.* 1972	de Jager *et al.* 1974	
5.088 (5)	4.108	5.4 (8)	3.969 (6)		Jenkins *et al.* 1970*a*	Engfer *et al.* 1974	
			4.044 (18)		Neuhausen *et al.* 1972	de Jager *et al.* 1974	
5.255 (4)	4.107	6.5 (7)	4.104 (4)		Bardin *et al.* 1965	Engfer *et al.* 1974	
			4.17 (2)		Fivozinsky *et al.* 1974		
5.414 (3)			4.227 (3)		Ehrlich 1968	Engfer *et al.* 1974	
			4.27 (2)		Fivozinsky *et al.* 1974		
5.438 (3)	4.099	5.6 (3)	4.239 (7)		Kessler *et al.* 1970	Engfer *et al.* 1974	
			4.274 (22)	4.244 (26)	Fajardo *et al.* 1971		2
			4.309 (22)		Fajardo *et al.* 1971		
			4.300 (22)		Fajardo *et al.* 1971		2
			4.332 (22)		Fajardo *et al.* 1971		
			4.396 (22)		Fajardo *et al.* 1971		
			4.31		Shevchenko *et al.* 1967*a*		
5.537 (1)	4.099	5.62 (5)	4.324 (2)		Cheng *et al.* 1971		

Table 6.3 continued

Nucleus	Particle	Momentum transfer (fm^{-1})	Shape	c (fm)	a (fm)	w	α	k_1
^{92}Mo	e	0.6-2.0	G3	4.538 (13)	2.5445 (27)	0.304 (18)		
	μ		F1	4.9748 (15)	0.5234		0.096	2.156
^{94}Mo	e	0.6-2.0	G3	4.517 (13)	2.5874 (36)	0.330 (18)		
^{96}Mo	e	0.6-2.0	G3	4.534 (14)	2.6235 (36)	0.306 (18)		
	μ		F1	5.071 (4)	0.5234		0.096	2.161
^{98}Mo	e	0.6-2.0	G3	4.562 (14)	2.6501 (36)	0.297 (18)		
^{100}Mo	e	0.6-2.0	G3	4.559 (14)	2.6723 (36)	0.339 (19)		
^{103}Rh	μ		F1	5.231 (60)	0.5234		0.100	2.171
^{nat}Pd	μ		F1	5.266 (44)	0.5234		0.102	2.173
^{107}Ag	μ		F1	5.2985 (24)	0.5234		0.103	2.177
^{109}Ag	μ		F1	5.3284 (24)	0.5234		0.103	2.179
^{110}Cd	e	0.2-1.1	F2	5.261 (17)	0.5907 (7)			
^{112}Cd	e	0.2-1.1	F2	5.316 (14)	0.5907 (8)			
^{114}Cd	e	0.2-1.1	F2	5.334 (15)	0.6005 (9)			
	μ		F1	5.27 (11)	0.58 (5)		0.105	2.184
^{116}Cd	e	0.2-1.1	F2	5.356 (16)	0.5973 (9)			
^{nat}In	e	0.5-1.6	F2	5.24 (10)	0.52 (5)			
	μ		F1	5.38 (19)	0.53 (12)		0.106	2.185
^{112}Sn	e,μ	0.6-2.4	G6	4.962 (7)	2.638 (3)	0.285 (12)		

R_{k_1} (fm)	k_2	R_{k_2} (fm)	$\langle r^2\rangle^{1/2}$ (fm) MD	MI	References: Experiment	Analysis	Notes
			4.317 (4)	4.296 (26)	Dreher *et al.* 1973		3
5.526 (2)	4.096	5.74 (9)	4.317 (2)		Macagno 1968	Engfer *et al.* 1974	
			4.358 (4)	4.334 (26)	Dreher *et al.* 1973		3
			4.390 (4)	4.364 (26)	Dreher *et al.* 1973		3
5.619 (5)			4.383 (4)		Macagno 1968	Engfer *et al.* 1974	
			4.423 (4)	4.391 (26)	Dreher *et al.* 1973		3
			4.461 (4)	4.431 (26)	Dreher *et al.* 1973		3
5.75 (5)			4.49 (4)		Backenstoss *et al.* 1965	Engfer *et al.* 1974	
5.79 (4)			4.52 (4)		Backenstoss *et al.* 1965	Engfer *et al.* 1974	
5.815 (3)			4.542 (3)		Carrigan *et al.* 1968	Engfer *et al.* 1974	
5.841 (3)			4.563 (3)		Carrigan *et al.* 1968	Engfer *et al.* 1974	
			4.63 (1)		Gillespie *et al.* 1973		
			4.67 (1)		Gillespie *et al.* 1973		
			4.70 (1)		Gillespie *et al.* 1973		
5.905 (1)			4.624 (8)		Kast *et al.* 1971		
			4.70 (1)		Gillespie *et al.* 1973		
			4.50 (9)		Hahn *et al.* 1956		
5.907 (1)	4.102	6.15 (8)	4.619 (15)		Kast *et al.* 1971		
			4.586 (5)		Ficenec *et al.* 1972		4

Table 6.3 continued

Nucleus	Particle	Momentum transfer (fm^{-1})	Shape	c (fm)	a (fm)	w	α	k_1
^{114}Sn	e,μ	0.6-2.4	G6	4.971 (7)	2.636 (3)	0.320 (12)		
^{116}Sn	e,μ	0.6-2.4	G6	5.062 (7)	2.625 (3)	0.272 (12)		
^{117}Sn	e,μ	0.6-2.4	G6	5.058 (7)	2.625 (3)	0.295 (12)		
^{118}Sn	e,μ	0.6-2.4	G6	5.072 (7)	2.623 (3)	0.304 (12)		
^{119}Sn	e,μ	0.6-2.4	G6	5.100 (7)	2.618 (3)	0.290 (12)		
^{120}Sn	e,μ	0.6-2.4	G6	5.110 (7)	2.619 (3)	0.292 (12)		
^{122}Sn	e,μ	0.6-2.4	G6	5.088 (7)	2.611 (3)	0.378 (12)		
^{124}Sn	e,μ	0.6-2.4	G6	5.155 (7)	2.615 (3)	0.311 (12)		
^{126}Te	μ		F2	5.43 (7)	0.59 (3)		0.110	2.197
^{127}I	μ		F1	5.5964 (5)	0.5234		0.112	2.201
^{133}Cs	μ		F1	5.674 (1)	0.5234		0.115	2.208
^{138}Ba	e	0.6-2.8	F6	5.338	2.678	0.375		
	μ		F2	5.798 (57)	2.11 (15)		0.116	2.212
^{142}Nd	e,μ	0.6-3.0	F2	5.6838	0.5868		0.122	2.226
^{144}Nd	e,μ	0.6-3.0	F2	5.6256	0.6178		0.122	2.228
^{146}Nd	e,μ	0.6-3.0	F2	5.6541	0.6321		0.122	2.231

R_{k_1} (fm)	k_2	R_{k_2} (fm)	$\langle r^2\rangle^{1/2}$ (fm) MD	$\langle r^2\rangle^{1/2}$ (fm) MI	References: Experiment	References: Analysis	Notes
			4.602 (5)		Ficenec *et al.* 1972		4
			4.619 (5)		Ficenec *et al.* 1972		4
			4.625 (5)		Ficenec *et al.* 1972		4
			4.634 (5)		Ficenec *et al.* 1972		4
			4.639 (5)		Ficenec *et al.* 1972		4
			4.646 (5)		Ficenec *et al.* 1972		4
			4.658 (5)		Ficenec *et al.* 1972		4
			4.670 (5)		Ficenec *et al.* 1972		4
6.055 (1)	4.103	6.24 (3)	4.774 (6)		Kast *et al.* 1971		
6.0795 (6)	4.105	6.19 (2)	4.737 (7)		Acker *et al.* 1966	Engfer *et al.* 1974	
6.150 (1)	4.105	6.20 (4)	4.806 (1)		Lee *et al.* 1969	Engfer *et al.* 1974	
			4.836		Heisenberg *et al.* 1971*a*		
6.1916 (7)	4.108	6.34 (4)	4.839 (1)		Thompson 1969		
6.2856	4.107	6.429	4.913		Heisenberg *et al.* 1971*b*		
					μ Macagno *et al.* 1970		
6.3213 (10)	4.110	6.469 (14)	4.926		Heisenberg *et al.* 1971*b*		
					μ Macagno *et al.* 1970		
6.3558 (9)	4.113	6.505 (11)	4.970		Heisenberg *et al.* 1971*b*		
					μ Macagno *et al.* 1970		

Table 6.3 continued

Nucleus	Particle	Momentum transfer (fm^{-1})	Shape	c (fm)	a (fm)	w	α	k_1
^{148}Nd	e,μ	0.6-3.0	F2	5.6703	0.644		0.122	2.233
^{197}Au	e	0.6-1.4	F2	6.38 (6)	0.535 (27)			
	μ		F1	6.5542 (4)	0.5234		0.149	2.318
^{nat}Hg	μ		F1	6.616 (11)	0.5234		0.150	2.325
^{205}TI	μ		F2	6.629 (7)	0.5193		0.152	2.328
^{204}Pb	μ		F2	6.627 (14)	0.5204 (77)		0.153	2.331
^{206}Pb	e	0.2-0.9	F2	6.61 (5)	0.545 (8)			
	μ		F2	6.649 (19)	0.5170 (102)		0.153	2.332
^{207}Pb	e	0.2-0.9	F2	6.62 (6)	0.546 (10)			
	μ		F2	6.647 (5)	0.5232 (27)		0.153	2.332
^{208}Pb	e	0.2-0.9	F2	6.624 (35)	0.549 (8)			
	μ		F2	6.650 (7)	0.5258 (36)		0.153	2.333
	e,μ	0.2-2.7	MI					
	e,μ	0.4-2.7	G6	6.3032	2.8882	0.3379		
^{209}Bi	e	0.7-2.8	G6	6.315 (10)	2.881 (8)	0.39 (6)		
	μ		F2	6.682 (5)	0.512 (34)		0.154	2.339

R_k (fm	k_2	R_k (fm)	$\langle r^2\rangle^{1/2}$ (fm) MD	MI	References Experiment	Analysis	Notes
6.3940 (10)	4.116	6.559 (14)	5.002		Heisenberg *et al.* 1971*b*		
					μ Macagno *et al.* 1970		
			5.33 (5)		Hahn *et al.* 1956		
6.9466 (7)	4.148	7.068 (5)	5.436 (2)		Robert-Tissot 1975	Engfer *et al.* 1974	
6.997 (15)	4.153	7.18 (6)	5.48 (1)		Acker *et al.* 1966	Engfer *et al.* 1974	
7.0039 (10)	4.153	7.13 (3)	5.484 (6)		Backe *et al.* 1972		
7.0044 (6)	4.152	7.128 (1)	5.486 (1)		Kessler *et al.* 1975	Kessler *et al.* 1975	
			5.509 (29)		de Jager 1973		
7.0175 (6)	4.153	7.1442 (5)	5.497 (2)		Kessler *et al.* 1975	Kessler *et al.* 1975	
			5.513 (32)		de Jager 1973		
7.0231 (6)	4.154	7.1509 (8)	5.504 (1)		Kessler *et al.* 1975	Kessler *et al.* 1975	
			5.521		de Jager 1973		
7.0317 (6)	4.155	7.1611 (5)	5.510 (1)		Kessler *et al.* 1975	Kessler *et al.* 1975	
			5.502 (6)		See note 5	Friar & Negele 1973	
			5.501		Heisenberg *et al.* 1969		
			5.521 (2)		Sick 1973*a*		
7.050 (3)	4.159	7.240 (2)			Bardin *et al.* 1967		

Notes

For explanation of the abbreviations used in the column 'Shape', see Table 6.2, notes 3 and 11.

1. Using corrected ^{12}C radius of Jansen *et al.* (1972).
2. Muonic X-ray energies for ^{90}Zr and ^{92}Zr from Ehrlich (1968). Simultaneous analysis of muonic X-ray data and electron scattering by Fajardo *et al.* (1971).
3. Model-independent analysis using a sum of gaussians. Note that the value of $\langle r^2 \rangle^{1/2}$ is affected by the assumption of a maximum radius for $\rho(r)$.
4. Electron-scattering cross-sections analysed simultaneously with muonic X-ray results of Macagno *et al.* (1970) and Ehrlich (1968) together with optical isotope shifts of Stacey (1964) normalized to electronic K X-ray shifts of Chesler and Boehm (1968).
5. The analysis of Friar and Negele (1973) is the most extensive published investigation of a charge density incorporating as much model independence as possible and using nuclear theory to constrain the density where this could not be done from the data. The generous error on the r.m.s. radius includes possible systematic errors and theoretical uncertainties in the muonic-atom corrections. The analysis was simultaneously carried out on the muonic-atom data of Anderson *et al.* (1969), Jenkins *et al.* (1971), Martin (1972), and Kessler (1971), and the electron-scattering data of van Niftrik (1969), Friedrich and Lenz (1972), and Heisenberg (1973).

at low q (where $F(q) \approx 1$).

The model-independence of $\langle r^2 \rangle^{1/2}$ determinations from muonic X-ray measurements decreases with Z because the property of the nucleus which determines the finite-size effect in muonic transitions (see Fig. 4.1) is the generalized moment f_{ij} defined in §4.2. Table 6.3 gives values of α, k_1, and k_2 for the $2p_{3/2}$-1s and $3d_{5/2}$-$2p_{3/2}$ transitions respectively, together with the equivalent uniform radii R_{k_1} and R_{k_2}. These particular transitions are selected from the vast number of experimental measurements, and the values of k and R_k for other transitions may be found in the tables of Engfer *et al.*

(1974).

6.1.4. *Isotope shifts*

Many of the difficulties of extracting the charge distribution of an individual nucleus from experimental results are not present in the analysis of *differences* in charge distributions of neighbouring nuclei, especially isotopes. Moreover, the vast body of results from optical and electronic X-ray transitions is available to provide information on differences in charge radii, particularly for nuclei which are too unstable for electron-scattering and muonic X-ray experiments. In addition the resulting charge-density differences can be used as a sensitive test of the validity of calculations of nuclear wavefunctions. In such calculations, which are almost always based on the use of effective nucleon-nucleon forces, one would expect the differences between charge densities to be more accurate than the absolute density.

Analyses of the actual shapes of the differences in terms of model densities have almost no significance since the unknown effects of model dependence are very much exaggerated. On the other hand, the analysis in terms of a complete set of functions would be expected to be more model independent for isotope differences than for absolute charge densities. This is because the addition of a small number of neutrons would not be expected to affect appreciably either the long-range behaviour of the density or the amount of very short wavelength oscillation (whose magnitude is not determined by electron scattering). Such an analysis has been carried out on the isotopes of iron (Friar and Negele 1975*a*).

The isotope shift between ^{40}Ca and ^{48}Ca is particularly interesting because the measurements indicate a *decrease* in the r.m.s. charge radius of about 0.004 fm when the neutrons are added, while Hartree-Fock calculations indicate that the proton radius increases. This puzzle was solved by Bertozzi *et al.* (1972*a*) who calculated the contribution from the neutron charge form factor and the spin-orbit interaction between the electron and the $1f_{7/2}$ neutrons in ^{48}Ca. The spin-orbit interaction can be represented by an effective charge which, together with the charge distribution due to the $1f_{7/2}$ neutrons produces a negative shift of 0.021 fm, indicating that the proton density r.m.s. radius increases by 0.017 fm. This is approximately the value obtained by calculation (Campi and Sprung 1972, Negele 1970).

In Table 6.4 the values of the shifts $\delta\langle r^2\rangle$ in the mean-square charge radius are compared for a number of sets of isotopes. Additional results may be found in compilations by de Jager *et al.* (1974) (from electron scattering), Engfer *et al.* (1974) (from muonic X-rays), Lee and Boehm (1973) (from electronic K X-rays), and Heilig and Steudel (1974) (from optical isotope shifts). In the optical and electronic X-ray columns the values quoted for $\delta\langle r^2\rangle$ are in fact (except where noted) the values of the generalized moment $\delta\langle f\rangle$ where

$$f = r^2 + ar^4 + br^6.$$

The coefficients a and b are small for heavy nuclei and negligible for light nuclei and, as noted in 5.1, are very nearly the same for both optical transitions and electronic K X-rays Since the coefficient of r^4 is negative the actual value of $\delta\langle r^2\rangle$ is larger by an amount which depends on the shape of the change $\delta\rho$ in the charge distribution; for a uniform distribution of radius $1.2\,A^{1/3}$ fm the correction varies from about 1 per cent for $Z = 30$ to 6 per cent for $Z = 82$.

For nuclei with $Z \leqslant 40$ the entries in the column for optical shifts are relative values, while for $Z > 40$ they are absolute shifts based on calculations of the magnitude of the atomic wavefunctions and the specific mass shifts. Further details of these calculations are given by Heilig and Steudel (1974). For most of the nuclei the agreement between the different atomic methods (i.e. all except the very model-dependent electron scattering) is well within the experimental errors. In a number of cases the muonic values are higher, which could be due to the fact that for Fermi distributions $\delta\rho_{ch}$ tends to be peaked near the nuclear surface. The remarkably extensive results for Hg isotopes show some extremely interesting features, notably the sudden decrease in $\langle r^2\rangle$

at about A=190, as illustrated in Fig. 6.3. The suggestion that this is because ^{181}Hg and ^{185} Hg are 'bubble nuclei' (see §2.4.2) has not been confirmed by calculations.

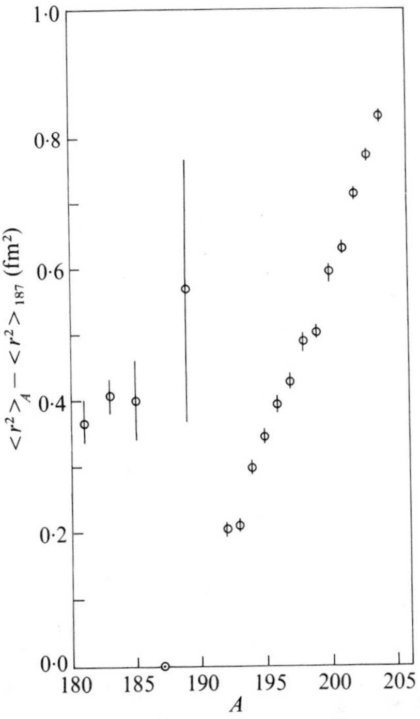

FIG. 6.3. Values of the mean-square charge radius $\langle r^2 \rangle_A$ for different Hg isotopes relative to that for ^{187}Hg. The graph is based on the values of $\delta\langle r^2 \rangle$ given in Table 6.4.

6.1.5. *Isotone shifts*

In Table 6.5 the values of $\Delta\langle r^2 \rangle$ from isotone-shift measurements are given. Most of the shift in energy is due to the change in the electric field due to the change in nuclear charge. We define the *volume shift* by the equation

$$\delta E_{\text{volume}} = \delta E_{\text{total}} - \delta E_Z - \delta E_{\text{nucl.pol.}}$$

where δE_Z is defined to be the shift calculated using the same charge density (normalized to unity) for the heavier isotone (mass number A_2) as for the lighter (mass number A_1). This

TABLE

Isotope

Element	Mass numbers		$\left(\frac{(\delta E)_{\text{field}}}{(\delta E)_{\text{std.}}}\right)^{\dagger}$ Muonic	$\delta\langle r^2\rangle$ (fm^2) Standard	Muonic	Optical	Electronic X-rays‡
						Relative	
Ca	40	42	0.64(5)	0.291	0.180(14)	1	
	40	44	0.49(2)	0.577	0.271(12)	1.3	
	40	48	-0.02(2)	1.138	-0.024(28)	-1.0	
Cr	50	52	-0.37(4)	0.271	-0.098(11)	-0.37(17)	
	52	54	1.16(10)	0.267	0.302(27)	1	
Ni	58	60	1.10(5)	0.258	0.278(12)		
	58	62	0.92(11)	0.513	0.458(21)		
	58	64		0.765			
Zr	90	92	1.38(7)	0.223	0.306(15)	1	
	92	94		0.222		0.69	
	94	96		0.220		0.54	
						Absolute	
Mo	92	96	1.33(2)	0.442	0.588(7)	0.278(62)	
	94	96		0.220		0.220(29)	
	96	98	0.96(11)	0.218	0.209(23)	0.169(25)	
	98	100		0.217		0.256(33)	
	94	95		0.110		0.186(28)	
	96	97	0.35(3)	0.109	0.038(3)	0.031(8)	

6.4

shifts

	References				Notes
Electron[§] Scattering	Muonic	Optical	Electronic X-rays	Electron scattering	
0.21	Ehrlich 1968	Epstein & Davis 1971		Frosch *et al.* 1968	
0.20	Macagno *et al.* 1970	Epstein & Davis 1971		Frosch *et al.* 1968	
-0.07	Ehrlich 1968	Epstein & Davis 1971		Frosch *et al.* 1968	
	Macagno *et al.* 1970	Bruch *et al.* 1969			
0.39	Ehrlich 1968			Khvastunov *et al.* 1970	
0.46	Ehrlich 1968			Litvinenko *et al.* 1972	
0.81				Khvastunov *et al.* 1970	
0.22	Ehrlich 1968	Heilig *et al.* 1963		Fajardo *et al.* 1971	
0.28		Heilig *et al.* 1963		Fajardo *et al.* 1971	
0.56		Heilig *et al.* 1963		Fajardo *et al.* 1971	
0.59	Macagno *et al.* 1970	Aufmuth *et al.* 1971		Dreher *et al.* 1973	
0.24		Aufmuth *et al.* 1971		Dreher *et al.* 1973	
0.25	Chasman *et al.* 1965	Aufmuth *et al.* 1971		Dreher *et al.* 1973	
0.34		Aufmuth *et al.* 1971		Dreher *et al.* 1973	
		Aufmuth *et al.* 1971			
	Macagno *et al.* 1970	Aufmuth *et al.* 1971			

Table 6.4 continued

Element	Mass numbers	$\left(\frac{(\delta E)_{\text{field}}}{(\delta E)_{\text{std.}}}\right)^{\dagger}$	$\delta\langle r^2\rangle$ (fm^2)			
		Muonic	Standard	Muonic	Optical	Electronic X-rays‡
					Absolute	
Sn	112 114		0.208		0.129(9)	
	114 116		0.206		0.133(9)	
	116 118	0.72(1)	0.205	0.148(2)	0.124(9)	
	118 120	0.62(1)	0.204	0.127(2)	0.112(8)	
	120 122	0.53(2)	0.203	0.108(4)	0.101(7)	
	122 124	0.53(1)	0.202	0.107(2)	0.091(6)	
	114 115		0.103		0.047(7)	
	116 117	0.55(2)	0.103	0.057(2)	0.049(7)	
	118 119	0.44(4)	0.102	0.044(4)	0.045(7)	
	116 124	0.60(1)	0.814	0.495(5)	0.42(6)	0.43(2)
Ba	134 136		0.195		0.02(1)	-0.022(13)
	136 138	0.32(3)	0.195	0.062(5)	0.05(1)	0.057(13)
	134 135		0.098		-0.02(1)	-0.040(13)
	136 137		0.097	0.002(12)	-0.02(1)	-0.001(13)
	138 140		0.194		0.242(41)	
Nd	142 144	1.43(2)	0.192	0.276(3)	0.28(1)	0.29(1)
	144 146	1.38(2)	0.191	0.264(3)	0.26(1)	0.23(3)
	146 148	1.56(2)	0.190	0.295(3)	0.29(1)	0.28(3)
Nd	148 150	2.09(2)	0.189	0.390(4)	0.38(2)	0.40(2)

	References				
Electron[§] Scattering	Muonic	Optical	Electronic X-rays	Electron scattering	Notes
0.15		Goble *et al.* 1974		Ficenec *et al.* 1972	
0.16		Silver & Stacey 1973		Ficenec *et al.* 1972	
0.14	Macagno *et al.* 1970	Silver & Stacey 1973		Ficenec *et al.* 1972	
0.11	Macagno *et al.* 1970	Silver & Stacey 1973		Ficenec *et al.* 1972	
0.11	Macagno *et al.* 1970	Silver & Stacey 1973		Ficenec *et al.* 1972	
0.11	Macagno *et al.* 1970	Silver & Stacey 1973		Ficenec *et al.* 1972	
		Goble & Silver 1973		Ficenec *et al.* 1972	
0.055	Macagno *et al.* 1970	Silver & Stacey 1973		Ficenec *et al.* 1972	1
0.055	Macagno *et al.* 1970	Silver & Stacey 1973		Ficenec *et al.* 1972	
0.47	Macagno *et al.* 1970	Silver & Stacey 1973	Boehm 1974	Ficenec *et al.* 1972	2
		Fischer *et al.* 1971	Sumbaev *et al.* 1969		
	Thompson 1969	Fischer *et al.* 1974	Sumbaev *et al.* 1969		
		Fischer *et al.* 1974	Sumbaev *et al.* 1969		
	Thompson 1969	Fischer *et al.* 1974	Sumbaev *et al.* 1969		
		Fischer *et al.* 1974			
	Macagno *et al.* 1970	King *et al.* 1973	Lee & Boehm 1973		
	Macagno *et al.* 1970	King *et al.* 1973	Lee & Boehm 1973		
	Macagno *et al.* 1970	King *et al.* 1973	Lee & Boehm 1973		
	Macagno *et al.* 1970	King *et al.* 1973	Lee & Boehm 1973		

Table 6.4 continued

Element	Mass numbers	$\left\{\frac{(\delta E)_{\text{field}}}{(\delta E)_{\text{std.}}}\right\}^{\dagger}$	$\delta\langle r^2\rangle$ (fm^2)			
		Muonic	Standard	Muonic	Optical	Electronic X-rays‡
					Absolute	
Nd	142 143	1.03(5)	0.096	0.102(5)	0.13(1)	0.12(2)
	144 145	1.03(9)	0.096	0.098(8)	0.11(1)	0.13(2)
	142 146	1.41(3)	0.382	0.540(4)	0.54(2)	0.52(3)
	146 150	1.82(3)	0.379	0.685(5)	0.67(3)	0.68(4)
Yb	168 170		0.182		0.128(19)	
	170 172	0.79(2)	0.181	0.145(4)	0.118(16)	0.167(16)
	172 174	0.63(3)	0.180	0.115(5)	0.092(15)	0.138(12)
	174 176	0.59(3)	0.179	0.107(5)	0.087(13)	0.104(11)
	170 171	0.67(6)	0.090	0.062(6)	0.041(10)	0.077(32)
	172 173	0.55(8)	0.090	0.050(7)	0.041(10)	0.050(27)
W	180 182		0.177		0.064(12)	
	182 184	0.67(3)	0.177	0.120(5)	0.091(16)	0.102(12)
	184 186	0.52(3)	0.176	0.092(5)	0.080(14)	0.066(9)
	182 183		0.088		0.048(10)	
Hg	181 183		0.177		0.042(28)	
	183 185		0.176		-0.007(21)	
	185 187		0.176		-0.400(61)	
	187 189		0.175		0.57(22)	
	192 194		0.174		0.095(16)	

Electron[§] Scattering	References				Notes
	Muonic	Optical	Electronic X-rays	Electron scattering	
	Macagno *et al.* 1970	King *et al.* 1973	Lee & Boehm 1973		
	Macagno *et al.* 1970	King *et al.* 1973	Lee & Boehm 1973		
0.54(2)	Macagno *et al.* 1970	King *et al.* 1973	Lee & Boehm 1973	Maas & de Jager 1974	
0.69(3)	Macagno *et al.* 1970	King *et al.* 1973	Lee & Boehm 1973	Maas & de Jager 1974	
		Champeau *et al.* 1973			
	Zehnder 1973	Champeau *et al.* 1973	Lee & Boehm 1973		
	Zehnder 1973	Champeau *et al.* 1973	Lee & Boehm 1969		
	Zehnder 1973	Champeau *et al.* 1973	Lee & Boehm 1973		
	Zehnder 1973	Champeau *et al.* 1973	Lee & Boehm 1973		
	Zehnder 1973	Champeau *et al.* 1973	Lee & Boehm 1973		
		Champeau & Miladi 1974			
	Hiltin *et al.* 1970	Champeau & Miladi 1974	Chesler & Boehm 1968		
	Hiltin *et al.* 1970	Champeau & Miladi 1974	Chesler & Boehm 1968		
		Champeau & Miladi 1974			
		Bonn *et al.* 1973			
		Bonn *et al.* 1973			
		Bonn *et al.* 1973			
		Tomlinson & Stroke 1974			

Table 6.4 continued

Element	Mass numbers	$\left(\frac{(\delta E)_{\text{field}}}{(\delta E)_{\text{std.}}}\right)^{\dagger}$ muonic	$\delta\langle r^2\rangle$ (fm^2) Standard	Muonic	Optical	Electronic X-rays‡
					Absolute	
Hg	194 196		0.173		0.096(16)	
	196 198		0.172		0.092(12)	
	198 200		0.172		0.108(13)	
	200 202		0.171		0.119(13)	
	202 204		0.171		0.117(13)	
	192 193		0.087		0.006(11)	
	194 195		0.087		0.046(12)	
	196 197		0.086		0.031(7)	
	198 199		0.086		0.014(2)	
	200 201		0.086		0.036(4)	
	202 203		0.085		0.038(22)	
	204 205		0.085		0.039(14)	
Pb	204 208	0.682(4)	0.340	0.264(1)	0.237(18)	0.205(3)
	206 208	0.722(7)	0.170	0.140(1)	0.125(14)	0.099(10)
	208 210		0.169		0.218(28)	
	206 207	0.56(3)	0.085	0.073(1)	0.047(6)	0.027(11)

†The standard shift is based on a uniform charge distribution of magnitude $1.2A^{1/3}$ fm.

§The electron scattering $\delta\langle r^2\rangle$ values are extremely model dependent.

Electron[§] scattering	References: Muonic	Optical	Electronic X-rays	Electron scattering	Notes
		Tomlinson & Stroke 1964			
		Tomlinson & Stroke 1964			
		Tomlinson & Stroke 1964			
		Tomlinson & Stroke 1964			
		Tomlinson & Stroke 1964			
		Tomlinson & Stroke 1964			
		Tomlinson & Stroke 1964			
		Tomlinson & Stroke 1964			
		Tomlinson & Stroke 1964			
		Tomlinson & Stroke 1964			
		Tomlinson & Stroke 1964			
		Tomlinson & Stroke 1964			
	Kessler *et al.* 1975	Blaise 1958	Lee & Boehm 1973		
0.013(4)	Kessler *et al.* 1975	Blaise 1958	Lee & Boehm 1973	de Jager 1973	
		Blaise 1958			
0.005(7)	Kessler *et al.* 1975	Blaise 1958	Lee & Boehm 1973	de Jager 1973	

‡For all isotopes heavier than Sn, the shifts given for electronic X-rays are the quantity $\delta\langle r^2\rangle + a\ \delta\langle r^4\rangle + b\ \delta\langle r^6\rangle$ where the last two terms contribute about 5 per cent in the region of Z = 60.

Notes

1. The electron-scattering results were analysed with densities constrained to fit results of other experiments, especially electronic K X-rays.

2. The e^- K X-ray value of $\delta\langle r^2\rangle$ is obtained from the value $C_1\delta\langle r^2\rangle + C_2\delta\langle r^4\rangle + C_3\ \delta r^6$ quoted by Boehm (1973) by assuming a uniform distribution. See Table 5.1 for values of the coefficients C_i.)

3. The muonic-atom shifts were calculated from shifts in the 1s level rather than shifts in the 2p - 1s energies in order to eliminate the uncertainty in the 2p level due to the nuclear polarization.

depends somewhat on the model used to represent the shape of the density of A_1. The values quoted here are based on a two-parameter Fermi distribution (with parameters c_1, a_1) which fits the transition energies. The value of $\delta E_{\text{standard}}$ is the change in the transition energy for isotone A_2 calculated for a change δc in the half-density radius given by

$$\delta c = c_1\{(A_2/A_1)^{1/3} - 1\}.$$

Just as in the case of isotope pairs, we would expect the charge-density differences between isotones to be given more reliably than the absolute densities, provided that a model-independent analysis is carried out. (Model-dependent analyses are almost as misleading here as in the case of isotope shifts.) As a test of theoretical nuclear wavefunctions the isotone shifts are particularly interesting, since in the single-particle model the change in density is equal to the sum of the squares of the wavefunctions of the added protons. The density difference between ^{64}Zn and ^{62}Ni obtained from electron scattering (Wohlfahrt *et al.* 1973) is compared with the Hartree-Fock prediction (Negele and Rinker 1975) in Fig. 6.4. A model-independent analysis has also been carried out by Sick (1975) on the electron results of Sinha *et al.* (1971) and the muonic spectrum for Ca measured by Suzuki (1967), and the difference between the charge densities of ^{209}Bi and ^{208}Pb has been extensively analysed by Sprung, Martorell and Campi

TABLE 6.5

Isotone shifts

N	Nuclei	$\left[\frac{(\delta E)_{\text{vol.}}}{(\delta E)_{\text{std.}}}\right]^{\dagger}$	$\Delta\langle r^2\rangle$		References		Notes
		Muonic	Muonic	Electron scattering	Muonic	Electron Scattering	
30	^{56}Fe ^{58}Ni			0.13(7)		Litvinenko *et al.* 1972	
50	^{88}Sr ^{90}Zr			0.44(6)		Schmitt 1973	
50	^{90}Zr ^{92}Mo			0.42(4)		Dreher *et al.* 1973	
66	^{114}Cd ^{115}In	0.23(12)	-0.05(5)		Kast *et al.* 1971		1
66	^{115}In ^{116}Sn	0.91(11)	0.21(5)		Kast *et al.* 1971		1
70	^{120}Sn ^{121}Sb	2.8(9)			Quitmann 1967		
72	^{123}Sb ^{124}Te	2.7(6)			Quitmann 1967		
74	^{124}Sn ^{126}Te	1.99(4)	0.43(5)		Kast 1970		1
74	^{126}Te ^{127}I	1.80(20)	0.18(5)		Kast 1970		1
126	^{208}Pb ^{209}Bi	2.77		0.23	See note 2	Sick 1973*b*	2

†See text for definitions of volume and standard shifts.

Notes

1. The numbers quoted are obtained by averaging results from $2p_{1/2}$-1s and $2p_{3/2}$-1s transition measurements and include an estimate of the error due to uncertainty in the nuclear polarization.
2. The muonic result is from the shift in the $2p_{3/2}$-1s energy obtained from the measurements of Kessler *et al.* (1971) for ^{208}Pb and Powers (1968) for ^{209}Bi.

(1976).

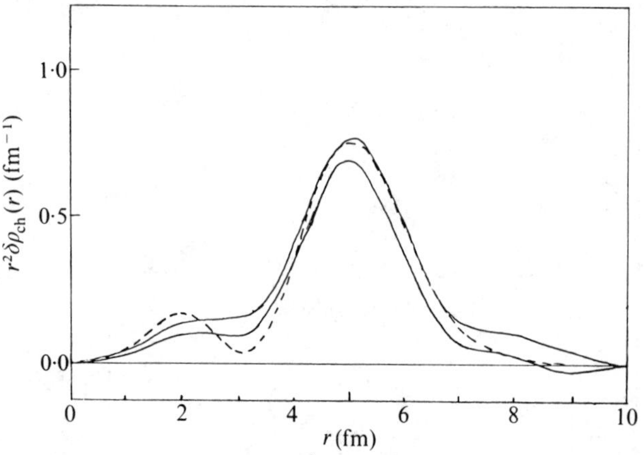

FIG. 6.4. The difference between the charge densities of the isotones ^{64}Zn and ^{62}Ni multiplied by r^2. The full curves show the range of densities obtained in the model-independent analysis of Sick (1975). The broken curve shows the density obtained from the Hartree-Fock calculation of Negele and Rinker (1975).

6.1.6. *Isomer shifts*

A selection of values obtained from muonic, Mössbauer, and optical measurements is given in Table 6.6. The optical measurements are more reliable than optical-isotope-shift measurements because of the absence of mass terms. The table gives a selection of values of the change $\Delta\langle r^2\rangle$ in the mean-square charge radius when the nucleus is excited. A table of radius changes from Mössbauer isomer shifts has been compiled by Kalvius and Shenoy (1974). Values for a few other nuclei are given in the tabulations of Engfer *et al*. (1974) (from muonic-atom data) and Heilig and Steudel (1974) (from optical data).

The optical method is applicable to real isomers, i.e. high angular momentum states whose lifetime is measured in minutes or hours. The muonic-atom and Mössbauer methods are applicable to nuclear states whose lifetime is very much shorter (down to about 10^{-11} s) and have been used to study

TABLE 6.6. *Isomer shifts*

Nucleus	Excited state		$\Delta\langle r^2\rangle$ (fm^2)			References		
	I^{π}	E(keV)	Muonic	Mössbauer	Optical†	Muonic	Mössbauer	Optical
^{150}Nd	2+	130.17	0.014(2)			Wu & Wilets 1969		
^{152}Sm	2+	121.78	0.014(1)	0.009		Wu & Wilets 1969	Yeboah-Amankwah *et al.* 1967	
^{154}Gd	2+	123.07	0.016(2)	0.019(6)		Wu & Wilets 1969	Kienle *et al.* 1968	
^{158}Gd	2+	79.5	-0.004(1)			Engfer *et al.* 1974		
^{182}W	2+	100.10	-0.0006(12)	0.0004(2)		Wu & Wilets 1969	Kienle *et al.* 1968	
^{184}W	2+	111.20	-0.0007(14)			Wu & Wilets 1969		
^{186}W	2+	122.57	-0.0007(14)					
^{193}Hg	13/2+				0.017(12)			Tomlinson and Stroke 1964
^{195}Hg	13/2+				-0.015(5)			Tomlinson and Stroke 1964
^{197}Hg	13/2+				0.014(4)			Tomlinson and Stroke 1964
^{199}Hg	13/2+				0.060(4)			Tomlinson and Stroke 1964
^{209}Bi	13/2+	1608.5	0.037(6)			Engfer *et al.* 1974		
	9/2+	2564.5	0.066(5)			Engfer *et al.* 1974		
	15/2+	2741.0	0.072(5)			Engfer *et al.* 1974		

†The electron density at the Hg nucleus used to obtain $\Delta\langle r^2\rangle$ was an average value from Heilig and Steudel (1974).

many states which decay by E2 transitions. The agreement between these two methods is within experimental error.

6.1.7. *Deformed nuclei*

The problems of deriving the charge density in spherical nuclei are magnified many times for deformed nuclei and there is little hope of obtaining the shape of the multipole transition densities $\rho_l(r)$. The model densities which are used in analysis have more parameters and are more model dependent than those used for spherical nuclei. Nevertheless, the quadrupole and hexadecapole moments and the deformation parameters obtained from Coulomb excitation, muonic hyperfine spectra, and inelastic electron scattering agree very well as shown in Table 6.7. The intrinsic quadrupole and hexadecapole moments are defined as in §3.5:

$$Q = Z\left(\frac{16\pi}{5}\right)^{1/2} \int r^2\, Y_{20}\, \rho_{\mathrm{ch}}^{\mathrm{int}}(\underset{\sim}{r})\,\mathrm{d}^3 r = Z\left(\frac{16\pi}{5}\right)^{1/2} \int \rho_2(r) r^4 \mathrm{d}r$$

$$\Pi = Z\left(\frac{4\pi}{9}\right)^{1/2} \int r^4\, Y_{40}\, \rho_{\mathrm{ch}}^{\mathrm{int}}(\underset{\sim}{r})\,\mathrm{d}^3 r = Z\left(\frac{4\pi}{9}\right)^{1/2} \int \rho_4(r) r^6 \mathrm{d}r.$$

The Coulomb excitation measurements are independent of the shapes of $\rho_2(r)$ and $\rho_4(r)$ and measure only the integral quantities Q and Π. Other methods give different integral moments of these functions. The moments of $\rho_2(r)$ obtained from muonic hyperfine measurements have been calculated by Zehnder (1974).

Apart from the quantities tabulated here these techniques give excited-state multipole moments, and transition moments between excited states, which are of particular value in the understanding of the validity and shortcomings of the collective model (Hitlin *et al.* 1970). Excited-state quadrupole moments obtained by measurement of the reorientation effect have been tabulated by Smilansky (1970), and they can also be measured by Mössbauer techniques (Kienle 1968).

6.2. MAGNETIC MOMENT DISTRIBUTIONS

Most of the available information on magnetic-moment distributions comes from low-q backward-angle electron scattering, where the effect of the M1 (magnetic dipole) moment dominates. The measurements provide magnetic form factors which can be used to obtain the mean-square radii of the M1 distributions and can be compared with calculations from theoretical wavefunctions as shown in Fig. 3.2. The finite size of the M1 distribution also affects hyperfine spectra (the Bohr-Weisskopf effect), but the effect is small, both for ordinary atoms (where the orbit is far out) and for muonic atoms (owing to the small magnetic moment of the muon).

In §6.1 we were concerned with nuclear ground-state charge distributions which are almost the same as proton density distributions (after due account has been taken of the proton size). The charge form factor of the neutron introduces a very small correction. In the case of magnetic-moment distributions the effect is due to valence nucleons, and the magnetic moment of the neutron is almost as large as that of the proton so that magnetic scattering is an important tool for studying the wavefunctions of both neutron and proton valence particles. In principle the determination of magnetic form factors could lead to the magnetic-moment density in configuration space just as is the case for charge densities, but the analysis is more complicated and the quantity and accuracy of data available mean that this is not possible. The measurements are extremely valuable, however, in checking the validity of calculated wavefunctions.

Instead of reproducing graphs of magnetic form factors, we refer readers to reviews of magnetic scattering by Donnelly and Walecka (1973, 1975). A number of results from M1 scattering from light nuclei have been given by de Jager *et al.* (1974) who fit the form factor F_1^{mag} (defined in eqn 3.70) with a linear combination of the $l = 0$ and $l = 2$ spherical Bessel transforms of the square of the radial wavefunction

$$F_1^{\text{mag}}(q) = \int R_{nl}^2(r)\, j_0(qr)\mathrm{d}^3r + \gamma \int R_{nl}^2(r)\, j_2(qr)\mathrm{d}^3r \qquad (6.3)$$

where the coefficient γ is often used as a free parameter. The value of γ predicted from the nuclear wavefunctions depends on the amount of configuration mixing, and the fitted values

TABLE

Transition quadrupole and hexadecapole moments

Nucleus	Coulomb excitation				Muonic hyperfine	
	$\|Q\|$ $(\mathrm{fm})^2$	$\|\Pi\|$ $(\mathrm{fm})^4$	β_2	β_4	Q $(\mathrm{fm})^2$	Π $(\mathrm{fm})^4$
^{12}C	21(2)		0.60			
^{20}Ne	69(7)		0.87			
^{24}Mg	72(6)		0.65			
^{28}Si	56(3)		0.40			
^{150}Nd	516(19)		0.279		515(10)	
^{152}Sm	584(3)	4100(800)	0.286(6)	0.066(24)	578(10)	
^{154}Sm	658(3)	5100(900)	0.317(7)			
^{158}Gd	707(10)	4000(1300)	0.33(1)			
^{160}Gd	726(6)	4300(1100)	0.34(1)			
^{162}Dy	735(3)	3200(1100)	0.33(1)		736(10)	
^{164}Dy	748(3)	3300(1300)	0.34(1)		742(10)	
^{165}Ho					744(7)	5200(1000)
^{166}Er	761(6)	700(1700)	0.35(1)			
^{168}Er	761(6)	2300(1700)	0.33(1)		777(10)	
^{170}Er	764(6)	2800(1700)	0.33(1)		775(10)	
^{170}Yb	746(17)		0.319		780(4)	
^{171}Yb	796(15)		0.339		795(4)	
^{172}Yb	770(13)		0.327		791(4)	
^{173}Yb	794(30)		0.336		792(5)	
^{174}Yb	766(8)		0.321		782(5)	
^{176}Yb	725(8)		0.302		759(5)	
^{182}W	646(16)		0.250		657(8)	
^{184}W	606(12)		0.232		627(8)	
^{186}W	597(17)		0.227		590(8)	

† For a uniform distribution of radius $R_0(1+\beta_2 Y_{20}+\beta_4 Y_{40})$ the intrinsic moments are given in terms of the deformations up to second order in β_2 and first order in β_4 by

$$Q = 3(1/5\pi)^{1/2} Z R_0^{\,2} \beta_2 (1+0.36\beta_2 + 0.97\beta_4)$$

$$\Pi = (1/4\pi)^{1/2} Z R_0^{\,4} (\beta_4 + 0.73\beta_2^{\,2} + 0.98\beta_2\beta_4).$$

6.7

and deformations of charge distributions†

structure		Inelastic electron scattering				Notes		
β_2	β_4	Q $(\mathrm{fm})^2$	Π $(\mathrm{fm})^4$	β_2	β_4	CE	μ^-	(e,e')
		-20	25	0.57	0.07	1,3		1,9
		58(3)	249(27)	0.40	0.19	1,3		1,10
		69(3)	48(16)	0.45	0.03	1,3		1,10
		-64(2)	205(33)	-0.39	0.10	1,3		1,10
0.278				0.279		1,3	2,6	2,13
0.296		582(6)	4300(200)	0.286(6)	0.058(22)	2,4	2,6	2,11,14
				0.315(7)	0.066(24)	2,4		2,11
				0.33(1)	0.030(25)	2,4		2,11
						2,4		
0.338						2,4	2,6	
0.334						2,4	2,6	
							2,7	
						2,4,5		
						2,4,5	2,6	
0.326						2,4,5	2,6	
						1	8	
						1	8	
						1	8	
						1	8	
						1	8	
		733		0.310	0.335	1	8	2,12
0.248						2,4,5	2,6	
0.237						2,4	2,6	
0.222						2,4	2,6	

Notes

1. Deformations based on uniform distribution with $R_0 = 1.2A^{1/3}$ fm
2. Deformations based on deformed Fermi distribution, $\rho = \rho_0/(1+e^x)$ with $x = \{r-R(\theta)\}/a$.
3. Results taken from tabulation of Stelson and Grodzins (1965).
4. Measurement and analysis by Erb *et al.* (1972). The diffuseness

parameter was taken to be 0.6 fm and the value of R_0 chosen to give a constant central density.

5. The error estimates given for Π are the mean of upper and lower limits. The upper value quoted by Erb *et al.* was 20 per cent smaller and the lower limit 20 per cent larger.
6. Experiments and analysis of Hitlin *et al.* (1970).
7. Experiment and analysis of Powers *et al.* (1975).
8. Results taken from Zehnder (1974) who does not quote deformation parameters. For ^{173}Yb the hyperfine splitting in the 3d states was used to obtain a model-independent value of Q. For the other isotopes the error is somewhat model dependent and the errors should be increased by a factor of 10 to account for this.
9. Results and analysis from Nakada *et al.* (1971).
10. Results and analysis from Horikawa *et al.* (1971).
11. Results and analysis from Bertozzi *et al.* (1972*a*).
12. Results of Heisenberg (1974) quoted by Zehnder (1974).
13. Results and analysis of Maas and de Jager (1974).
14. Coupled-channel calculations by Mercer and Ravenhall (1974) show that the two-step process for excitation of the second excited state does not have a negligible contribution so that the value of Π deduced from the cross-section would be modified by an amount which is of the order of 10 per cent.

differ substantially from the single-particle-model values. de Jager *et al.* also give r.m.s. radii $\langle r^2\rangle_M^{1/2}$ of the dipole distributions which can be deduced from the slope at $q^2 = 0$ of the form factor whose expansion in powers of q^2 is

$$F_1^{\text{mag}}(q) = 1 - \frac{1}{6}\langle r_1^{\,2}\rangle_M q^2 + \ldots . \tag{6.4}$$

Values of $\langle r_1^{\,2}\rangle_M$ for selected nuclei are shown in Table 6.8.

TABLE 6.8

Root-mean-square radii of magnetic dipole form factors

Nucleus	$\langle r_1^2 \rangle^{1/2}$ (fm)	References
n	0.79(15)	Hand *et al.* 1963
^1H	0.83(7)	Borkowski *et al.* 1974
^3H	1.70(5)	Collard *et al.* 1965
^3He	1.94(19)	Chertok *et al.* 1969
^6Li	3.00(47)	Rand *et al.* 1966
^7Li	2.69(13)	Rand *et al.* 1966
^9Be	2.64(16)	Rand *et al.* 1966
^{10}B	2.21(32)	Rand *et al.* 1966
^{11}B	2.42(10)	Rand *et al.* 1966
^{14}N	2.52(31)	Rand *et al.* 1966
^{27}Al	3.19(11)	Lapikás *et al.* 1973

7

NUCLEAR SCATTERING

7.1. THEORY OF SCATTERING FROM NUCLEI

Experimental measurements on nuclear scattering yield differential cross-sections for elastic scattering, total cross-sections for scattering and absorption, and a variety of other data. The role of scattering theory is to provide a framework for the interpretation of these data; the simplest forms of the theory provide a description of the data in terms of a limited number of model parameters while the more sophisticated versions connect more fundamental quantities.

7.1.1. *Partial wave methods*

The idea of a partial wave expansion forms the basis of many methods of making calculations for elastic scattering and reactions, and the replacement of an infinite sum over partial waves by an integration over impact parameters leads to many approximate formulae.

We consider non-relativistic two-body scattering. For a neutron and spin-zero projectile we have to find a steady-state solution of the Schrödinger equation

$$\nabla^2\psi + [k^2 - \bar{V}(r)]\psi = 0 \tag{7.1}$$

where

$$k^2 = \frac{2\mu}{\hbar^2} E, \qquad \bar{V}(r) = \frac{2\mu}{\hbar^2} V(r), \tag{7.2}$$

μ is the reduced mass, E is the energy in the centre-of-mass system, and $V(r)$ is a typical nuclear potential. The solution ψ must be regular at the origin and have the asymptotic form of a plane wave and an outgoing spherical wave, i.e.

$$\psi \to e^{ikz} + f(\theta)\, e^{ikr}/r \tag{7.3}$$

where θ is the scattering angle in the centre-of-mass system and $f(\theta)$ is the scattering amplitude. The asymptotic form of ψ can also be written as

$$\psi \rightarrow \frac{1}{2} \sum_{l} i^{l+1} (2l+1) \, P_l(\cos\theta) \, [\mathscr{I}_l(kr) - \eta_l \mathscr{O}_l(kr)] \tag{7.4}$$

where $\mathscr{I}_l$ and $\mathscr{O}_l$ represent incoming and outgoing spherical waves, respectively, and the reflection coefficient η_l is defined in terms of the phase shift δ_l as

$$\eta_l = \exp(2i\delta_l). \tag{7.5}$$

If $|\eta_l| = 1$ the intensity of the outgoing wave is equal to that of the incoming wave and reflection is complete, but if non-elastic processes, such as inelastic scattering and re-arrangement collisions, compete with elastic scattering the intensity of the outgoing elastically scattered wave must be less than that of the incident wave.

When a Coulomb potential is included for the scattering of charged particles the arguments given above must be modified, but a partial wave expansion can still be made and the scattering amplitude is given by

$$f(\theta) = f_c(\theta) + (2ik)^{-1} \sum_{l} (2l+1) \exp(2i\sigma_l) \, (\eta_l - 1) \, P_l(\cos\theta) \tag{7.6}$$

where $f_c(\theta)$ and σ_l are the scattering amplitude and phase shift, respectively, for pure Coulomb scattering.

If the projectile has spin s we must define a set of scattering amplitudes $f_{\nu\mu}$ for scattering from spin state (s,μ) to state (s,ν). If the final spin state is not observed and the spins in the incident beam are randomly oriented, the cross-section is obtained by taking the usual sum over final spin states and average over initial states. If the target has spin S there are $g = (2s+1)(2S+1)$ spin states, and it is convenient to construct the scattering matrix M of order $g \times g$ whose elements are the $f_{\nu\mu}$. For projectiles of spin $\frac{1}{2}$ scattering from spin-zero targets, or vice versa, it is found that M has the form

$$M = A(\theta) + C(\theta)\, \underset{\sim}{\sigma}.\hat{\underset{\sim}{n}} \tag{7.7}$$

where $\underset{\sim}{\sigma}$ is the Pauli spin operator and $\hat{\underset{\sim}{n}}$ is a unit vector in the direction perpendicular to the reaction plane defined by the initial momentum $\underset{\sim}{k}$ and final momentum $\underset{\sim}{k}'$, i.e.

$$\hat{\underset{\sim}{n}} = \frac{(\underset{\sim}{k} \wedge \underset{\sim}{k}')}{|\underset{\sim}{k} \wedge \underset{\sim}{k}'|} . \tag{7.8}$$

The appropriate form for nucleon-nucleon scattering is

$$M = A + B\,(\underset{\sim}{\sigma}_0.\hat{\underset{\sim}{n}})(\underset{\sim}{\sigma}_1.\hat{\underset{\sim}{n}}) + C\,(\underset{\sim}{\sigma}_0+\underset{\sim}{\sigma}_1).\hat{\underset{\sim}{n}}$$

$$+ E\,(\underset{\sim}{\sigma}_0.\hat{\underset{\sim}{q}})(\underset{\sim}{\sigma}_1.\hat{\underset{\sim}{q}}) + F\,(\underset{\sim}{\sigma}_0.\hat{\underset{\sim}{P}})(\underset{\sim}{\sigma}_1.\hat{\underset{\sim}{P}}) \tag{7.9}$$

where $\hat{\underset{\sim}{q}}$ is a unit vector in the direction of the momentum transfer $\underset{\sim}{q} = \underset{\sim}{k}-\underset{\sim}{k}'$ and $\hat{\underset{\sim}{P}}$ is the third unit vector required to make a right-handed set of axes. The coefficients A, B, etc. are functions of momentum transfer and energy. Because the scattering matrix is spin dependent, there is a net alignment of spins in the $\hat{\underset{\sim}{n}}$-direction as a result of the scattering process. This alignment is measured by the polarization vector

$$\underset{\sim}{P}(\theta) = \langle \sigma \rangle = \frac{\text{Trace}(M^+\, \underset{\sim}{\sigma}\, M)}{\text{Trace}(M^+ M)} . \tag{7.10}$$

The polarization arising from the scattering of spin-$\frac{1}{2}$ projectiles from a spin-zero target is obtained by substituting the expression (7.7) into eqn (7.10). This gives

$$P(\theta) = \underset{\sim}{P}(\theta).\hat{\underset{\sim}{n}} = \frac{2\text{Re}(AC^*)}{|A|^2 + |C|^2} . \tag{7.11}$$

It can be argued that the physically significant quantities are the phase shifts and that the potential may be regarded as just one of several parametrizations which lead to the phase shifts (see, for example, Frahn 1967). This point of view connects the description of nuclear scattering with the methods of direct phase shift analysis and dispersion theory which are used for nucleon-nucleon scattering and elementary particle processes. It is then natural to parametrize the phase shifts or the reflection coefficients directly.

The simplest case is the limit of complete absorption of all partial waves up to L-1 so that the reflection coefficients can be represented by (Blair 1954)

$$\left.\begin{aligned} \eta_l &= 0, \qquad l < L \\ \eta_l &= 1, \qquad l \geqslant L \end{aligned}\right\}. \tag{7.12}$$

This is known as the sharp cut-off model and has been applied mainly to α-particle scattering; smooth cut-off models have been developed (McIntyre, Wang, and Becker 1960, McIntyre, Baker, and Wang 1962, Alster and Conzett 1965) and usually involve four parameters. Still more elaborate parametrizations have been developed to describe a variety of scattering processes (Frahn and Venter 1963, 1964, Springer and Harvey 1965, Hufner and de Shalit 1965). The semiclassical relations $pb \simeq \hbar\{l(l+1)\}^{1/2}$, $p = \hbar k$, can be used to link the angular momentum L with a size parameter R, and, if the transition of η_l from zero to unity is a gradual one occurring over a range of angular momenta Δ, this parameter can also be linked with a surface diffuseness d. This gives

$$L = kR, \qquad \Delta = kd. \tag{7.13}$$

When Coulomb effects are taken into account these formulae become (Frahn 1967)

$$L = kR\{1 - (2n/kR)\}^{1/2} \tag{7.14a}$$

$$\Delta = kd\,\frac{1 - (n/kR)}{\{1 - (2n/kR)\}^{1/2}}. \tag{7.14b}$$

where n is the Coulomb parameter $\mu Z_1 Z_2 e^2/\hbar^2 k$.

7.1.2. *Multiple scattering and impulse approximation*

The scattering amplitude can also be written in the form

$$f(\theta) = f(\underset{\sim}{k}',\underset{\sim}{k})$$

$$= -\frac{\mu}{2\pi\hbar^2}\int \exp(-i\underset{\sim}{k}'.\underset{\sim}{r})\ V(\underset{\sim}{r})\ \psi^+(\underset{\sim}{k},\underset{\sim}{r})\ d^3r \quad (7.15)$$

where ψ^+ is the solution with outgoing boundary conditions, as defined in eqn (7.3). It is convenient to introduce the transition operators $T^{\pm}$:

$$T^{\pm} = V + V\ G_0^{\pm}\ T^{\pm} = V + V\ G^{\pm}\ V \quad (7.16)$$

where G_0 and G are the propagators of the projectile in free space† and in the nuclear medium, and the scattering amplitude then becomes

$$f(\underset{\sim}{k}',\underset{\sim}{k}) = -\frac{\mu}{2\pi\hbar^2}\langle\underset{\sim}{k}'|T|\underset{\sim}{k}\rangle. \quad (7.17)$$

In the context of scattering from a nucleus T is the operator for scattering from the nucleus as a whole. Since the nucleus is composed of a number of nucleons it is of interest to describe the nuclear scattering in terms of successive scatterings from single nucleons, each of which can be represented by an operator

$$t(j) = V(j) + V(j)\ G_0\ t(j) \quad (7.18)$$

where the label j indicates the jth nucleon in the nucleus and $V = \Sigma_j\ V(j)$. By iterating eqn (7.18) it can be seen that $t(j)$ sums the interaction $V(j)$ of the projectile and the jth nucleon to all orders. It can be shown (Watson 1953, 1957) that the relation between T and the $t(j)$ is

---

†G_0 includes the internal Hamiltonian for the target nucleus but excludes the interaction between the projectile and the nucleus.

$$T = \sum_j t(j) + \sum_j t(j)\, G_0 \sum_{k\neq j} t(k) + \sum_j t(j)\, G_0 \sum_{k\neq j} t(k)\, G_0 \sum_{l\neq k} t(l) + \ldots \tag{7.19}$$

This result is known as a multiple-scattering expansion, and the physical interpretation is that the projectile makes successive scatterings from different nucleons and propagates freely between them. Thus, the first term represents single scattering, the second double scattering, and so on.

The scattering from a nucleon in the nucleus differs from scattering from a free nucleon in a number of respects owing to the presence of the other nucleons. For example, in the nucleus the exclusion principle acts to block a number of energetically available scattering states. In certain circumstances, however, it is permissible to replace the $t(j)$ by the operator $\tau(j)$ defined by

$$\tau(j) = V(j) + V(j)\, g(j)\, \tau(j) \tag{7.20}$$

$$g(j) = (E_2 - H_0 - H(j) + i\varepsilon)^{-1} \tag{7.21}$$

where $H(j)$ is the Hamiltonian for a single nucleon and E_2 is the total energy in the centre-of-mass system of the two particles. For this impulse approximation to be valid the binding energy of the struck nucleon should be small compared with the incident kinetic energy and the mean free path λ between collisions should be long enough that the energy spread of the scattered particle $\Delta E \sim \hbar v/\lambda$ is small compared with the incident energy.

A potential U which describes the elastic scattering from a system of bound particles can be derived (Watson 1957, 1958) by introducing a slightly different operator

$$\tilde{t}(j) = V(j) + V(j)\, Q\, G_0\, \tilde{t}(j) \tag{7.22}$$

where Q is a projection operator which projects off the ground state of the target nucleus. This t-operator approaches the operator (7.20) in this high-energy limit. The multiple-scattering series for U is then given by

$$U = \langle 0|\tilde{U}|0\rangle$$

$$= \langle 0| \sum_j \tilde{t}(j) \, |0\rangle + \langle 0| \sum_{j\neq k} \tilde{t}(j) \, Q \, G_0 \, \tilde{t}(k) \, |0\rangle + \ldots \tag{7.23}$$

where $|0\rangle$ represents the ground state of the target nucleus. Apart from the presence of the operator Q, this expansion has the same form as eqn (7.19); however, the presence of Q implies that the intermediate states in the multiple-scattering process involve excited states of the target nucleus. The exact scattering amplitude for elastic scattering is then given in terms of the potential operator $\tilde{U}$ as

$$f_{\mathrm{el}}(\underset{\sim}{k},\underset{\sim}{k}') = -\frac{\mu}{2\pi\hbar^2}\langle 0,\underset{\sim}{k}'|\tilde{U} + \tilde{U}\, P \, G_0 \, \tilde{U} + \ldots |0,\underset{\sim}{k}\rangle \tag{7.24}$$

where $P = 1-Q$ projects on to the ground state of the target. Thus, exact solution of the Schrödinger equation with the potential U additionally takes into account all the multiple-scattering processes in which the intermediate state involves the ground state of the target.

7.1.3. *Approximate methods for elastic scattering*

In the WKB approximation the wavefunction is written as

$$\psi^+(\underset{\sim}{k},\underset{\sim}{r}) = \exp(\mathrm{i}S(\underset{\sim}{k},\underset{\sim}{r})). \tag{7.25}$$

The function $S(\underset{\sim}{k},\underset{\sim}{r})$ is then expanded in powers of $\hbar$, and the lowest-order term is

$$S^{(0)}(\underset{\sim}{k},\underset{\sim}{r}) = \int^{r} k(\underset{\sim}{r}')\,\mathrm{d}^3 r' \tag{7.26}$$

where the integral is along the classical trajectory and $k(r)$ is the local wavenumber

$$k^2(r) = k^2 - \bar{V}(r)$$

with k^2 and $\bar{V}$ defined as in eqn (7.2). If $E \gg |V|$, the integrand in eqn (7.26) can be expanded to give

$$S^{(0)}(\underset{\sim}{k},\underset{\sim}{r}) = \underset{\sim}{k}.\underset{\sim}{r} - \frac{\mu}{\hbar^2 k} \int^{r} V(\underset{\sim}{r}')\, d^3r'$$

and, if the condition $kd \gg 1$ also applies where d is the range of the potential, the trajectory will approximat to a straight line in the direction of incidence $\underset{\sim}{k}$, which we have taken to be the z-axis. This gives

$$r = \underset{\sim}{b} + \hat{\underset{\sim}{k}}z \tag{7.27}$$

where b is the impact parameter and

$$S^{(0)}(\underset{\sim}{k},\underset{\sim}{r}) = \underset{\sim}{k}.\underset{\sim}{r} - \frac{1}{\hbar v}\int_{-\infty}^{z} V(\underset{\sim}{b} + \hat{\underset{\sim}{k}}z')\, dz'. \tag{7.28}$$

This formula also holds in the relativistic case.

Using eqns (7.28) and (7.25) and substituting in eqn (7.15) we obtain the high-energy or eikonal approximation for the scattering amplitude:

$$f(\theta,\phi) = -\frac{\mu}{2\pi\hbar^2}\int \exp\{i(\underset{\sim}{k}-\underset{\sim}{k}').\underset{\sim}{r}\}\, V(\underset{\sim}{r})\, \exp\left(-\frac{i}{\hbar v}\int_{-\infty}^{z} V(\underset{\sim}{b} + \underset{\sim}{k}z')dz'\right) d^3r. \tag{7.29}$$

For elastic scattering and small scattering angles

$$(\underset{\sim}{k}-\underset{\sim}{k}').\underset{\sim}{r} \simeq (\underset{\sim}{k}-\underset{\sim}{k}').\underset{\sim}{b}$$

and the scattering amplitude can be reduced to

$$f(\theta,\phi) = \frac{k}{2\pi i}\int \exp\{i(\underset{\sim}{k}-\underset{\sim}{k}').\underset{\sim}{b}\}\,[\exp\{i\chi(b)\} - 1]\, d^2b \tag{7.30}$$

where

$$\chi(\underset{\sim}{b}) = -\frac{1}{\hbar v}\int_{-\infty}^{\infty} V(\underset{\sim}{b} + \hat{\underset{\sim}{k}}z)\, dz. \tag{7.31}$$

If the potential is spherically symmetric, $\chi(\underset{\sim}{b}) \equiv \chi(b)$ and for small angles $(\underset{\sim}{k}-\underset{\sim}{k}').\underset{\sim}{b} \simeq 2kb \sin \frac{1}{2}\theta \cos \phi$, and hence, using the formula

$$\int_0^{2\pi} \exp(i\lambda \cos \phi)\, d\phi = 2\pi\, J_0(\lambda), \tag{7.32}$$

the scattering amplitude becomes

$$f(\theta,\phi) = ik \int_0^\infty J_0(2kb \sin \tfrac{1}{2}\theta)\ [1 - \exp\{i\chi(b)\}]\ b\ db. \quad (7.33)$$

These formulae are valid only for small scattering angles, but additional terms can be derived to extend them to larger angles (Schiff 1956, 1968, Ross 1968).

The semiclassical formula can also be expanded in a multiple-scattering series. The basic assumption in this development is that the phase shift function $\chi(b)$ for scattering from the nucleus as a whole can be replaced by the sum of phase shifts due to scatterings from individual nucleons, i.e.

$$\chi \simeq -\frac{1}{\hbar v} \int_{-\infty}^{\infty} \sum_j V_j\, dz \simeq \sum_j \chi_j \qquad (7.34)$$

(Glauber 1959, Czyz 1971). Hence

$$1 - \exp i\chi(b) = 1 - \exp(i\Sigma\chi_j) = 1 - \Pi \exp(i\chi_j) \qquad (7.35a)$$

$$= 1 - \Pi(1 - \Gamma_j) \qquad (7.35b)$$

$$= \sum_j \Gamma_j - \sum_{j\neq k} \Gamma_j \Gamma_k + \sum_{j\neq k\neq l} \Gamma_j \Gamma_k \Gamma_l + \ldots \qquad (7.35c)$$

Equation (7.35*c*) represents the multiple-scattering expansion for the operator or profile function $\Gamma = 1 - \exp(i\Sigma\chi_j)$ and, when inserted into eqn (7.30), yields the multiple-scattering expansion for the scattering amplitude. It permits successive small-angle scattering from different nucleons but does not allow the large-angle scattering necessary for scattering more than once from the same nucleon, in contrast to the multiple-scattering expansion (7.29). This means that the semiclassical expansion contains a finite number of terms; in addition, each individual scattering is on the energy shell.

7.1.4. *Diffraction theory*

Using the Schrödinger equation, the exact expression (7.15) for the scattering amplitude can be transformed to

$$f(\theta,\phi) = -(4\pi)^{-1}\int \exp(-i\underset{\sim}{k}'.\underset{\sim}{r})\,(\nabla^2 + k^2)\,\psi^+(k,\underset{\sim}{r})\,d^3r$$

$$= -(4\pi)^{-1}\int [\exp(-i\underset{\sim}{k}'.\underset{\sim}{r})\,(\nabla^2\psi^+) - \{\nabla^2\exp(-i\underset{\sim}{k}'.\underset{\sim}{r})\}\,\psi^+]\,d^3r$$

where we have used the elastic scattering condition $k^2 = k'^2$. The integral is over the volume within which $V(\underset{\sim}{r})$ is non-zero, but the integral can be transformed to an integral over the surface bounding this volume. The amplitude for scattering from a black disc then becomes (Blair 1966, Frahn 1967)

$$f(\theta,\phi) = \frac{ik}{4\pi}(1 + \cos\theta)\int \exp\{i(\underset{\sim}{k}-\underset{\sim}{k}').\underset{\sim}{r}\}\,dS \qquad (7.36)$$

where the integration is around the boundary of the shadow region. If this is defined by two co-ordinates ρ and η, where ρ is a length and η is an angle in the scattering plane, the amplitude can be written as

$$f(\theta,\phi) = \frac{ik}{4\pi}(1 + \cos\theta)\int_0^{2\pi} d\eta \int_0^{\rho(\eta)} \exp(-ik\rho\, 2\sin\tfrac{1}{2}\theta\,\cos\eta)\,\rho d\rho$$

$$= \frac{ik}{2\pi}\int_0^{2\pi} d\eta \int_0^{\rho(\eta)} \exp(-ik\rho\,\theta\cos\eta)\,\rho d\rho$$

where the second line applies for small θ. For scattering from from a spherical nucleus of radius R_F this gives

$$f(\theta,\phi) \simeq \frac{ik}{2\pi}\int_0^{2\pi} d\eta \int_0^{R_F} \exp(-ik\rho\,\theta\cos\eta)\,\rho d\rho$$

which can be evaluated using eqn (7.49) to give

$$f(\theta) = ik\,R_F^2\left(\frac{J_1(k\,R_F\,\theta)}{k\,R_F\,\theta}\right) \qquad (7.37)$$

which is the same as the result obtained using the sharp cut-off model (7.12) for the reflection coefficients and the semi-classical relation $L + \frac{1}{2} = k\,R_F$. It is also identical to the Airy formula for Fraunhofer diffraction by a circular aperture, and for this reason we refer to the radius R_F as the diffraction radius or Fraunhofer radius. If the nucleus is sharp surfaced but not spherical the boundary can be represented by the formula

$$\rho(\eta) = R_F\{1 + \sum_{lm} \alpha_{lm} Y_l^m(\tfrac{\pi}{2},\eta)\} \tag{7.38}$$

where α_{lm} are the collective co-ordinates.

Diffraction theory can be extended to take account of Coulomb scattering and partial absorption through appropriate changes in the boundary conditions (Frahn 1967). Coulomb scattering is very important for heavy nuclei, but for light nuclei the Fraunhofer diffraction theory describes quite well the oscillatory patterns seen in certain differential cross-sections. The main deficiency in this diffraction theory arises from the assumption of a sharp surface, and this can be remedied (Inopin and Berezhnoy 1965) by introducing a shadow function which is the convolution of a sharp-edged function and a smoothing function. After more manipulation the scattering amplitude reduces to the product of the amplitude for a sharp-edged nucleus and a damping factor. This again introduces a surface diffuseness parameter as well as a radius parameter.

7.2. THE OPTICAL MODEL

Many very accurate measurements have been made of differential and total cross-sections for elastic scattering, and polarization, for a variety of nuclear projectiles from a wide range of target nuclei. In the majority of cases these data have been analysed in terms of the optical model using a complex potential whose real part describes the average potential energy inside the nucleus and whose imaginary part describes the effect of all processes which tend to deplete the flux in the elastic channel. Much of the work has been concerned with the optical model as a phenomenological model, and careful fits to the data have established the form of the potential and the change in the parameters as functions of basic variables such as the energy of the incident projectile and the mass number of the target. More recent work has been concerned with the connection between the phenomenological potential and the microscopic structure of the target nucleus and of the projectile, which has an important role in the interpretation of nuclear sizes from nuclear scattering.

7.2.1. *Formal theory of the optical potential*

There are a number of different ways of developing a formal theory of the optical potential. We outline here the approach due to Feshbach (Feshbach 1958, 1962, Lemmer 1962*b*).

We have to solve the Schrödinger equation $E\psi = H\psi$ where $H = H_0 + H_N + V$ and ψ is the total wavefunction which can be expanded in terms of the scattering states ψ of the projectile and the eigenstates Φ of the nuclear Hamiltonian H_N. When projection operators P and Q are introduced, as in §7.1.2, the Schrödinger equation reduces to two coupled equations

$$(E - H_{PP})\, P\Psi = H_{PQ}\, Q\Psi \tag{7.39a}$$

$$(E - H_{QQ})\, Q\Psi = H_{QP}\, P\Psi \tag{7.39b}$$

where $H_{PQ} = P\,H\,Q$, etc. Then, eliminating $Q\Psi$ and inserting a suitable form for P we obtain an equation for elastic scattering of the projectile in the form

$$[E - H_0 - \langle\Phi_0|V|\Phi_0\rangle - \langle\Phi_0|VQ(E-H_{QQ})^{-1}QV|\Phi_0\rangle]\,\psi_0 = 0 \tag{7.40}$$

where we have put the energy of the nuclear ground state equal to zero. Hence, the generalized optical potential may be defined as

$$V_{\text{opt}} = \langle\Phi_0|V|\Phi_0\rangle + \langle\Phi_0|VQ(E-H_{QQ})^{-1}QV|\Phi_0\rangle. \tag{7.41}$$

The second term in eqn (7.41) will vary rapidly with energy whenever E is in the vicinity of an eigenvalue of H_{QQ}. This will give rise to a marked fluctuation, i.e. a resonance, in the cross-section at this particular energy. The propagator $(E-H_{QQ})^{-1}$ can be expressed in terms of the eigenstates Ψ_t

of H_{QQ} so that, for scattering in the vicinity of an isolated resonance at $E = E_s$, the optical potential can be written as

$$V_{\text{opt}} = \langle \Phi_0 | V | \Phi_0 \rangle + \sum_{t \neq s} \frac{\langle \Phi_0 | VQ | \Psi_t)(\Psi_t | QV | \Phi_0 \rangle}{E - E_t}$$

$$+ \frac{\langle \Phi_0 | VQ | \Psi_s)(\Psi_s | QV | \Phi_0 \rangle}{E - E_s} \quad (7.42a)$$

$$= U_{\text{opt}} + \frac{\langle \Phi_0 | VQ | \Psi_s)(\Psi_s | QV | \Phi_0 \rangle}{E - E_s} \quad (7.42b)$$

where the parenthesis indicates integration and summation over the co-ordinates of the target nucleons and the projectile. Thus, the optical potential U_{opt}, and the corresponding cross-section, will vary smoothly with energy and include effects of distant resonances.

Using eqn (7.16), eqn (7.41) can be rewritten in terms of the T-matrix. This suggests the use of a multiple-scattering expansion for the potential, which leads directly to the Watson formula given in eqn (7.23) or to the slightly different formulation of Kerman, McManus, and Thaler (1959). These formal expressions provide a basis for various attempts to calculate the optical potential; at high energies the natural procedure is to use the impulse approximation and replace $\tilde{t}$, defined in eqn (7.22), by the free t-matrix τ for projectile-nucleon scattering, while at lower energies it may be more realistic to use an effective interaction, such as that defined in eqn (2.13), which is appropriate for the interaction of bound nucleons. It is evident from eqns (7.23) and (7.41) that the leading term in any such calculation must involve the nuclear matter distribution.

A nucleon in the nucleus which is involved in a collision with the projectile is initially bound and is moving in the nucleus with a certain momentum, usually called the Fermi momentum. Thus, the energy and momentum conditions for such a collision are different from those pertaining in a free projectile-nucleon collision, and so we need to know something about the projectile-nucleon interaction off the energy shell for free scattering. Such information can, in principle, be obtained from measurements on meson production and knock-out reactions, but these studies are still in progress. This suggests that, for the study of nuclear sizes, we should concentrate on scattering at an energy sufficiently high that the process is only slightly off-shell, or use credible nucleon-nucleon potentials and hope that the off-shell behaviour is reasonably correctly described.

7.2.2. *General properties of the optical potential*

In order to examine general properties of the optical potential for nucleon-nucleus scattering we use the first term in eqn (7.23) together with the impulse approximation, and neglect off-shell effects. The t-matrix for free nucleon-nucleon scattering can be expressed in terms of the scattering matrix defined by eqn (7.9). This gives

$$\tau(j) = -\frac{\hbar^2}{4\pi^2\mu_0} M(j) \tag{7.43}$$

where μ_0 is the reduced mass in the nucleon-nucleon centre-of-mass system. We have already noted that the coefficients, A, B, etc., of the scattering matrix are functions of momentum transfer. They also have a dependence on isobaric spin which may be expressed in the form

$$A(q) = A_\alpha(q) + A_\beta(q)\, \underset{\sim}{t}_0 \cdot \underset{\sim}{t}_j \tag{7.44}$$

where $\underset{\sim}{t}_0$, $\underset{\sim}{t}_j$ are the isospin operators for the incident nucleon and the jth nucleon in the nucleus. Neglecting Coulomb scattering and assuming the charge independence of nuclear forces, it can be shown that

$$A_\alpha = \tfrac{1}{2}(A_{pp} + A_{pn}), \qquad A_\beta = \tfrac{1}{2}(A_{pp} - A_{pn}) \tag{7.45}$$

where A_{pp}, A_{pn} are the relevant coefficients for p-p and p-n scattering. Neglecting off-shell effects and using eqn (7.23) we obtain the following expression for the optical potential for nucleon-nucleus scattering in monentum space:

$$U_{\text{opt}}(q) = -\frac{2\pi\hbar^2}{\mu_0}\langle 0,\underset{\sim}{k}'|\sum_j M(j)|0,\underset{\sim}{k}\rangle \tag{7.46}$$

and in configuration space

$$U_{\text{opt}}(r) = (2\pi)^{-3}\int \exp(-i\underset{\sim}{q}.\underset{\sim}{r})\; U(q)\; d^3q \tag{7.47}$$

For proton-nucleus scattering, the spin-independent part of the potential is given by

$$U_{\text{D}}(r) = -\frac{\hbar^2}{4\pi^2\mu_0}\int \{Z\,A_{\text{pp}}(q)\;F_{\text{p}}(q) + N\,A_{\text{pn}}(q)\;F_{\text{n}}(q)\}\exp(-i\underset{\sim}{q}.\underset{\sim}{r})\;d^3q \tag{7.48}$$

where F_{p}, F_{n} are the form factors of the proton and neutron distributions, and, if $F_{\text{p}} = F_{\text{n}} = F_{\text{m}}$, this reduces to

$$U_{\text{D}}(r) = -\frac{\hbar^2 A}{4\pi^2\mu_0}\int \exp(i\underset{\sim}{q}.\underset{\sim}{r})\;\bar{A}(q)\;F_{\text{m}}(q)\;d^3q \tag{7.49}$$

where

$$\bar{A}(q) = \tfrac{1}{2}(A_{\text{pp}} + A_{\text{pn}}) + \varepsilon\tfrac{1}{2}(A_{\text{pn}} - A_{\text{pp}}) \tag{7.50a}$$

$$= A_{\alpha} - \varepsilon A_{\beta} \tag{7.50b}$$

with

$$\varepsilon = (N-Z)/A. \tag{7.51}$$

The same result is obtained for neutron-nucleus scattering but with the opposite sign between the two terms in eqns (7.50a) and (7.50b). If we assume that $\bar{A}(q)$ varies slowly with q compared with $F_{\text{m}}(q)$, which is reasonably true for heavy nuclei, we find

$$U_{\text{D}}(r) \approx -\frac{2\pi\hbar^2 A}{\mu_0}\bar{A}(0)\;\rho_{\text{m}}(r) \tag{7.52}$$

so that this approximation predicts a potential proportional to the density. In general, however, the range and shape of the potential is determined by the folding integral in eqn (7.49). It follows from the properties of this integral that the r.m.s. radii of the potential, the matter distribution, and the two-nucleon force are related by

$$\langle r^2 \rangle_{\mathrm{opt}} = \langle r^2 \rangle_{\mathrm{m}} + \langle r^2 \rangle_{\mathrm{d}}. \qquad (7.53)$$

The real and imaginary parts of $\bar{A}(q)$ are energy dependent and for a given energy are differently dependent on momentum transfer q. Thus, these formulae predict a local potential whose real and imaginary parts have different radial behaviour, and both the strength and range are energy dependent.

For a spherical spin-zero nucleus the terms in $M(j)$ linear in σ_j give no contribution, so that the spin-dependent term is of the form

$$U_{\mathrm{so}}(r) = -\frac{\hbar^2 A}{4\pi^2\mu_0}(\underset{\sim}{\sigma}_0\cdot\hat{\underset{\sim}{n}})\int \exp(-\mathrm{i}\underset{\sim}{q}\cdot\underset{\sim}{r})\ \bar{C}(q)\ F(q)\ \mathrm{d}^3 q \qquad (7.54)$$

where $C(q)$ is defined in a similar way to $\bar{A}(q)$ in eqn (7.50a) and this can be shown to be equivalent to a potential of the form

$$b\left[V_{\mathrm{so}}\,\frac{1}{r}\,\frac{\mathrm{d}h(r)}{\mathrm{d}r} + \mathrm{i}\,W_{\mathrm{so}}\,\frac{1}{r}\,\frac{\mathrm{d}g(r)}{\mathrm{d}r}\right]\underset{\sim}{\sigma}\cdot\underset{\sim}{\ell} \qquad (7.55)$$

with $b = (\hbar/m_\pi c)^2$ and

$$b\,V_{\mathrm{so}}\,k^2\sin\theta\,h(r) = \frac{\hbar^2 A}{4\pi^2\mu_0}\int \exp(-\mathrm{i}\underset{\sim}{q}\cdot\underset{\sim}{r})\ F_{\mathrm{m}}(q)\ \mathrm{Im}\,\bar{C}(q)\ \mathrm{d}^3 q \qquad (7.56a)$$

$$b\,W_{\mathrm{so}}\,k^2\sin\theta\,g(r) = -\frac{\hbar^2 A}{4\pi^2\mu_0}\int \exp(-\mathrm{i}\underset{\sim}{q}\cdot\underset{\sim}{r})\ F_{\mathrm{m}}(q)\ \mathrm{Re}\,\bar{C}(q)\ \mathrm{d}^3 q. \qquad (7.56b)$$

The real part of the spin-dependent potential has a much smaller r.m.s. radius than is obtained for the spin-independent terms (Kerman *et al.* 1959). This result is connected with the short range of the two-nucleon spin-orbit force (Bryan and Scott 1964).

It is clear from eqn (7.50) that we expect a symmetry

term in the potential. The nature of the p-p and n-p interactions are such that the symmetry term deepens the real and imaginary parts of the proton potential at medium energies and has the reverse effect on the neutron potential (Slanina and McManus 1968, Satchler 1969), but there is an indication that the effect on the real potential is reversed above 200 MeV (Kerman *et al.* 1959). Eqns (7.49) and (7.50) suggest that the symmetry term should have the same radial shape as the isospin-independent part of the spin-independent potential. This is called the volume form. However, these results were based on the assumption that $F_p = F_n$, i.e. that $\rho_p = \rho_n$. If we allow the distributions to be different, but for simplicity replace $M(q)$ by $M(0)$, we obtain for proton-nucleus scattering

$$U_D(r) = -\frac{2\pi\hbar^2}{\mu_0}\left[\{A_{pp}(0) + A_{pn}(0)\}\, A\, \rho_m(r) + \{A_{pn}(0) - A_{pp}(0)\}\{N\,\rho_n(r) - Z\,\rho_p(r)\}\right]. \quad (7.57)$$

Thus, for $\rho_n \neq \rho_p$ the symmetry term can take a variety of shapes including the possibility of surface peaking and is likely to show the effects of shell structure.

We have presented this discussion of the general properties of the nucleon-nucleus optical potential in terms of the free nucleon-nucleon parameters using impulse approximation. The fact that we do not expect this approximation to be valid at low and medium energies in no way invalidates the qualitative behaviour deduced, and the same conclusions can be reached using a potential model (Satchler 1969, Greenlees, Pyle, and Tang 1968*a*).

7.2.3. *The phenomenological approach*

In many analyses the optical potential is taken to be a local potential with a specified functional form and the parameters of this potential are then varied to yield a good fit to the data. The form most commonly used is the Saxon-Woods type given by

$$V(r) = -U(1+e^{x})^{-1} - i\left(W - 4\,W_D\,\frac{d}{dx'}\right)(1+e^{x'})^{-1}$$

$$+ \left(\frac{\hbar}{m_\pi c}\right)^2 (V_{so} + i\,W_{so}) \frac{1}{r}\frac{d}{dr}(1+e^{x'})^{-1}\, \underset{\sim}{\sigma}\cdot\underset{\sim}{\ell} \qquad (7.58)$$

where

$$x = (r-R_0)/a, \qquad x' = (r-R')/a', \qquad x_s = (r-R_s)/a_s \quad (7.59)$$

and the halfway radii are often expressed as $R_s = r_s\, A^{1/3}$, etc. The shape chosen for the imaginary spin-independent term allows for volume or surface absorption or some intermediate situation.

From the discussion of the previous section it may be expected that several of the parameters will depend on the incident energy and on the symmetry coefficient $\varepsilon = (N-A)/A$. For this reason, the depth of the real part of the potential is frequently written as

$$U = U_0 \pm \alpha(N-Z)/A + \beta\, Z/A^{1/3} - \gamma E. \qquad (7.60)$$

Using an effective mass approximation with $m^* \simeq 0.7\, m$, the coefficient β may be estimated as ~ 0.4 MeV (see §2.1) and the coefficient $\gamma = 1 - m^*/m$ becomes ~ 0.3.

For target nuclei with non-zero spin it is possible that the potential can contain terms involving the nuclear spin I. In first order these could be proportional to $\underset{\sim}{\sigma}\cdot\underset{\sim}{I}$ and $\underset{\sim}{\ell}\cdot\underset{\sim}{I}$. The existence of a spin-spin interaction $V_{ss}(r)\ \underset{\sim}{\sigma}\cdot\underset{\sim}{I}$ has been looked for by several methods, and the most recent examination of nucleon scattering data (Batty 1971) yields strengths for V_{ss} of less than 1 MeV, assuming a radial shape of Saxon-Woods form, except for ^{9}Be which gave a somewhat larger value. Recent theoretical work using a microscopic model (Satchler 1971*c*) has suggested that V_{ss} might be less than 100 keV. Studies of the $\underset{\sim}{\ell}\cdot\underset{\sim}{I}$ potential produce estimates of similar magnitude (Love 1972, 1974, Rawitscher 1972).

For projectiles with spin $s > \frac{1}{2}$ a derivation similar to that given in the preceding section indicates the presence of tensor components in the optical potential even when the target has spin zero. Similarly, the optical potential for projectiles with isospin $T > \frac{1}{2}$ can contain isotensor terms.

Independent analyses of data for one type of target nucleus at various energies or for many different target nuclei at one energy will generally produce parameters which show wide fluctuations with mass number and energy. In order to obtain meaningful information it is necessary to impose some constraints on the parameters and then determine optimum values for the remaining parameters. In 'fixed-geometry' analyses the radial parameters are fixed and the strengths are varied. With large and fast computers it is now possible to carry out a simultaneous analysis of several sets of data for various target nuclei and various energies; this procedure is often called a 'global analysis'. The parameters which give the best fit to the elastic scattering data are obtained by minimizing the quantity

$$\chi_\sigma^2 = \frac{1}{N} \sum_{i=1}^{N} \left\{ \frac{\sigma_{th}(\theta_i) - \sigma_{exp}(\theta_i)}{\Delta\sigma_{exp}(\theta_i)} \right\}^2 \tag{7.61}$$

where N is the number of data points, σ_{th} and σ_{exp} are the calculated and measured differential cross-sections, respectively, at angle θ_i, and $\Delta\sigma_{exp}$ is the uncertainty associated with the measured value. The uncertainties $\Delta\sigma_{exp}$ may be chosen to weight a particular angular region, or to be the experimental values, or to be a constant percentage of σ_{exp}. The latter choice is useful when comparing the quality of fit for data from different target nuclei. A similar quantity χ_p^2 can be constructed for polarization data. In some cases the angular acceptance of the detectors is such that some averaging of data occurs. This can be important in the vicinity of deep minima in the differential cross-section and for polarization data, and can be taken into account by averaging the calculated results over an angular range comparable with the experimental angular acceptance.

The general formalism developed in §7.2.1 indicates the possibility of resonances in the elastic scattering. The corresponding elastic scattering amplitude consists of a non-resonant part, due to U_{opt}, which is the amplitude for potential or shape-elastic scattering, and a resonant part which represents compound elastic scattering due to re-emission into the elastic channel after formation of the compound nucleus. Thus, the cross-section for elastic scattering can be written as

$$\sigma_S = \sigma_{SE} + \sigma_{CE}. \tag{7.62}$$

The total cross-section σ_C for removal of the incident particle from the elastic channel by any process is the sum of the cross-section for formation of the compound nucleus and subsequent re-emission through the elastic channel, i.e. σ_{CE}, and the reaction cross-section, so that

$$\sigma_C = \sigma_{CE} + \sigma_R \tag{7.63}$$

and

$$\sigma_T = \sigma_{SE} + \sigma_C = \sigma_S + \sigma_R. \tag{7.64}$$

In the resonance region, the reflection coefficients are rapidly varying functions of energy. It is then convenient to assume that the reflection coefficients and cross-sections can be averaged over an energy interval I such that $D \leqslant I \leqslant E$ where D is the mean spacing of the resonances. If the energy-averaged reflection coefficient is now associated with the reflection coefficient calculated with the optical potential, it can be shown (Feshbach *et al.* 1954) that the elastic cross-sections predicted by the optical potential U_{opt} should be connected with the shape-elastic scattering. At sufficiently high incident energies the density of the compound states is very great and the widths become very much larger than the spacing so that energy averaging automatically occurs. For this reason, the theory of the optical potential at higher energies makes no reference to energy averaging and the predictions of the optical model can be compared directly with the data. At lower energies some estimate of the compound-elastic scattering must be made or the data must be averaged.

7.2.4. *Scattering in the Hartree-Fock approximation*

In some recent work the Hartree-Fock field derived from bound-state calculations has been used to calculate the continuum wavefunctions and hence the phase shifts and elastic cross-section and polarization. This has been done with the Brink-Boeker force (Vautherin and Veneroni 1968) and with the Skyrme force (Dover and Van Giai 1971). Such a procedure should give a reasonable description of the smoothly varying part of the cross-section and of resonances in low-energy scattering which are single-particle in character.

It is possible to define a local equivalent potential which yields the same phase shifts as the non-local HF field. For Skyrme's interaction the non-locality is specified through a nucleon effective mass $m^*(r)$, which is state dependent, but because $m^*(r)/m$ tends to unity for large r the potential is local at large distances. The HF approximation neglects coupling to inelastic and reaction channels so that the equivalent local potential is real; it may nevertheless be usefully compared with the real part of the optical potential. Dover and Van Giai (1972) write the equivalent local potential in the form

$$V_{\mathrm{n}} = V_{\mathrm{av}} + V_{\mathrm{sym}} \tag{7.65a}$$

$$V_{\mathrm{p}} = V_{\mathrm{av}} - V_{\mathrm{sym}} + \Delta V_{\mathrm{c}} + V_{\mathrm{c}} \tag{7.65b}$$

where ΔV_{c} is a modification to the Coulomb potential due to non-locality. The average potential and the symmetry potential are given by

$$V_{\mathrm{av}} = V_{\mathrm{av}}^{\mathrm{o}} + V_{\mathrm{av}}^{\mathrm{so}}\, \boldsymbol{\ell}\cdot\boldsymbol{\sigma} + V_{\mathrm{av}}^{\mathrm{E}} \tag{7.66a}$$

$$V_{\mathrm{sym}} = V_{\mathrm{sym}}^{\mathrm{o}} + V_{\mathrm{sym}}^{\mathrm{so}}\, \boldsymbol{\ell}\cdot\boldsymbol{\sigma} + V_{\mathrm{sym}}^{\mathrm{E}} \tag{7.66b}$$

where V^{E} represents the energy-dependent part of the potential. They have studied a range of spherical nuclei from $^{16}\mathrm{O}$ to $^{208}\mathrm{Pb}$ and find $V_{\mathrm{av}}^{\mathrm{o}}$ is close to a Saxon-Woods shape, except for $^{16}\mathrm{O}$, with a value of - 43 MeV for the central depth. The diffuseness parameter is constant over a range of nuclei with a value of 0.52 ± 0.02 fm and the difference between the halfway radius R_o of the potential and the halfway radius R of the matter distribution is constant at $R_0 - R = 0.75 \pm 0.05$ fm. The energy-dependent part of the central potential also resembles the Saxon-Woods shape but with $a^E \approx 0.55$ fm and $R_0^E = R + 0.30$ fm. Thus, the energy-dependent term is not of the same shape as the energy-independent term. The energy dependence is linear in E and predicts that $V_{av}^o + V_{av}^E \approx 0$ at $E = 100$ MeV. As we shall see later, this is not in accord with phenomenological fits to the data; this discrepancy is probably due to the absence of coupling to the non-elastic channels which also affects the real part of the potential at higher energies.

Dover and Van Giai have also examined the behaviour of the symmetry term. Neglecting derivatives of $m^*(r)$, it can be written as

$$V_{sym}^o \approx \frac{m_n^*(r)\; U_n(r) \; - \; m_p^*(r)\; U_p(r)}{2m} \tag{7.67}$$

which can be reduced to the form

$$V_{sym}^o \approx \frac{b_0\{N\rho_n(r) \; - \; Z\rho_p(r)\}}{\{1 + C\,A\,\rho_m(r)\}^2} \tag{7.68}$$

where C is a constant. Thus V_{sym}^o should follow the behaviour of the neutron excess in the interior region but its behaviour is more complicated in the surface region. The energy-independent part of the symmetry potential varies from nucleus to nucleus, and could not readily be represented by a simple universal form factor. For ^{40}Ca, for which $N = Z$, V_{sym}^o does not vanish identically but is small and oscillates in sign. This result is due to the non-equivalence of the neutron and proton wavefunctions. The spin-orbit potential is found to resemble the usual Thomas form for spherical nuclei but is proportional to $(1/r)(d\rho/dr)$ rather than the derivative of the potential. The peak of this potential occurs at

$$R_{so} \approx R + 0.15 \text{ fm} \approx R_0 - 0.6 \text{ fm}.$$

This is in accord with the result noted in the previous section from the impulse approximation, and so too is the presence of a spin-orbit contribution to the symmetry term. The non-locality contribution to the Coulomb potential has the form

$$\Delta V_{\mathrm{C}} \approx -\,(0.8\ Z/A^{1/3})\ g(r) \tag{7.69}$$

where the radial behaviour is the same as for $V^{\mathrm{E}}_{\mathrm{av}}$.

It is clear that the local equivalent potential derived from the HF approximation may not readily be expressed in terms of $(N-Z)/A$. However, for heavy nuclei the central depth for proton scattering may be roughly approximated by

$$U_0 + 0.8\ Z/A^{1/3} + 14(N-Z)/A - \gamma E. \tag{7.70}$$

It may be noted that, because the average value of m^*/m is ~ 0.57 in the HF calculation with Skyrme's potential, the coefficient of the Coulomb correction is larger than that commonly used in the phenomenological potential, i.e. as given in eqn (7.60), but it is consistent with the value obtained in bound-state calculations as can be seen by comparison with eqn (2.10). The symmetry term is also different, but the sum of the Coulomb and symmetry terms in eqn (7.70) is comparable with the sum in eqn (7.60) although the two terms are separately different.

7.3. NUCLEON-NUCLEUS SCATTERING

We now examine nucleon-nucleus scattering and its interpretation, with particular reference to the study of nuclear sizes. It is convenient to classify the data according to incident energy. Such a classification is inevitably somewhat arbitrary, but we use here the following division according to the incident kinetic energy E_{lab}:

low energy	$E_{\mathrm{lab}} < 19$ MeV	$\lambda > 6.5$ fm
medium energy	$19 < E_{\mathrm{lab}} < 100$ MeV	$3 < \lambda < 6.5$ fm

intermediate energy	$100 < E_{\text{lab}} < 800$ MeV	$0.9 < \lambda < 3$ fm
high energy	$0.8 < E_{\text{lab}} < 70$ GeV	$\lambda < 0.9$ fm.

The wavelength λ corresponding to a free nucleon of kinetic energy E_{lab} is given to indicate the ability of the incident nucleon to probe the structure of the target nucleus. It must be remembered, however, that the wavelength inside the nucleus is changed by the refractive effect of the nuclear potential.

7.3.1. *Low-energy region*

At very low energies, up to a few keV, only s-wave neutrons interact with the nucleus. Below the inelastic threshold the compound nucleus can decay only through re-emission of the incident particle or by γ-emission, but as the energy is increased up to a few MeV the inelastic channels increase in importance. The low-energy data are usually presented in terms of the scattering length or hard-sphere radius R_{a} and the s-wave strength function. The former is the radius of a hard sphere which would give the same shape elastic scattering in the low-energy limit, i.e.

$$\sigma_{\text{SE}} = 4\pi\, R_{\text{a}}^2, \qquad E \to 0 \tag{7.71}$$

$$= \frac{\pi}{k^2}\, |1 - \langle \eta_0 \rangle|^2, \qquad k \to 0 \tag{7.72}$$

where $\langle\ \rangle$ denotes an energy-averaged quantity. Using a Breit-Wigner form for the resonance behaviour it can be shown that

$$\sigma_{\text{c}} = \frac{\pi}{k^2}\, \frac{2\pi\, \bar{\Gamma}_{\text{n}}}{D} = \frac{\pi}{k^2}\, (1 - |\langle \eta_0 \rangle|^2) \tag{7.73}$$

where $\bar{\Gamma}_{\text{n}}$ is the average width for a given spin and parity and D is the mean spacing of all levels of the same spin and parity. In order to compare data at different energies it is usual to define an energy-independent quantity through the relation

$$\frac{\bar{\Gamma}_{\text{n}}^{(0)}}{D} = \left(\frac{E_0}{E}\right)^{1/2} \frac{\bar{\Gamma}_{\text{n}}}{D} \tag{7.74}$$

where E_0 is an arbitrary energy, usually taken to be 1 eV. The quantity $\bar{\Gamma}_n^{(0)}/D$ is known as the s-wave strength function and is determined by averaging over the observed resonances or obtained directly from a poor resolution experiment. A p-wave strength function can be defined in a similar way (Saplakoglu, Bollinger, and Coté 1958). It follows from the definitions in eqns (7.71)-(7.74) that the averaged total cross-section is given by

$$\langle \sigma_T \rangle = a + b\, E^{-1/2} \tag{7.75}$$

where

$$a = 4\pi R_a^2 \tag{7.76}$$

$$b = 2\pi^2\, E_0^{-1/2}\, \frac{E}{k^2}\, \frac{\bar{\Gamma}_n{}^{(0)}}{D}\,. \tag{7.77}$$

The two quantities R_a and $\bar{\Gamma}_n^{(0)}/D$ are not sufficient to determine a multi-parameter optical potential of the kind defined in eqn (7.59). Nevertheless, the implication that a quantity such as the s-wave strength function can be described by an optical model whose parameters vary smoothly with mass number is a strong test of the basic concepts of the optical model and energy averaging. In addition, a simple analysis predicts that a giant resonance or size resonance will occur whenever an integral number of half-wavelengths fit into the real potential well. For s-wave resonances in a square well of depth V_0 and radius R_0 this occurs when $K\, R_0 = (n+\frac{1}{2})\pi$, where $K^2 \simeq 2\mu\, V_0/\hbar^2$ for small E, and it occurs for p-waves when $K\, R_0 \simeq n\pi$. If it is assumed that V_0 is independent of mass number while $R_0 \propto A^{1/3}$, the position of the resonances can be located as a function of A. An example of this is shown in Fig. 7.1. The resonances are sharpest for a square well and become broader as the potential surface becomes more diffuse (Vogt 1968). The widths of the resonances also increase as the strength of the imaginary part of the potential increases until the width eventually becomes comparable with the spacing, but for weak absorption the average cross-

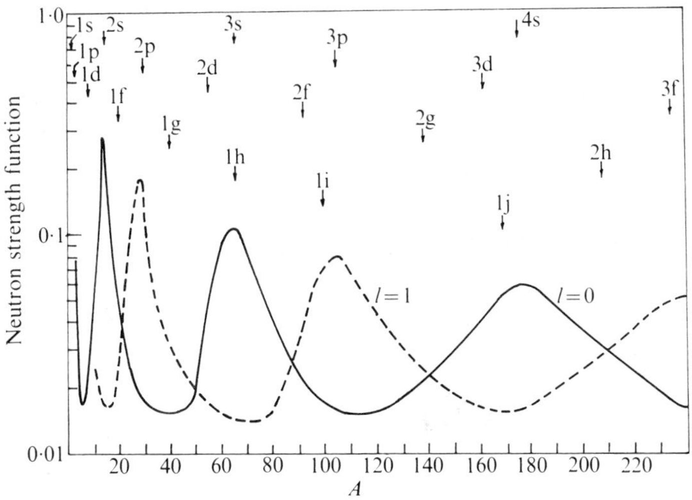

FIG. 7.1. The neutron strength function at zero energy as a function of the mass number A for a square well of depth 51 MeV. The s-wave and p-wave strength functions are shown as solid and broken lines, respectively, while the positions of the giant resonance in the other partial waves are indicated by arrows. (From Vogt 1968.)

sections do display the phenomenon of size resonance.

The experimental data for s-wave strength functions and scattering lengths are shown in Figs. 7.2 and 7.3, and both indicate the presents of the 3s and 4s size resonances. Early optical model analyses using a spherically symmetric potential and volume absorption fitted the data moderately well, but did not reproduce the splitting of the 4s resonance and the very low values of $\bar{\Gamma}_n^{(0)}/D$ for $A \sim 100$-120. The first of these discrepancies can be integrated in terms of the known deformation of nuclei in this region of mass number which leads to coupling between the 4s resonance and the nearby 3d resonance (Margolis and Troubetskoy 1957). Detailed calculations using the rotational model have yielded satisfactory agreement with the data in this region (Margolis and Troubetskoy 1957, Chase, Wilets, and Edmonds 1958, Buck and Perey 1962, Jain 1964), and calculations using a vibrational model have explained fine structure in the 3s and 3p resonances (Buck and Perey 1962, Jain 1964).

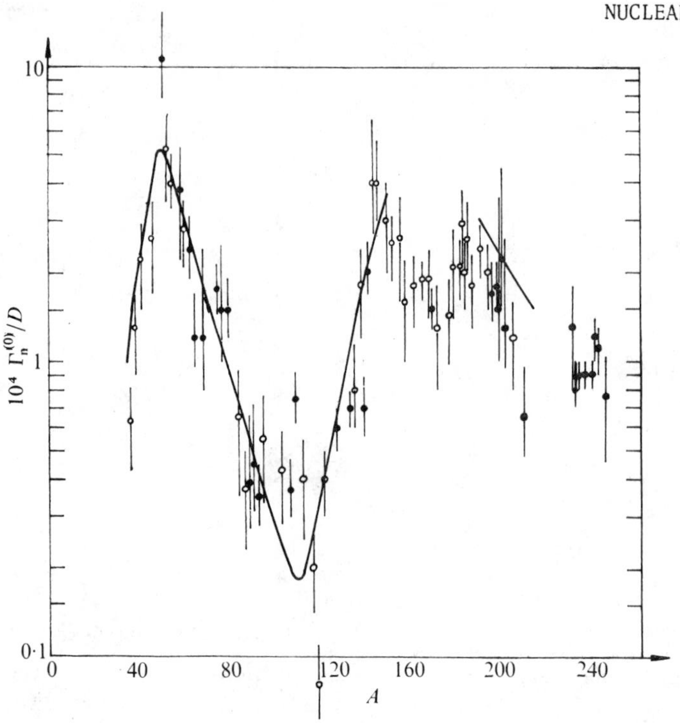

FIG. 7.2. s-wave strength functions. The solid line is calculated from Moldauer's potential. (From Morgenstern *et al.* 1969.)

The presence of the deep minimum in the s-wave strength functions yields more basic information about the optical potential. The connection between the behaviour of the strength function and the parameters of the optical potential has been examined by Vogt (1962, 1968) and Moldauer (1962, 1963). This can be done by using the relation between the imaginary part of the optical potential and the strength function noted by Porter (1955). For a beam with a flux of $v = \hbar k/\mu$ particles per second, the cross-section for removal from the beam can be written as

$$\sigma_C = -\frac{1}{v}\int \operatorname{div} \underset{\sim}{j}\, d^3r$$

Using the relation $\operatorname{div} \underset{\sim}{j} = -(2/\hbar)\, W\, |\psi|^2$, this becomes

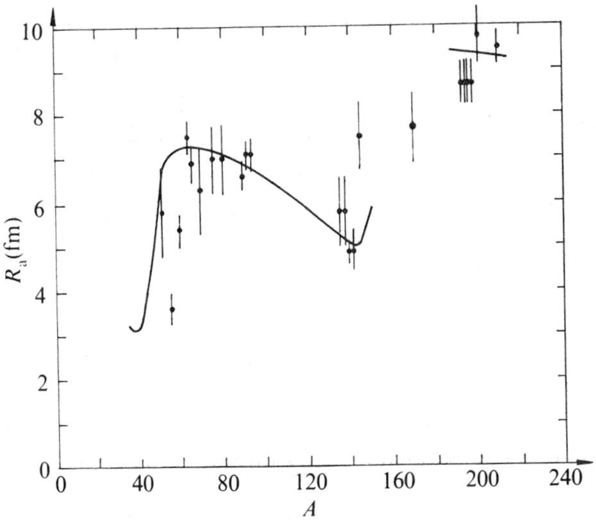

FIG. 7.3. The hard-sphere radius R_a deduced using Moldauer's potential. (From Morgenstern *et al.* 1969.)

$$\sigma_C = \frac{2\mu}{\hbar^2 k} \int W(r) \; |\psi(r)|^2 \; d^3r$$

and finally, using eqns (7.73) and (7.74), the s-wave strength function becomes

$$\frac{\bar{\Gamma}_n^{(0)}}{D} = \left(\frac{E_0}{E}\right)^{\frac{1}{2}} \frac{k\mu}{\pi^2\hbar^2} \int W(r) \; |\psi(r)|^2 \; d^3r \tag{7.78}$$

For volume absorption the magnitude of the strength function depends essentially on the magnitude of W and of $|\psi|^2$ inside the nucleus, and for weak absorption the latter is determined essentially by the size of the nucleus. Thus, if ψ has an antinode at the surface of the nucleus the interior amplitude is comparable with the exterior amplitude and $\bar{\Gamma}_n^{(0)}/D$ is large, while if ψ has a node at the surface the interior amplitude is small compared with the exterior amplitude and $\bar{\Gamma}_n^{(0)}/D$ has a minimum, but a deep minimum can be obtained only if W is very small and this disturbs the fit to the strength function maxima and to other data. If, on the other hand, the absorption is concentrated in a small region in the nuclear surface the strength function minimum can be made very deep even for a relatively large peak value for W (Moldauer 1963).

Moldauer (1963) has fitted the s-wave strength functions, total cross-sections from 0.2 MeV to 1.0 MeV, and elastic scattering cross-sections for even-mass nuclei with $40 < A < 150$ using an optical potential whose real central and spin-orbit terms were of Saxon-Woods shape with

$$R_0 = r_0\, A^{1/3} + r_1. \tag{7.79}$$

The surface absorption was described by a gaussian function

$$g(r) = \exp\left\{\frac{-(r-R_0-c)^2}{b^2}\right\}. \tag{7.80}$$

The optimum set of parameters was found to be

$$\left.\begin{array}{lll} U = 46 \text{ MeV} & W = 14 \text{ MeV}, & V_{so} = 7 \text{ MeV} \\ r_0 = 1.16 \text{ fm}, & r_1 = 0.6 \text{ fm}, & c = 0.5 \text{ fm} \\ a = 0.62 \text{ fm}, & b = 0.5 \text{ fm} & \end{array}\right\} \tag{7.81}$$

and fits to the data are shown in Figs. 7.2, 7.3, and 7.4. It should be noted that Figs. 7.2 and 7.3 include data (Seth 1966, Morgenstern *et al.* 1969) more recent than those used in Moldauer's analysis, but the agreement remains good. Fig. 7.5 compares results for volume and surface absorption and also illustrates the effect of the parameter c which shifts the peak of the surface absorption term out into the fringe of the nucleus, beyond the halfway radius of the real potential.

The surface peaking of the absorptive part of the optical potential can be justified on theoretical grounds, since at very low energies the exclusion principle inhibits interactions in the nuclear interior. Several calculations have reproduced the observed behaviour of the imaginary part (Gomes 1959, Lemmer, Maris, and Tang 1959, Shaw 1959), including the result that the imaginary potential peaks outside the halfway radius of the potential.

Further support for the surface-absorption model has

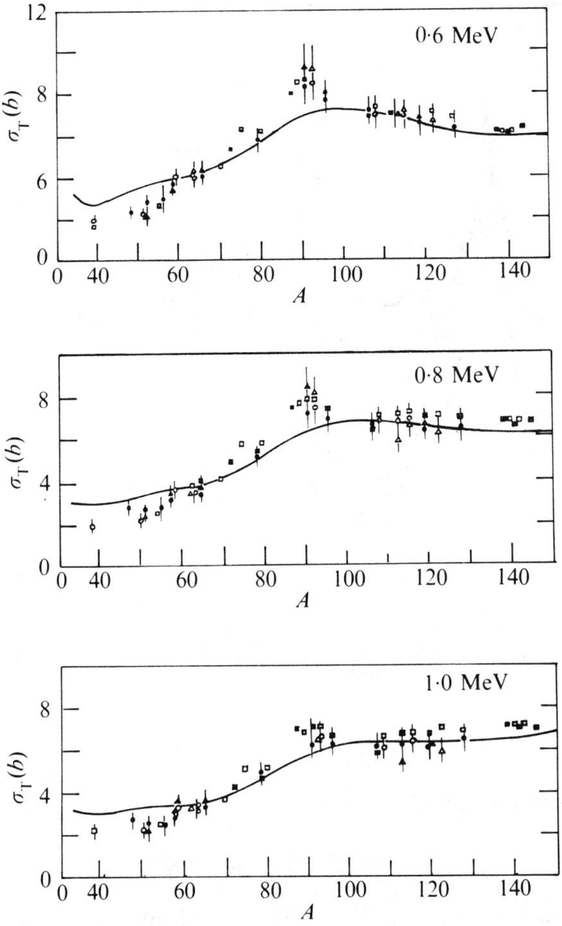

FIG. 7.4. Neutron total cross-sections at 0.6, 0.8, and 1.0 MeV (see Moldauer (1963) for details of the experimental data). (From Moldauer 1963.)

arisen in several analyses of differential cross-section and polarization data in the energy region 1-24 MeV using a surface derivative form as defined in eqn (7.59). The results are listed and discussed in detail by Green, Sawada, and Saxon (1968) and Satchler (1969). The data can be fitted with surface absorption only, and a constrained form for the spin-orbit term, i.e. $R_0 = R_s$ and $a = a_s$, but some difference between a and a' has been noted. If the halfway radius is

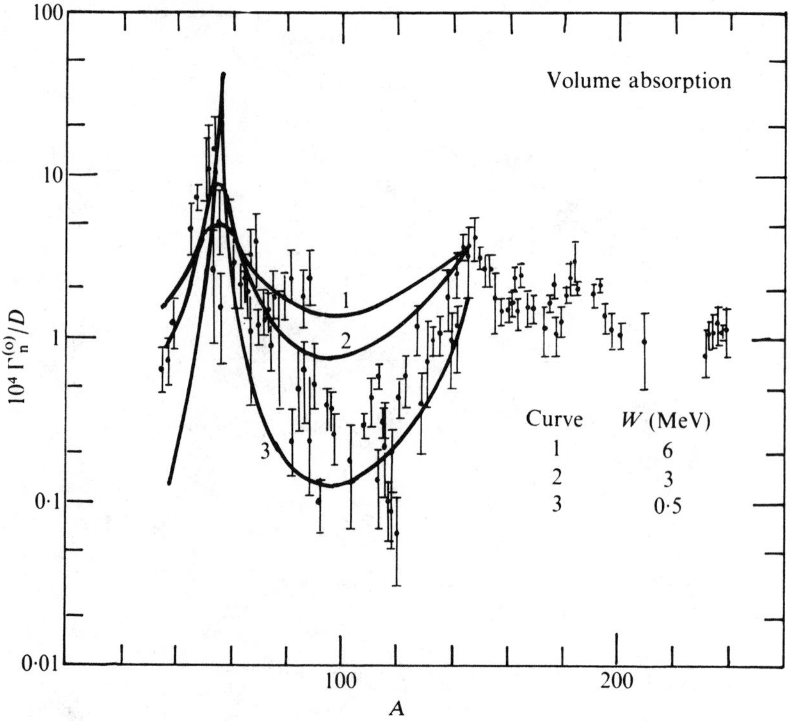

FIG. 7.5(a). Neutron strength functions. (See page 218 for details.)

written as $R_0 = r_0\ A^{1/3}$ a value of 1.25-1.27 fm is obtained for r_0. The radial parameters obtained by Perey (1963) by fitting data between 9 MeV and 22 MeV for a wide range of targets were

$$R_0 = R' = R_s = 1.25\ A^{1/3}\ \text{fm},$$

$$a = a_s = 0.65\ \text{fm}, \qquad a' = 0.47\ \text{fm}. \tag{7.82}$$

Satchler (1969) has examined the validity of the formula (7.61) for the depth of the real potential and the magnitudes of the coefficients α and β. For proton scattering it appears that if β is taken to be 0.4 MeV then α = 25 ± 10 MeV. For neutron scattering, the study by Perey and Buck (1962) using a non-local potential indicated no variation with $(N-Z)/A$ at

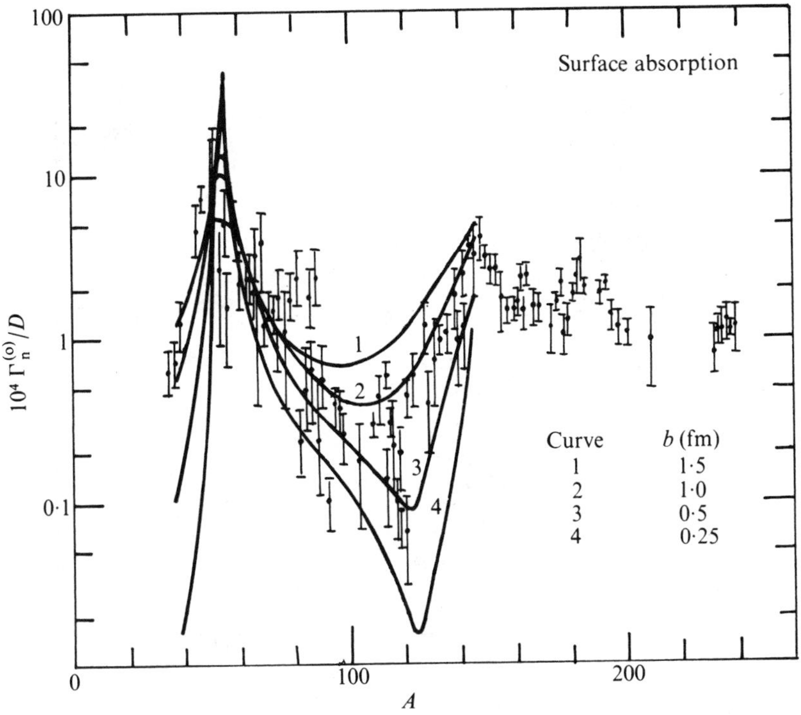

FIG. 7.5(b). Neutron strength functions. (See page 218 for details.)

all; other studies have indicated a range of values and Satchler quotes $\alpha = 15 \pm 15$ MeV as an average value.

One of the remarkable recent developments in the use of the optical model at low and medium energies has been its application to nucleon scattering from very light nuclei. Because the optical model, like the shell model, is based on the concept of an averaged interaction between the incident nucleon and the nucleons in the nucleus, it is usually anticipated that these methods will break down for very light systems. The work of Devries, Perrenoud, and Slaus (1972) has shown that it is possible to fit the elastic cross-section for p+d in the energy range 17-46 MeV using a conventional optical potential with surface absorption plus a nucleon-exchange term, but that $n+{}^3H$ scattering at 18 MeV can be fitted out to large angles using the optical potential alone.

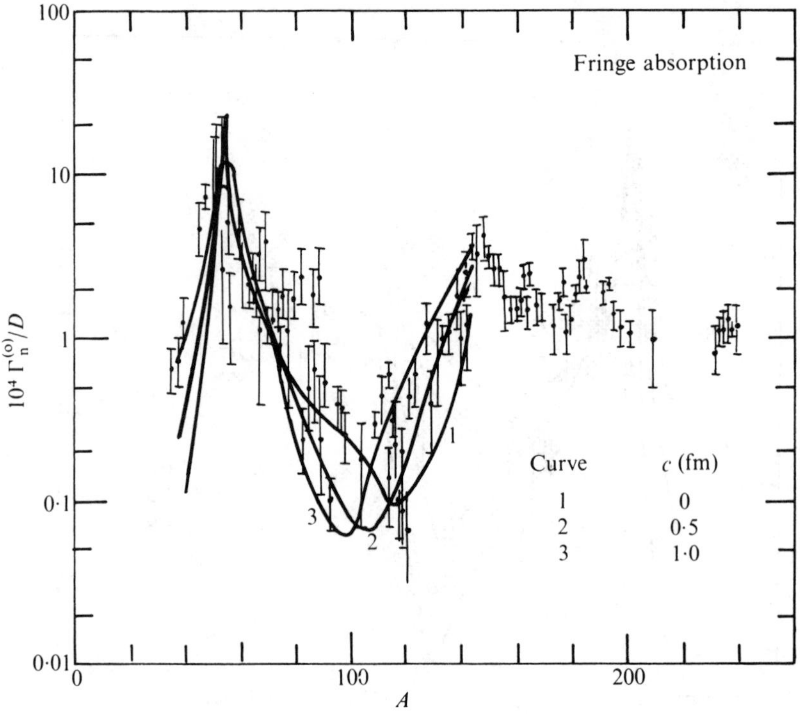

FIG. 7.5(c). Neutron strength functions for spherical nuclei calculated from Moldauer's potential with U = 46 MeV, r_0 = 1.3 fm, r_1 = 0, and a = 0.58 fm. (a) Volume absorption, W varied as shown. (b) Surface absorption with W = 7 MeV, c = 0, and b varied as shown. (c) Fringe absorption with W = 7 MeV, b = 0.5 fm, and c varied as shown. (From Moldauer 1963.)

Further analysis of n+^{3}H cross-sections and polarization over the range 6-23 MeV (Sherif and Podmore 1972) confirms this conclusion but indicates that volume absorption is preferred. Proton and neutron scattering from ^{4}He has been studied over the range 2-22 MeV (Satchler *et al.* 1968, Mailandt, Lilley, and Greenlees 1972) and the study of proton scattering has been extended up to 55 MeV (Thompson, Epstein, and Sawada 1970). Some of the parameters obtained are listed in Table 7.1. It is difficult to deduce anything about the size of these light targets from such analyses, but these results are of considerable importance in the microscopic analysis of scattering of these light complex nuclei from heavier nuclei, as explained in §7.4.4.

7.3.2. *Medium-energy region*

Extensive analyses of proton scattering data at 29-31 MeV, 30, 40, and 50 MeV have been carried out in terms of phenomenological potentials. Fricke and Satchler (1965) found a set of average parameters which gave good agreement with the cross-section data for a range of nuclei at 40 MeV. The radius parameters are

$$\left.\begin{aligned} &R_0 = R_s = 1.18\,A^{1/3}\ \text{fm}, \qquad a = a' = a_s = 0.7\ \text{fm} \\ &R' = 1.40\,A^{1/3}\ \text{(volume absorption)} \\ \textit{or}\quad &R' = 1.04\,A^{1/3}\ \text{(surface absorption)} \end{aligned}\right\} \quad (7.83)$$

which are significantly different from those obtained in the low-energy region. They obtained equally good fits with volume or surface absorption or a mixture of both provided that the potentials were similar in the surface region, for example beyond 5 fm in ^{60}Ni, and suggested that elastic scattering at this energy is rather insensitive to the amount of absorption in the interior region. However, subsequent analyses have shown that this conclusion is connected with the limited angular range of the earlier data, and that a mixture of volume and surface absorption is needed in the medium-energy range (Satchler 1967*c*, Fricke *et al.* 1967, Calderbank *et al.* 1967).

In connection with an analysis of 30 MeV data, Satchler (1967*c*) has compared the results obtained when cross-section and polarization data are analysed separately or simultaneously. It appears that cross-section data alone can be fitted with $R_0 = R_s$ and $a = a_s$, but inclusion of polarization data requires that these parameters should take different values, particularly in order to fit the data at forward angles. Minimizing χ^2_σ and χ^2_p separately leads to contradictory results, but simultaneous minimization indicates that $R_s < R_0$

TABLE

Optical-model parameters for nucleon

	E_{lab} (MeV)	R_0 (fm)	R' (fm)
p + ^{2}H Devries *et al.* 1972	17-46	$1.0A^{1/3}$	$1.9A^{1/3}$
n + ^{3}H Devries *et al.* 1972	18	$1.49A^{1/3}$	$1.74A^{1/3}$
n + ^{3}H Sherif and Podmore 1972	6-19.5 21-23	$1.488A^{1/3}$ $1.48A^{1/3}$	$1.501A^{1/3}$ $1.501A^{1/3}$
p + ^{4}He Mailandt *et al.* 1972	<12	$(1.48-0.009E_{\text{lab}})$ $A^{1/3}$	-
p + ^{4}He Satchler *et al.* 1968	2-22	$(1.50-0.01E_{\text{lab}})$ $\times (M_\alpha/M_p)^{1/3}$	
p + ^{4}He Thompson *et al.* 1970	31 40 55	$1.1A^{1/3}$ $1.1A^{1/3}$ $1.1A^{1/3}$	$2.56A^{1/3}$ $2.39A^{1/3}$ $2.30A^{1/3}$
n + ^{4}He Mailandt *et al.* 1972	<10	$(1.48-0.009E_{\text{lab}})$ $\times A^{1/3}$	-
n + ^{4}He Satchler *et al.* 1968	2-22	$(1.50-0.01E_{\text{lab}})$ $\times (M_\alpha/M)^{1/3}$	-

7.1

scattering from light nuclei

R_{so} (fm)	a (fm)	a' (fm)	a_s (fm)	U (MeV)	V_{so} (MeV)
—	0.402	0.961	—	$76.9-0.55E_{lab}$	—
—	0.144	0.378	—	37.4	—
$1.049A^{1/3}$	0.144	0.378	0.289	$53.5-0.60E_{lab}$	—
$1.049A^{1/3}$	0.144	0.378	0.289	$52.8-0.50E_{lab}$	
$1.117A^{1/3}$	0.34	—	0.34	42.5	$3.86+0.032E_{lab}$
$1.0(M_\alpha/M_p)^{1/3}$	0.25	—	0.25	43.0	$2.7+0.1E_{lab}$
$1.1A^{1/3}$	0.35	0.10	0.35	56.2	4.26
$1.1A^{1/3}$	0.35	0.10	0.35	51.0	2.75
$1.1A^{1/3}$	0.35	0.10	0.35	42.6	2.63
$1.117A^{1/3}$	0.34	—	0.34	42.5	$2.50+0.26E_{lab}$
$1.0(M_\alpha/M_p)^{1/3}$	0.25	—	0.25	41.8	$3.0+0.1E_{lab}$

and $a_s < a$ are required. Satchler also showed that with a constrained spin-orbit term, i.e. $R_s = R$ and $a_s = a$, an average geometry potential can be obtained with the following parameters:

$$\left.\begin{aligned} \text{potential 1:}\quad & R_0 = 1.12\,A^{1/3}\text{ fm}\quad R' = 1.33\,A^{1/3}\text{ fm}\quad a = 0.75\text{ fm}\\ & a' = 0.58\text{ fm (Fe, Ni, Co, Cu)}\\ & = 0.65\text{ fm (Sn)}\\ & = 0.75\text{ fm (Pb)}\\ & U = 47.5 + 0.4\,Z/A^{1/3} + 30(N-Z)/A\text{ MeV}\\ & W = 3\text{ MeV}\\ & W_D = 4.5 + 16(N-Z)/A\text{ MeV.} \end{aligned}\right\} \qquad (7.84)$$

This potential was obtained by fixing the radius parameters and varying U, W_D, and V_{so}. A second potential was obtained by fixing V_{so} and U and varying $r_0 = R_0\,A^{-1/3}$ and W_D. This gives

$$\left.\begin{aligned} \text{potential 2:}\quad & U = 51 + 0.4\,Z/A^{1/3}\text{ MeV}\\ & W_D = 4.25 + 16(N-Z)/A\text{ MeV}\\ & r_0 = 1.09 + 0.25(N-Z)/A\text{ fm.} \end{aligned}\right\} \qquad (7.85)$$

It is not surprising that optimum parameters can be obtained by varying either U or r_0, since it is well known that there is a potential ambiguity such that small changes in U or r_0 keeping the product $U\,r_0^n$ constant will not change the cross-section. The values of U and r_0 given above imply $n \sim 2.6$. Because of this ambiguity and the dependence of U, W_D, and r_0 on $(N-Z)/A$, very great care must be taken in comparing values of R_0 for different nuclei and in attempting any deductions about nuclear sizes from phenomenological analyses at these energies.

A further analysis at 40 MeV (Fricke *et al.* 1967) has given the following average geometry parameters:

$$\left.\begin{aligned} &R_0 = 1.16\ A^{1/3}\ \text{fm},\ R' = 1.37\ A^{1/3}\ \text{fm},\ R_s = 1.064\ A^{1/3}\ \text{fm} \\ &a = 0.75\ \text{fm},\ a' = 0.63\ \text{fm},\ a_s = 0.738\ \text{fm}. \end{aligned}\right\} \quad (7.86)$$

This geometry also gives good fits to the data at 50 MeV (Burge *et al.* 1967, Calderbank *et al.* 1967) and at 61 MeV (Fulmer *et al.* 1967). The real strength has the form given in eqn (7.60) with α = 26.4 MeV at 40 MeV, α = 38 ± 8 MeV at 50 MeV, and α = 27.2 MeV at 61 MeV, and γ = 0.22 ± 0.03.

A global analysis has been carried out by Becchetti (1968, Becchetti and Greenlees 1969) for proton scattering in the energy range 10-40 MeV and for a wide range of target nuclei. The parameters obtained are

$$\left.\begin{aligned} &R_0 = 1.17\ A^{1/3}\ \text{fm},\ R' = 1.32\ A^{1/3}\ \text{fm},\ R_s = 1.01\ A^{1/3}\ \text{fm} \\ &a = 0.75\ \text{fm},\ a' = 0.51 + 0.7(N-Z)/A,\ a_s = 0.75\ \text{fm} \\ &V_{so} = 6.2\ \text{MeV},\ W = 0.22E - 2.7\ \text{MeV} \\ &U = 54.0 + 24.0(N-Z)/A + 0.4Z/A^{1/3} - 0.32E\ \text{MeV} \\ &W_D = 11.8 + 12(N-Z)/A - 0.25E\ \text{MeV}. \end{aligned}\right\} \quad (7.87)$$

These parameters are rather similar to the set (7.80) obtained at 40 MeV except for the variation of the imaginary diffuseness a' which improved the fits to the total absorption cross-sections σ_R. The increase in a' for heavy nuclei is consistent with the parameters (7.84) and with other studies of scattering from heavy nuclei (Perey 1963, 1964), and also with an analysis of σ_R by the Oak Ridge group (Menet *et al.* 1971) which gave the parameters

$$\left.\begin{aligned} a' &= 0.75 + 1.0(N-Z)/A - 0.008E \\ W_D &= 4.2 + 15.5(N-Z)/A - 0.05E \\ W &= 1.2 + 0.09E \end{aligned}\right\} \quad (7.88)$$

Thus, there is good evidence that both the strength and the width of the surface absorption term increase in heavier nuclei, although the parametrization in terms of the symmetry parameter $(N-Z)/A$ is largely a matter of convenience. It also appears that the reaction cross-section σ_R is particularly sensitive to the imaginary diffuseness a' (Perey 1966, Menet *et al.* 1971). The situation for neutron scattering has been examined by Hodgson (1971) and Satchler (1969), but as yet it is difficult to reach precise conclusions.

A number of studies of proton scattering from isotopes have been carried out. In those cases where the data have been analysed using a phenomenological model the principal interest has been in the symmetry dependence of the parameters. This procedure contains certain fundamental theoretical difficulties since the optical model is not intended to take into account specific differences between nuclei which might arise, for example, from coupling of the ground state to collective excitations of various types and strengths, or from changes in pairing effects or deformations as a neutron shell is filled. Thomas and Burge (1969) have proposed that some of the experimental difficulties can be removed by minimizing fits to the difference functions

$$D_\sigma(\theta) = \frac{\sigma_1(\theta) - \sigma_2(\theta)}{\sigma_1(\theta) + \sigma_2(\theta)} \quad (7.89)$$

$$D_P(\theta) = P_1(\theta) - P_2(\theta). \quad (7.90)$$

They have shown that this method can be used to eliminate normalization errors and reject data points of low relative accuracy, and obtain some information about trends in the parameter shifts as a function of energy but conclude that the existing data are not sufficiently accurate or complete to permit firm conclusions about the behaviour of symmetry-dependent terms.

A phenomenological analysis of proton scattering from ^{40}Ca and ^{48}Ca has been carried out by Maggiore *et al.* (1970) using $R_0 = R'$ and $a = a'$; the central strengths were either related by

$$U(^{48}\text{Ca}) = U(^{40}\text{Ca}) + 26.4(N-Z)/A$$

or taken to be equal. Good fits to the ratios of the cross-sections were obtained at incident energies of 25, 30, 35, and 40 MeV, but the sign and magnitude of the shifts in R_0 and a depended on the incident energy and on the presence or otherwise of the symmetry term in the real potential. The average result was a difference in the r.m.s. radii of 0.15 fm.

In order to obtain more fundamental information about nuclear sizes from medium-energy nuclear scattering we need to know which region of the potential or, more precisely, which moments are well determined in the analysis of the data. Also, we need to examine whether the potentials can be reliably connected with the nuclear matter distributions. An approach to these problems has been developed by Greenlees and collaborators. In the initial work by Greenlees *et al.* (1968*a*) the phenomenological form for the central potential for proton scattering from nuclei with spin zero was replaced by the potential

$$U_{\text{RS}} = U_{\text{R}} + U_{\text{S}} \tag{7.91}$$

where

$$U_{\text{R}}(r) = \int A\, \rho_{\text{m}}(\underset{\sim}{r}')\, u_{\text{d}}(|\underset{\sim}{r}-\underset{\sim}{r}'|)\, \text{d}^3r' \tag{7.92}$$

$$U_{\text{S}}(r) = \int \{Z\, \rho_{\text{p}}(\underset{\sim}{r}') - N\, \rho_{\text{n}}(\underset{\sim}{r}')\}\, u_{\tau}(|\underset{\sim}{r}-\underset{\sim}{r}'|)\, \text{d}^3r' \tag{7.93}$$

and u_{d}, u_{τ} are terms in an effective nucleon-nucleon potential which can be written as

$$u(S) = u_d(S) + u_\tau(S)\ \underset{\sim}{\tau}.\tau_i + \text{spin-dependent and tensor terms,} \tag{7.94}$$

In the special case, when $\rho_p \equiv \rho_n$ and $u_\tau = -\zeta\, u_d$, the potential (7.91) reduces to

$$U_{RS}(\underset{\sim}{r}) = \left\{1 + \frac{\zeta(N-Z)}{A}\right\} U_R(\underset{\sim}{r}). \tag{7.95}$$

The relation between the r.m.s. radii and volume integrals J are given by

$$\langle r^2\rangle_{RS} = \langle r^2\rangle_m + \langle r^2\rangle_d \tag{7.96}$$

$$J_{RS} = A\, J_d \left\{1 + \frac{\zeta(N-Z)}{A}\right\}. \tag{7.97}$$

In the original analysis the real central term of the phenomenological potential (7.59) was replaced by that obtained from eqns (7.92) and (7.95) with

$$\rho_m(r) \propto \left\{1 + \exp\left(\frac{r-R_m}{a_m}\right)\right\}^{-1} \tag{7.98}$$

$$u_d(r) \propto e^{-\mu r}/\mu r. \tag{7.99}$$

The overall strength of V_{RS} of $U_{RS}(r)$ is treated as a parameter in the fitting procedure. The imaginary part of the potential was treated phenomenologically as in eqn (7.59) and the spin-orbit term was treated similarly, except that the radial parameters were taken to be R_m and a_m. This gives a total of eight parameters for the model, V_{RS}, W, W_D, V_{so}, R_m, a_m, R', and a', assuming that a value for μ (the range of the nucleon-nucleon interaction) is known or can be precisely determined.

For the Yukawa function (7.99) the mean-square radius is $\langle r^2\rangle_d = 6/\mu^2$. Studies of the variation of χ^2 against $\langle r^2\rangle_d$ for best fits to 30.3 MeV proton scattering showed that a rather wide range of $\langle r^2\rangle_d$ is acceptable although more accurate polarization data would improve the situation. However, it is found that $\langle r^2\rangle_{RS}$ is quite well determined, as can be seen from Table 7.2 and Fig. 7.6. This result has been confirmed by studies of neutron scattering (Pyle and Greenlees 1969) and further studies of proton scattering in which the restraint $\rho_p \equiv \rho_n$ was relaxed, the value of ζ was varied, and the spin-orbit potential was also derived by a folding procedure (Greenlees, Makofske, and Pyle 1970); the variation of best-fit χ^2 values with $\langle r^2 \rangle_d$ is shown in Fig. 7.7. These studies indicate that the volume integral J_{RS} is also well determined, and this is supported by many other studies of proton scattering in the range 16-50 MeV (Boyd and Greenlees 1968, Woollam *et al.* 1970, Boyd *et al.* 1971, Hnizdo *et al.* 1971, Lombardi *et al.* 1972) which have used both the conventional optical model and the folding procedure.

Thus, it appears that analyses of medium-energy nucleon scattering, at least up to 50 MeV, yield essentially the volume integral and $\langle r^2 \rangle$ moment of the *potential* for a given target nucleus at a given energy. This important observation has often appeared as a subsidiary result in attempts to derive information on the nuclear matter distribution, and has not received sufficient systematic attention. It would be illuminating to know, for example, what is the energy dependence of these quantities, whether sensitivity to $\langle r^4 \rangle$ occurs at higher energies, and whether the constancy of J_{RS} and $\langle r^2 \rangle_{RS}$ really persists when a much wider variety of potential shapes is used. At present there is some indication that J_{RS}/A is constant at a given energy, as shown by the results given in Table 7.3, and this has rather disconcerting implications about the nature and existence of the symmetry potential (Greenlees *et al.* 1968*a*, Satchler 1969).

The folding integral (7.92) can be fairly regarded as an alternative, and relatively meaningful, way of parametrizing the optical potential, but attempts to derive $\langle r^2 \rangle_m$ or the difference between proton and neutron distributions depend on the validity of this first-order approximation to the optical potential and confidence to the choice of u_d. The dependence of the difference between the neutron and proton r.m.s. radii on the choice of the nucleon-nucleon interaction is shown in Table 7.4. These results were derived from $\langle r^2 \rangle_{RS}$ using eqn.

TABLE 7.2

Sets of parameters giving best fits to elastic scattering data for a Yukawa nucleon-nucleon interaction and various values of $\langle r^2 \rangle_d$ †

^{59}Co	30.3 MeV proton scattering					
$\langle r^2 \rangle_d$ (fm)	0.01	0.75	1.5	3.0	4.5	6.0
V_{RS} (MeV)	47.8	48.3	50.7	55.1	59.9	63.8
$R' A^{-1/3}$ (fm)	1.217	1.266	1.278	1.360	1.404	1.428
a' (fm)	0.643	0.658	0.645	0.581	0.588	0.507
$\langle r^2 \rangle_{RS}$ (fm^2)	17.97	19.12	19.08	19.34	19.78	20.47
χ^2_σ	29.7	11.7	7.5	5.6	9.9	15.7
χ^2_p	33.3	12.6	10.6	9.0	7.4	7.2
^{208}Pb	30.3 MeV proton scattering					
$\langle r^2 \rangle_d$ (fm)	0.1	1.0	2.0	3.0	5.0	6.0
V_{RS} (MeV)	51.21	51.56	51.87	52.63	54.07	54.33
$R' A^{-1/3}$ (fm)	1.243	1.239	1.241	1.260	1.310	1.320
a' (fm)	0.742	0.746	0.740	0.734	0.714	0.686
$\langle r^2 \rangle_{RS}$ (fm^2)	36.00	36.00	36.12	36.00	35.52	35.64
χ^2_σ	1.3	1.3	1.4	1.6	2.3	2.7
χ^2_p	3.4	2.9	2.6	3.1	4.9	6.7
^{208}Pb	14.5 MeV neutron scattering					
$\langle r^2 \rangle_d$ (fm)	0.1	1.0	2.0	3.0	5.0	6.0
V_{RS} (MeV)	45.43	45.50	45.68	46.09	46.94	47.29
$R' A^{-1/3}$ (fm)	1.211	1.197	1.205	1.229	1.280	1.299
a' (fm)	0.698	0.713	0.704	0.682	0.620	0.577
$\langle r^2 \rangle_{RS}$ (fm^2)	36.97	36.97	37.09	37.21	37.95	38.56
χ^2_σ	0.43	0.42	0.44	0.51	0.81	1.03

† From Greenlees *et al.* 1968*a*, 1970, Pyle and Greenlees 1969.

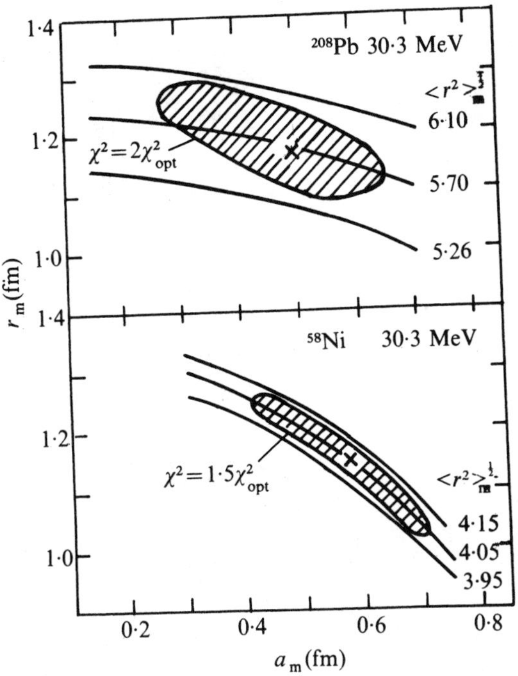

FIG. 7.6. Contours of constant best-fit χ^2 for variations of diffuseness and halfway radius of the matter distribution. The shaded areas enclose regions where the fits to data are visibly indistinguishable. (From Greenlees *et al.* 1968*a*.)

(7.96) and the parameters of Acker *et al.* (1966) for the proton distribution. It can be seen that the results obtained with the gaussian interaction are nearest to the predicted values given in Tables 2.1 and 2.2.

Calculations of the real part of the optical potential using more realistic effective interactions have been carried out by Slanina and McManus (1968), Friedman (1969), Kidwai and Rook (1971), and Thomas and Sinha (1971). The effective interaction can be written in the form

$$t(r) = t_D(r) + t_E(r) \tag{7.100}$$

where it is assumed that t is an effective local interaction

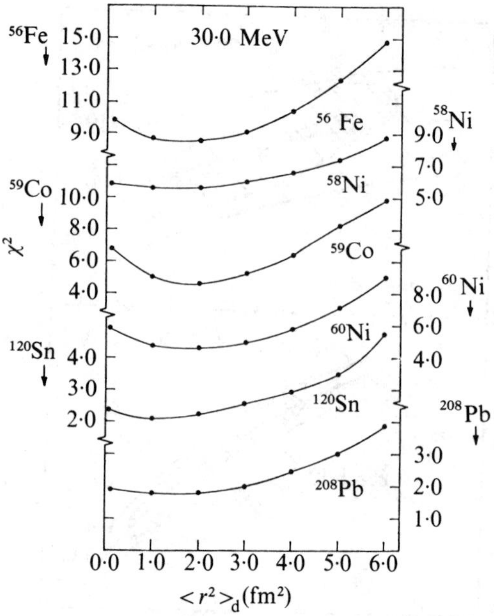

FIG. 7.7. Variation of best-fit χ^2 values against the mean-square radius $\langle r^2 \rangle_d$ of a Yukawa two-body interaction, using $\langle r^2 \rangle_{ls} = 0.5$ fm^2 and $\zeta = 0.48$. (From Greenlees *et al.* 1970.)

and that t_D, t_E are, respectively, the direct and exchange operators. If t operates only in relative even states it can be written as

$$t(r) = \bar{t}(r)[1 + P] \tag{7.101}$$

where P is a space-exchange operator,

$$\bar{t}(r) = \frac{3}{16}(V_S + V_T) - \frac{N-Z}{16A}(V_S - V_T)$$

and V_S, V_T are the singlet-even and triplet-even parts of the interaction. Slanina and McManus have used the Kuo-Brown interaction (Kuo and Brown 1966)

$$V_S(r) = 0, \; r < d_S; \; V_S(r) = v_{c\ell}(r), \; r > d_S$$

TABLE 7.3

Values of $\langle r^2 \rangle_{RS}$ *and* J_{RS}/A *obtained from various analyses of data for 30.3 MeV proton elastic scattering*†

	J_{RS}/A (MeV fm^3)						
	a	*b*	*c*	*d*	*e*	*f*	*g*
^{56}Fe	401	401	408	407	379	—	—
^{58}Ni	400	402	408	406	386	409	395
^{59}Co	402	403	411	409	384	411	398
^{60}N	403	404	413	409	387	413	396
^{120}Sn	399	397	402	399	392	406	399
^{208}Pb	408	411	411	402	395	411	399
	$\langle r^2 \rangle^{1/2}_{RS}$ (fm)						
^{56}Fe	4.32	4.33	4.39	4.39	4.21	—	—
^{58}Ni	4.39	4.40	4.46	4.43	4.28	4.38	4.35
^{59}Co	4.35	4.35	4.39	4.42	4.26	4.37	4.36
^{60}Ni	4.38	4.42	4.46	4.47	4.32	4.42	4.34
^{120}Sn	5.21	5.20	5.22	5.23	5.14	5.24	5.21
^{208}Pb	6.01	6.02	6.00	5.97	5.92	6.03	5.95

a Yukawa: $\langle r^2 \rangle_d = 2.25\ \text{fm}^2$, $\langle r^2 \rangle_{ls} = 0.5\ \text{fm}^2$, $\zeta = 0.48$

b Yukawa: $\langle r^2 \rangle_d = 2.25\ \text{fm}^2$, $\langle r^2 \rangle_{ls} = 0.5\ \text{fm}^2$, $\zeta = 0$

c Yukawa: $\langle r^2 \rangle_d = 0.1\ \text{fm}^2$, $\langle r^2 \rangle_{ls} = 0.5\ \text{fm}^2$, $\zeta = 0.48$

d Yukawa: $\langle r^2 \rangle_d = 4.27\ \text{fm}^2$, $\langle r^2 \rangle_{ls} = 0.5\ \text{fm}^2$, $\zeta = 0.48$

e Gaussian: $\langle r^2 \rangle_d = 4.27\ \text{fm}^2$, $\langle r^2 \rangle_{ls} = 0.5\ \text{fm}^2$, $\zeta = 0.48$

f Yukawa: $\langle r^2 \rangle_d = 2.25\ \text{fm}^2$, $\rho_p = \rho_n$, phenomenological *ls*

g Phenomenological optical model: Satchler (1967)

† From Greenlees *et al*. 1970. Typical errors are ±15 MeV fm^3 and ±0.15 fm.

TABLE 7.4

The difference in neutron and proton r.m.s. radii obtained from various analyses of data for 30.3 MeV elastic proton scattering†

	$\langle r^2\rangle_n^{1/2} - \langle r^2\rangle_p^{1/2}$ (fm)				
	a	*b*	*c*	*d*	Error
^{56}Fe	—	0.45	0.22	-0.04	± 0.15
^{58}Ni	0.71	0.50	0.24	0.01	± 0.18
^{59}Co	0.64	0.43	0.20	-0.03	± 0.16
^{60}Ni	0.70	0.48	0.25	0.03	± 0.16
^{120}Sn	0.71	0.49	0.27	0.15	± 0.19
^{208}Pb	0.64	0.46	0.19	0.13	± 0.25

a Yukawa: $\langle r^2\rangle_d = 2.25$ fm^2, $\rho_p = \rho_n$, phenomenological *ls*

b Yukawa: $\langle r^2\rangle_d = 2.25$ fm^2, $\langle r^2\rangle_{ls} = 0.5$ fm^2, $\zeta = 0.48$

c Yukawa: $\langle r^2\rangle_d = 4.27$ fm^2, $\langle r^2\rangle_{ls} = 0.5$ fm^2, $\zeta = 0.48$

d Gaussian: $\langle r^2\rangle_d = 4.27$ fm^2, $\langle r^2\rangle_{ls} = 0.5$ fm^2, $\zeta = 0.48$

†From Greenlees *et al.* 1970.

$$V_T(r) = 0,\ r < d_T;\ V_T(r) = v_{c\ell}(r) - 8\, v_{t\ell}^2(r)/240,\ r > d_T$$

where $v_{c\ell}$ and $v_{t\ell}$ are the central and tensor components of the long-range parts of the Hamada-Johnston potential. For a laboratory energy of 40 MeV, separation distances are $d_S = 1.05$ fm, $d_T = 1.07$ fm. They also use Green's density-dependent interaction (Green 1967)

$$V_S(r) = C_S(1 - a_S\,\rho^{2/3})\ V_S^{KK}$$

$$V_T(r) = C_T(1 - a_T\,\rho^{2/3})\ V_T^{KK}$$

where KK specifies the Kallio-Kolltveit interaction (Kallio

and Kolltveit 1964)

$$V_S(r) = 0,\ r < d_S;\ V_S(r) = -330.8\ \exp\{-2.4021(r-0.4)\},\ r > d_S$$

$$V_T(r) = 0,\ r < d_T;\ V_T(r) = -475.0\ \exp\{-2.5214(r-0.4)\},\ r > d_T$$

and the separation distances for 40 MeV are $d_S = 1.046$ fm and $d_T = 0.924$ fm. Two versions of Green's interaction with different parameters a_S, a_T, C_S, C_T correspond to weak density dependence (WG) and to strong density dependence (SG). Thomas and Sinha (1971) also use Green's interaction while Friedman (1969) used a linear density-dependent interaction

$$V_{S(T)} = \beta_{S(T)}(1 - \alpha_{S(T)}\ \rho)\ V^{KK}_{S(T)}$$

due to Lande *et al.* (1968). Thomas, Sinha, and Duggan (1973) have compared results using several of these interactions. Slanina and McManus also fitted the free nucleon-nucleon t-matrix to obtain a pseudopotential in the form of a Yukawa or a sum of Yukawa terms in order to carry out a calculation in impulse approximation. For the single Yukawa this led to $\langle r^2 \rangle_d \sim 3.26$ fm compared with $\langle r^2 \rangle_d = 2.25 \pm 0.6$ fm^2 which was the preferred value obtained by Greenlees *et al.* (1968*a*). Antisymmetrization leads to a non-local optical potential which can be represented by an effective local potential for purposes of comparison with phenomenological potentials. Examples of fits to the data achieved in this type of calculation are given in Fig. 7.8.

The results obtained for $\langle r^2 \rangle_{RS}$ and J_{RS} obtained by Slanina and McManus are shown in Table 7.5. Since the calculations are carried out with $\langle r^2 \rangle_p = \langle r^2 \rangle_n$, it appears that a major part of the neutron-proton difference inferred in the early work by Greenlees *et al.* is removed when realistic effective interactions are used. A similar conclusion was reached by Friedman. Slanina and McManus find that the coefficient of the symmetry term is 20-30 MeV, and that the energy dependence can be represented by $\gamma = 0.21$ but that 80 per cent of this energy dependence comes from the exchange

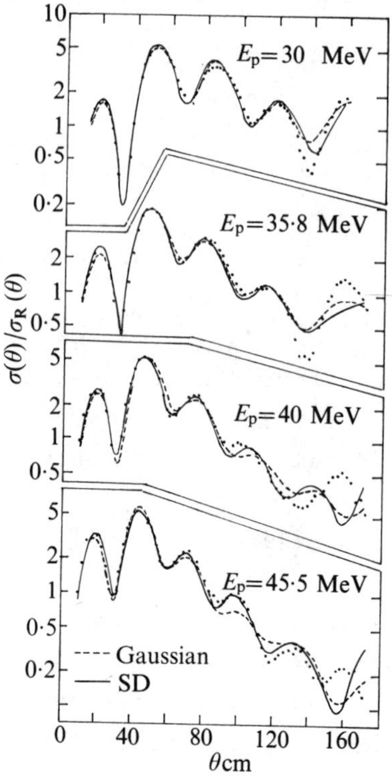

FIG. 7.8. Microscopic fits to proton scattering from ^{40}Ca for a gaussian two-body interaction with mean square radius of 4.5 fm^2 and for Green's strong density-dependent force compared with experimental data at 30 MeV (dots). (From Thomas and Sinha 1971.)

term. The actual effect of the exchange term is small at 40 MeV; it smooths the surface of the total potential and adds 0.4 fm^2 to the mean square radius. A further study of exchange effects has been carried out by Owen and Satchler (1970) using the long-range part of the S-wave Hamada-Johnston potential. In this case the scattering from the non-local potential was calculated exactly and the resulting cross-sections were then fitted using local Saxon-Woods potentials. At 30 MeV the exchange effects are quite important leading to a *reduction* in the r.m.s. radius of the potential of a few per cent for heavy target nuclei but as much as 28 per cent for ^{16}O. The exchange contribution also appears to enhance

TABLE 7.5

Moments of the direct term of the real central potential calculated with various effective interactions†

	^{40}Ca $\langle r^2\rangle_m = 11.36$ fm^2			^{58}Ni $\langle r^2\rangle_m = 14.81$ fm^2		
	J_{RS} (MeV fm^3)	$\langle r^2\rangle_{RS}$ (fm^2)	$\langle r^2\rangle_d$ (fm^2)	J_{RS} (MeV fm^3)	$\langle r^2\rangle_{RS}$ (fm^2)	$\langle r^2\rangle_d$ (fm^2)
Impulse approx.	12031	14.52	3.16	17470	17.98	3.15
Kuo-Brown	14560	18.03	6.76	21410	21.45	6.75
WG	12910	15.12	3.76	19020	18.75	3.94
SG	15690	15.50	4.14	23240	19.26	4.45
Greenlees *et al.*‡	17160	16.78	23690	23690	19.72	
Fricke *et al.*§	15330	16.43		21770	19.51	
	^{120}Sn $\langle r^2\rangle_m = 20.86$ fm^2			^{208}Pb $\langle r^2\rangle_m = 29.54$ fm^2		
Impulse approx.	36370	24.08	3.22	63180	32.74	3.20
Kuo-Brown	46590	27.40	6.71	82070	35.95	6.70
WG	39570	24.95	4.09	68960	33.80	4.26
SG	47440	25.57	4.71	82390	34.56	5.02
Greenlees *et al.*‡	50120	27.79		88090	37.10	
Fricke *et al.*§	45140	27.60		79230	37.19	

†From Slanina and McManus 1968.

‡Greenlees and Pyle 1966.

§Fricke, Gross, and Zucker 1967.

the asymmetry dependence, but this result depends on assumptions about the distribution of the excess neutrons. The energy dependence of the equivalent local potential found by Owen and Satchler yields values for γ of 0.30 (^{16}O), 0.27 (^{40}Ca) and 0.30 (^{208}Pb). This A-dependence of the coefficient γ is also observed by Van Oers and Haw (1973) in an empirical study of proton optical potentials.

Some more recent studies of the real part of the optical potential relate the equivalent uniform radius of the potential U_V (see §1.2.3) to that of the nuclear matter distribution U_ρ (Myers 1973, Srivastava, Ganguly, and Hodgson 1974). Srivastava *et al.* (1974) calculated the direct part of the potential for 40 MeV protons using the Kallio-Kolltveit force and found $U_V = U_\rho$, as predicted by Myers (1973). When the exchange term calculated with the same force was included, they found $U_V = U_\rho + 0.12$ fm. However, when Green's strong density-dependent force is used the relations become $U_V = U_\rho + 0.40$ fm (without exchange) and $U_V = U_\rho + 0.55$ fm (with exchange). This result is again consistent with the predictions of Myers (1973) based on the saturation property of the density-dependent force. It is suggested that the uniform radius of the potential obeys a general formula

$$U_V = 1.13A^{1/3} + 0.55 \text{ fm}.$$

Jeukenne, Lejeune, and Mahaux (1974) derive an optical potential from the leading term of a low-density expansion of the mass operator, using Reid's hard-core nucleon-nucleon interaction. The energy dependence of the real part of the potential is given by $56-0.3E$ MeV in the energy range 20-150 MeV, and the potential changes sign at $E \sim 200$ MeV. The imaginary part has a volume shape at lower energies and magnitude $0.19E + 1.9$ MeV. At low energy the real part of the symmetry energy is $\sim 14(N - Z)/A$ but it changes sign at $E \sim 110$ MeV.

These results obtained with realistic effective interactions can be regarded as justifying the essential idea behind the folding procedure for constructing the real central potential, but they also demonstrate the dangers in attempting to derive information about the nuclear matter distribution or the nucleon-nucleon interaction from a simplified calculation.

The impulse approximation, as normally used, is not expected to be valid at 40 MeV owing to the importance of off-shell effects and exchange. Lerner and Redish (1972) have investigated this point by constructing a potential from a fully off-shell t-matrix and including exchange between the incident and struck nucleons though neglecting other exchange terms. The t-matrix was calculated from the Reid soft-core potential for the nucleon-nucleon interaction. The target nucleus was described by a single-particle model so that the contribution from each target nucleon could be treated separately, and the single-particle contributions were then summed to give an equivalent local potential. Only the real central part of the potential was calculated in this way and it was found that the mean square radius exceeded $\langle r^2 \rangle_m$ by 4 fm^2 which is comparable with the results of the simple folding calculations. As can be seen from Fig. 7.9, the calculated potential for 65 MeV scattering is substantially deeper than the phenomenological potential but is in excellent agreement with it in the surface region. In contrast, a potential calculated from an on-shell t-matrix falls off much too rapidly; this is consistent with the results of Slanina and McManus listed in Table 7.5 where the impulse approximation gives the smallest value for the volume integral and r.m.s. radius of the potential at 40 MeV. When used with a phenomenological imaginary and spin-orbit potential, the calculated potentials give good agreement with elastic scattering from the oxygen isotopes at 65 MeV (Lerner and Marion 1972).

7.3.3. *Intermediate- and high-energy region*

The essential features of elastic scattering in the 100-200 MeV region are the smooth fall in the differential cross-sections with increasing scattering angle with very little of the oscillatory behaviour seen at lower energies and the

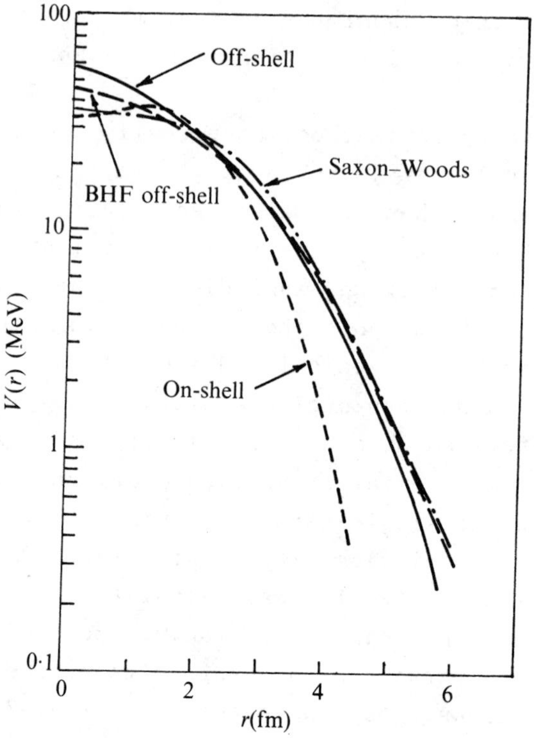

FIG. 7.9. Calculated optical potentials using a fully off-shell t-matrix and an on-shell approximation compared with a phenomenological Saxon-Woods potential. (From Lerner and Redish 1972.)

marked oscillations in the polarization data. In the early analyses of data at 182 MeV it was found (Johansson, Svanberg, and Hodgson 1961, Hodgson 1961) that these data could be fitted using the phenomenological potential (7.59) with volume absorption provided that $R' > R_0$. An imaginary term in the spin-orbit potential is required. Early studies of neutron total cross-sections over the range 15-120 MeV (Bowen *et al.* 1961) indicated that the transition from surface to volume absorption occurs at an incident energy of ~ 80 MeV. Later analysis of the 182 MeV data (Satchler and Haybron 1964) and studies of data at 75, 100, 155, and 160 MeV (Satchler 1965, Roos and Wall 1965, Rolland *et al.* 1966, Geoffrion *et al.* 1968, Willis *et al.* 1968, Seth 1969, Van Oers 1971, Horowitz

1972) confirm the need for an imaginary spin-orbit potential above 100 MeV and for volume or mixed surface and volume absorption in the higher energy range, except possibly for very light nuclei. Some of the parameters obtained by these authors are listed in Table 7.6.

The small-angle data, i.e. <30°, can be reasonably well represented using potentials calculated from the on-shell impulse approximation. A number of estimates of multiple-scattering corrections have been carried out for nucleon scattering in the 100-300 MeV region. Most of these calculations (Kerman *et al.* 1959, McDonald and Hull 1966, Chalmers and Saperstein 1967, 1968, Tatischeff 1967) use rather simple approximations to treat the two-nucleon correlation function which appears in the double-scattering term, so that these calculations indicate clearly the importance of multiple scattering but do not substantially improve agreement with the data. In a more recent paper (Johnson and Martin 1972) a method has been presented for treating exactly the non-locality of the second term in the multiple-scattering expansion of the potential and of incorporating the Pauli correlations obtained from a realistic single-particle model of the nucleus. An important effect seen in all these calculations is the significant reduction in the magnitude of the imaginary part of the potential for incident energies below ~ 200 MeV and the rather smaller increase in the imaginary part at higher energies; this leads to a reduction in the predicted values of the reaction cross-sections below ~ 200 MeV.

In the phenomenological analyses it was noted that, when R' was increased to damp the oscillations in $d\sigma/d\Omega$, the predicted values of σ_R became too large, except for heavy nuclei. Seth (1969) suggests that the values of σ_R are strongly dependent on a' and that an energy dependence of the form

$$a' = 0.736 - 0.0012\,E \tag{7.103}$$

should provide satisfactory agreement with the data. This is consistent with the results (7.88) obtained by fitting the reaction cross-section in the medium-energy region. Horowitz

TABLE

Optical-model parameters for

	E_{lab}(MeV)	U(MeV)	$R_0A^{-1/3}$(fm)	a(fm)	W(MeV)	W_D(MeV)
^{6}Li	155[a]	18.7	1.07	0.51	0	8.3
^{7}Li	155[a]	20.5	1.10	0.42	0	8.5
	183[b]	10.3	1.25	0.46	7.9	0
	183[b]	13.5	1.02	0.44	0	11.1
^{9}Be	160[c]	-5.40	1.02	0.49	14.17	0
	160[d]	16.2	1.0	0.39	14.0	0
	183[b]	0.9	1.12	0.57	9.7	0
^{11}B	155[a]	14.2	1.44	0.44	5.79	0
^{12}C	75[e]	25.4	1.13	0.49	0	8.5
	152[e]	18.7	1.13	0.72	10.4	0
	156[j]	12.8	1.40	0.52	25.4	0
	183[b]	19.4	0.90	0.45	15.6	0
^{14}N	155[a]	21.5	1.22	0.57	10.2	0
^{24}Mg	100[h]	22.1	1.27	0.68	7.2	0
	155[f]	18.3	1.27	0.74	10.3	0
^{27}Al	156[j]	12.4	1.39	0.55	16.4	0
	183[b]	11.9	1.01	0.54	11.8	0
^{28}Si	100[h]	21.7	1.27	0.68	6.2	0
	155[f]	22.7	1.24	0.75	11.0	0
^{32}S	155[f]	19.0	1.26	0.74	10.6	0
^{40}Ca	75[e]	23.0	1.20	0.53	0	13.2
	75[g]	31.8	1.20	0.43	5.1	9.4
	75[i]	33.5	1.15	0.71	11.7	0
	153[e]	15.0	1.01	0.47	15.0	0
	153[g]	18.7	1.13	0.52	13.6	2.8
	153[i]	10.8	1.01	0.31	16.0	0
	155[f]	19.4	1.24	0.74	12.6	0
	156[j]	12.3	1.42	0.55	13.1	0
	160[c]	17.5	1.17	0.59	8.2	0

7.6

75-200 MeV proton scattering

$R'A^{-1/3}$(fm)	a'(fm)	V_{so}(MeV)	W_{so}(MeV)	$R_{so}A^{-1/3}$(fm)	a_s(fm)	$R_cA^{-1/3}$(fm)
1.18	0.52	0.63	-2.08	1.10	0.62	1.56
1.23	0.50	0.90	-1.90	1.18	0.48	1.49
1.72	0.40	3.02	-1.03	1.25	0.46	1.90
1.06	0.55	2.74	-2.09	1.02	0.44	1.90
1.63	0.39	2.31	-4.82	1.02	0.49	1.89
1.55	0.40	2.50	-1.0	1.0	0.39	1.3
1.68	0.45	4.61	-2.25	1.12	0.57	1.89
1.18	0.85	2.27	-1.74	0.97	0.48	1.28
1.45	0.45	0.72	0	1.86	0.40	1.33
1.18	0.85	3.52	-2.11	0.93	0.50	1.33
0.81	0.71	1.84	-2.29	0.92	0.45	1.42
1.19	0.56	4.12	-0.10	0.90	0.45	1.33
1.30	0.52	3.28	-2.66	0.93	0.47	1.30
1.50	0.53	9.9	0	1.00	0.60	
1.20	0.86	2.44	-2.49	0.97	0.62	1.33
1.05	0.75	3.01	-1.72	1.01	0.57	1.34
1.44	0.46	4.60	-1.86	1.01	0.54	1.30
1.55	0.42	9.54	0	1.08	0.61	
1.26	0.67	2.78	-3.35	0.95	0.62	1.29
1.25	0.81	2.96	-2.30	0.99	0.64	1.30
1.20	0.56	0.91	0	1.69	0.39	1.32
1.30	0.65	3.32	0	1.10	0.65	1.25
1.32	0.55	5.26	0	1.04	0.51	1.32
1.40	0.52	2.11	-2.65	1.12	0.61	1.32
1.37	0.55	1.42	-3.5	1.10	0.65	1.25
1.37	0.51	3.55	2.7	1.07	0.61	1.32
1.23	0.84	2.87	-1.80	1.02	0.58	1.32
1.22	0.57	3.48	-0.94	1.03	0.64	1.34
1.54	0.48	4.04	-0.04	1.17	0.59	1.32

TABLE 7.1 continued

	E_{lab}(MeV)	U(MeV)	$R_0A^{-1/3}$(fm)	a(fm)	W(MeV)	W_D(MeV)
^{40}Ca	160[d]	14.5	1.17	0.51	10.0	0
	160[g]	17.8	1.12	0.53	14.3	2.1
	160[i]	17.1	1.10	0.49	10.7	0
	182[g]	15.5	1.10	0.55	16.47	0
	182[b]	19.3	1.01	0.55	16.4	0
	182[i]	18.7	1.00	0.33	14.9	0
^{58}Ni	160[c]	11.3	1.13	0.65	9.9	0
	160[d]	14.2	1.20	0.52	10.0	0
	160[g]	17.8	1.14	0.33	14.3	2.1
^{90}Zr	156[j]	12.0	1.41	0.56	12.0	0
^{115}In	182[b]	11.0	1.15	0.71	8.2	0
	182[g]	15.5	1.15	0.71	16.5	0
^{120}Sn	156[j]	18.2	1.29	0.72	7.3	0
	160[c]	37.6	1.00	0.71	16.1	0
	160[d]	16.2	1.18	0.55	10.0	0
	160[g]	17.8	1.16	0.73	14.3	2.1
^{140}Ce	76[e]	27.4	1.23	0.56	0	12.0
	76[g]	31.8	1.22	0.62	5.1	9.4
	153[g]	18.7	1.17	0.71	13.6	2.7
^{208}Pb	156[j]	26.5	1.12	0.79	16.2	0
	160[c]	28.7	1.12	0.70	17.6	0
	160[d]	17.0	1.25	0.69	9.0	0
	160[g]	17.8	1.18	0.80	14.3	2.1
^{209}Bi	78[e]	26.9	1.22	0.58	0	13.1
	78[g]	31.4	1.22	0.70	5.3	9.3
	153[e]	22.5	1.11	0.73	10.8	0
	153[g]	18.7	1.18	0.79	2.7	13.6

[a] Geoffrion *et al.* (1968).
[b] Satchler and Haybron (1964).
[c] Satchler (1965).
[d] Roos and Wall (1965).
[e] Rolland *et al.* (1966).
[f] Willis *et al.* (1968).

$R'A^{-1/3}$(fm)	a'(fm)	V_{so}(MeV)	W_{so}(MeV)	$R_{so}A^{-1/3}$(fm)	a_s(fm)	$R_cA^{-1/3}$(fm)
1.50	0.50	2.45	-1.0	1.17	0.51	1.30
1.38	0.54	1.32	-3.5	1.10	0.65	1.25
1.47	0.48	3.39	1.0	1.04	0.51	1.32
1.40	0.52	1.06	-3.5	1.10	0.65	1.25
1.36	0.54	4.06	-2.0	1.01	0.55	1.32
1.38	0.54	4.00	1.8	1.09	0.64	1.32
1.49	0.32	3.41	-4.1	1.13	0.65	1.25
1.45	0.50	2.45	-1.0	1.20	0.52	1.30
1.36	0.54	1.32	-3.5	1.10	0.65	1.25
1.25	0.43	2.03	-1.65	1.08	0.71	1.25
1.42	0.43	3.2	-3.0	1.15	0.71	1.20
1.35	0.52	1.1	-3.5	1.10	0.65	1.25
1.32	0.54	2.5	-1.3	1.11	0.65	1.23
1.34	0.51	4.5	-2.2	1.00	0.71	1.20
1.40	0.50	2.5	-1.0	1.18	0.55	1.30
1.34	0.54	1.3	-3.5	1.10	0.54	1.25
1.22	0.76	1.0	0	1.52	0.45	1.20
1.28	0.65	3.3	0	1.10	0.65	1.25
1.33	0.55	1.4	-3.5	1.10	0.65	1.25
1.31	0.48	1.2	-2.3	1.15	0.72	1.20
1.31	0.53	0.4	-3.0	1.12	0.70	1.20
1.37	0.70	2.5	-1.0	1.25	0.69	1.30
1.32	0.54	1.3	-3.5	1.10	0.65	1.25
1.21	0.68	0.9	0	1.48	0.43	1.20
1.28	0.64	3.2	0	1.10	0.65	1.25
1.29	0.59	1.7	-3.3	1.10	0.75	1.20
1.32	0.55	1.4	-3.5	1.10	0.65	1.25

[g] Seth (1969).
[h] Horowitz (1972).
[i] Van Oers (1971).
[j] Comparat *et al.* (1974).

(1972) noted that, at 100 MeV, no potential with surface absorption alone could be found to fit both the differential and the reaction cross-sections for ^{24}Mg and ^{28}Si. Reasonable agreement can be found with volume absorption or a mixture of volume and surface terms with different radial parameters. Thus, an explanation of the problem of fitting σ_R in terms of the behaviour of the imaginary part of the potential is consistent with observations at lower energies and with more fundamental considerations relating to the effect of multiple scattering. An alternative explanation has been given in terms of the behaviour of the real part of the potential (Elton 1966, Jeukenne, Lejeune and Mahaux 1974).

Systematic analyses of differential cross-sections, polarization, and reaction cross-sections for a wide range of nuclei over a wide range of incident energies are rather rare in the intermediate- and high-energy region owing to the general scarcity of data. Data over an extensive energy range exists only for scattering from ^{12}C and has been studied in the 100 MeV-1 GeV range by Batty (1961) and Shah (Shah 1971, Jackson and Shah 1969). Accurate measurements of differential cross-sections for proton scattering from light nuclei in the region of Coulomb-nuclear interference (2°-20°) have recently been made at 144 MeV (Jarvis, Whitehead, and Shah 1972), and have been combined with other measurements of proton and neutron differential and total cross-sections and polarization in the range 144-155 MeV for an optical model analysis.

The optical potential used by Jarvis *et al.* was constructed from the free nucleon-nucleon amplitudes following Kerman *et al.* (1959) using essentially eqns (7.50) and (7.57). The dependence of the nucleon-nucleon amplitudes on momentum transfer was expressed as

$$\bar{A}(q) = A_R(0)\exp(-\tfrac{1}{4}\alpha_1^2 q^2) + i\,A_I(0)\exp(-\tfrac{1}{4}\alpha_2^2 q^2) \quad (7.104a)$$

$$\bar{C}(q) = C_R(0)\exp(-\tfrac{1}{4}\alpha_3^2 q^2) + i\,C_I(0)\exp(-\tfrac{1}{4}\alpha_4^2 q^2) \quad (7.104b)$$

and the nuclear density distribution taken to be of gaussian form, which yields a nuclear form factor also of gaussian form so that the potential has four radial parameters given by $a_i^2 = a_N^2 + \alpha_i^2$. The Coulomb potential was taken to be $(Ze^2/r)\ \mathrm{erf}(r/a_c)$, which is the correct form for a Gaussian charge distribution, where a_c includes the finite electromagnetic size of the proton. The predicted parameters are given in Table 7.7, together with the best-fit parameters obtained from a search procedure. The predicted parameters for the lithium isotopes give quite good agreement with the data without modification; the parameters obtained from the search, though apparently rather different, yield the same values for the volume integral Va^3. For the other nuclei a significant change in the imaginary potential is required to fit the small-angle data which also has the effect of reducing the total cross-sections to agree with the experimental values.

The semiclassical analysis of the ^{12}C data by Batty (1961) also used potentials derived in a similar manner but took the density distribution to be

$$\rho(r) = \left(1 + \frac{4}{3}\frac{r^2}{a_N^2}\right)\exp\left(-\frac{r^2}{a_N^2}\right). \tag{7.105}$$

This is the form derived for ^{12}C from oscillator functions and can be shown to fit elastic electron scattering up to $q \sim 2.5\ \mathrm{fm}^{-1}$ with $a_N = 1.64$ fm. All terms in the potential were taken to have the same radius parameter a_1 which was allowed to vary. This gave a total of five parameters, the four potential strengths and a_1. The important results obtained from this analysis are that the real part of the potential changes sign at an incident proton energy of ~ 450 MeV, as shown in Fig. 7.10, and that the radius parameter a_1 tends to a_N at high energy.

The analysis for ^{12}C was repeated by Shah (1971) using the same data and such new data as had become available in the meantime in the same energy region. The usual phenomenological potential (7.59) was used both in a semiclassical calculation and in an optical model search code. In the latter

TABLE

Optical-model parameters for

	a_N(fm)	U(MeV)	a_1(fm)	Ua_1^3(MeV fm^3)	W(MeV)	a_2(fm)
^{4}He	1.15	20.6	1.82	124	16.7	1.83
		10.3	2.27	121	23.5	1.55
^{6}Li	2.02	13.8	2.46	205	11.3	2.48
		9.7	2.76	205	21.2	2.00
^{7}Li	2.02	16.4	2.46	244	12.9	2.48
		15.7	2.49	242	20.2	2.17
^{9}Be	1.82	26.8	2.30	326	21.4	2.32
		18.3	2.64	335	12.6	2.64
^{12}C	2.01	30.9	2.45	455	25.4	2.46
		19.5	2.77	415	28.2	2.23

† From Jarvis *et al.* 1972. For each nucleus the first line gives the predicted parameters and the second line gives the best-fit parameters.

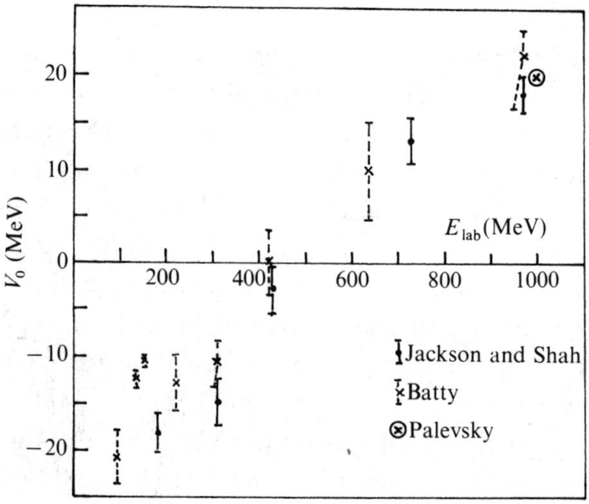

FIG. 7.10. Variation of the depth of the real part of the nucleon optical potential with incident proton energy. (From Jackson and Shah 1969.)

7.7
143-155 MeV proton scattering†

Wa_2^3(MeV fm^3)	V_{so}(MeV)	a_3	W_{so}(MeV)	a_4(fm)	σ_T(mb)	σ_T^{exp}(mb)
102	4.34	1.55	-0.52	1.92	156	119 ± 4
88	2.57	1.75	-0.82	1.92	116	
172	2.13	2.28	-0.35	2.54	238	
170	1.97	2.28	0.26	2.54	193	
197	2.55	2.28	-0.34	2.54	268	
206	1.78	2.28	-0.57	2.54	233	
290	4.22	2.10	-0.56	2.48	348	284 ± 14
231	4.08	2.10	-0.61	2.48	282	
378	4.42	2.26	-0.72	2.52	487	344 ± 5
325	4.52	2.26	-0.35	2.52	344	

analysis the diffuseness parameters were held fixed and equal to 0.5 fm. It was found that for incident energies below 500 MeV, the data require $R' > R_0$. The agreement with σ_R is generally good; the agreement with $d\sigma/d\Omega$ is good at 310, 424, and 970 MeV, is good at large angles at 180 MeV, and is good up to and slightly beyond the first diffraction minimum at 1 GeV, while the agreement with $P(\theta)$ is good at 310 and 424 MeV and is reasonable at 180, 725, and 970 MeV (Shah 1971). The variation of the depth of the real potential with energy is compared with Batty's results in Fig. 7.10. The energy dependence displayed in this figure can be represented approximately by eqn (7.60) with $\gamma = 0.04 \pm 0.01$ and $U_0 \sim 20$ MeV. It is valid only over a limited energy range, since these parameters are quite different from those obtained at lower energies.

The variation of a radius parameter of the real potential with incident energy is shown in Fig. 7.11. In Batty's

'optical limit' of Glauber theory. These comparisons show that this transformation is valid for nuclei as light as ^{12}C and ^{16}O but not for ^{4}He.

Very precise data for 1 GeV proton scattering from ^{12}C, ^{58}Ni, and ^{208}Pb have been published by the Saclay group (Bertini *et al.* 1973). They have analysed their data in an eikonal approximation with the optical potential given by impulse approximation in the form (7.49) or (7.52). This yields satisfactory agreement with the data except that the diffraction minima are too deep, and this appears to be a general defect of the eikonal method (Clark *et al.* 1973). Auger and Lombard (1973*a*) have analysed the same data in a simple Glauber or semiclassical model, i.e. using eqns (7.31), (7.33), and (7.49). The nuclear densities are taken from Hartree-Fock calculations so that there are no adjustable parameters in the calculation. This yields reasonable agreement with the data but the fits to the positions of the maxima and minima are not exact, particularly for ^{208}Pb. Reasonable agreement with the data for ^{12}C has been obtained using the optical limit of Glauber theory and a deformed density distribution (Starodubsky and Domchenkov 1972).

Additional data at 1 GeV have been obtained at Leningrad and at Saclay (Alkhazov *et al.* 1972, 1975, 1976, Chaumeaux *et al.* 1976, Thirion 1973). The data appear to be consistent with a difference between the proton and neutron radii which is zero or less than 0.15 fm, except possibly for ^{48}Ca. The 1 GeV data for ^{40}Ca have been analysed by Brissaud and Brussel (1976*a*) using the model independent method of Sick (see §3.4.2) to generate the matter distribution. The best defined moments are for k = 2-4 and the uncertainty on M_2 is comparable with that obtained in model-dependent analyses of electron and proton scattering (Lombard and Wilkin 1975).

The elastic scattering data in the intermediate-energy region can also be fitted by means of direct parametrization of the reflection coefficients (Venter and Frahn 1964*a*). At 180 MeV excellent fits to the data can be achieved with five parameters: the semiclassical radius R and diffuseness d, as defined in eqn (7.13), a spin-orbit parameter ν, a real-phase or refraction parameter μ, and a transparency parameter ε which allows for incomplete absorption of the low partial waves. Good overall agreement with the data of Johansson *et al.* (1961) can be obtained with standard parameters $R = 1.15\ A^{1/3}$, $d = 0.66$ fm, $\mu = 0.85$ kd, and $\nu = 1.50$ kd. The parameter ε decreases with mass number from ~ 0.5 for lithium to ~ 0.02 for gold, and is determined principally by the reaction cross-sections. In the formula for the cross-section the refractive term $(\mu/kR)^2\{J_0(kR\theta)\}^2$ adds to the term $\{J_1(kR\theta)/kR\theta\}^2$ which arises in the sharp cut-off model, and consequently damps out the oscillatory pattern.

Some elastic scattering data exist at incident proton momenta of 20 GeV/c. This has been analysed by a sharp cut-off model (Bellettini *et al.* 1966), a sharp cut-off model with inclusion of inelastic excitation of low-lying states (Matthiae 1967), and a smooth diffraction model with refraction included (Frahn and Wiechers 1966). The latter two calculations yield filling-in of the minima of the differential cross-section which are too deep in the sharp cut-off model. Frahn and Wiechers invert the semiclassical expression (7.31) to obtain a potential and show that this yields a real part which is repulsive, at least in the nuclear surface.

In addition to the measurements of neutron total cross-sections by Bowen *et al.* (1961) for C, Al, Cu, Cd, Pb, and U in the energy range 15-120 MeV, there have been high-energy measurements of neutron total and reaction cross-sections for C, Al, Cu and Pb in the energy range 0.3-4.5 GeV (Coor *et al.* 1955, Atkinson *et al.* 1959). More recently, there have been measurements of proton reaction cross-sections on Be, C, Al, Fe, Cu, Ge, Sn, and Pb in the energy range 220-570 MeV (Renberg *et al.* 1972), of proton total cross-sections on light nuclei in the energy range 180-560 MeV (Schwaller *et al.* 1972), of neutron total cross-sections on Be, C, Al, Fe, Cu, and Pb in the energy range 0.4-26 GeV (Engler *et al.* 1968, Lakin *et al.* 1970, Parker *et al.* 1970, Jones *et al.* 1971 Schimmerling *et al.* 1971), and of neutron total cross-sections on C, O, Al, Cu, Sn, and Pb over an energy range of 28-54 GeV (Babaev *et al.* 1974). The earlier neutron data have been

wavefunction or the nature of the interaction in the vicinity of the target nucleus.

This uncertainty about the physical content of the potential theory for composite projectiles has led to considerable emphasis on the interpretation of the behaviour of the reflection coefficients. The method of direct parametrization was outlined in §7.1.1 and is taken further in §7.4.3 where we consider the relation between the size parameters determined in this way with the size parameters arising in the optical model. A similar motivation has led to extensive use of diffraction theory, described in §7.1.4, and this leads to additional size parameters.

In phenomenological analyses the optical potential for tritons and helions is taken to be of the same form as the nucleon-nucleus potential (7.59), and this form is also used for α-particles except that the spin-orbit term is absent. For the deuteron, which has spin 1, the potential can in principle contain tensor terms of the form (Satchler 1960)

$$T_{\mathrm{R}}(r) = \{(\mathbf{s}.\mathbf{r})^2/r^2 - \tfrac{2}{3}\}\, f_{\mathrm{R}}(r) \tag{7.106a}$$

$$T_{\mathrm{L}}(r) = \{(\mathbf{l}.\mathbf{s})^2 + \tfrac{1}{2}(\mathbf{l}.\mathbf{s}) - \tfrac{2}{3}\,\mathbf{l}^2\}\, f_{\mathrm{L}}(r) \tag{7.106b}$$

$$T_{\mathrm{p}}(r) = \{(\mathbf{s}.\mathbf{p})^2 - \tfrac{2}{3}\,\mathbf{p}^2\}\, f_{\mathrm{p}}(r). \tag{7.106c}$$

There is now evidence from polarization data for a contribution from such terms (Glashauser and Thirion 1969, Johnson 1971, Haerberli and Knutson 1973), but they are frequently omitted in analyses of differential cross-sections.

A convenient way of comparing the expected behaviour of the optical potential for composite projectiles with the nucleon potential is to use the folding model originally proposed by Watanabe (1958). For the deuteron this yields a potential of the form

$$V_{\mathrm{d}}(\mathbf{r}) = \langle \phi_{\mathrm{d}}(\mathbf{R})|V_{\mathrm{n}}(|\mathbf{r}+\tfrac{1}{2}\mathbf{R}|) + V_{\mathrm{p}}(|\mathbf{r}-\tfrac{1}{2}\mathbf{R}|)|\phi_{\mathrm{d}}(\mathbf{R})\rangle \tag{7.107}$$

and the extension to the three-body system is obvious.

Several authors have evaluated the expression (7.107) taking the S-state part of the deuteron wavefunction as a Hulthen function (Rook 1965, Abul-Magd and El-Nadi 1966, Perey and Satchler 1967, Johnson and Soper 1970). For a nucleon-nucleus potential whose real part has parameters r_0 = 1.25 fm, a = 0.65 fm Perey and Satchler (1967) obtain a deuteron-nucleus potential for medium and heavy nuclei whose real part has a shape similar to the original Saxon-Woods form but with a much larger diffuseness. For $60 \lesssim A \lesssim 200$ they find that the halfway radius of the potential for 11.8 MeV deuterons is given by $(1.174\ A^{1/3} + 0.01\ A^{2/3})$fm and the diffuseness is ~ 0.88. (For light nuclei the potential is not flat for small values of r and the potential is not symmetric about the halfway point.) The central depth is given by

$$U_{\mathrm{d}}(0) = C|V_{\mathrm{n}}(0) + V_{\mathrm{p}}(0)|,$$

with $C \sim 0.96$ for medium-mass nuclei. Using the surface-peaked form for the imaginary part of the nucleon-nucleus potential with depth W_{D}, r_0' = 1.25 fm, and a' = 0.47 fm, Perey and Satchler found an imaginary part for the deuteron potential which peaks at $\sim 1.23\ A^{1/3}$ fm with a width given by $a' \simeq 0.73$ fm and a maximum value of $0.64\ (W_{\mathrm{Dp}} + W_{\mathrm{Dn}})$. Comparison with elastic scattering data in the region 10-25 MeV shows that the prescription (7.107) yields a real potential in good agreement with the data provided that the strength is reduced by 10-20 per cent, but it yields an imaginary part which is too weak and does not give enough absorption at large distances. A similar study has been made for 52 MeV deuterons by Hinterberger *et al.* (1968).

For deuteron scattering from ^{90}Zr at 5.5 MeV, Haerberli and Knutson (1973) have fitted the differential cross-section together with the three tensor and one vector polarization functions using the spin-orbit potential predicted by the Watanabe model and also the tensor term T_{R} generated by the D-state of the deuteron. They did not adjust the parameters of these terms, but treated the real and imaginary central terms of the potential phenomenologically. Using the folded

TABLE 7.8

Average parameter sets for deuteron optical potentials

	27.5, 52 MeV[a]	11-52 MeV[b]	34.4 MeV[c]	11-27 MeV[d]	11-27 MeV[e]
U (MeV)	$79+1.5Z/A^{1/3}-0.35E$	$100+2.5Z/A^{1/3}-0.5E$	$90.2+0.89Z/A^{1/3}$	$81+2.0Z/A^{1/3}-0.22E$	$75+1.14Z/A^{1/3}-0.42E$
R_0 (fm)	$1.25A^{1/3}$	$1.05A^{1/3}$	$0.968A^{1/3}+0.029A^{2/3}$	$1.15A^{1/3}$	$1.30A^{1/3}$
a (fm)	$0.81-0.024A^{1/3}$	$0.71+0.04A^{1/3}$	0.814	0.81	0.73
W_D (MeV)	13	$5+2A^{1/3}$	$4.33+2.2A^{1/3}$	f	f
R' (fm)	$1.25A^{1/3}$	$1.28A^{1/3}$	$1.09A^{1/3}+0.8$	$1.34A^{1/3}$	$1.34A^{1/3}$
a' (fm)	$0.51+0.076A^{1/3}$	$0.71+0.02A^{1/3}$	$0.554+0.059A^{1/3}$	0.68	0.65
V_{so} (MeV)	6	6	7	0	0

[a] Hinterberger *et al.* (1968), set 1.
[b] Hinterberger *et al.* (1968), set 2
[c] Newman *et al.* (1967)
[d] Perey and Perey (1963), set B.
[e] Perey and Perey (1963), set D.
[f] No systematic trend with A.

Watanabe model. A global analysis at 38 MeV for nine target nuclei in the range A = 46-91 confirmed preference for a surface-peaked imaginary potential (Urone, Put, and Ridley 1972), but no clear evidence for a symmetry term was observed. Fulmer and Hafele (1973*a*) find a spin-orbit strength about 1 MeV greater for targets with non-zero spin. These values appear to be weakly dependent on target mass number and incident energy.

The validity of the Watanabe model for ^{3}He scattering has been examined in the energy range 14-84 MeV. Marchese, Clarke, and Griffiths (1972) found that the volume integral of the real part of the observed phenomenological potential increases with energy up to $E \sim 40$ MeV and then probably tends to a constant value, whereas the prediction of the Watanabe model is that the volume integral should decrease linearly with energy owing to the energy dependence of the nucleon-nucleus potential. Rather simplified model calculations by Simbel (1974), following the adiabatic method of Johnson and Soper (1970) to include break-up of the projectile, give qualitative agreement with the phenomenological results and suggest that at low energies the disintegration of the projectile has an important effect.

The elastic scattering data for 40-50 MeV α-particles on medium and heavy nuclei for scattering angles up to ~ 70° can be fitted with volume absorption and four variable parameters (Drisko, Satchler, and Bassel 1963, Fernandez and Blair 1970), but for large-angle data and higher energies a six-parameter potential with $R' > R_0$ is usually required (Jackson and Morgan 1968, Hauser *et al.* 1969, Tatischeff and Brissaud 1970). In the energy region of 15-30 MeV there is considerable difficulty in fitting the large-angle data, particularly for light nuclei, unless very unusual optical potentials are introduced. This has led to a variety of explanations (see, for example, Hodgson 1971, Agassi and Wall 1972) in terms of α-clustering in the target nucleus, glory scattering, angular momentum mismatch, and exchange effects. However it has recently been shown that the optical model can describe the data for A = 40 provided the potential yields sufficient reflection at the

changes necessary to satisfy eqn (7.10). For each of these discrete sets of parameters there are continuous ambiguities which give rise to a family of potentials and have the effect of keeping the surface of the potential unchanged. Extensive studies of these ambiguities have been made, and more recent studies have looked for situations in which they are removed.

Two extensive studies of ^{3}He scattering from medium-mass nuclei have been carried out at 38 MeV with data extending to $\theta \sim 155°$ (Urone *et al.* 1971*b*) and at 33 MeV with data out to $\theta \sim 175°$ (Cage *et al.* 1972). Both studies show that at these energies the discrete ambiguities are not removed by the existence of large-angle data, although it appears that the existence of large-angle data does change the values obtained for the η_l, as shown in Fig. 7.13. A further study by Baugh (1969) at 29 MeV has shown that inclusion of σ_R also does not serve to remove the ambiguities. It is evident from these papers that the volume integral and r.m.s. radius are well-determined for a given family of potentials, but that these

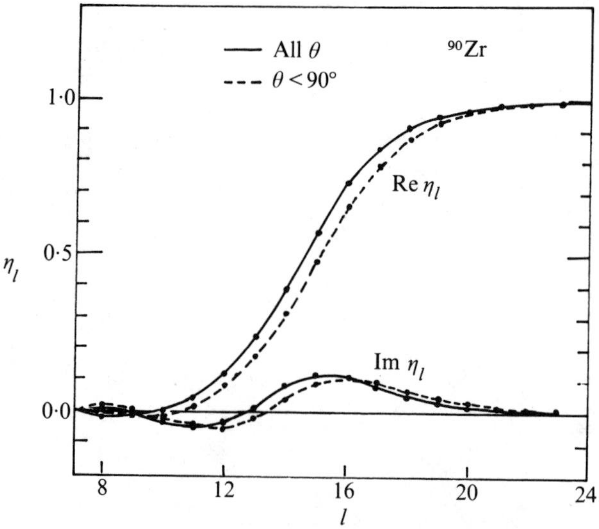

FIG. 7.13. Reflection coefficients obtained by fitting full angular range (solid line) compared with those obtained by fitting a restricted angular range. (From Urone *et al.* 1971*b*.)

values change abruptly from one family to another owing to the discrete ambiguity. Some results are given in Table 7.9. Thus the volume integral may usefully be used as a label for a potential family, and is to be preferred to the real depth for this purpose, but at these energies there are no unique values for these moments, in contrast to the situation for nucleon scattering. The volume integral of the imaginary potential is not so well determined. Fulmer and Hafele (1972, 1973*b*,*c*) have extended measurements of helion scattering to higher incident energies. They find that fits to the higher energy data at 59.8 and 71.1 MeV yield one unique potential family with $U \sim 130$ MeV and $J_{\mathrm{R}}^{\mathrm{P}} \sim 330$ MeV fm^3. They also obtain an energy dependence for the potential parameters of the form (Fulmer and Hafele 1973*c*)

$$U = 133.9 - 0.14\, E$$

$$W_{\mathrm{D}} = 22.4 - 0.04\, E.$$

Alpha-particle scattering has been studied extensively in the range 20-64 MeV and $\theta \lesssim 80°$. At 30-44 MeV, Jackson and Morgan (1968) found that the invariance of the potential in the extreme surface region, often expressed for Saxon-Woods potentials in terms of the Igo conditions (Igo 1958, 1959),

$$U \exp(R_0/a) = \text{const.} \tag{7.111a}$$

$$W \exp(R_0'/a') = \text{const.} \tag{7.111b}$$

is valid only for rather small variations of the diffuseness, and that a better criterion is the constancy of the potential in the vicinity of the strong absorption radius $R_{1/2}$, which is defined by the relation

$$k\, R_{1/2} = n + \{n^2 + L_{1/2}(L_{1/2} + 1)\}^{1/2} \tag{7.112a}$$

where n is the Coulomb parameter and $L_{1/2}$ is the (non-integer) angular momentum for which

1976*a*) that the relation between the real and imaginary parts of the potential in the surface region is not unimportant. A similar effect has been noted in heavy-ion scattering (Satchler 1975).

It appears from Table 7.9, and from similar results for α-particle scattering given in Table 7.10, that the volume integral per interacting nucleon pair, i.e.

$$J_R^p = J_R/A\ A_p, \tag{7.113}$$

has the same order of magnitude for different composite projectiles. Cage, Cole, and Pyle (1973) have made this comparison more precise by choosing data for which the centre-of-mass energy per projectile nucleon, E/A_p, is constant. They find that there exist families of potentials for proton, deuteron, helion, and α-particle scattering from the same target nucleus which have a constant volume integral $J_R^p \sim 480$ MeV fm^3, although other families of potentials can also be found.

Data for 10.8 MeV α-particle scattering from ^{24}Mg over the whole angular range were studied by Thompson, Crawford, and Davis (1967). They found that the existence of large-angle data did not remove the discrete ambiguities. The separate fits to the forward-angle or backward-angle data were somewhat superior than those for the whole angular range, and fits to the forward-angle data gave rise to a value of W about twice that obtained from the whole angular range.

For α-particles, data at higher energies are becoming available. Singh *et al.* (1969) have compared analyses of data on ^{24}Mg at 40 MeV and 80 MeV. At the lower energy several sets of parameters were obtained, but at 80 MeV only one acceptable minimum was found, and the corresponding parameters gave an excellent fit up to ~ 170°. When the 80 MeV data were restricted to $\theta \lesssim 70°$, four families of potentials were obtained. Thus, the discrete ambiguity was removed but the continuous ambiguities remained, such that the product VR_0^4 and WR'^4 remained constant for good fits. The r.m.s. radii of the real and imaginary potentials also remained constant.

TABLE 7.10

Optical-model parameters for α-particle scattering from ^{58}Ni at 50 MeV†

U (MeV)	$R_0 A^{-1/3}$ (fm)	a (fm)	J_R^P (MeV fm^3)	$\langle r^2 \rangle_R^{1/2}$ (fm)	W (MeV)	$R' A^{-1/3}$ (fm)	a' (fm)
57.2	1.652	0.487	285	5.272	18.6	1.717	0.221
95.1	1.555	0.516	402	5.040	19.6	1.692	0.246
132.9	1.505	0.518	511	4.905	20.8	1.678	0.250
178.4	1.450	0.540	621	4.787	20.7	1.687	0.251
225.6	1.424	0.531	745	4.703	25.3	1.671	0.253
269.0	1.394	0.533	836	4.624	26.0	1.655	0.259
327.3	1.366	0.530	961	4.544	27.7	1.634	0.272

†From Weisser *et al.* 1970.

$$U(\mathrm{E}) = 160\ (1 - 0.003\ E) \qquad (7.114c)$$

where the uncertainties on $U(0)$ and α are $\pm$ 20 MeV and $\pm$ 0.0015 MeV^{-1}, respectively. For $^{90}\mathrm{Zr}$, Paans, Put, and Malfliet (1973) have studied the energy dependence over the range 40-118 MeV and obtain

$$U(E) = 160 - 0.26\ E = 160\ (1 - 0.0016\ E). \qquad (7.114d)$$

Elastic scattering of helions at 217 MeV has recently been studied for a wide range of target nuclei (Willis *et al.* 1973). The angular range of the data extends to $\theta_{\mathrm{cm}} \sim 40°$ and this appears to be sufficient to remove the discrete ambiguity. Good fits to the data were obtained, except for the lightest target nuclei, using a six-parameter Saxon-Woods potential without a spin-orbit term. The diffuseness parameters of the real potential lie in the range a = 0.77-0.86 fm and are consistent with values given in Table 7.9, but the other parameters of the real potential lie in the range U = 65-78 MeV and r_0 = 1.24-1.30 fm. These parameters lead to a volume integral per pair of interacting nucleons of 255 MeV fm^3.

Thus, it appears that a combination of high-energy and large-angle data serves to eliminate the discrete ambiguity. Goldberg and Smith (1972) have proposed criteria based on the semiclassical description of scattering; these are that the energy is high enough for the cross-section to exhibit an exponential decrease beyond a certain angle θ and that the measurements must be continued beyond this angle. The angle θ increases with increasing A and decreases with increasing E, so that for large A and small E the situation of an exponential fall in the cross-section may not occur at all, and in this case discrete ambiutities can always be expected. This result is consistent with the work of Hauser *et al.*, and it also agrees with the appearance of the α-particle data of Tatischeff and Brissaud but not with their interpretation of their data. An alternative approach, suggested by Parks *et al.* (1972), is the study of scattering from aligned targets. They have studied α-particle scattering from aligned $^{165}\mathrm{Ho}$ in the energy range 14-23 MeV at fixed large angles, and show that parameter sets which give equally good fits for unaligned nuclei give sufficiently different results for the aligned case. This method requires independent knowledge of the nuclear deformation. Cross-sections for ^{16}O scattering from ^{28}Si at 145 MeV and 215 MeV do not show the structureless exponential fall observed for light-ion scattering (Cramer *et al.* 1976). The analyses of these cross-sections together with data in the range 33-81 MeV shows that satisfactory fits can be obtained with an energy-independent optical potential. The depth of the best-fit real potential they obtain is only 10 MeV and the depth of the imaginary potential exceeds this by a factor of ~2. The energy independence assumed in this analysis is consistent with the predictions of the microscopic model (see §7.4.4 and eqn 7.128).

7.4.3. *Strong absorption radii*

Although an optical model analysis of elastic scattering data determines a radius parameter R_0, the existence of ambiguities in the potential parameters implies that the radius parameter is a function of the strength of the potential. This means that for strongly absorbed projectiles R_0 is not a significant size parameter, and comparisons of such parameters for different nuclei should be treated with considerable caution.

Fortunately, it has proved possible to define a number of radius parameters for strongly absorbed projectiles which are well-determined and reproducible. Diffraction theory leads immediately to a single radius parameter R_F, defined in eqn (7.37), which can be determined quite accurately by fitting the positions of maxima or minima in the cross-section. For comparison of results for a range of nuclei it is necessary to use a Coulomb-corrected radius R_{FC} (Fernandez and Blair 1970). However, the sharp cut-off model does not give good agreement with the data unless a smoothing factor is introduced, either in configuration space or in angular momentum space. The usual procedure is to parametrize the reflection coefficients in l-space and to define a strong absorption radius (s.a.r.) through eqns (7.112) and (7.113). An

From these data, in particular, it can be seen that the radii increase approximately monotonically with $A^{1/3}$, except for ^{48}Ca which is anomalously low, but that the rate of increase for different isotope sequences is clearly not constant.

Blair has discussed the relation between the strong absorption parameters and the potential parameters of the optical model (Blair 1966, Fernandez and Blair 1970), and in particular has shown that the radius r_b at which the real potential barrier is just surmounted in a grazing collision and the magnitude of the real part of the optical potential $U(r_b)$ at this radius are related by

$$U(r_b) \simeq \left(\frac{a}{r_b - 2a}\right)\left(E - \frac{Z Z_p e^2}{r_b^2}\right). \qquad (7.118)$$

From this it follows that, for a fixed incident energy and diffuseness parameter, $U(r_b)$ decreases with increasing mass number, and hence the average behaviour of the strong absorption radii increases more rapidly with $A^{1/3}$ than does the halfway radius of the optical potential or of the nuclear matter distribution. A study of optical model wavefunctions (Jackson and Morgan 1968) has shown that the strong absorption radius can be interpreted as a measure of the distance at which the process of absorption begins to be effective. This picture is valid for the illuminated side of the nucleus and is consistent with diffraction theory, but on the dark side the focus effect (McCarthy 1959) is observed owing to refraction.

For heavy-ion scattering and for low energy α-particle scattering on heavy nuclei it is necessary to take notice of the importance of Coulomb scattering. The scattering angle θ_c at which the distance of closest approach for pure Coulomb scattering is equal to the strong-absorption radius is given by

$$k R_{1/2} = n(1 + \operatorname{cosec} \tfrac{1}{2} \theta_c). \qquad (7.119)$$

It can be shown (Frahn 1966, 1971) that for $L_{1/2} \lesssim 10$ non-diffractive scattering occurs, while for $L_{1/2} \gtrsim 10$ and

$k\ R_{1/2} \gg n$ forward-angle diffraction patterns of Fraunhofer type are observed for $\theta > \theta_c$. For $L_{1/2} \gtrsim 10$ and $k\ R_{1/2} \simeq n$ diffraction patterns of Fresnel type are observed in the vicinity of $\theta \simeq \theta_c$. An example of Fresnel scattering is shown in Fig. 7.17.

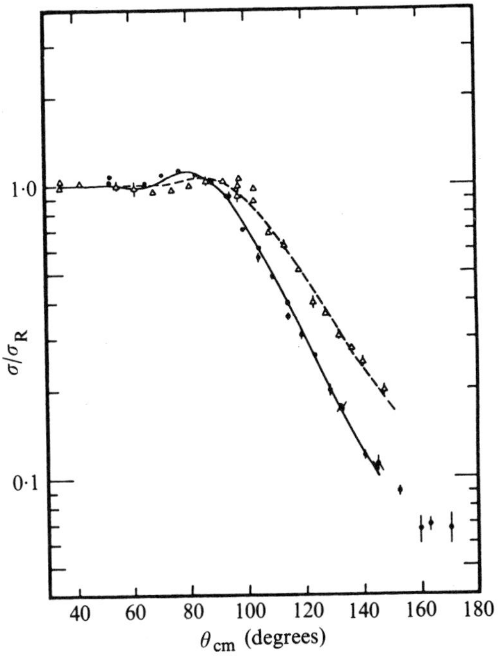

FIG. 7.17. Cross-sections for elastic scattering of ^{16}O on ^{40}Ca and ^{48}Ca near the Coulomb barrier: ● ^{48}Ca, experimental results at 33.23 MeV; Δ ^{40}Ca, experimental results at 34.37 MeV; solid curve, ^{48}Ca, theoretical results at 33.23 MeV, broken curve, ^{40}Ca, theoretical results at 34.37 MeV. (After Blair *et al.* 1972.)

Elastic scattering of heavy ions has been analysed in terms of parametrized phase shift models (McIntyre *et al.* 1960, 1962, Venter and Frahn 1964*b*, Baker and McIntyre 1967). Strong absorption radii determined from such analyses have been tabulated by Anni and Taffara (1970). An analysis of the scattering of 500 MeV ^{84}Kr ions from ^{208}Pb and ^{232}Th has been carried out (Colombani *et al.* 1972) to determine $R_{1/2}$ from

heavy ion scattering from ^{208}Pb are given in Table 7.13; in this case the strong absorption radius and the value of the real potential at this radius are very well-determined and are considered to be of greater significance than the barrier height and position.

A distance of closest approach in heavy ion reactions has also been studied by determination of the energy threshold E_F for reactions in which the heavy projectile and target nucleus form a compound nucleus which subsequently de-excites by emission of neutrons. The distance of closest approach is written as $r_0(A_1^{1/3} + A_2^{1/3})$ and the data for ^{40}Ar + Dy yield r_0 = 1.45 fm (Le Beyec *et al.* 1971) while that for ^{84}Kr + ^{116}Cd and ^{84}Kr + ^{72}Ge yields r_0 = 1.32 fm (Gauvin *et al.* 1972). A further analysis of these data for reactions with ^{40}Ar and ^{84}Kr and additional data for ^{32}S on a range of target nuclei from ^{24}Mg to ^{58}Ni have been presented (Gutbrod, Winn, and Blann 1973) in terms of the interaction barrier radius R_F and height V_F. These quantities are defined by the relations

$$\frac{\mathrm{d}}{\mathrm{d}r} V(r) \bigg|_{r=R_F} = 0 \qquad (7.120\ )$$

$$E_F = V_F = V(R_F). \qquad (7.120b)$$

The interaction barrier is also used to define an effective radius through the Coulomb interaction, i.e.

$$R_{\text{eff}} = Z_1 Z_2\, e^2/V_F \qquad (7.120c)$$

where $R_{\text{eff}} > R_F$. It is found that R_{eff} can be represented by the formula

$$R_{\text{eff}} = R_1 + R_2 + d \qquad (7.120d)$$

where R_1, R_2 are the equivalent uniform or quadratic radii of the heavy ion and the target respectively and $d \simeq 1.7$ fm. The fact that d is a constant within the experimental uncertainties of ± 0.2-0.3 fm implies that there is no experimental evidence for dynamic effects on the radii larger than 2-3 per cent. A similar analysis by Lefort *et al.* (1972) gave a value of d which decreased with increasing projectile mass. The difference between the two calculations is that Lefort *et al.* took $R_1 + R_2 = 1.26(A_1^{1/3} + A_2^{1/3})$ whereas Gutbrod *et al.* took the values of R_1 and R_2 from electromagnetic measurements. Wong (1972*a*) calculated V_F assuming a Saxon-Woods form for the nuclear potential with fixed depth and diffuseness and a radius parameter $r_0 = 1.2$ fm, and allowing both the nuclear and Coulomb potentials to be deformed when appropriate. He then derived values of $r_e = R_{eff}(A_1^{1/3} + A_2^{1/3})^{-1}$. These values decrease with increasing Z_1 or Z_2 and increase if the target or projectile is permanently deformed. This behaviour for r_e is consistent with the result that the nuclear quadratic radius does not increase precisely with $A^{1/3}$ (see §1.2.3 and Chapter 3). This use of formula $r_0(A_1^{1/3} + A_2^{1/3})$ for a projectile and target of widely different mass numbers is liable to lead to error, and the approach of Gutbrod *et al.* (1973) is to be preferred.

Tamain *et al.* (1975) have now given a definition of complete fusion reactions. They write $\sigma_R = \sigma_{CF} + \sigma_{NCF}$ where σ_{NCF} is the total cross-section for reactions not leading to complete fusion and the reaction cross-section and complete fusion cross-section are given by

$$\sigma_R = \pi\lambda\!\!\!^{-2} \sum_{l=0}^{l_{max}} (2l+1), \qquad \sigma_{CF} = \pi\lambda\!\!\!^{-2} \sum_{l=0}^{l_{crit}} (2l+1).$$

Classically, $\sigma_R = \pi(R_1 + R_2)^2\,(1 - V_F/E)$. For ^{40}Ar reactions on Mo, Sb, Ho, Bi, and U, they find that the critical angular momentum l_{crit} for complete fusion increases with the mass and energy of the projectile. The corresponding radius parameter is $R_c = r_{crit}(A_1^{1/3} + A_2^{1/3})$ with $r_{crit} = 1.03 \pm 0.08$ fm.

7.4.4. *Microscopic methods*

In a microscopic model we seek to describe the scattering of a composite projectile from the nucleus in terms of the more fundamental interactions between the nucleons, and the result that the volume integral per interacting pair is approximately constant suggests that some sort of folding model could be appropriate. For composite projectiles we have two alternative single-folding methods, but in both cases the simple folding integral is the first term in an expansion so that the best choice is the one which minimizes the importance of the higher-order terms or allows reliable estimation of these terms. For the deuteron the Watanabe model given by eqn (7.106) has proved useful while the extension of the microscopic model used for nucleon scattering does not appear to yield physically meaningful results (Winsborrow *et al.* 1972). For other projectiles the microscopic treatment of the optical potential has been used in order to relate the nuclear matter distribution to strong interaction parameters.

In the single-folding, microscopic procedure the real part of the optical potential for α-particle scattering has been generated from the expression

$$U(r) = \int V_{\mathrm{n}\alpha}(|\underset{\sim}{r} - \underset{\sim}{R}|)\ \rho(R)\ \mathrm{d}^3R. \qquad (7.121)$$

Comparison of the calculated potentials with phenomenological potentials (Jackson 1964, 1969*a*, Lilley 1971) and examination of the behaviour of the integrand (Batty, Friedman, and Jackson 1971) shows that eqn (7.121) can produce the required behaviour for the surface region of the potential and that this region of the potential is determined mainly by the 50-10 per cent region of the matter distribution. A more detailed and qualitative use of this method depends on the precision with which we can define the interaction $V_{\mathrm{n}\alpha}$ between the free α-particle and a target nucleon. Since $V_{\mathrm{n}\alpha}$ is an effective interaction whose formal properties are determined by the many-body aspects of the problem, it may be argued that $V_{\mathrm{n}\alpha}$ should be treated as a phenomenological interaction which simulates some of the many-body effects and whose parameters may therefore be determined by fitting the data on α-particle scattering by nuclei (Morgan and Jackson 1969, Tatischeff and Brissaud 1970, Bernstein and Seidler 1971). Alternatively, it may be argued that the interaction of the α-particle with the nucleus is confined to a region where the nucleon density is low, so that, unless substantial clustering occurs, the effect of exchange and multiple scattering may be very much reduced compared with the situation for nucleon-nucleus scattering. This would imply that $V_{n\alpha}$ should be the same or very similar to the free n-α interaction (Madsen and Tobocman 1965, Lilley 1971, Mailandt *et al.* 1972). Yet another approach is to use a double-folding procedure and derive the n-α interaction by folding the n-n interaction into the ground-state distribution of the α-particle (Bernstein 1969, Budzanowski *et al.* 1970, Batty and Friedman 1971). The double-folding procedure has been developed by Budzanowski *et al.* (1974) who use the nucleon-nucleon effective interaction of Slanina and McManus (1968) and Fermi distributions for the α-particle and the target nucleus. They examine 27 MeV α-particle scattering from a range of nuclei from $A = 45$ to $A = 208$ and find reasonable agreement with experiment, even though this method completely neglects all exchange effects. The real part of the potential has a similar shape to the Saxon-Woods potential but has a shorter tail.

These various approaches have been compared by Batty *et al.* (1971) who conclude that the best choice for a simple, local, effective interaction is the gaussian form

$$V_{n\alpha} = -V_0 \exp\{-K^2(\underset{\sim}{r} - \underset{\sim}{R})^2\} \tag{7.122}$$

and that parameters V_0 and K can be found to give reasonable agreement with free n-α scattering and good agreement with low low- and medium-energy elastic and inelastic α-particle scattering. A Saxon-Woods form gives superior agreement with the n-α scattering (Satchler *et al.* 1968, Mailandt *et al.* 1972, 1973) but is less satisfactory for inelastic scattering[†] (Satchler 1971*b*).

Mailandt *et al.* (1972, 1973) have used their Saxon-Woods form for $V_{n\alpha}$, whose parameters are given in Table 7.1, to calculate the real part of the optical potential for 27 MeV

[†]See §8.1.3.

α-particle scattering from ^{45}Sc, ^{65}Cu, ^{90}Zr, and ^{122}Sn by folding $V_{n\alpha}$ into a nuclear matter distribution of Fermi shape. They take the imaginary part of the potential to have the standard phenomenological form and fit the data over an angular range from 15° to 175°. By comparison with the parameters for the proton distributions obtained from an analysis of muonic X-ray data (Acker *et al.* 1966) they compare the r.m.s. radii of the neutron and proton distributions. For ^{45}Sc, ^{65}Cu, and ^{90}Zr the results are consistent with

$$\langle r^2 \rangle_n^{1/2} - \langle r^2 \rangle_p^{1/2} \simeq 0$$

while for ^{122}Sn they obtain

$$\langle r^2 \rangle_n^{1/2} - \langle r^2 \rangle_p^{1/2} = 0.22 \pm 0.009 \text{ fm.}$$

Other authors prefer to construct both the real and the imaginary parts of the optical potential using the folding procedure. Since the free n-α interaction is real at the relative energies associated with medium-energy α-particle scattering from nuclei (see Table 7.1), use of the free interaction would then require the whole of the imaginary part of the potential to come from second- and higher-order terms in the expansion for the potential. However, the effective interaction should in principle be complex. (This can be seen from eqn (7.23) and from §7.2.1.) Also it is known that for medium energies and a restricted angular range the data are not very sensitive to the details of the imaginary potential and that fits to the data can be obtained with four parameters. This encourages the view that the folded potential may be written as

$$V(r) = -(V_0 + iW_0) \int \exp\{-K^2(\underset{\sim}{r} - \underset{\sim}{R})^2\}\, \rho(R)\, d^3R \quad (7.123)$$

giving an imaginary potential of the same radial form as the real potential. It has been shown that this calculated potential leads to excellent agreement with the data at 22 MeV (Batty and Friedman 1971), 42 MeV (Morgan and Jackson 1969),

104 MeV (Bernstein and Seidler 1971, 1972, Rebel *et al.* 1972*a*), and fair agreement at 166 MeV (Tatischeff and Brissaud 1970).

The potential (7.122) has been used to fit the 42 MeV data of Fernandez and Blair (1970) for the calcium and nickel isotopes (Jackson 1970*b*, Batty *et al.* 1971). For the calcium isotopes both the ES and ZD distributions were used, while for the nickel isotopes the BG and ZD distributions were chosen,† and two-parameter fits to the data were obtained by varying V_0 and W_0 for fixed K, or by varying W_0 and K for fixed V_0. Results for the strong absorption radii are shown in Fig. 7.18 compared with the values obtained by other methods. It is clear that the microscopic model can reproduce the isotopic behaviour of the strong absorption radii and that no *ad hoc* explanation of this behaviour is required.

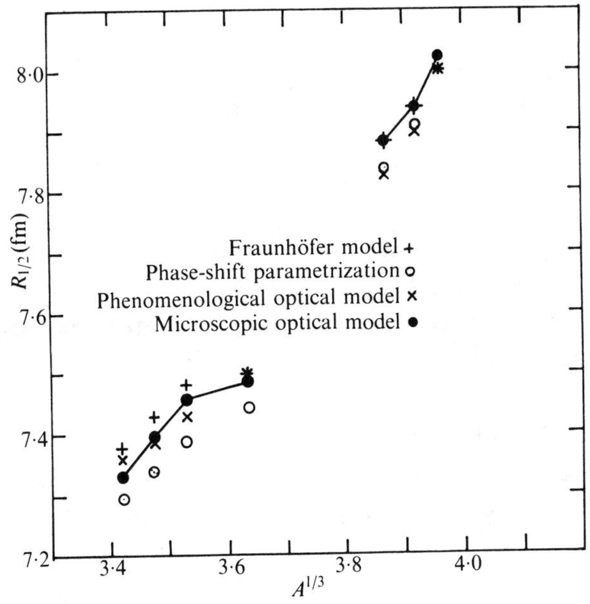

FIG. 7.18. Comparison of strong absorption radii obtained by microscopic and phenomenological methods. (From Jackson 1970*b*.)

†For details of these distributions, see §2.2 and Table 1.1.

A typical microscopic fit to the 42 MeV data is shown in Fig. 7.19. Examination of the corresponding potentials

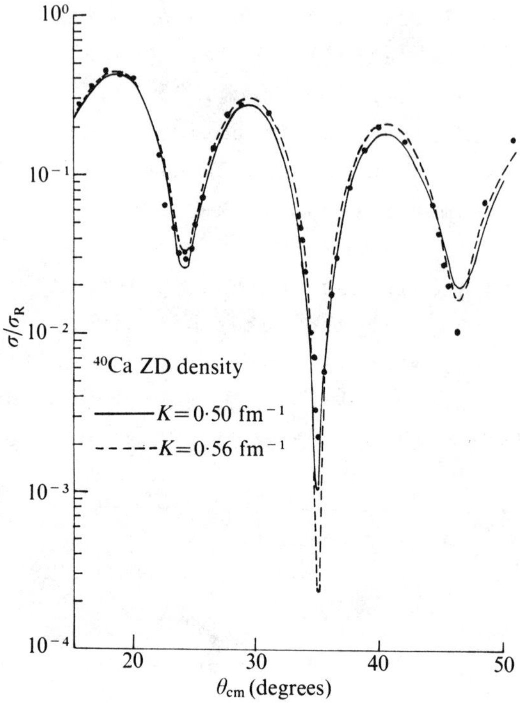

FIG. 7.19. Microscopic results for 42 MeV α-particle calculated using the ZD matter distribution compared with the data of Fernandez and Blair. (From Batty *et al.* 1971.)

shows that the search procedure adjusts the free parameters to give the same real potential in the vicinity of $R_{1/2}$ which, as we have seen earlier, leads to equally good fits to the data. This leads to an ambiguity of the form

$$V_0K^{-6} = \text{constant} \tag{7.124}$$

where the constant depends on the type of nuclear distribution used, and is $(2.6 \pm 0.2) \times 10^3$ MeV fm^6 for the ZD distribution (Batty *et al.* 1971). Using the 42 MeV data it is not possible to distinguish between the various distributions unless some specific choice is made for K, but for the 22 MeV data of Goldring *et al*. (1970) on the lead and bismuth isotopes Batty and Friedman (1971) have shown that the BG distributions will not yield the necessary values of r_b and $U(r_b)$ for any variation of the parameters of the gaussian effective interaction. The ZD distributions do yield satisfactory results. This result again indicates that a large difference in the r.m.s. radii of the neutron and proton distributions is not consistent with the data. Differential cross-sections for ^{208}Pb and ^{209}Bi at 16-24 MeV together with (α,n) cross-sections have been analysed by Barnett and Lilley (1974) using the effective interaction of Mailandt *et al*. (1972). They obtain

$$\langle r^2 \rangle_m^{1/2} = 5.48 \text{ fm}$$

for ^{208}Pb which is consistent with a difference of essentially zero between the neutron and proton radii. More recent theoretical work at this energy (Jackson and Rhoades-Brown 1976) has shown that the quality of fit obtained with any real microscopic potential is dependent on some variation of the imaginary part.

Sumner (1974) has taken new data at 42 MeV on isotopes of Ca, Ni, Cu, Zn, Mo, and Pb. He has unfolded various effective interactions from the pheneomenological potentials which fit the data in order to obtain matter distributions in the extreme surface region and compare them with theoretical predictions.

Bernstein and Seidler (1971, 1972) have taken the view that nuclei with $N = Z$ should be used to calibrate the effective interaction which can then be used to predict cross-sections for other nuclei. Using the gaussian interaction with $K = 0.5$ fm^{-1} they fit the cross-section for 104 MeV α-particle scattering from ^{40}Ca and then predict the cross-sections for ^{16}O and ^{28}Si without further variation of the parameters and taking the proton and neutron distributions to be identical (Bernstein and Seidler 1971). They subsequently examined scattering from ^{90}Zr and ^{208}Pb at the same energy (Bernstein and Seidler 1972) and determined parameters of a nuclear matter distribution of Fermi shape. Combining these parameters with those for the proton distribution taken from fits to electron scattering they obtain

$$\langle r^2\rangle_n^{1/2} - \langle r^2\rangle_p^{1/2} = 0.20 \pm 0.13 \text{ fm}$$

for ^{90}Zr and 0.26 ± 0.13 fm for ^{208}Pb. The data at 104 MeV have also been used to test Hartree-Fock wavefunctions for ^{20}Ne and ^{24}Mg (Mackintosh 1972) but a good fit to the data was not obtained. Extensive microscopic calculations have also been carried out at 166 MeV using eqn (7.123) with $K = 0.5 \text{ fm}^{-1}$ (Tatischeff, Brissaud, and Bimbot 1972, Brissaud *et al.* 1972*a*) in order to deduce parameters of neutron distributions using Fermi or modified Fermi distributions. The results are generally consistent with

$$\langle r^2\rangle_n^{1/2} - \langle r^2\rangle_p^{1/2} \simeq 0.15 \pm 0.1 \text{ fm}.$$

The fits to the data with the fully microscopic model (7.123) are satisfactory only up to $\theta \sim 22°$ but are somewhat improved when the imaginary part of the potential is replaced by a microscopic Saxon-Woods form.

Lerner *et al.* (1975) have obtained data at high relative accuracy for 79 MeV α-particle scattering from ^{40}Ca and ^{48}Ca. Following the method of Bernstein and Seidler, they find that the difference function $D_\sigma(\theta)$ defined by eqn (7.89) is sensitive primarily to the difference in the r.m.s. radii of the matter distributions of the two isotopes; they obtain a difference of 0.05 ± 0.04 fm which is in disagreement with most of the HF results given in Table 2.1. Using Negele's distributions (Negele 1970) they obtain good agreement with the cross-section for ^{40}Ca but not for ^{48}Ca. By combining their results with other information they obtain

$$\langle r^2\rangle_n^{1/2} - \langle r^2\rangle_p^{1/2} = 0.03 \pm 0.08 \text{ fm}$$

for ^{48}Ca. This again is in disagreement with the HF predictions. Because of the folding procedure, contributions to the icroscopic potential in the vicinity of the strong absorption adius comes from regions of the nucleus at somewhat smaller istances but definitely not from the interior. For α-article scattering at 104 MeV from ^{208}Pb, Gils and Rebel 1976) found little sensitivity to the matter distribution for < 6.4 fm. Brissaud and Brussel (1976*b*) have carried out a odel independent analysis on the higher energy data at 166 eV for ^{40}Ca. For this lighter nucleus they found that the oments with k = 2-4 were well-determined, but the matter disribution in the interior region for $r < 1$ fm was essentially ndetermined.

By analysing data for a range of nuclei, Tatischeff and rissaud (1970) obtained average values of V_0 and W_0 and howed that V_0 decreases roughly linearly with incident article energy. Lerner *et al.* (1972) have studied the energy ependence of α-particle scattering from ^{40}Ca between 39.6 MeV nd 115.4 MeV and have interpreted the data using the microscopic model of Bernstein and Seidler. They find that the oefficient V_0 decreases linearly with energy while the coefficient W_0 increases. For V_0 they find

$$V_0(E) = V_0 - \gamma E \tag{7.125}$$

with $V_0 = 44 \pm 4$ MeV and $\gamma = 0.11 \pm 0.04$. This result can also be written as

$$V_0(E) = V_0(1 - \alpha E) \tag{7.126}$$

with $\alpha = 0.0025 \pm 0.001$ MeV^{-1}, so that this energy dependence is consistent with the results given in eqn (7.114*b*,*c*,*d*) for phenomenological analyses.

The magnitude of the coefficient α can be related to the energy dependence of the n-α interaction. The results of Thompson *et al.* (1970) over the range 31-55 MeV (see Table 7.1) give

$$V_{p\alpha}(E) = 74 - 0.57\, E_{p}^{L}. \tag{7.127}$$

For simplicity we take the mass of the target nucleus to be infinitely heavy. In order to make the total energy the same in the free n-α scattering and in the α-nucleus scattering, we need $E_\alpha = 4E_p^L$, where E_p^L is the lab energy of the nucleon. Hence

$$V_{p\alpha}(E) = 74\ (1 - 0.0019\ E_\alpha). \tag{7.127a}$$

Satchler *et al.* (1968) give two expressions for the n-α interaction for different Saxon-Woods geometries, viz.

$$V_{n\alpha}(E) = 53.5 - 0.69\ E_n^L = 53.5\ (1 - 0.0032\ E_\alpha) \tag{7.127b}$$

$$V_{n\alpha}(E) = 45.1 - 0.43\ E_n^L = 45.1\ (1 - 0.0024\ E_\alpha). \tag{7.127c}$$

These coefficients for the energy dependence are consistent with that given by eqn (7.126). According to Thompson and Tang (1971) the energy dependence of the effective n-α interaction derived from a resonating-group calculation arises mainly from the exchange term, and over the range E_n = 50-100 MeV the energy dependence is not linear.

The energy dependence of the potential for scattering of a composite projectile can be examined in a general way (Jackson and Johnson 1974). The potential for the interaction between nucleus 1 and nucleus 2 can be written as

$$V_{12} = A_2\langle\psi_2|V_{1n}|\psi_2\rangle \tag{7.128}$$

where ψ_2 is the ground-state wavefunction for nucleus 2 and V_{1n} is the interaction between nucleus 1 and a nucleon. It is assumed that V_{1n} can be represented by an energy-independent, non-local potential of Perey-Buck form with non-locality parameter β_{1n}. This yields a non-local potential V_{12} with non-locality parameter β_{12} given by

$$\beta_{12} = \frac{A_1+A_2}{A_2(A_1+1)}\ \beta_{1n}. \tag{7.129}$$

This result is an extension of the work of Johnson and Soper (1972) for deuteron-nucleus scattering.

If we put $A_1 = 4$, $A_2 = A$, we find from eqn (7.129) that $\beta_{\alpha A} = (A + 4)\beta_{\alpha n}/5A$. This corresponds to the folding model (7.121), and shows that in this model the non-locality of the α-nucleus potential arises from the non-locality of the α-n interaction. Alternatively, we may put $A_1 = A$, $A_2 = 4$ which gives $\beta_{A\alpha} = (A + 4)\beta_{An}/4(A + 1)$ which is approximately $\beta_{A\alpha} \sim \frac{1}{4}\beta_{An}$ for large A. Similarly for $A_2 = 2$ we have $\beta_{Ad} \sim \frac{1}{2}\beta_{An}$, as shown by Johnson and Soper. In this case we have the Watanabe model, and the non-locality comes from the nucleon-nucleus potential. The depth of the energy-dependent equivalent local potential is given in an effective mass approximation by

$$V_A = V_0(1 - \frac{1}{4} a_{An} E_\alpha) = V_0(1 - a_{A\alpha}E_\alpha)$$

where $a_{Ax} = \mu_{Ax}\beta_{Ax}^2/2\hbar^2$ and μ is the reduced mass. Using the nucleon-nucleus energy dependence from eqn (7.87), we find $a_{A\alpha} \sim 0.0015\ \mathrm{MeV}^{-1}$. Allowing for experimental uncertainties this is consistent with the observed values.

Eqn (7.129) was obtained by neglecting the variation of ψ_2 over the range of the non-locality. If this approximation is improved by taking ψ_2 to be a gaussian function with range parameter α_2, the non-locality parameter becomes

$$(\bar{\beta}_{12})^{-2} = (\beta_{12})^{-2} + (\bar{\alpha}_2)^{-2} \tag{7.130}$$

$$\bar{\alpha}_2 = 2(A_1 + A_2)\alpha_2/A_1A_2 \tag{7.131}$$

and the potential is multiplied by the factor $(\bar{\beta}_{12}/\beta_{12})^3$. The effect of this factor is to reduce the magnitude of the potential so that it is always less than A_2 times the nucleon-nucleus potential. The reduction has been estimated to be ~6 per cent for deuteron scattering (Johnson and Soper 1972) and ~15 per cent for α-particle scattering (Jackson and Johnson 1974). The observed reductions in the potentials required to fit the data are somewhat larger than these estimates (Perey and Satchler 1967, Goldberg *et al.* 1974).

The use of microscopic models for heavy-ion scattering

has recently been reviewed by Siemssen (1972). The folding model (7.128) yields a potential of the form

$$U(\underset{\sim}{r}) = A_2 \int \rho_2(\underset{\sim}{R})\, V_{1n}\, (|\underset{\sim}{r} - \underset{\sim}{R}|)\, d^3R \tag{7.132}$$

where ρ_2 is the matter distribution for nucleus 2 and V_{1n} is the nucleon-nucleus optical potential for nucleus 1. Potentials of this type have been used by McIntosh *et al.* (1964), Eisen (1971), Brink and Rowley (1973), and by Ball *et al.* (1975). Such potentials are very deep in the interior region, probably unrealistically so, but may be used when the combination of strong absorption and strong Coulomb repulsion causes the scattering to be sensitive only to the long-range behaviour of the potential. Brink and Rowley (1974) have successfully predicted the position and height of the barrier for various combinations of projectile and target and, using the same method, Rowley (1974) has reproduced the modified Fresnel cross-sections for 500 MeV ^{84}Kr scattering observed by Colombani *et al.* (1972). The double-folding procedure gives a potential of the form

$$U(\underset{\sim}{r}) = \iint d^3R_1\, d^3R_2\, \rho_1\, (\underset{\sim}{R}_1)\, \rho_2\, (\underset{\sim}{R}_2)\, u\, (|\underset{\sim}{r} + \underset{\sim}{R}_1 - \underset{\sim}{R}_2|) \tag{7.133}$$

where u is the nucleon-nucleon interaction appropriately averaged over spin and iso-spin. A zero-range and a gaussian interaction gave quite good agreement (see Table 7.13) with data for ^{12}C and ^{16}O scattering from ^{208}Pb (Ball *et al.* 1975). In contrast, the long-range part of the Hamada-Johnston potential was not satisfactory owing to the long-range OPEP component, although it is acceptable for nucleon-nucleus scattering which depends more on volume properties (see §§7.3.2 and 8.1.3).

Other calculations of the heavy ion potentials yield shallow potentials which may be repulsive in the interior region owing to exchange effects or to the compression of nuclear matter above the equilibrium density. The theory which underlies these calculations is under rapid development and at the present time nuclear size information may be most reliably obtained from diffraction models, from direct phase shift parametrization, or from studies of barrier properties.

TABLE 7.13

Parameters for heavy ion scattering from ^{208}Pb *(Ball et al. 1975)*

Projectile	E_L (MeV)	E_L/A_p (MeV/nucleon)	$L_{1/2}$	$R_{1/2}$ (fm)	$-U(R_{1/2})$ (MeV)	$\frac{W(R_{1/2})}{U(R_{1/2})}$	r_b (fm)	$ReV(r_b)$ (MeV)
^{4}He	42	10	20.5	10.5	2.33	0.49	10.9	20.3
^{11}B	72	6.5	36.7	11.9	2.10	0.20	11.4	48.2
^{12}C	96	8	51.7	12.2	1.40	0.64	11.3	58.2
^{12}C	116	10	64.1	12.1	1.64	0.65	11.4	58.0
^{16}O	129.5	8	72.5	12.7	1.16	0.79	11.5	76.4
^{16}O	192	12	106.7	12.5	1.28	0.98	11.3	77.4
^{20}Ne	161	8	92.2	13.1	0.90	1.03	11.35	95.5

8
NUCLEAR REACTIONS

In this chapter we are concerned with the interpretation of nuclear size information from a wide range of reactions such as inelastic scattering, rearrangement collisions, particle production, and decay processes. Those processes which lead to definite final states of the residual nucleus are primarily studied for the sake of spectroscopic information related to the overlap of the initial and final states, whereas processes which lead to a sum of unresolved final states give information about average nuclear properties. In all cases we restrict the discussion to direct nuclear reactions and assume that the projectile energy is above the region in which formation of a compound nucleus has a significant effect. Unfortunately, this does not mean that many-body effects, such as core excitation, are necessarily unimportant, and the influence of such effects on the interpretation of the data must be considered.

Throughout this chapter the standard formulae will be quoted rather than derived, and the reader is referred to books on nuclear reaction theory for derivations of the formulae and discussions of their validity (McCarthy 1968, Jackson 1970*a*, Austern 1970).

8.1. INELASTIC SCATTERING TO RESOLVED FINAL STATES

8.1.1. *Nuclear transition densities*

Using the notation introduced in §7.1.2, the scattering amplitude for inelastic scattering from an initial state i to a definite final state f can be written as

$$f(\underset{\sim}{k}',\underset{\sim}{k}) = -\left(\frac{\mu}{2\pi\hbar^2}\right) T_{fi}(\underset{\sim}{k}',\underset{\sim}{k}). \tag{8.1}$$

The interaction potential can be separated into two parts, one of which is the usual optical potential for elastic scattering. The transition matrix element is then given by

$$T_{fi} = \langle \chi_f^- \Phi_f | t^+ | \chi_i^+ \Phi_i \rangle \tag{8.2}$$

where t is the transition operator in the nuclear medium represented by the optical potential, χ_i^+, χ_f^- are the solutions for the optical potential with outgoing and incoming boundary conditions respectively, and Φ_i, Φ_f are the initial and final nuclear wavefunctions. The multiple-scattering expansion for t has the same form as that for T given by eqn (7.19) with G_0 replaced by a propagator containing the optical potential. Replacing t by the first iteration gives the distorted-wave Born approximation (DWBA), while replacing t by the sum of the free two-body interactions $\Sigma\tau(j)$ gives the distorted-wave impulse approximation (DWIA).

In the impulse approximation, use of the expression (7.43) for the free nucleon-nucleon interaction leads to a scattering amplitude for nucleon-nucleus scattering of the form

$$f(\underset{\sim}{k}', \underset{\sim}{k}) = \left(\frac{\mu}{\mu_0}\right) G(q) \tag{8.3}$$

$$G(q) = \langle \chi_f^- \Phi_f | \sum_j M(j) | \chi_i^+ \Phi_i \rangle . \tag{8.4a}$$

If we neglect spin-orbit terms in the distorting potentials, $G(q)$ becomes

$$G(q) = \bar{M}(q) \int \chi_f^{-*}(\underset{\sim}{k}', \underset{\sim}{r}) \rho_{if}(\underset{\sim}{r}) \chi_i^+(\underset{\sim}{k}, \underset{\sim}{r}) \, d^3r \tag{8.4b}$$

where the coefficients of $M(j)$ are taken to be functions of the incident energy and momentum transfer $\underset{\sim}{q} = \underset{\sim}{k} - \underset{\sim}{k}'$ as in eqn (7.44), and $\bar{M}$ represents the matrix of the spin and isospin operators in these coefficients. The nuclear transition density ρ_{if} is defined as

$$\rho_{if}(\underset{\sim}{r}) = \langle f | \sum_j \delta(\underset{\sim}{r} - \underset{\sim}{r}_j) | i \rangle . \tag{8.5}$$

For convenience, we have *not* normalized this quantity to unity. If the transition takes place between substates M_i and M_f, the transition density (8.5) must be labelled by these quantum numbers. The cross-section for inelastic scattering of unpolarized nucleons from an unoriented target initially in spin state I_i is then given by

$$\frac{d\sigma}{d\Omega} = \frac{1}{2(2I_i+1)} \sum_{M_i M_f} |f^{M_i M_f}(\underset{\sim}{k}', \underset{\sim}{k})|^2 \tag{8.6a}$$

$$= \frac{\mu}{\mu_0} \frac{1}{2(2I_i+1)} \text{Trace } (G^+ G). \tag{8.6b}$$

In a plane-wave approximation the function $G(q)$ reduces to the form

$$G(q) \propto \sum_{LM} Y_L^{M*}(\theta,\phi) \int j_L(qr)\, \rho_{if}(\underset{\sim}{r})\, Y_L^M(\hat{\underset{\sim}{r}})\, d^3r \tag{8.7}$$

with the selection rules

$$|I_i - I_f| \leqslant L \leqslant I_i + I_f, \qquad M_f = M + M_i \tag{8.8}$$

so that a connection can be made with the corresponding cross-section for inelastic electron scattering through the Coulomb or longitudinal part of the electromagnetic interaction. The relevant formulae are given in eqns (3.33) and (3.37), and we may, for convenience, write (3.33) in the form

$$\frac{d\sigma}{d\Omega} = \sum_{L=0}^{\infty} \frac{d\sigma}{d\Omega}(CL) + \sum_{L=1}^{\infty} \frac{d\sigma}{d\Omega}(EL) + \sum_{L=1}^{\infty} \frac{d\sigma}{d\Omega}(ML) \tag{8.9}$$

to distinguish between the longitudinal, electric transverse, and magnetic transverse components. The lower limits of multipolarity L for the EL and ML transitions are determined by conditions explained in §3.2, and the actual values for the longitudinal transitions are determined by the selection rules (8.8). This connection provides a method of combined analysis of inelastic electron scattering and inelastic nucleon scattering from light nuclei at intermediate energies (Haybron, Johnson, and Metzger 1967). If the data for excitation by electron scattering are fitted using a parametrization of the transition density, this density may be used for a phenomenological analysis of inelastic nucleon scattering to the same state, provided of course that $A = 2Z$, and it may be assumed that the distribution of protons and neutrons are identical. For heavier nuclei a distorted-wave treatment is necessary for electron scattering (Griffy *et al.* 1962, Onley, Reynolds, and Wright 1964, Schucan 1965). Knoll (1973) has used some of the ground state charge distributions for ^{208}Pb obtained by Friedrich and Lenz (1972) to show that the variation in these distributions does not significantly influence the inelastic calculations.

Knoll (1973) has applied the δ-shell method of Friedrich and Lenz to construct a set of model-independent transition densities of the form

$$\rho_{\mathrm{if}}^{L}(r) = \sum_{n=1}^{N} \frac{P_{nL}}{R_{nL}^{2}}\,\delta(r - R_{nL}) = \sum_{n=1}^{N} P_{nL}\,\rho_{\mathrm{if}}^{nL}(r). \quad (8.10a)$$

The Fourier transforms of such transition densities which fit computed 'data' for a C2 transition over the range 1-2.5 fm^{-1} in momentum transfer q are shown in Fig. 8.1. It appears that

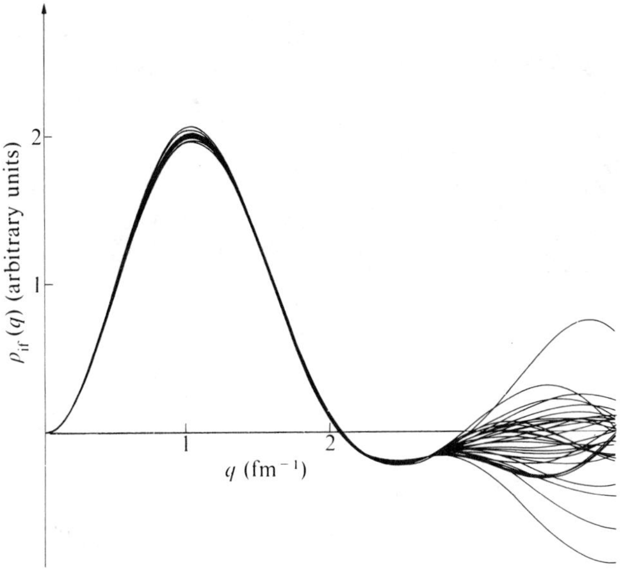

FIG. 8.1. Fourier transforms of various transition densities which fit the same inelastic cross-section for a C2 transition over the range of momentum transfer 1.0-2.5 fm^{-1}. (From Knoll 1973.)

the transition density can be determined over the range of q for which the inelastic cross-section has been measured. An alternative representation of a model-independent transition density is given by the Fourier-Bessel expansion

$$\rho_{if}(r) = \sum_{\nu} a_{\lambda\nu}\, j_\lambda(q_{\lambda\nu}\, r), \qquad r \leqslant R$$
$$= 0 \qquad\qquad , \qquad r > R \tag{8.10}$$

where the coefficients $q_{\lambda\nu}$ are determined from the requirement that $j_\lambda(q_{\lambda\nu}\, R) = 0$. The expansion for $\lambda = 0$ has also been used to represent the difference in charge density between isotopes or isotones (Neuhausen 1975). A more realistic expression could be obtained by expanding in terms of functions which fall exponentially at large r, such as the Weinberg states (Jackson, Hilton, and Roberts 1976).

Tibell (1969) has shown that for inelastic proton scattering at 185 MeV a plot of $\log\{(d\sigma_{exp}/d\Omega)/q^{2L}\}$ against q^2 is a straight line with negative gradient, at least for light nuclei and small values of momentum transfer $\lesssim 1.2\ \mathrm{fm}^{-1}$, as shown in Fig. 8.2. Now, using oscillator wavefunctions to construct shell model transition densities and a plane wave approximation, the theoretical cross-section for a transition of multipolarity L is given by

$$\frac{d\sigma}{d\Omega} \propto (qb)^{2L} \exp(-\tfrac{1}{2}q^2b^2)$$

where b is the oscillator-length parameter. The agreement between this simple prediction and Tibell's empirical result suggests that inelastic nucleon scattering in the intermediate-energy region is not strongly sensitive to the radial shape of the transition density or to distortion effects.† Provided the multipolarity is known and the transition density has the correct peak position or 'effective radius', the overall shape

---

†The insensitivity to distortion is not maintained near 0° where an additional maximum may be introduced by distortion effects alone.

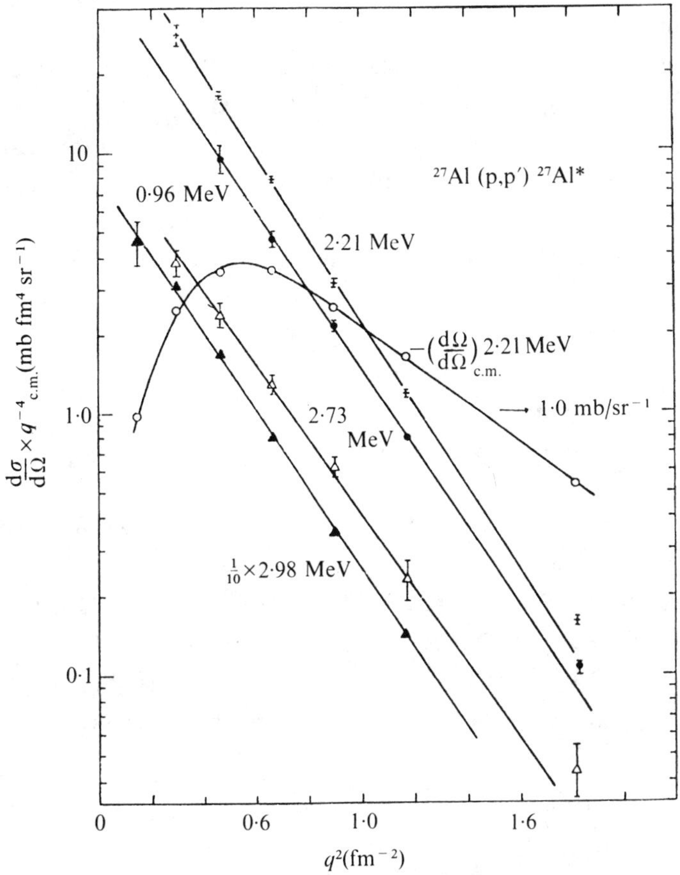

FIG. 8.2. Tibell plots of cross-sections for excitation of various states in ^{27}Al. (From Tibell 1969.)

of the cross-section will be given adequately. However, extrapolation of the Tibell plots to $q = 0$ gives information on the ratio of values of the reduced transition probability B_L (see below, §8.1.2.).

Many analyses of inelastic electron scattering and inelastic nucleon scattering at intermediate energies have been carried out. The main concern has been to obtain the correct strength for the nuclear transition density, which is related to the choice of nuclear model or coupling scheme. In general, the simplest single-particle transitions do not yield sufficient strength and it is necessary to include the contribution from many particle-hole pairs via RPA or some similar method (Haybron and McManus 1965, Gillet and Melkanoff 1964) or to compensate for truncation of the shell model space by the addition of core excitation (Love and Satchler 1967, Love 1968). The same conclusion was obtained from a study of inelastic proton scattering at 1 GeV (Brissaud *et al.* 1974). These high-energy data are restricted to low momentum transfer and consequently give essentially the same information as that derived from medium-energy data.

8.1.2. *Collective parameters*

The DWBA matrix element can be written in the form

$$T_{\mathrm{fi}} = \int \chi_{\mathrm{f}}^{-*}(\underset{\sim}{k}',\underset{\sim}{r})\langle\phi_{\mathrm{f}}(\xi)|V(r,\xi)|\phi_{\mathrm{i}}(\xi)\rangle\chi_{\mathrm{i}}^{+}(\underset{\sim}{k},\underset{\sim}{r})\,\mathrm{d}^3r \quad (8.11)$$

where ξ represents all the relevant internal co-ordinates of the target nucleus. It is convenient to make a multipole expansion of $V(r,\xi)$ so that, with suitable assumptions about the interaction, the nuclear matrix element becomes

$$\langle\phi_f|V|\phi_i\rangle = \sum_{LM}\langle I_fM_f|V_{LM}|I_iM_i\rangle\{i^LY_L^M(\hat{\underset{\sim}{r}})\}^* \quad (8.12a)$$

$$= \sum_{LM}(I_iM_iLM|I_fM_f)\langle I_f\|V\|I_i\rangle\{i^LY_L^M(\hat{\underset{\sim}{r}})\}^*. \quad (8.12b)$$

The reduced matrix element[†] is a function of r only and is usually written in the form

$$\langle I_f\|V\|I_i\rangle = A_LF_L(r). \quad (8.13)$$

The differential cross-section is obtained by taking the sum over final states M_f and average over initial states M_i and becomes

---

[†]The definition of the reduced matrix element used here differs from the definition used in Chapters 3 and 6 by a factor of $(2I_f+1)^{1/2}$, as can be seen by comparing eqns (3.41) and (8.12*b*).

$$\frac{d\varepsilon}{d\Omega} = \left(\frac{\mu}{2\pi\hbar^2}\right) \frac{k'}{k} \sum_{LM} \frac{2I_f+1}{(2I_i+1)(2L+1)} |T_{LM}|^2 \tag{8.14}$$

where

$$T_{LM} = i^{-L} A_L \int \chi_f^{-*}(\underset{\sim}{k}',\underset{\sim}{r})\, F_L(r)\, Y_L^{M*}(\hat{\underset{\sim}{r}})\, \chi_i^{+}(\underset{\sim}{k},\underset{\sim}{r})\, d^3r. \tag{8.15}$$

This formalism lends itself to a macroscopic description of the nucleus in which it is assumed that the interaction potential or generalized optical potential follows the shape of the nuclear surface and so becomes a function of the collective co-ordinates. The non-spherical terms induce nuclear excitations associated with a change in the state of vibration of the nucleus about a mean spherical shape or a change in the state of rotation of a permanently deformed nucleus. A simple procedure is to assume that the nuclear surface can be deformed and take the potential to be $V\{r - R(\theta,\phi)\}$. A Taylor expansion about $R = R_0$ yields

$$V(r - R) = V(r - R_0) - \delta R \frac{d}{dr} V(r - R_0) + \frac{1}{2}(\delta R)^2 \frac{d^2}{dr^2} V(r - R_0) - \ldots. \tag{8.16}$$

The lowest term of this expansion can be associated with the usual optical potential while the higher terms give the inelastic scattering and some corrections to elastic scattering. For a vibrational nucleus we have

$$R = R_0\{1 + \sum_{Rq} \alpha^*_{kq}\, Y_k^{q*}(\theta,\phi)\} \tag{8.17}$$

which yields, for the first-order excitation,

$$V_{LM} = -i^L R_0\, \alpha^*_{LM} \frac{dV}{dr} \tag{8.18}$$

so that

$$A_L F_L(r) = -i^L \left(\frac{\hbar\omega_L}{2C_L}\right)^{1/2} R_0 \frac{dV}{dr} \tag{8.19a}$$

where $\hbar\omega_L$ and C_L are the parameters of the vibration (Davidson 1968). Similarly for a rotational excitation we have

$$A_L F_L(r) = -i^L \beta_L R_0 (2L+1)^{-1/2} \frac{dV}{dr} \tag{8.19b}$$

where β_L is the usual deformation parameter. Thus the form factor for the transition is determined from the potential which fits elastic scattering, so that the only free parameters are β_L or $(\hbar\omega_L/2C_L)^{1/2}$ which may be determined by fitting the inelastic cross-section.

It has been pointed out by Satchler (1967*b*, 1972*c*) that other types of shape oscillation are possible and physically reasonable. The form factors $F_L(r)$ for these other modes of oscillation in general have different shapes from the surface oscillation described above.

For inelastic proton scattering at medium and intermediate energies it is necessary to deform the imaginary part of the optical potential as well as the real part in order to obtain good angular distributions (Haybron 1966, Fulling and Satchler 1968), and the same is true for medium-energy deuterons and helions (Dickens *et al.* 1965 , Gray *et al.* 1966). In order to fit the inelastic asymmetries for protons in the range 30-155 MeV it has also proved necessary to deform the spin-orbit term in the potential (Fricke *et al.* 1967, Sherif and Blair 1968, 1970, Sherif 1969).

The exact form of the matrix element (8.11) can be written as

$$T_{fi} = \langle \phi_f \Phi_f | V + V(E - H_0 - H(\xi) - V + i\varepsilon)^{-1} V | \phi_i \Phi_i \rangle \tag{8.20}$$

where ϕ_i, Φ_f are plane wavefunctions, $H(\xi)$ is the nuclear Hamiltonian, and ξ represents the nuclear co-ordinates. This expression can be approximated by neglecting $H(\xi)$ in the propagator and neglecting the energy loss of the projectile, so that the excited states of the nucleus are assumed to be degenerate and the nucleus is frozen during the time of the collision. The scattering amplitude can then be written in the form

$$f(\underset{\sim}{k}', \underset{\sim}{k}) = \langle \Phi_f | f(\xi, \theta, \phi) | \Phi_i \rangle \tag{8.21}$$

where $f(\xi,\theta,\phi)$ may be regarded as a generalized scattering amplitude from which both elastic and inelastic scattering can be derived. In the diffraction theory this amplitude is given by eqn (7.37*a*) where the integration extends over the projection of the nuclear surface whose boundary is defined by the formula (7.38). This model leads to the following expressions (Blair 1966) for the cross-sections for single excitation with multipolarity L:

$$\frac{d\sigma}{d\Omega}(L, I_i \to I_f) = \frac{2I_f+1}{(2I_i+1)(2I+1)} \frac{d\sigma}{d\Omega}(0^+ \to L) \qquad \text{vibrational model} \qquad (8.22)$$

$$= (I_i K_i L0 | I_f K_f)^2 \frac{d\sigma}{d\Omega}(0^+ \to L). \qquad \text{rotational model} \qquad (8.23)$$

where

$$\frac{d\sigma}{d\Omega}(0^+ \to L) = (kR_F^2)^2 \sum_{\substack{M \\ L+M \text{ even}}} (4\pi)^{-1}(2L+1)D_L(i^{-L}[L:M])^2[J_{|M|}(kR_F\theta)]^2 \qquad (8.24)$$

$$D_L = \hbar\omega_L/2C_L \qquad \text{vibrational model} \qquad (8.25a)$$

$$= \beta_L{}^2 \qquad \text{rotational model} \qquad (8.25b)$$

$$i^{-L}[L:M] = \frac{\{(L-M)!(L+M)!\}^{1/2}}{(L-M)!!(L+M)!!}.$$

As noted in §7.1.4, the sharp-edged diffraction model may be much improved by the introduction of a smoothing function (Inopin and Berezhnoy 1965).

The values of the collective parameters obtained by fitting data for excitation of the nucleus by strongly interacting projectiles may be compared with values derived from the study of electromagnetic transitions. For Coulomb excitation of a nucleus of charge Ze by a projectile of charge $Z'e$ the multipole operator is

$$V_{LM}^C = i^L \frac{4\pi Z'e^2}{(2L+1)r^{L+1}} \langle I_f \| \mathscr{O}_L \| I_i \rangle \qquad (8.26)$$

$$\mathscr{O}_{LM} = \sum_j r_j^L \, Y_L^M(\hat{\underset{\sim}{r}}_j) \qquad (8.27)$$

where the sum runs over the protons in the nucleus. For a Taylor expansion of the deformed charge distribution similar to that for a potential given in eqn (8.16) we have

$$A_L^C\, F_L^C(r) = \mathrm{i}^L \frac{4\pi ZZ'e^2}{(2L+1)^{3/2} r^{L+1}} \beta_L^C\, c \int \frac{\mathrm{d}\rho}{\mathrm{d}s} s^{L+2}\, \mathrm{d}s \qquad (8.28)$$

where c is the halfway radius. For a uniformly charge sphere of radius U, we take

$$\frac{\mathrm{d}\rho}{\mathrm{d}s} = \frac{3}{4\pi U^3} \delta(s - U) \qquad (8.29)$$

so that in this case, with $c = U$,

$$A_L\, F_L^C(r) = \mathrm{i}^L \frac{3ZZ'e^2}{(2L+1)^{3/2} r^{L+1}} \beta_L^C\, U^L. \qquad (8.30)$$

These quantities may be linked to the reduced transition probability B which is usually (though not universally)† defined as (Alder *et al.* 1956)

$$B_L(I_\mathrm{i} \rightarrow I_\mathrm{f}) = \sum_{MM_\mathrm{f}} |\langle I_\mathrm{f} M_\mathrm{f} | \mathcal{O}_{LM} | I_\mathrm{i} M_\mathrm{i} \rangle|^2 \qquad (8.31a)$$

$$= (2I_\mathrm{i}+1)^{-1} |\langle I_\mathrm{f} \| \mathcal{O}_L \| I_\mathrm{f} \rangle|^2 \qquad (8.31b)$$

Hence, for Coulomb excitation and the transition density (8.29), we have

$$B_L^C(I_i \rightarrow I_f) = (2I_i+1)^{-1}(2L+1)^{-1}(Z\, \beta_L^C\, U^L). \qquad (8.32)$$

The radiative width for decay of an excited state by γ-emission is given by (Alder *et al.* 1956)

---

†The definition of the reduced matrix element in eqns (8.31*a*,*b*) is the same as that used in Chapter 3 but differs from that given in eqns (8.12*a*,*b*). This situation arises from the different definitions used respectively in electromagnetic theory and the DWBA formalism for strongly interacting projectiles.

$$\Gamma_L = \frac{8\pi(L+1)}{L\{(2L+1)!!\}} \left(\frac{E_\gamma}{\hbar c}\right)^{2L+1} B_L^{E(M)}(I_f \rightarrow I_i) \tag{8.33}$$

where E_γ is the nuclear excitation energy and

$$B_L(I_f \rightarrow I_i) = (2I_i+1)(2I_f+1)^{-1} B_L(I_i \rightarrow I_f). \tag{8.34}$$

For inelastic electron scattering the relevant Coulomb operator is $(2L+1)!!q^{-L}M_{LM}^C(q)$ where $M_{LM}^C(q)$ is defined in eqn (3.38) so that the reduced transition probability is

$$B_L^C(q,I_i \rightarrow I_f) = (2I_i+1)^{-1}\{q^{-L}(2L+1)!! \sum_M \int \rho_{if}(\underset{\sim}{r}) j_L(qr) Y_L^M(\hat{\underset{\sim}{r}})\, d^3r\}^2 \tag{8.35}$$

and in the limit $qr \rightarrow 0$ when $j_L(qr) \rightarrow (qr)^L/(2L+1)!!$ this reduces to the previous expression.† The form factor for inelastic excitation via the Coulomb operator is given by

$$Z^2|F_L^C(q)|^2 = 4\pi q^{2L}\{(2L+1)!!\}^{-2} B_L^C(q,I_i \rightarrow I_f) \tag{8.36}$$

while the form factors for transverse electric or magnetic excitation are

$$Z^2|F_L^{E(M)}(q)|^2 = 4\pi\left(\frac{L+1}{L}\right)\left\{\frac{q^L}{(2L+1)!!}\right\}^2 B_L^{E(M)}(q,I_i \rightarrow I_f). \tag{8.37}$$

A variety of expressions have been used for the macroscopic transition density for inelastic electron scattering. A form due to Helm (1956) gives the form factor

$$F_L(q) \propto j_L(qr) \exp(-\tfrac{1}{2}g^2q^2) \tag{8.38}$$

where g is a parameter connected with the width of the nuclear surface. The hydrodynamical model yields (Tassie 1956, 1957)

$$\rho_{if} \propto r^{L-1} \frac{d\rho}{dr} \tag{8.39}$$

---

†In inelastic scattering the limit for q is $q \rightarrow \omega$ where $\omega = E/\hbar c$, and it is at this limit that the magnitudes of B_L should be compared.

where ρ is normally taken to be a Fermi distribution. This form can conveniently be used in a DWBA calculation (Ziegler 1967, Onley, Griffy and Reynolds 1963, Onley *et al.* 1964).

Thus, we have a wide variety of formulae available for the analysis of inelastic scattering of various projectiles and these may be used to extract and compare collective parameters. However, the assumption that the deformation of the optical potential is the same as that of the matter or charge distribution is not obvious, and some microscopic calculations discussed in §8.1.3 suggest that it is not justified. It has been suggested that this difficulty can be circumvented by noting that the normalization factor between theory and experiment is $\beta_L R$ (Austern and Blair 1965) and hence that this quantity should be compared for different projectiles.† Since the parameter R has a different physical significance in the various formulae, the corresponding parameter β_L may take different values and is clearly model dependent. Alternatively, the intrinsic multipole moments can be calculated from the deformation parameters. These moments are given by

$$Q_L = 2Z\left(\frac{4\pi}{2L+1}\right)^{1/2} \int r^L\, Y_L^0(\hat{\mathbf{r}})\, \rho_{\mathrm{if}}(\mathbf{r})\, \mathrm{d}^3 r. \tag{8.40}$$

It is also assumed that we may compare the transition rate for a process involving an electromagnetic operator, such as that defined in eqn (8.27) which involves only the nuclear protons, with the rate for a process in which protons and neutrons contribute equally. Bernstein (1969) has made this assumption explicit by introducing an isoscalar transition rate

$$B_L^{\mathrm{IS}}(I_{\mathrm{i}} \rightarrow I_{\mathrm{f}}) = (2I_{\mathrm{i}}+1)^{-1}\,(Z/A)^2 |\langle I_{\mathrm{f}} \| \mathcal{O}_L \| I_{\mathrm{i}} \rangle|^2 \tag{8.41}$$

where the operator $\mathcal{O}_{LM}$ has the same form as (8.27) but the sum

---

†There is some divergence of opinion on what to do when the real and imaginary parts of the potential are both deformed but have different halfway radii.

runs over all nucleons, and has shown that this quantity can be extracted from inelastic α-particle scattering. It is convenient to present the transition rates in Weisskopf units by dividing by the single-particle estimate for B_L (Blatt and Weisskopf 1952). This gives

$$G_L = \frac{B_L(I_i \to I_f)}{B_{SP}(0 \to L)} \tag{8.42}$$

$$B_{SP}(0 \to L) = \frac{2L+1}{4\pi}\left(\frac{3}{3+L}\right)^2 (r_0\ A^{1/3})^{2L} \tag{8.43}$$

where r_0 is usually taken as 1.2 fm. A further complication arises when the coupling is strong owing to the contribution of higher order excitations which are not taken into account in DWBA. These contributions can be taken into account in standard coupled-channels calculations or using the Austern-Blair approach (Austern and Blair 1965).

An extensive tabulation of B_2 and β_2 values obtained from Coulomb excitation has been given by Stelson and Grodzins (1965), and a further tabulation of B_2 $(0 \to 2)$ and Q_2 values measured from the reorientation effect in Coulomb excitation has been published by Smilansky (1970).†

The most complete analysis of results obtained from inelastic scattering is that by Bernstein (1969) who demonstrates the equivalence of the isoscalar and electromagnetic transition rates. The agreement for light $T = 0$ nuclei, shown in Fig. 8.3, is not surprising, but that for medium and heavy nuclei with $N > Z$, shown in Fig. 8.4, is quite remarkable. Bernstein also gives a systematic analysis of transition rates, mainly in terms of the G_L defined in eqn (8.42), for 2^+, 3^-, 4^+, and 5^- states in a wide range of nuclei. Since Bernstein's review was published an analysis of inelastic α-particle scattering at 44 MeV from a range of nuclei near

---

†Values of β_2, etc., obtained from electromagnetic studies of deformed ground states are discussed in Chapter 6. Table 6.7 gives a comparison of values obtained by different methods.

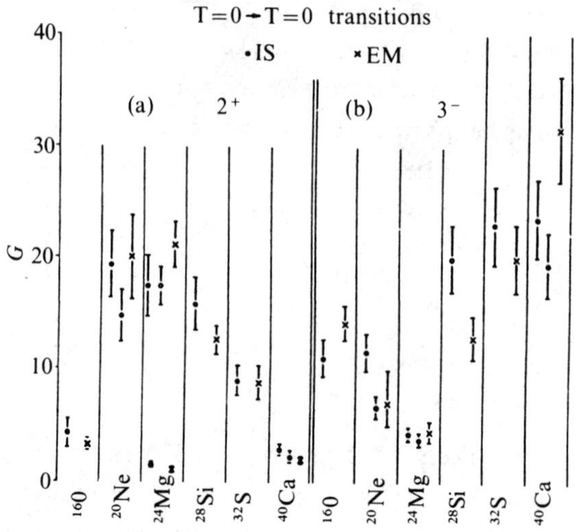

FIG. 8.3. Comparison of isoscalar (IS) and electromagnetic (EM) transition strengths G_L for $T = 0$ excited states in $N = Z$ nuclei. (a) 2^+ states; (b) 3^- states. (From Bernstein 1969.)

the closed shells with 28 neutrons or 20, 28, 50, and 82 protons has appeared (Bruge *et al.* 1970), and the collective parameters obtained are compared with earlier results for α-particle scattering, proton, deuteron, and ^{3}He scattering, and electromagnetic processes. Fulling and Satchler (1968) have studied 30 MeV inelastic proton scattering from a range of nuclei and have tabulated β_L values. The inelastic scatter-scattering of 50 MeV protons from ^{40}Ca, ^{45}Sc, and ^{56,58}Fe has been studied in a similar manner (Mani 1971*a*,*b*,*c*, Mani and Jacques 1971). Results for $\beta_L R$ and Q_L have been obtained from 104 MeV α-particle scattering from ^{12}C, ^{20}Ne, and ^{64}Ni (Specht *et al.* 1970) and compared with values given by other methods, while results from a more detailed analysis for ^{12}C have been compared with theoretical predictions (Specht *et al.* 1971). A study of excitation of ^{58,60,62,64}Ni at the same α-particle energy (Rebel *et al.* 1972*a*) has given values of $\beta_L R$ and G_L for 2^+ and 3^- states. Rebel *et al.* (1972*b*) have also presented a detailed study of deformation in the sd-shell using new data for 104 MeV α-particle scattering, and the

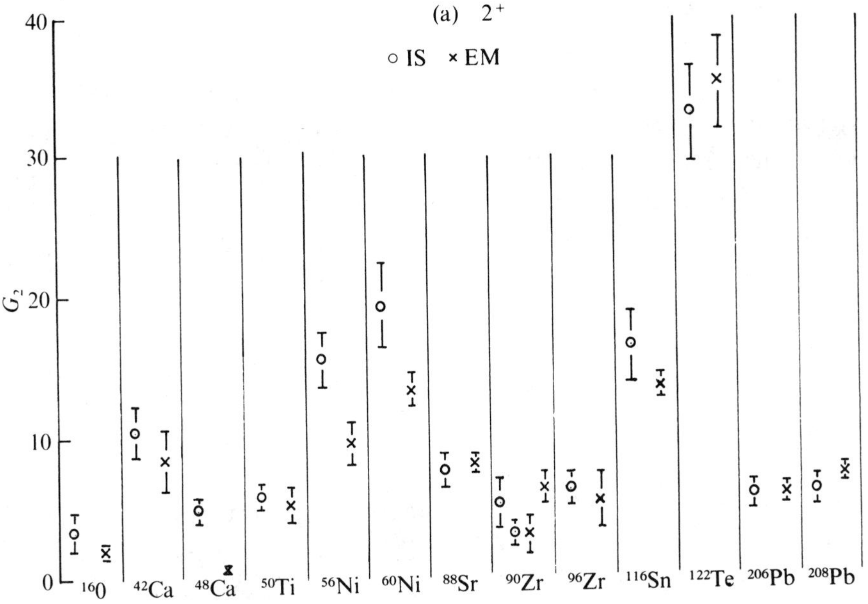

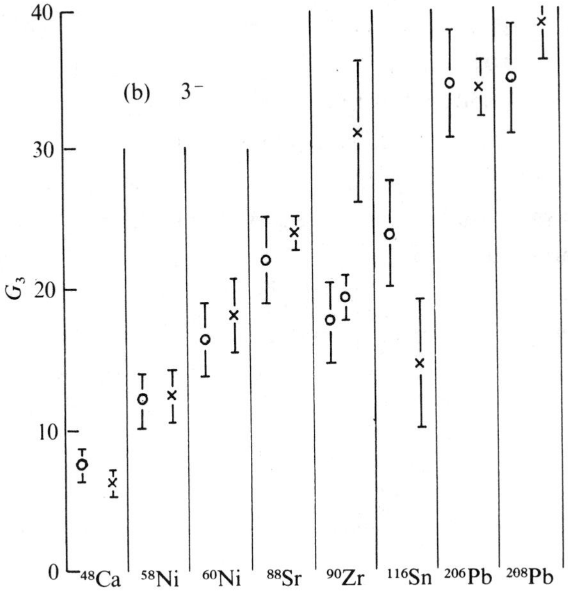

FIG. 8.4. Comparison of isoscalar (IS) and electromagnetic (EM) transition strengths G_L for $N > Z$ nuclei. Part (a) refers to 2^+ states and part (b) refers to 3^- states. (From Bernstein 1969.)

results have been compared with those from other methods and with theoretical predictions.

Similar analyses have been carried out for inelastic electron scattering. In this case there is greater sensitivity to the features of the transition density, and the more recent analyses indicate that the Tassie model (8.39) does not yield a satisfactory shape fit to the data. Consequently, a phenomenological form has been adopted with

$$\rho_{\text{if}} \propto r^{L-1} \frac{\mathrm{d}}{\mathrm{d}r} \left\{ 1 + \exp\left(\frac{r - c_{\text{tr}}}{a_{\text{tr}}} \right) \right\}^{-1} \tag{8.44a}$$

$$c_{\text{tr}} = (RV)c, \qquad t_{\text{tr}} = (SV)t, \qquad t_{\text{tr}} = 4.39\, a_{\text{tr}} \tag{8.44b}$$

where the factors RV and SV indicate the departure from the Tassie model which has $c_{\text{tr}} = c$, $a_{\text{tr}} = a$. Heisenberg, McCarthy, and Sick (1971*a*) have studied the excitation of 2^+, 3^-, 4^+, and 5^- levels in ^{40,42,44}Ca, ^{46,48,50}Ti, and ^{56}Fe over a range of momentum transfer q = 0.4–2.6 fm. They find that for the 2^+ states the Tassie model gives a poor shape fit to the data and yields very large B_2 values, whereas fits to the data with the modified transition density (8.44) yield B_2 values in agreement with Coulomb excitation and strong-interaction scattering. This requires a value for SV of 2.5–3. The Tassie model gives satisfactory agreement with data at low momentum transfer for 2^+ and 3^- states in ^{58,60,62}Ni (Duguay *et al.* 1967), but in a study of the excitation of the lowest 2^+ and 3^- states in ^{116,118,120}Sn Curtis *et al.* (1969) found that the Tassie model does not give a satisfactory shape fit over a larger range of q, although it does appear that this model yields B_2 values in better agreement with results from Coulomb excitation. Some fits to the data for Sn isotopes are shown in Fig. 8.5. Ziegler and Peterson (1968) have studied collective 3^- states in ^{206,207,208}Pb and ^{209}Bi. The range of momentum transfer was 0.25–0.6 fm^{-1} and in this range the Tassie model proved adequate, but this model was later shown to be inadequate for ^{208}Pb at higher q (Heisenberg and Sick 1970). Friedrich (1972) has studied the lowest 3^- level in ^{208}Pb and two 5^-

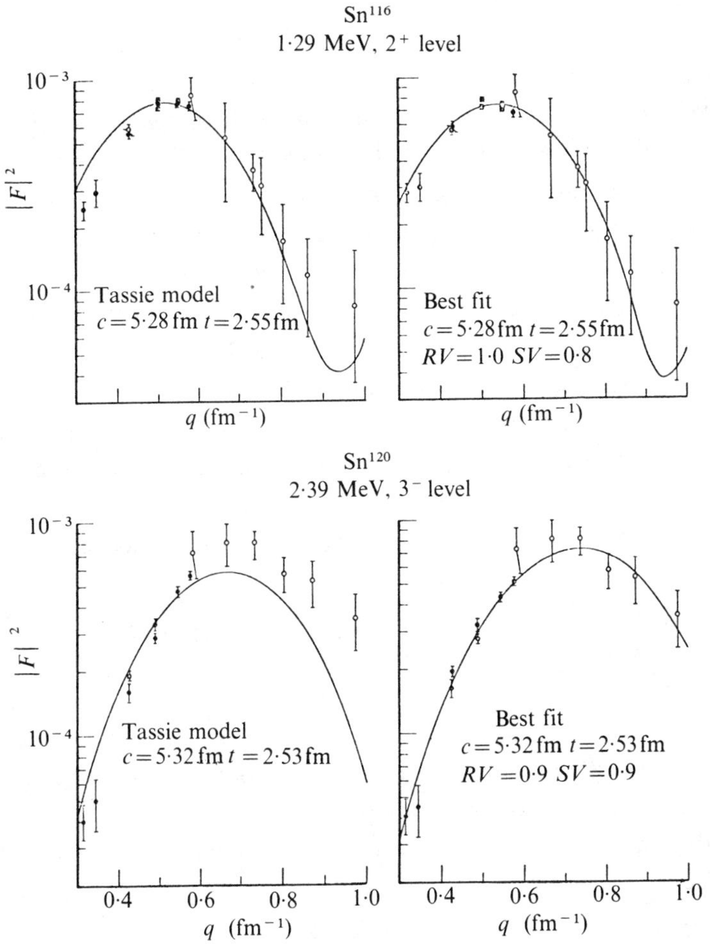

FIG. 8.5. Form factors for inelastic electron scattering from ^{116}Sn and ^{120}Sn: for details of the experimental data, see Curtis *et al.* (1969). (After Curtis *et al.* 1969.)

levels in the same nucleus, and concludes that it is not possible to obtain a good fit to data for the 3^- level over the range of momentum transfer q = 0.48–1.54 fm^{-1} even with the modified transition density, although a satisfactory fit is obtained with the sum of two gaussians. The modified Tassie model is satisfactory for the 5^- levels.

These studies confirm that inelastic scattering of

strongly interacting projectiles to resolved final states can provide information on nuclear shapes in terms of the deformation lengths and reduced transition probabilities of the nuclear collective model. Electron scattering shows a greater sensitivity to the details of the transition density so that a macroscopic model is barely adequate. The degree of consistency between the various methods can be inferred from Table 8.1 which gives parameters deduced for the 2^+ state in ^{58}Ni at 1.45 MeV and the 3^- state at 4.47 MeV. Recent work on the excitation of low-lying states in strongly deformed nuclei has been reviewed by Donnelly and Walecka (1975), who anticipate considerable progress in this region from experiments utilising the new high-energy high-resolution electron linacs.

For low momentum transfer, it is possible to expand the Bessel function in eqn (8.35) to obtain the relation

$$\left\{\frac{B_L^C\ (q)}{B_L^C\ (q)}\right\}^{1/2} = 1 - A_L q^2 \frac{\langle r^{L+2}\rangle}{\langle r^L\rangle} + B_L q^4 \frac{\langle r^{L+4}\rangle}{\langle r^L\rangle} \cdots \qquad (8.45)$$

where A_L, B_L are constants. Similar expansions can be obtained for transverse electric and magnetic excitations. It is possible to define a transition radius R_{tr} (Drechsel 1968) through the formulae

$$R_{\text{tr}}^2 = \langle r^{L+2}\rangle/\langle r^L\rangle, \qquad L \geqslant 1$$

$$= \langle r^4\rangle/\langle r^2\rangle, \qquad L = 0.$$

For simple models, such as the homogeneous liquid drop, R_{tr} has a simple interpretation, but in general it is not necessarily associated with a peak in or localization of the transition density. Nevertheless, R_{tr} has proved to be a reasonably model-independent parameter which can be deduced from inelastic electron scattering on light nuclei at low momentum transfer (Drechsel 1968, Theissen 1972). Values obtained from many analyses for nuclei from ^{4}He to ^{39}K have been tabulated by Theissen (1972).

TABLE 8.1

Values of $\beta_L R$ and G_L for the 2^+ (1.45 MeV) and the 3^- (4.47 MeV) states in ^{58}Ni

Projectile	E_{lab} (MeV)	$\beta_2 R$ (fm)	G_2 (W units)	$\beta_3 R$ (fm)	G_3 (W units)	Reference
α	104	0.77	9.2±0.5	0.69	9.0±0.4	Rebel *et al.* 1972*b*
α	50.2	1.00	15.8±1.6	0.73	7.8±1.0	Jarvis *et al.* 1967
α	44	0.91	12.7±0.6	0.60	6.8	Bruge *et al.* 1970
α	34	1.07	18.0±2.7	0.89		Inoue 1968
^{3}He	51.3	1.05		0.88		Gibson *et al.* 1967
^{3}He	37.7	0.90		0.75		Bingham and Halbert 1968
p	40	0.86	11.5	0.80	12.3	Lingappa & Greenlees 1970
p	17.9	1.00		0.76		Jarvis *et al.* 1967
e					13.3±2.0	Duguay *et al.* 1967
Coulomb excitation			9.9±1.5			Stelson and Grodzins 1965

8.1.3. *Microscopic methods*

In recent years powerful techniques have been developed to handle the nuclear many-body problem, and many of the macroscopic properties of nuclei can be understood in microscopic terms. A microscopic theory of medium-energy nuclear scattering has been developed simultaneously and provides more detailed information about the mechanism of nuclear excitation.

Using the projection operator formalism introduced in §7.2.1 an effective interaction can be defined as

$$V_{\text{eff}} = V + V \frac{Q}{E + H_{QQ} + \mathrm{i}\varepsilon} V \tag{8.46}$$

so that the transition matrix element in distorted-wave theory is

$$T_{fi} = \langle \chi_{\mathrm{f}}^{-} \Phi_{\mathrm{f}} | V_{\text{eff}} | \chi_{\mathrm{i}}^{+} \Phi_{\mathrm{i}} \rangle \tag{8.47}$$

and the generalized optical potential is

$$V_{\text{opt}} = \langle \Phi_i | V_{\text{eff}} | \Phi_i \rangle. \tag{8.48}$$

Apart from the projection operator Q, the expression for V_{eff} resembles the eqn (7.16) for the transition operator T, but because of the presence of Q the effective interaction is related to the Brueckner reaction matrix (2.15) for bound-state problems (McVoy and Romo 1969). In the form (8.46), V_{eff} is still a many-body operator. The essential assumption of the microscopic theory is that V_{eff} can be replaced by a sum of one-body operators, i.e.

$$V_{\text{eff}} = \sum_j t(j). \tag{8.49}$$

At sufficiently high energies $t(j)$ approximates to the free two-body interaction and we have the DWIA discussed in §8.1.1. At medium energies $t(j)$ represents an interaction in the nuclear medium, where the exclusion principle blocks intermediate states which are occupied by other nucleons. From the formal expressions (8.46) and (7.22), we may expect $t(j)$ to be energy dependent, non-local, and possibly density dependent.

The development of the microscopic theory for inelastic nucleon scattering up to 1967 has been described by Satchler (1966, 1967*a*) and Glendenning (1967). In this early work, the nucleon-nucleon interaction at medium energies was taken to be real, local, and a function of the separation distance between the projectile and the bound nucleon. Some spin and isospin dependence was incorporated by writing $t(j)$ in the form

$$t(j) = -(V_0 + V_1\, \boldsymbol{\sigma}\cdot\boldsymbol{\sigma}_j)\; g(|\mathbf{r} - \boldsymbol{\xi}_j|) \tag{8.50a}$$

with

$$V_0 = V_{0\alpha} + V_{0\beta}\, \boldsymbol{\tau}\cdot\boldsymbol{\tau}_j \tag{8.50b}$$

and similarly for V_1. The radial function g was taken to be of Yukawa or gaussian shape. Reasonable agreement with the data for inelastic proton scattering at medium energies is obtained with a Yukawa interaction of range 1 fm and strength $V_{0\alpha}$ of around 200 MeV or somewhat less for light nuclei. The spin and isospin strengths are small and the energy dependence of the interaction is weak.

Many of the early calculations did not take account of exchange effects. The contribution from exchange forces is implicitly taken into account when the effective interaction is written in the form (8.50) so that the additional exchange terms on the inelastic amplitude appear as 'knock-on' terms arising from antisymmetrization between the projectile and the target nucleon (Amos, McCarthy, and Madsen 1967, Schaeffer 1969*a*, Atkinson and Madsen 1970). Extensive studies of these exchange contributions show that they are equally important when the simple Yukawa or gaussian interactions are replaced by a more realistic interaction such as the long-range part of the Hamada-Johnston interaction (Love and Satchler 1970). The importance of exchange increases as the multipolarity L increases and the incident energy decreases (Schaeffer 1969*a*, Satchler 1973); the effect is extremely important over the range 10-60 MeV, as can be seen from Fig. 8.6 and Table 8.2. At 155 MeV the exchange amplitudes are small and interfere destructively with the direct terms. Similar calculations have recently been carried out using effective interactions derived from Skyrme's force (Davies and Satchler 1974).

Another important feature of recent calculations arises from the realization that the contribution from core excitation is often large and that pure single-particle excitations are rarely sufficient to reproduce the observed strengths. Core polarization can be incorporated by separating the target nucleons into two groups, the core and the valence nucleons (Love and Satchler 1970). A simple collective model is used to allow excitations of the core, while the interaction between the projectile and the valence nucleon is treated accurately and antisymmetrized. Recent studies (Satchler 1970*a*, 1971*a*, Terrien 1973) have shown that a complex interaction is required for inelastic proton scattering; this is consistent with the collective model for the generalized

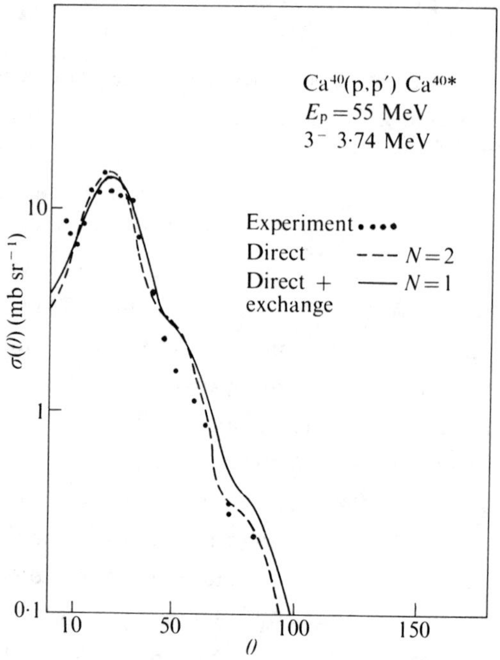

FIG. 8.6. Comparison of calculated cross-sections for 55 MeV inelastic proton scattering from ^{40}Ca. The factor N is the normalization required to bring the calculated curve into agreement with the peak of the experimental cross-section. (From Schaeffer 1969*a*.)

optical potential and with the formal definition of the effective interaction.

For closed-shell nuclei or those with closed subshells; the RPA offers a description of the excitations which contains substantial enhancement over the TDA or the simple single-particle model. The RPA state vectors for levels in ^{12}C, ^{16}O, ^{40}Ca, and ^{208}Pb have been used by Petrovich (1970) in a study of medium-energy proton scattering using a variety of nucleon-nucleon interactions, including the Kallio-Kolltveit force. Schaeffer (1969*a*) has also studied ^{40}Ca using RPA and the Blatt-Jackson force with Yukawa shape, while Satchler (1973) used RPA for ^{40}Ca with the long-range part of the Hamada-Johnston force. All these authors use the RPA state vectors

TABLE 8.2

Ratios of the integrated cross-section for inelastic proton scattering from the lowest 3^- and 5^- states in ^{40}Ca with and without the exchange term[†]

E_p (MeV)	Ratio $\sigma(D+E)/\sigma(D)$[‡]				
	17.3	20.3	30	40	50
3^- (BJ)	2.7	3.3	2.9	2.5	2.3
3^- (HJ)		3.5	2.8	2.4	2.0
5^- (BJ)	6.8	7.9	6.4	4.6	3.6
5^- (HJ)		11.2	7.2	5.2	3.9

[†]Calculated with the Blatt-Jackson (BJ) or Hamada-Johnston (HJ) force (Schaeffer 1969*a*, Satchler 1973).
[‡]*D+E*, with exchange term; *D*, without exchange term.

of Gillet and Sanderson (1967) for the lowest 3^- and 5^- states. Petrovich (1970) and Schaeffer (1969*a*) use also the GS transition densities derived from oscillator functions which give a satisfactory description of inelastic electron scattering (Gillet and Melkanoff 1964) and of inelastic proton scattering at 155 MeV in DWIA (Haybron and McManus 1965), while Satchler constructs transition densities from single-particle wavefunctions in a Saxon-Woods potential and renormalizes the transition densities to reproduce B_3 and B_5 values. These transition densities are in reasonable agreement with those deduced from electron scattering although the transition radius for the 3^- state is too small. In general, agreement with experiment is good but deteriorates below 30 MeV. The exchange contribution is essential to this agreement, as noted earlier and illustrated in Fig. 8.6.

These analyses of medium-energy nucleon scattering by means of the microscopic DWBA indicate that the method provides an important test of microscopic bound-state wavefunctions, particularly when combined with studies of intermediate-energy proton scattering and electron scattering.† However, this test must be seen as an indication of the overall validity of the nuclear wavefunctions, since fine details of the angular distributions can be modified either through the effective interaction or through the transition densities and it is not yet possible to define the former with complete precision. Consequently, the macroscopic model sometimes gives better agreement with the data. This is the case for the lowest 3^- state in ^{40}Ca at 30 MeV, for which the transition potentials are compared in Fig. 8.7.

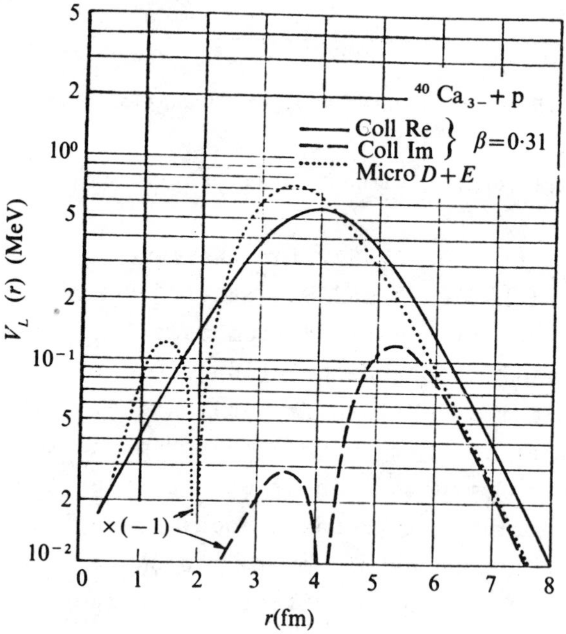

FIG. 8.7. Real and imaginary parts of the transition potentials for excitation of the 3^- state in ^{40}Ca by proton scattering compared with the transition potential calculated in the microscopic model including direct and exchange parts. (From Satchler 1973.)

---

†See also Hammerstein, Howell, and Petrovich (1973).

Picard *et al.* (1969) and Schaeffer (1969*b*) have studied inelastic proton scattering from ^{88}Sr and ^{90}Zr (closed neutron shells) and from ^{60,62}Ni (closed proton shells) using a BCS + RPA description of the nuclear states which includes two-quasiparticle excitation of the open shell and 1p-1h excitation of the core (Gillet, Giraud, and Rho 1969). The proton and neutron amplitudes are renormalized by λ_p and λ_n respectively to fit the (p,p') data at medium energies. The factor λ_p may be compared with the value required to fit B_L values. Table 8.3 indicates that these wavefunctions give a

TABLE 8.3

Enhancement factors λ_p and λ_n for the proton and neutron amplitudes for inelastic proton scattering derived from BCS + RPA wavefunctions†

	2^+ states					
	^{60}Ni	^{60}Ni	^{62}Ni	^{88}Sr	^{90}Zr	^{90}Zr
E_p(MeV)	40	13	11	20	20.3	12.7
λ_p	3.0	3.0	2.6	2.0	2.0	2.0
λ_n	0.9	1.2	0.9	5.5	5.5	5.0

	3^- states				
	^{60}Ni	^{60}Ni	^{88}Sr	^{90}Zr	^{90}Zr
E_p(MeV)	40	13	20	20.3	12.7
λ_p	1.9	1.9	1.3	1.4	1.4
λ_n	0.9	1.1	1.3	1.2	1.3

†Schaeffer 1967*b*, Picard *et al.* 1969.

good description of 3^- states but the model does not give a good description of the 2^+ states, possibly owing to insufficient neutron core excitation. Studies of core excitation have also been carried out by Petrovich (1970) for ^{50}Ti and ^{90}Zr which have closed neutron shells and two valence protons and for ^{89}Y and ^{209}Bi which have one valence proton. He describes the core by a macroscopic vibrational model or microscopically with RPA. The microscopic transition densities have a large contribution from neutron core excitations in ^{50}Ti and ^{90}Zr, but the excess neutrons are less important in ^{89}Y. Further studies of ^{89}Y and ^{90}Zr have been reported by Whiten, Scott, and Satchler (1972) who use a macroscopic description of the core excitation. They give values for the deformation of the core and also of the effective charge (Mottelson 1960). Compared with the results of Whiten *et al.* and the electromagnetic values, the effective charges given by the microscopic model of Petrovich are too small, particularly for $L = 2$. This suggests that the imbalance of proton and neutron core excitations in this model is not consistent with the data, and that an isoscalar core excitation may be closer to reality.

The excitation of collective 2^+, 3^-, and 4^+ states by electron scattering has been observed by Phan-Xuan-Ho *et al.* (1973) for ^{50}Ti, ^{52}Cr, and ^{54}Fe (N = 28 closed shell), ^{90}Zr and ^{92}Mo (N = 50 closed shell), and ^{116,120,124}Sn (Z = 50 closed shell). The range of momentum transfer is 0.5-1.7 fm^{-1}. From the analysis it is again concluded that the BCS + RPA wavefunctions of Gillet *et al.* (1969) give a good description of the 3^- states, but not of the 2^+ states. In the latter case not only is the strength insufficient but the shape of the transition density is systematically incorrect.

A similar microscopic approach can be applied to the inelastic scattering of composite projectiles although a further complication now arises concerning the nature of the effective interaction.† For medium-energy deuteron, helion, and triton scattering, the effective interaction has been obtained by folding a realistic nucleon-nucleon force into the projectile density distribution. In this case the knock-on exchange amplitude arising from exchange between a projectile nucleon

---

†Aspects of this problem are discussed in connection with elastic scattering in §7.4.4.

and a target nucleon with which it is important (Schaeffer 1970, Park and Satchler 1971, Satchler 1971*d*). Spectator exchange is not taken into account. Also, for these projectiles a large imaginary interaction is expected owing to the importance of projectile break-up and other processes; this has been taken into account through a collective imaginary form factor.

Calculations have been carried out for medium-energy deuteron and helion scattering from ^{40}Ca and ^{208}Pb, and for triton and helion scattering from ^{90}Zr (Park and Satchler 1971, Satchler 1971*d*). An essential feature of these processes is that the excitation occurs at large distances, $r \gtrsim 6$ fm in ^{40}Ca, $r \gtrsim 9$ fm in ^{208}Pb; consequently the collective imaginary term which is associated with a large halfway radius gives an important contribution and, because the microscopic density due to the valence nucleons falls more rapidly at large distances, the core excitation terms dominate. This localization at large distances leads to a much greater contribution from the neutron transition density than from the proton density, so that the inelastic scattering of composite projectiles is in principle a method for studying neutron distributions, although a number of uncertainties must be removed before this becomes a practical possibility.

The microscopic approach to inelastic scattering has been used extensively for medium-energy α-particles (Jackson 1964, Wall 1964, Madsen and Tobocman 1964, Alster, Shreve, and Peterson 1966, Yntema and Satchler 1967, Bernstein 1969). In most of these calculations, the gaussian n-α interaction (7.121) has been used. In so far as this interaction fits low-energy n-α scattering it may be assumed to take account of exchange between the projectile and an interacting nucleon but not of spectator exchange (Batty *et al*. 1971). The contribution from the imaginary part of the interaction is small (Satchler 1971*b*), and this is again consistent with results obtained from the macroscopic generalized optical model. Owing to the strong absorption of the projectile there is very marked localization in the outer region of the nucleus so that the process probes only the tail of the nuclear wavefunctions

(Austern and Blair 1965, Jackson 1969*a*). This is illustrated in Figs. 8.8 and 8.9. At these energies the phenomenological optical potential for α-particle scattering shows discrete ambiguities, as discussed in §7.4.2, and raises some uncertainty over which potential should be used to generate distorted wavefunctions for inelastic scattering. Calculations at 42 MeV (Morgan and Jackson 1969) indicate that not all these potentials lead to equally good fits in the microscopic DWBA although they do give good fits in the macroscopic DWBA or coupled-channels method. In the macroscopic case the diagonal and off-diagonal matrix elements are generated consistently, whereas in the microscopic case there is no connection

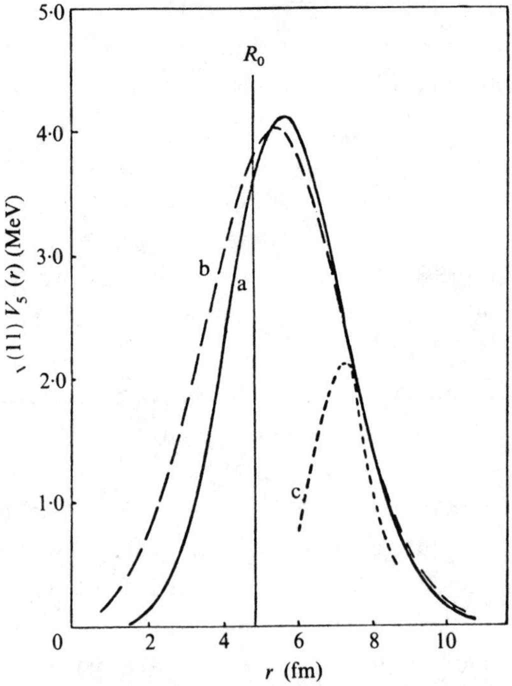

FIG. 8.8. The effective interaction for the excitation of the $9/2^+$ state in ^{89}Y by α-particle scattering. Curve (a) is calculated directly by a microscopic model, curve (b) is the derivative of the optical potential calculated by the microscopic model, and curve (c) is the derivative of a Saxon-Woods optical potential. For curves (b) and (c) $\beta_5 = 0.0125$. (From Jackson 1969*a*.)

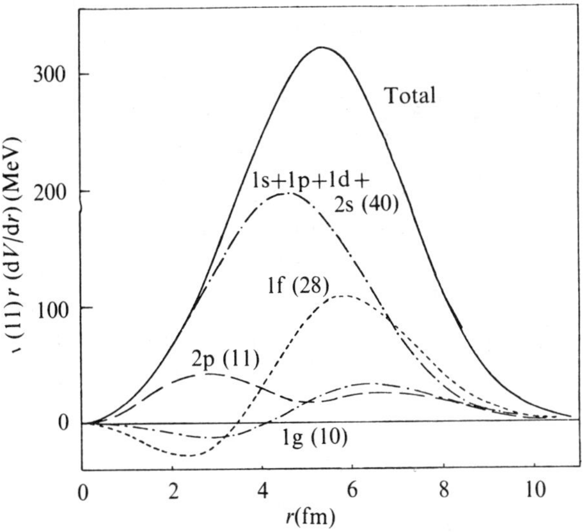

FIG. 8.9. The solid curve is curve (b) of Fig. 8.8 divided by β_5. The broken curves show the contribution from various single-particle states, and the number associated with each curve gives the number of nucleons contributing to that curve. (From Jackson 1969*a*).

between the diagonal (elastic) and off-diagonal (inelastic) matrix elements (Jackson 1964). When the microscopic DWBA is used with a consistent derivation of the diagonal and off-diagonal elements, good shape agreement with the data is achieved (Morgan and Jackson 1969). The inelastic form factors in the microscopic model are substantially different in shape to those given by the macroscopic model (Glendenning and Veneroni 1966, Rawitscher and Spicuzza 1971), as can be seen from Fig. 8.8.

Studies of inelastic α-particle scattering in the intermediate-energy region have also been carried out. At 166 MeV Tatischeff and Brissaud (1970) have measured angular distributions for the first 2^+ levels in ^{12}C, ^{28}Si, and ^{120}Sn and also the 3^- (9.64 MeV) and 0^+ (7.66 MeV) levels in ^{12}C, Brissaud *et al.* (1972*b*) have measured the angular distributions for the lowest 2^+ and 3^- levels in $^{116,118,120,124}Sn$, and Bimbot *et al.* (1973) have studied the excitation of the first collective level in 22 nuclei. The results for the tin isotopes have been studied in detail using the BCS + RPA wavefunctions of Clement and Baranger (1968). The shape of the angular distributions is not sensitive to the details of the transition densities, so the agreement between theory and experiment is determined from the enhancement coefficients λ_p and λ_n, as for proton scattering, and it appears that the BCS + RPA gives the best description of the 3^- states but the TDA calculation is superior for the 2^+ states because it includes more configurations. A variety of wavefunctions have been tested for the other nuclei, again in terms of the factors λ_p and λ_n.

The inelastic α-particle scattering data at 104 MeV, discussed in the previous section, has been reanalysed in terms of microscopic models for the deformed nuclei in the sd shell. Mackintosh (1973) has used a phenomenological deformed density with the gaussian α-n interaction to generate the inelastic form factor for ^{24}Mg. The values of B_2 and $\beta_2 R$ required to fit the data are substantially larger than those for the deformed potential obtained by Rebel *et al.* (1972), because of the smoothing effect of folding in the α-n interaction. Rebel and Schwemer (1973) have also used a deformed density for ^{20}Ne and ^{28}Si. The lower strengths required for the α-n interaction in these calculations are consistent with the energy dependence discussed in §7.4.4.

Mackintosh and Tassie (1974) have studied different ways of deforming the nuclear density for ^{20}Ne and ^{24}Mg, and have compared fits to the inelastic scattering data at 104 MeV. The deformation parameters β_2 and β_4 extracted show a considerable variation and a strong dependence on the deformation method, but the intrinsic moments Q_L, defined in eqn (8.40), are more consistently determined. The fits obtained with Vautherin's deformed Hartree-Fock solutions are quite good for ^{24}Mg but poor for ^{20}Ne. Mackintosh and de Swiniarski (1975) have studied inelastic scattering of ^3He from ^{20}Ne at 68 MeV and find that, in this case also, the moments Q_L are well determined and the fit with the deformed HF distribution is poor, but the quality of fit to the angular distribution shows somewhat greater sensitivity to the parametrization of the nuclear deformation than for α-particle scattering.

Satchler (1972) has shown that, for the usual kind of folding model, the moments of the potential Q_L^V and of the density Q_L^ρ obey the relation

$$Q_L^V/J = Q_L^\rho/A \tag{8.51}$$

where J is the volume integral of the potential. By comparing the values of Q_L^V determined by scattering of nuclear projectiles with values of Q_L^ρ obtained by electromagnetic methods, Mackintosh (1976) has shown that the relation (8.51) is reasonably well satisfied for sd-shell nuclei. For certain heavy nuclei the limited evidence available suggests that the relation can be satisfied only if the neutron moments of the deformation are less than the proton moments.

8.2. INELASTIC SCATTERING TO SUMMED FINAL STATES

Fig. 8.10 gives an idealized representation of the excitation spectrum for inelastic scattering. The sharp spikes correspond to excitation of discrete low-lying levels in the target nucleus. There then appear broader peaks corresponding to collective excitation of closely spaced levels. These levels are usually above the threshold for particle decay;

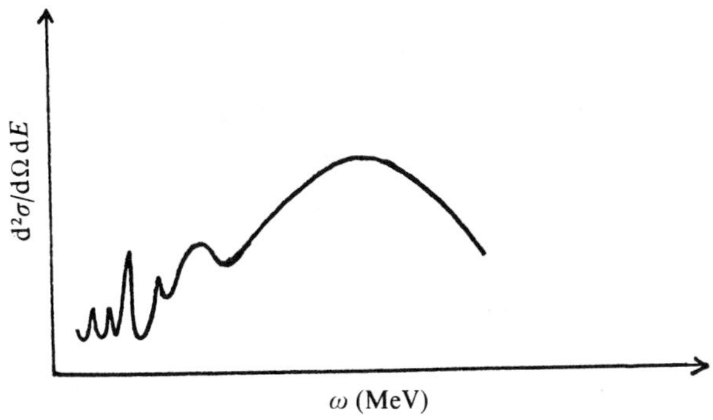

FIG. 8.10. A summed inelastic cross-section as a function of energy loss ω.

consequently they are broadened and generally not resolved. The giant electric dipole resonance appears in this region at an excitation energy of 13-25 MeV (Spicer 1969), and other giant resonances may also occur (deForest and Walecka 1966, Satchler 1972*a*). Beyond this region is observed a broad quasi-elastic peak, which corresponds to ejection of individual nucleons from the target nucleus.

The information which can be obtained from electron excitation in the giant resonance region has recently been reviewed by Bellicard (1975) and Donnelly and Walecka (1975), while the earlier work was reviewed by deForest and Walecka (1966). The location and strength of states predicted by appropriate nuclear models can be compared with the excitation spectrum, as shown in Fig. 8.11. Comparison of the theoretical and experimental form factors yields information on the strength and shape of the transition density, as is the case for single resolved states. For measurements at large angles and relatively high q the higher multipoles of the transverse magnetic operator tend to dominate. This means that from a $T = 0$ ground state the high-spin $T = 1$ states can be excited; for example 4^- states in ^{12}C and ^{16}O and a 6^- state in ^{28}Si have been observed.

Some measurements have been made for excitation of the giant resonance region in ^{12}C and ^{16}O by 185 MeV protons (Tyren and Maris 1957, Sundberg and Tibell 1969) and the early data for ^{12}C have been analysed in DWIA using a particle-hole model (Sanderson 1961, 1962).

If the quasi-elastic peak is due to single-nucleon knockout a simple relation can be found between the energy loss ω and the momentum transfer q. Using non-relativistic kinematics and dropping terms of order $1/A$ we have

$$\omega = \frac{\hbar^2 q^2}{2m} + \frac{\hbar^2}{m}\,\underset{\sim}{q}\cdot\underset{\sim}{P} + \frac{\hbar^2 P^2}{2m} + E_s \qquad (8.52)$$

where E_s is the separation energy for the ejected nucleon and P is its Fermi momentum in the nucleus. Hence for a free nucleon at rest we have $\omega = \hbar^2 q^2/2m$, but for a bound nucleon ω can take a range of values for fixed momentum transfer and

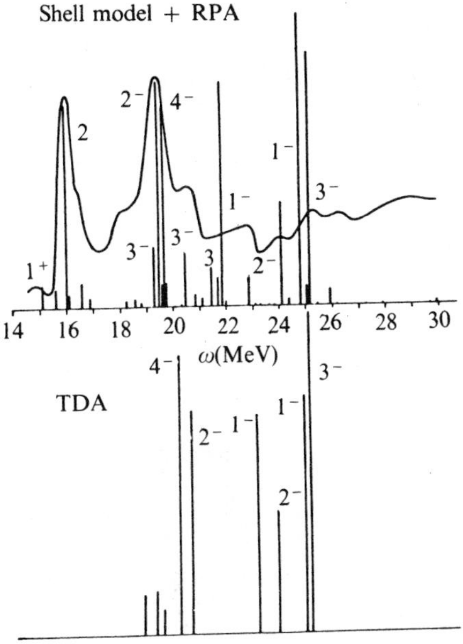

FIG. 8.11. The smooth curve shows the form factor for summed inelastic electron scattering from ^{12}C at $\theta = 135°$ and $q = 241$ MeV/c as a function of excitation energy ω. For comparison the $T = 1$ odd-parity states predicted in the Tamm-Dancoff approximation are shown and also the $T = 0$ and $T = 1$ states given in a shell model plus RPA calculation. (From Donnelly 1972.)

this gives a finite width to the peak. Measurements of the summed inelastic spectrum have been made for protons at 160 MeV (Wall and Roos 1966) and at 1 GeV (Corley 1968, Corley *et al.* 1972), and yield widths for the quasi-elastic peak of ~80 MeV. Measurements for electron scattering have been published by Bishop, Isabelle, and Bétourné (1964) and by Berthot *et al.* (1969); they show similar peaks which are narrower for the lighter nuclei.

For electron scattering the cross-section can be written in the form (see §3.2)

$$\frac{d^2\sigma}{d\Omega dE'} = K_e \left(\frac{d\sigma}{d\Omega}\right)_M |f(q_p^2)|^2 \left[\left(\frac{q_\mu^2}{q^2}\right)^2 R_L(q,\omega) + \left\{\frac{1}{2}\left(\frac{q_\mu^2}{q^2}\right) + \tan^2 \frac{1}{2}\theta\right\} R_T(q, \ )\right] \tag{8.53}$$

where E' is the energy of the scattered electron, $\omega = E - E'$ is the energy loss and $q_\mu = (\underset{\sim}{q}, i\omega)$ is the electron four-momentum. The coefficient K absorbs kinematic factors including the recoil factor (3.22), $f(q_\mu^2)$ is the nucleon form factor, and the $R(q,\omega)$ are the response functions (deForest and Walecka 1966)

$$R(q,\omega) = \sum_{\lambda\neq\alpha} |\langle\Phi_\lambda|\mathscr{O}(q)|\Phi_\alpha\rangle|^2 \,\delta(E_\lambda - E_\alpha - \omega) \tag{8.54}$$

where $\mathscr{O}(q) = \Sigma_j \mathscr{O}_j(q)$ is the operator for transverse electric or magnetic transitions or the Coulomb operator (including the Darwin-Foldy correction). These response functions have been evaluated in a simple single-particle model for ^{12}C and ^{16}O (deForest 1969) and for ^{12}C and ^{40}Ca (Donnelly 1970). The width and position of the peak is found to depend quite strongly on the momentum transfer q, as can be seen from Fig. 8.12, while Fig. 8.13 shows that contribution of higher multipoles increases for increasing q and so does the importance of the Mλ transitions. These calculations yield reasonable agreement with the data except for the lower values of ω.

A similar description can be given for inelastic proton scattering using DWIA in the form

$$\frac{d^2\sigma}{d\Omega dE'} = K_p \frac{d\sigma^{AV}}{d\Omega} \sum_{\lambda\neq\alpha} |\langle\chi^-(\underset{\sim}{k}',\underset{\sim}{r})\Phi_\lambda| \sum_{j=1}^{A} \delta(\underset{\sim}{r}-\underset{\sim}{r}_j)|\chi^+(\underset{\sim}{k},\underset{\sim}{r})\Phi_\alpha\rangle|^2 \,\delta(E_\lambda^{AV} - E_\alpha - \omega) \tag{8.55}$$

where the free nucleon-nucleon cross-section $d\sigma^{AV}/d\Omega$ is to be evaluated at some suitable average momentum transfer and we have assumed that the excitation energies in the δ-function can be replaced by an average value. The cross-section then depends on the sum of the transition densities $\Sigma\rho_{\lambda\alpha}(\underset{\sim}{r})$, where $\rho_{\lambda\alpha}$ is defined in eqn (8.5), and in a single-particle model for closed-shell nuclei with single-particle states μ occupied up to $\mu = \varepsilon$ this is given by

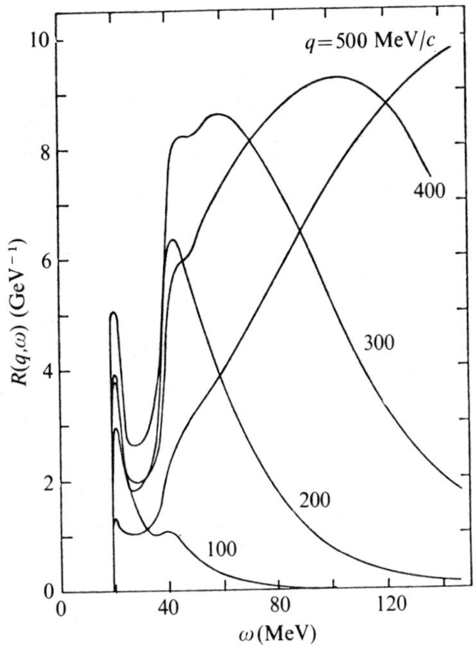

FIG. 8.12. The response function $R(q,\omega)$ for ^{12}C as a function of excitation energy ω for several values of momentum transfer q. (From Donnelly 1970.)

$$\sum_{\lambda \neq \alpha} \rho_{\lambda\alpha}(\underset{\sim}{r}) = \sum_{\mu \leqslant \varepsilon} \left\{ \sum_{\nu > \varepsilon} \phi_\nu^*(\underset{\sim}{r})\ \phi_\mu(\underset{\sim}{r}) + \int d^3p\ \phi_{\underset{\sim}{p}}^*(\underset{\sim}{r})\ \phi_\mu(\underset{\sim}{r}) \right\} \tag{8.56}$$

where $\phi_{\underset{\sim}{p}}$ is a continuum state for the ejected nucleon. The cross-section now becomes (Jackson 1971*a*)

$$\frac{d^2\sigma}{d\Omega dE'} = K_p \frac{d\sigma^{AV}}{d\Omega}\ \delta(E_\lambda^{AV} - E_\alpha - \omega)$$
$$\times \sum_{\mu \leqslant \varepsilon} \iint G(\underset{\sim}{k},\underset{\sim}{k}',\underset{\sim}{r})\ G^*(\underset{\sim}{k},\underset{\sim}{k}',\underset{\sim}{r}')\ \phi_\mu(\underset{\sim}{r})\ \phi_\mu^*(\underset{\sim}{r}')\ F(\underset{\sim}{r},\underset{\sim}{r}')\ d^3r\ d^3r' \tag{8.57}$$

where

$$G(\underset{\sim}{k},\underset{\sim}{k}',\underset{\sim}{r}) = \chi^+(\underset{\sim}{k},\underset{\sim}{r})\ \chi^{-*}(\underset{\sim}{k}',\underset{\sim}{r}) \tag{8.58}$$

$$F(\underset{\sim}{r},\underset{\sim}{r}') = \sum_{\nu > \varepsilon} \phi_\nu^*(\underset{\sim}{r})\ \phi_\nu(\underset{\sim}{r}') + \int d^3p\ \phi_{\underset{\sim}{p}}^*(\underset{\sim}{r})\ \phi_p(r') \tag{8.59}$$

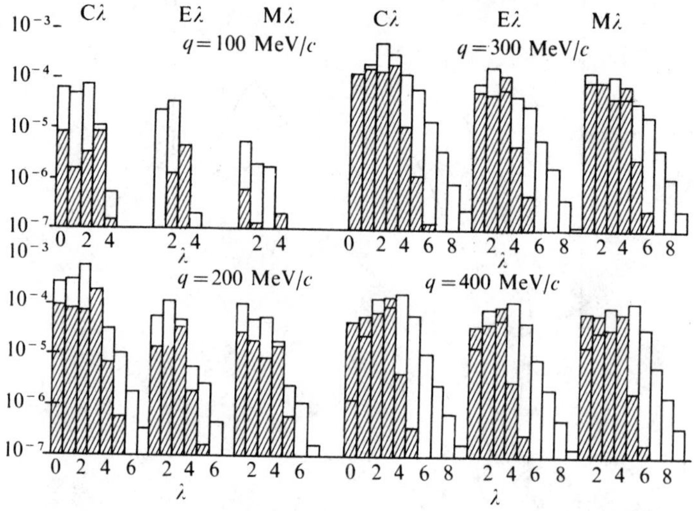

FIG. 8.13. The response function $R(q,\omega)$ decomposed into the contributions from the various multipoles Cλ, Eλ, Mλ for ^{12}C at an excitation energy of 60 MeV and the values of momentum transfer shown. (From Donnelly 1970.)

and the limits of the integration over $\underset{\sim}{p}$ are determined by energy conservation.

Since ϕ_ν, $\phi_{\underset{\sim}{p}}$ should be orthogonal and form a complete set, a suitable approximation, valid at high incident energies and high energy loss, should be

$$F^{(0)}(\underset{\sim}{r},\underset{\sim}{r}') = \delta(\underset{\sim}{r} - \underset{\sim}{r}') \tag{8.60}$$

in which case the cross-section becomes

$$\frac{d^2\sigma}{d\Omega dE'} = K_p \frac{d\sigma^{AV}}{d\Omega} N(A)\ \delta(E_\lambda^{AV} - E_\alpha - \omega) \tag{8.61}$$

where $N(A)$ is the effective nucleon number, also denoted by A_{eff} (see §8.5.4),

$$N(A) = \sum_{\mu\leqslant\varepsilon} \int G(\underset{\sim}{k},\underset{\sim}{k}',\underset{\sim}{r})\ G^*(\underset{\sim}{k},\underset{\sim}{k}',\underset{\sim}{r})|\phi_\mu(\underset{\sim}{r})|^2\ d^3r \tag{8.62a}$$

$$= A \int G(\underset{\sim}{k},\underset{\sim}{k}',\underset{\sim}{r})\ G^*(\underset{\sim}{k},\underset{\sim}{k}',\underset{\sim}{r})\ \rho(\underset{\sim}{r})\ d^3r. \tag{8.62b}$$

In a semiclassical treatment of the distortion this expression for $N(A)$ reduces to eqn (10.20), while in a plane-wave approximation $N(A) = A$. Thus, the cross-section (8.61) represents the sum of incoherent scatterings from each nucleon and is proportional to the free nucleon-nucleon cross-section. Cross-sections with this characteristic at high momentum transfer have been observed by Bellettini *et al.* (1966) for 19 GeV protons on light- and medium-mass nuclei, and the data have been analysed using the expression (8.61) (Glauber 1967). At this high energy the integrand of (8.62*b*) peaks about 0.5 fm beyond the halfway radius (Glauber 1967), while at 1 GeV the integrand spans the nuclear transition region (Jackson 1969*b*). If it is assumed that $\rho_p = \rho_n$, then the variation in $N(A)$ for Fermi distributions which fit low- and medium-energy electron scattering data is small, as can be seen from Table 8.4, but

TABLE 8.4

Values of the effective nucleon number $N(A)$ calculated using the semiclassical approximation and a Fermi distribution for $\rho_m(r)$†

A	$cA^{-1/3}$	a	1 GeV protons, σ=44 mb		2 GeV pions, σ=27 mb	
	(fm)	(fm)	$N(A)$	$A^{-1/3}N(A)$	$N(A)$	$A^{-1/3}N(A)$
16	1.03	0.409	2.71	1.08	4.58	1.82
40	1.05	0.568	5.08	1.49	8.84	2.58
88	1.08	0.523	5.69	1.28	10.66	2.40
107	1.10	0.569	6.68	1.41	12.46	2.52
208	1.09	0.523			13.52	2.28
	1.10	0.523	7.16	1.21	13.64	2.30
209	1.12	0.545			14.69	2.48
	1.14	0.455			13.14	2.22

†From Jackson 1969*b*.

significant changes in $N(A)$ are introduced if the r.m.s. radius of the neutron distribution is changed by a few per cent. This effect provides a method of studying the matter distribution in the surface region,[†] but it has not been exploited to any great extent.

An improved approximation for the kernel $F(\underset{\sim}{r},\underset{\sim}{r}')$ is given by

$$F^{(1)}(\underset{\sim}{r},\underset{\sim}{r}') = \delta(\underset{\sim}{r} - \underset{\sim}{r}') - \sum_{\nu\leqslant\varepsilon} \phi_\nu^*(\underset{\sim}{r})\, \phi_\nu(\underset{\sim}{r}') \tag{8.63}$$

where the second term gives the correction due to the exclusion principle which blocks scattering into states with $\nu \leqslant \varepsilon$. In this simple single-particle model the one-nucleon density function and the two-nucleon density function (eqn 1.7) are given by

$$A\rho(\underset{\sim}{r}) = \sum_{\mu\leqslant\varepsilon} |\phi_\mu(r)|^2 \tag{8.64}$$

$$A(A-1)\rho(\underset{\sim}{r},\underset{\sim}{r}') = A^2\rho(\underset{\sim}{r})\rho(\underset{\sim}{r}') - A\rho_{\mathrm{ex}}(\underset{\sim}{r},\underset{\sim}{r}') \tag{8.65}$$

$$A\rho_{\mathrm{ex}}(\underset{\sim}{r},\underset{\sim}{r}') = \sum_{\mu\leqslant\varepsilon}\sum_{\nu\leqslant\varepsilon} \phi_\mu^*(\underset{\sim}{r}')\, \phi_\nu^*(\underset{\sim}{r})\, \phi_\nu(\underset{\sim}{r}')\, \phi_\mu(\underset{\sim}{r}) \tag{8.66}$$

where ρ_{ex} is the exchange part of the two-nucleon density function. Hence the cross-section (8.57) with the approximate kernel (8.63) becomes

$$\frac{\mathrm{d}^2\sigma}{\mathrm{d}\Omega\mathrm{d}E'} = K_{\mathrm{p}} \frac{\mathrm{d}\sigma}{\mathrm{d}\Omega}^{\mathrm{AV}}\Big\{ A \int G(\underset{\sim}{k},\underset{\sim}{k}',\underset{\sim}{r})G^*(\underset{\sim}{k},\underset{\sim}{k}',\underset{\sim}{r})\rho(\underset{\sim}{r})\, \mathrm{d}^3r$$

$$- A \iint G(\underset{\sim}{k},\underset{\sim}{k}',\underset{\sim}{r})\, G^*(\underset{\sim}{k},\underset{\sim}{k}',\underset{\sim}{r}')\, \rho_{\mathrm{ex}}(\underset{\sim}{r},\underset{\sim}{r}')\, \mathrm{d}^3r\, \mathrm{d}^3r' \Big\}. \tag{8.67}$$

By introducing the nucleon-nucleon correlation function $C(\mathrm{r},\mathrm{r}')$ defined as

$$C(\underset{\sim}{r},\underset{\sim}{r}') = \rho(\underset{\sim}{r},\underset{\sim}{r}') - \rho(\underset{\sim}{r})\, \rho(\underset{\sim}{r}') \tag{8.68}$$

[†]See also §10.5.

[‡]There are many different definitions and notations for the correlation function. Some of these are summarized and related by Jackson and Murugesu (1970).

the cross-section (8.68) can be written in a more general form as

$$\frac{d^2\sigma}{d\Omega dE'} = K_p \frac{d\sigma}{d\Omega}^{AV'}\left\{ A \int G(\underset{\sim}{k},\underset{\sim}{k}',\underset{\sim}{r})\, G^*(\underset{\sim}{k},\underset{\sim}{k}',\underset{\sim}{r})\, \rho(\underset{\sim}{r})\, d^3r \right.$$
$$+ A(A-1) \iint G(\underset{\sim}{k},\underset{\sim}{k}',\underset{\sim}{r})\, G^*(\underset{\sim}{k},\underset{\sim}{k}',\underset{\sim}{r}')\, C(\underset{\sim}{r},\underset{\sim}{r}')\, d^3r\, d^3r'$$
$$\left. - A \iint G(\underset{\sim}{k},\underset{\sim}{k}',\underset{\sim}{r})\, G^*(\underset{\sim}{k},\underset{\sim}{k}',\underset{\sim}{r}')\, \rho(\underset{\sim}{r})\, \rho(\underset{\sim}{r}')\, d^3r\, d^3r' \right\}. \tag{8.69a}$$

In a plane-wave approximation this reduces to

$$\frac{d^2\sigma}{d\Omega dE} = K_p \frac{d\sigma}{d\Omega}^{AV} A\{1 + (A - 1)\, C(q) - F^2(q)\} \tag{8.69b}$$

where $F(q)$ is the usual form factor and

$$C(q) = \iint \exp\{i\underset{\sim}{q}.(\underset{\sim}{r}-\underset{\sim}{r}')\}\, C(\underset{\sim}{r},\underset{\sim}{r}')\, d^3r\, d^3r'. \tag{8.70}$$

When $q \to 0$, $F^2(q) \to 1$ and $C(q) \to 0$, but, for large q, $F^2(q) \to 0$ and $C(q) \to 0$, so that we regain the incoherent scattering approximation. The dependence of the cross-section on the correlation function has led to considerable interest in the possibility of studying dynamical correlations by summed inelastic scattering of both protons and electrons (McVoy and Van Hove 1962, Czyz 1963, 1971, Glauber 1959, Simenog and Sitenko 1966), although recent estimates seem somewhat unfavourable (Kofoed-Hansen and Wilkin 1971, Kofoed-Hansen 1973).

For electron energies in the GeV region the summed inelastic cross-section is dominated by the quasi-elastic peak and the peak for resonant pion production (Stanfield *et al.* 1971), Heimlich *et al.* 1974). The quasi-elastic part can be fitted using a Fermi gas model to obtain values for the Fermi momentum k_F and the mean separation energy. It is doubtful whether this procedure yields any new information for light nuclei which can be studied in more detail by other means, although it does provide a further check on predictions of the nuclear momentum distribution. There may be some value in studies of heavier nuclei, for which there is still considerable difficulty in separating final states in coincidence experiments.

8.3. TRANSFER REACTIONS

There is an important group of direct reactions which involve the transfer of a single nucleon or a group of nucleons between the projectile and the target. In a stripping reaction the projectile is stripped of a group of nucleons so that the target nucleus has the x nucleons added to it; thus the reaction A(a,b)C where A is the target nucleus can be represented as

$$\mathrm{a(b+x)} + \mathrm{A} \rightarrow \mathrm{b} + \mathrm{C(A+x)}. \tag{8.71a}$$

In a pick-up reaction the converse occurs as the projectile removes the nucleons x from the target nucleus, i.e.

$$\mathrm{a} + \mathrm{A} \rightarrow \mathrm{b(a+x)} + \mathrm{C(A-x)}. \tag{8.71b}$$

Thus measurement of the energy spectrum for a stripping reaction yields information about the states of the system x in the residual nucleus C, while a similar measurement for a pick-up reaction yields information about the states in the target nucleus A. The total angular momentum of the state in the residual nucleus must be given by

$$\mathbf{J}_{\mathrm{C}} = \mathbf{J}_{\mathrm{A}} + \mathbf{j}$$

where $\mathbf{j}$ is the total angular momentum transferred, or, since $\mathbf{j} = \boldsymbol{\ell} + \mathbf{s}$,

$$J_{\mathrm{A}} + l + s \geqslant J_{\mathrm{C}} \geqslant |J_{\mathrm{A}} - |l - s|| \tag{8.72}$$

and hence identification of l and j from the shape of the angular distribution leads to information about the spin and parity of excited states in the residual nucleus.

8.3.1. *The matrix element and overlap integral*

For the two-body transfer reaction A(a,b)C the differential cross-section for unpolarized projectiles and unoriented target nuclei is given by

$$\frac{d\sigma}{d\Omega} = \frac{\mu_{aA}\,\mu_{bC}}{4\pi^2\hbar^4}\,\frac{k_b}{k_a}\,\frac{1}{(2J_A+1)(2s_a+1)} \sum_{m_a m_b M_A M_C} |T_{fi}|^2 \tag{8.73}$$

where μ_{aA}, μ_{bC} are the reduced masses for the initial and final channels and $\underset{\sim}{k}_a$, $\underset{\sim}{k}_b$ are the incident and outgoing momenta in the centre-of-mass frame. In DWBA the matrix element for the pick-up reaction is given in the prior form by

$$T_{fi} = \langle \chi_b^- \, \phi_b \, \Phi_C | V_{ax} | \chi_a^+ \, \phi_a \, \Phi_A \rangle \tag{8.74}$$

where ϕ_a, ϕ_b, Φ_A, Φ_C represent bound-state wavefunctions and χ_a^+, χ_b^- represent the distorted wavefunctions. The derivation of this formula includes the assumption that the core C is inert; there is now considerable experimental evidence for core excitation in transfer reactions at medium energies (Pugh *et al.* 1965, Dupont and Chabre 1968).

In order to evaluate the matrix element (8.74) we denote the internal co-ordinates of the residual nucleus C, including spin co-ordinates, by $\underset{\sim}{\xi}_C$ and the internal co-ordinates of the system x by $\underset{\sim}{\xi}_x$. The target nucleus is then described by the co-ordinates ξ_C and ξ_x together with the relative co-ordinate $\underset{\sim}{r}_{xC}$. The internal co-ordinates of the projectile a are denoted by $\underset{\sim}{\sigma}_a$ so that the reactions product b is described by the co-ordinates $\underset{\sim}{\sigma}_a$, $\underset{\sim}{\xi}_x$ and the relative co-ordinate $\underset{\sim}{r}_{ax}$. If we neglect spin-orbit terms in the optical potentials, the distorted wavefunctions are functions of the separation distances $\underset{\sim}{r}_{aA}$ and $\underset{\sim}{r}_{bc}$ which are connected to the other space co-ordinates by the relations (see Fig. 8.14)

$$\underset{\sim}{r}_{aA} = \underset{\sim}{r}_{ax} + \gamma \underset{\sim}{r}_{xC}, \qquad \underset{\sim}{r}_{bC} = \underset{\sim}{r}_{xC} + \lambda \underset{\sim}{r}_{ax} \tag{8.75}$$

where $\gamma = M_C/M_A$ and $\lambda = M_a/M_b$.† We may also introduce the

---

†We have used the same symbols for mass and for projection quantum numbers, but this should not present any difficulty to the reader.

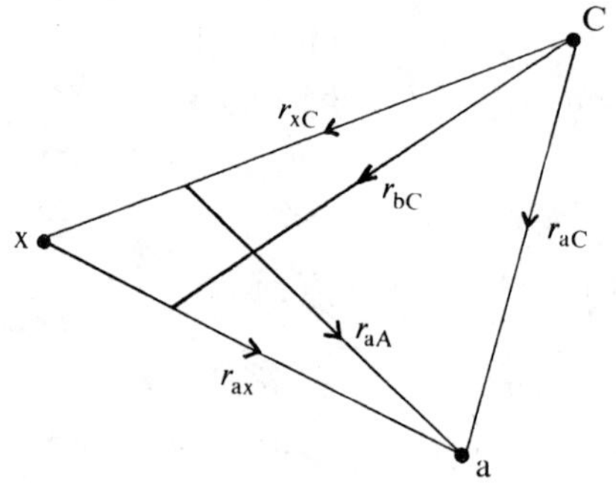

FIG. 8.14. Co-ordinate system for the analysis of transfer reactions.

overlap integral which is defined by the equation

$$\int d^3\xi_C\ \Phi_C^*(\underset{\sim}{\xi}_C)\ \Phi_A(\underset{\sim}{\xi}_C,\underset{\sim}{\xi}_x,\underset{\sim}{r}_{xC}) = \psi_x(\underset{\sim}{r}_{xC},\underset{\sim}{\xi}_x). \qquad (8.76)$$

If it is assumed that the systems a and x are in a relative s-state in b, the result of integration over $\underset{\sim}{\xi}_x$ and $\underset{\sim}{\sigma}_a$ can be expressed in a simplified form as

$$\int d^3\sigma_a\ d^3\xi_x\ \phi_b^*(\underset{\sim}{r}_{ax},\underset{\sim}{\xi}_x,\underset{\sim}{\sigma}_a)\ \phi_a(\underset{\sim}{\sigma}_a)\ \psi_x(\underset{\sim}{r}_{xC},\underset{\sim}{\xi}_x) = C_{ax}\ \psi_x(\underset{\sim}{r}_{xC})\ \phi_b^*(\underset{\sim}{r}_{ax}) \qquad (8.77)$$

where C_{ax} is a constant introduced so that ϕ_b can be normalized to unity. Using eqns (8.75), (8.76), and (8.77), we find that the matrix element (8.74) becomes

$$T_{fi} = C_{ax}\int d^3r_{ax}\ d^3r_{xC}\ \chi_b^{-*}(\underset{\sim}{k}_b,\underset{\sim}{r}_{xC}+\lambda\underset{\sim}{r}_{ax})\ \phi_b^*(\underset{\sim}{r}_{ax})\ V_{ax}(\underset{\sim}{r}_{ax})\ \times$$
$$\times\ \psi_x(\underset{\sim}{r}_{xC})\ \psi_a^+(\underset{\sim}{k}_a,\underset{\sim}{r}_{ax}+\lambda\underset{\sim}{r}_{xC}). \qquad (8.78)$$

The matrix element (8.78) is a six-dimensional integral. It can be reduced to a more simple form through the zero-range approximation (Bassel, Drisko, and Satchler 1962)

$$V_{ax}(\underset{\sim}{r}_{ax})\ \phi_b^*(\underset{\sim}{r}_{ax}) = D_0\ \delta(\underset{\sim}{r}_{ax}) = D_0\ \delta(\underset{\sim}{r}_{aA} - \gamma\underset{\sim}{r}_{xC}). \qquad (8.79)$$

Alternatively the local-energy approximation (Buttle and

Goldfarb 1964) may be used to give an approximate treatment of the finite-range calculation. Exact finite-range calculations (Austern *et al.* 1964, Dickens *et al.* 1965*a*) show that there is an important connection between the effect of distortion and of the finite-range interaction, and that the finite-range correction tends to reduce the contribution to the matrix element from the nuclear interior. In general finite-range effects are not too important in (p,d) and (d,p) reactions at medium energies and low Q-values. They are much more important when the momentum transfer is large, for example, for (p,d) at intermediate energies and for heavy-particle stripping (Robson 1963, Towner 1967).

The matrix element for heavy-ion transfer reactions is frequently written in the post form, for which the principal term in the interaction is $V_{\mathrm{bC}}(\mathbf{r}_{\mathrm{bC}})$. Using the relations

$$\begin{aligned} \mathbf{r}_{\mathrm{aC}} &= \mathbf{r}, \qquad \mathbf{r}_{\mathrm{ax}} = \mathbf{r}', \qquad \mathbf{r}_{\mathrm{xC}} = \mathbf{r} - \mathbf{r}', \\ \mathbf{r}_{\mathrm{aA}} &= \left(1 - \frac{M_{\mathrm{x}}}{M_{\mathrm{A}}}\right)\mathbf{r} + \frac{M_{\mathrm{x}}}{M_{\mathrm{A}}}\mathbf{r}', \qquad \mathbf{r}_{\mathrm{bC}} = \mathbf{r} - \frac{M_{\mathrm{x}}}{M_{\mathrm{b}}}\mathbf{r}' \end{aligned} \tag{8.80}$$

the DWBA matrix element becomes (Buttle and Goldfarb 1966, 1971)

$$T_{\mathrm{fi}} = \int \mathrm{d}^3 r\, \mathrm{d}^3 r'\, \chi_{\mathrm{f}}^{-*}\left(\mathbf{k}_{\mathrm{b}}, \mathbf{r} - \frac{M_{\mathrm{x}}}{M_{\mathrm{b}}}\mathbf{r}'\right) \phi_{\mathrm{b}}^{*}(\mathbf{r}')\, V_{\mathrm{bC}}(\mathbf{r} - \mathbf{r}')\, \times$$

$$\times\, \psi_{\mathrm{x}}(\mathbf{r} - \mathbf{r}')\, \chi_{\mathrm{i}}^{+}\left(\mathbf{k}_{\mathrm{a}}, \frac{M_{\mathrm{A}} - M_{\mathrm{x}}}{M_{\mathrm{A}}}\mathbf{r} + \frac{M_{\mathrm{x}}}{M_{\mathrm{A}}}\mathbf{r}'\right). \tag{8.81}$$

In order to simplify this expression, it can be argued that when x consists of a single nucleon or a few nucleons the coefficients $M_{\mathrm{x}}/M_{\mathrm{b}}$ and $M_{\mathrm{x}}/M_{\mathrm{A}}$ are of order A^{-1} and can therefore be neglected. With this approximation the matrix element for single-nucleon transfer becomes

$$T_{\mathrm{fi}} = \int \mathrm{d}^3 r\, \chi_{\mathrm{f}}^{-*}(\mathbf{k}_{\mathrm{b}}, \mathbf{r})\, G_{\mathrm{if}}(\mathbf{r})\, \chi_{\mathrm{i}}^{+}\left(\mathbf{k}_{\mathrm{a}}, \frac{A-1}{A}\mathbf{r}\right) \tag{8.82a}$$

$$G_{\mathrm{if}}(\mathbf{r}) = \int \mathrm{d}^3 r'\, \psi_{\mathrm{f}}^{*}(\mathbf{r}')\, V_{\mathrm{bC}}(\mathbf{r} - \mathbf{r}')\, \psi_{\mathrm{i}}(\mathbf{r} - \mathbf{r}') \tag{8.82b}$$

where ψ_i, ψ_f are the bound-state wavefunctions for the single nucleon on the initial and final systems. This is known as neglect of recoil terms and is valid at low energies, but for higher energies and heavy targets with large effective radii the recoil effects can be important (Greider 1970, Buttle and Goldfarb 1971).

8.3.2. *Spectroscopic factors and sum rules*

It is customary to make a fractional parentage expansion of the wavefunction of the target nucleus in the case of a pick-up reaction or of the residual nucleus in the case of a stripping reaction. For the pick-up reaction this gives

$$\Phi_{J_A}^{M_A}(\underset{\sim}{\xi}_C,\underset{\sim}{\xi}_x,\underset{\sim}{r}_{xC}) = \sum_{jmJ_P} (J_P M_P jm|J_A M_A)\mathscr{I}_{J_A J_P}(j)\Phi_{J_P}^{M_P}(\underset{\sim}{\xi}_C)\psi_j^m(\underset{\sim}{\xi}_x,\underset{\sim}{r}_{xC}) \tag{8.83}$$

where the Φ_{J_P} constitute a complete set of wavefunctions for the parent states J_P in the residual nucleus and $\mathscr{I}$ is the fractional parentage coefficient. The overlap integral defined by eqn (8.76) now becomes

$$\psi_x(\underset{\sim}{\xi}_x,\underset{\sim}{r}_{xC}) = \sum_{jm} (J_C M_C jm|J_A M_A)\mathscr{I}_{J_A J_C}(j)\psi_j^m(\underset{\sim}{\xi}_x,\underset{\sim}{r}_{xC}).$$

We assume that particle x moves within nucleus A with total angular momentum $\underset{\sim}{j}$ composed of orbital angular momentum $\underset{\sim}{\ell}$ and spin $\underset{\sim}{s}$. If x is a single nucleon $\underset{\sim}{s}$ is the intrinsic spin, but if x is a group of nucleons or cluster then $\underset{\sim}{s}$ is the total internal angular momentum of the system and may take several values. With this notation the overlap integral may be written as

$$\psi_x(\underset{\sim}{\xi}_x,\underset{\sim}{r}_{xC}) = \sum_{\substack{jls\\ m\lambda\mu}} (J_C M_C jm|J_A M_A)(l\lambda\ \mu|jm)\mathscr{I}_{J_A J_C}(j)$$
$$\times\ i^l R_{nlj}(r_{xC})\ Y_l^\lambda(\hat{\underset{\sim}{r}}_{xC})\ \psi_s^\mu(\underset{\sim}{\xi}_x). \tag{8.84}$$

The parentage coefficient $\mathscr{I}$ in these expansions selects a particular nucleon or cluster. If there are N_{lj} identical particles x within A, the cross-section is proportional to the *spectroscopic factor* S, where

$$S_{J_A J_C}(lj) = N_{lj}\,\{\mathscr{I}_{J_A J_C}(lj)\}^2. \tag{8.85}$$

The wavefunction $\phi_b(\underset{\sim}{r}_{ax},\underset{\sim}{\xi}_x,\underset{\sim}{\sigma}_a)$ for particle b can be expanded in a similar way in terms of the total internal angular momenta s_a and s_b for particles a and b. For simplicity we write eqn (8.84) as

$$\psi_x(\underset{\sim}{\xi}_x,\underset{\sim}{r}_{xC}) = \sum \psi_j^m(\underset{\sim}{r}_{xC})\;\psi_s^\mu(\underset{\sim}{\xi}_x)$$

so that the reduced overlap integral (8.77) becomes

$$\int d^3\sigma_a\, d^3\xi_x\, \phi_b^*(\underset{\sim}{r}_{ax},\underset{\sim}{\xi}_x,\underset{\sim}{\sigma}_a)\;\phi_a(\underset{\sim}{\sigma}_a)\;\psi_x(\underset{\sim}{\xi}_x,\underset{\sim}{r}_{xC})$$
$$= \sum (j_a\nu_a s\mu|s_b m_b)\mathscr{I}_{j_a s_b}(s)(s_a m_a \Lambda\lambda|j_a\nu_a)\phi_\Lambda^{\lambda *}(\underset{\sim}{r}_{ax})\;\psi_j^m(\underset{\sim}{r}_{xC}).$$

If a and x are in a relative s-state in b, so that $\Lambda = 0$, this reduces to

$$\int d^3\sigma_a\; d^3\xi_x\;\phi_b^*\;\phi_a\;\psi_x = \sum (s_a m_a s\mu|s_b m_b)\;\mathscr{I}_{s_a s_b}(s)\phi_0^*(\underset{\sim}{r}_{ax})\psi_j^m(\underset{\sim}{r}_{xC})$$

and, if there are n identical particles x within b, the cross-section is proportional to $n\{\mathscr{I}_{s_a s_b}(s)\}^2$. By comparison with eqn (8.77) we have

$$C_{ax}^2 = S_{s_a s_b}(s) = n\{\mathscr{I}_{s_a s_b}(s)\}^2. \tag{8.86}$$

In this formalism, for single-nucleon transfer the number N_{lj} is the number of neutrons or protons in the subshell. Alternatively, the isospin formalism can be used (Macfarlane and French 1960), in which case the fractional parentage coefficient and spectroscopic factor are labelled by T_A and T_B and the isospin Clebsch-Gordan coefficients $(t_a m_{ta} t m_t|t_b m_{tb})$ and $(T_B M_{TB} t m_t|T_A M_{TA})$ appear in the equations above. The value values of S and $\mathscr{I}$ are different in the two cases. For single-nucleon transfer N_{lj} is now the number of active nucleons in the subshell.

The experimental spectroscopic factor is determined by fitting the shape of the angular distribution with a calculated cross-section and then finding the magnitude of the spectroscopic factor required to give absolute agreement. For single-nucleon transfer reactions involving a proton and deuteron as projectile and emitted particle, or vice versa, the spectroscopic factor can in many cases be obtained with an accuracy of 20-30 per cent for incident energies in the range 10-60 MeV. At higher incident energies complications in the theory and limitations in the energy resolution have so far reduced the accuracy with which the spectroscopic factor can be obtained. For single-nucleon transfer with light and heavy ions the second spectroscopic factor, defined in (8.86), enters the formulae. For light ions this factor is usually calculated theoretically from model wavefunctions, but for heavy-ion transfer reactions this factor is taken from other experiments.

When the summation over initial and final states is carried out, the differential cross-section for single-nucleon transfer has the form (Bassel *et al.* 1962)

$$\frac{\mathrm{d}\sigma}{\mathrm{d}\Omega} = \sum_{lj} S_{J_\mathrm{A}J_\mathrm{C}}(lj)\ \sigma_{lj}(\theta) \tag{8.87}$$

where the function $\sigma_{lj} = \Sigma_\Lambda |B^\Lambda_{lj}|^2$ contains spherical harmonics and integrals over radial functions, and depends on the scattering angle θ, the incident energy, and the Q-value. Although σ_{lj} depends on the initial and final states it does so only weakly through the properties of the overlap integral. This supports the view that the spectroscopic factor contains the nuclear structure information for the transition. In contrast, the summation over initial and final states in a two-nucleon transfer reaction does not lead to such simple factorization but gives a cross-section of the form (Towner and Hardy 1969)

$$\frac{\mathrm{d}\sigma}{\mathrm{d}\Omega} = \sum_{LSJ\Lambda T} f(S,T)\left|\sum_{l_1 l_2}\{S(l_1 l_2;LSJT)\}^{1/2} B^\Lambda_{LJ}\right|^2 \tag{8.88}$$

where l_1 and l_2 are the angular momenta of the two nucleons transferred, $LSJT$ are the resulting quantum numbers of the pair of nucleons, and $S(l_1 l_2;LSJT)$ is the corresponding spectroscopic factor. The factor $f(S,T)$ depends on the particular reaction. As before the spectroscopic factor contains the essential nuclear structure information, but the sum over l_1 and l_2 introduces several terms which contribute coherently to the cross-section. This means that the spectroscopic factor cannot be determined directly from experiment but, if some model predictions are available, two-nucleon transfer tests the magnitude and phase of the predicted quantities very sensitively.

The theoretical spectroscopic factors are always model dependent and, within the framework of a given model, they may depend on the coupling scheme (Macfarlane and French 1960). This makes possible a test of various nuclear models for a given nucleus or for similar states in a different nuclei. The model may also be used to calculate the total particle or hole strength for given lj values. This is the monopole *sum rule* and is again a model-dependent quantity. Comparison of the sum of the observed spectroscopic factors with the predicted sum rule indicates whether there is a substantial amount of missing hole or particle strength at high excitations or how much of the sum rule is exhausted by a particular transition. More general multipole sum rules have been defined (French 1964, 1965).

In certain special cases the spectroscopic factors for single-nucleon transfer take a particularly simple form. In jj-coupling each subshell is labelled by the quantum numbers lj and the number of neutrons or protons in the subshell is N_{lj}. If a single nucleon is removed from the outer shell of a closed-shell-plus-one target, the coefficient of fractional parentage (cfp) is unity and so is the spectroscopic factor. Similarly, if one nucleon is added to a closed-shell nucleus the cfp and the spectroscopic factor are both unity. In jj-coupling, the sum rule for the single-nucleon pick-up reaction A(p,d)B is (Macfarlane and French 1960)

$$\sum_{J_\mathrm{P}} S_{J_\mathrm{A}J_\mathrm{P}}(lj) = N_{lj} \tag{8.89a}$$

while the sum rule for the stripping reaction B(d,p)A is

$$\sum_{J_\mathrm{P}} (2J_\mathrm{P}+1)\, S_{J_\mathrm{B}J_\mathrm{P}}(lj) = (2j+1-N_{lj}+1)(2J_\mathrm{B}+1) \tag{8.89b}$$

i.e. nucleus A has N_{lj} particles and $2j+1-N_{lj}$ holes in the unfilled shell, whereas nucleus B has $N_{lj}-1$ particles and $2j+1-N_{lj}+1$ holes in the same shell. If we compare pick-up and stripping on the same nucleus we have

$$\mathrm{A(p,d)B:}\quad \sum_{J_\mathrm{P}} S_{J_\mathrm{A}J_\mathrm{P}}(lj) = N_{lj} \tag{8.89d}$$

$$\mathrm{A(d,p)C:}\quad \sum_{J_\mathrm{P}} (2J_\mathrm{P}+1)\, S_{J_\mathrm{A}J_\mathrm{P}}(lj) = (2j+1-N_{lj})(2J_\mathrm{A}+1). \tag{8.89d}$$

The sum rules have particularly simple forms for spin-zero target nuclei. In this case the spin of the final nucleus is equal to j and conservation of parity permits only one l-value. Eqns (8.89*c*) and (8.89*d*) then yield

$$\sum \{(2j+1)\, S^{+}(lj) + S^{-}(lj)\} = 2j+1 \tag{8.90a}$$

where S^{+} is the spectroscopic factor for stripping and S^{-} is the spectroscopic factor for pick-up. This result can be interpreted as the sum rule for particles and holes. We may also write

$$U^2 = \sum S^{+}(lj) = \frac{2j+1-N_{lj}}{2j+1} \tag{8.90b}$$

$$V^2 = \frac{\sum S^{-}(lj)}{2j+1} = \frac{N_{lj}}{2j+1} \tag{8.90c}$$

so that

$$U^2 + V^2 = 1. \tag{8.90d}$$

Thus U^2 measures the probability of the subshell lj being empty and V^2 measures the probability that the subshell lj is occupied. Thus comparison of (d,p) and (d,t) reactions on the same spin-zero target nucleus yields information about the partial filling of states near the Fermi surface. Fig. 8.15 shows a comparison of the experimental values for a deformed heavy nucleus compared with BCS theory (Elbek and Tjom 1969). For a target with odd A and spin j, the (d,p) and (d,t) cross-sections are proportional to V^2 and U^2 respectively (Cohen and Price 1961).

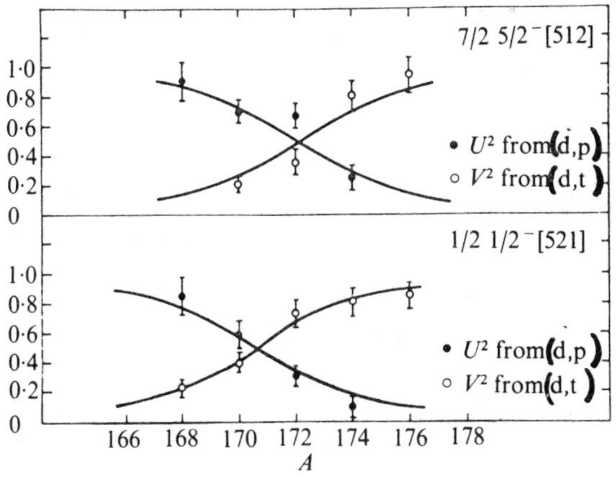

FIG. 8.15. The factors U^2 and V^2 for even Yb nuclei. (From Elbek and Tjom 1969.)

8.3.3. *Single-nucleon transfer*

In single-nucleon transfer reactions the system x is a single proton or neutron so that the overlap integral (8.76) reduces to

$$\psi_x(\underset{\sim}{r}) = \int d^3\xi_C\ \Phi_C^*(\underset{\sim}{\xi}_C)\ \Phi_A(\underset{\sim}{\xi}_C,\underset{\sim}{r})$$

$$= \sum_{jm} (J_C M_C j m | J_A M_A)\ \mathscr{F}_{J_A J_C}(j)\ \psi_j^m(\underset{\sim}{r}) \tag{8.91}$$

where ψ_j^m is a normalized single-particle wavefunction. The asymptotic behaviour of the overlap integral can readily be established (Berggren 1965, Pinkston and Satchler 1965) by writing the Hamiltonian H_A for the nucleus A as $H_C + T_{xC} + V_{xC}$

where H_C is the Hamiltonian for the core C and V_{xC} is the nucleon-core interaction. Hence

$$(H_C + T_{xC} + V_{xC})\Phi_A^i = E_A^i \Phi_A^i$$

and multiplying on the left by Φ_C^* and integrating over $\underset{\sim}{\xi}_C$ we have

$$(T_{xC} + V_{xC})\psi_x(\underset{\sim}{r}) = (E_A^i - E_C^f)\psi_x(\underset{\sim}{r}). \qquad (8.92)$$

Consequently the asymptotic behaviour of ψ_x, outside the range of V_{xC} and any residual interaction, is given by

$$R_{nlj}(r) \to e^{-\kappa r}/r \qquad (8.93)$$

where

$$\kappa^2 = \frac{2\mu_{xC}(E_C^f - E_A^i)}{\hbar^2} \qquad (8.94)$$

Thus the asymptotic behaviour of the overlap integral is determined by the separation energy E_s^{if}, defined in eqn (2.34), for the break-up of nucleus A into the residual nucleus C plus a proton or neutron and leaving C in the definite final state f. This result justifies the use of the separation energy procedure (SEP) in which the radial part of ψ_j^m is generated in a spherically symmetric Saxon-Woods potential whose parameters are chosen to give a nucleon binding energy equal to the separation energy. The number of nodes in the radial wavefunction is usually determined by the principal quantum number of the active nucleons in the subshell lj, although a more accurate treatment would imply summation over this and higher values of n in order to take account of configuration mixing. This procedure is satisfactory when the nucleus A is spherical or nearly so, but for strongly deformed nuclei the potential in which the ψ_j^m are generated is also deformed. Methods have been developed (Rost 1967, Anderson, Back, and Bang 1970) for calculating the required single-particle wavefunctions in a deformed Saxon-Woods potential and have been applied to the analysis of transfer reactions on deformed nuclei.

For energies below the Coulomb barrier transfer reactions are localized in the outer regions of the nucleus beyond the distance of closest approach. The corresponding cross-sections for the (d,p) and (t,d) reactions are insensitive to the l-value but are strongly energy dependent (Goldfarb 1965). In this situation it is a reasonable approximation to replace the radial part of the overlap integral by its asymptotic form (8.93) multiplied by a normalization factor. Many of the uncertainties due to nuclear effects are diminished or removed so that the spectroscopic factors can be deduced with greater accuracy. A similar approach has been taken to heavy-ion transfer reactions by Buttle and Goldfarb (1966, 1971) who write the single-nucleon wavefunction ψ_f in eqn (8.82b) as

$$\psi_f(\underset{\sim}{r}') = N\, h_l^{(1)}(i\kappa r')\; Y_l^{\lambda}(\hat{\underset{\sim}{r}}') \tag{8.95}$$

where κ is given by eqn (8.94). This function can be expanded in terms of Hankel functions and spherical Bessel functions so that the integral (8.82b) can be evaluated without difficulty. Rapaport and Kerman (1968) have introduced a dimensionless constant, called the reduced normalization, through the relation

$$\Lambda_{lj} = \frac{N^2\, S(lj)}{\kappa^3} \tag{8.96}$$

This constant has the important characteristic that it is independent of the interior behaviour of the nucleon wavefunction to which the Hankel function must be matched to determine the normalization N. This has been verified for sub-Coulomb (d,p) reactions (Rapaport and Kerman 1968, Rapaport, Sperduto, and Saloma 1972, Kent, Morgan, and Seyler 1972) by varying the halfway radius and diffuseness of the single-particle potential used to generate the nucleon wavefunction. Some results are shown in Fig. 8.16. It is apparently possible to determine a quantity from sub-Coulomb reactions which is parameter independent, but this quantity can provide no nuclear size information. On the other hand.

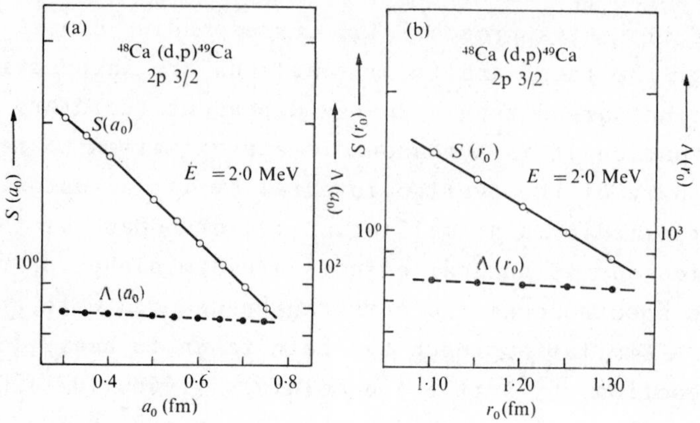

FIG. 8.16. The variation of the spectroscopic factor S and the reduced normalization Λ with parameters of the neutron bound-state potential: (a) variation with diffuseness for a fixed radius r_0 = 1.25 fm, and (b) variation with radius r_0 for a fixed diffuseness r = 0.65 fm. (After Rapaport *et al.* 1972.)

determination of the magnitude of the tail of the wavefunction, as defined in equation (8.95) together with the nucleon separation energy yields the r.m.s. radius of the nucleon orbit to within quite narrow limits, if a single-particle model is assumed. Such measurements have been made for light ion (Korner and Schiffer 1971, Friedman *et al.* 1972, Schiffer and Korner 1973) and heavy ion (G.D. Jones *et al.* 1974) transfer reactions below the Coulomb barrier. The results obtained for the r.m.s. radii of the neutron excess orbits can be compared with calculations for Coulomb energy differences (see §10.2).

It is customary to refer to ψ_j^m as a single-particle wavefunction. This is correct in the sense that it is generated in a one-body potential, but it is not a shell model wavefunction in the usual sense (Berggren 1965, Pinkston and

Satchler 1965). In constructing the overlap integral we have described the nucleus in terms of the relative co-ordinates ξ_C and r_{xC}, and the Hamiltonian H_A is translation invariant which is not the case when the shell model is used. The co-ordinate r_{xC} is referred to the centre of mass of the core C and not to the centre of mass of the target nucleus A. In addition, if the initial state J_A has more than one parent state J_p in the residual nucleus, the separation energy $E_C^f - E_A^i$ for a transition to a definite final state is different from the Hartree-Fock energy, defined in eqn (2.35) as $-(E_A^i - \Sigma_p \mathscr{S}_p^2 E_C^p)$. Comparison with Hartree-Fock calculations is nevertheless useful, and the extent to which the difference in energies and wavefunctions is significant depends on the strength of excitation of the various parent states and the energy splitting between them compared with the energy resolution of the experiment.

Although the SEP yields an overlap integral with the correct asymptotic radial behaviour it may not yield the correct interior behaviour or the correct magnitude in the asymptotic behaviour. Improvement of the description of the overlap integral can be achieved by inclusion of the residual interaction which leads to a set of coupled equations. A review of methods used to handle these equations has been given by Philpott, Pinkston, and Satchler (1968).

Most of the available information about neutron single-particle and single-hole states comes from the (p,d) and (d,t) reactions. The shapes of the angular distributions are characteristic of the orbital angular momentum l although the difference in shape diminishes for increasing incident energy. At medium energies the angular distributions and polarization also show j-dependence (Schiffer 1968), i.e. there is a difference between the shapes for $j = l + \frac{1}{2}$ and $j = l - \frac{1}{2}$. This effect, illustrated in Figs. 8.17 and 8.18, allows identification of both l and j. It appears that the j-dependence observed at large angles in $l = 1$ transitions is due to the spin-orbit terms in the optical potential. The j-dependence of the main peak seen in $l = 3$ transitions arises from the D-state of the deuteron (Johnson and Santos 1967, 1971) and from

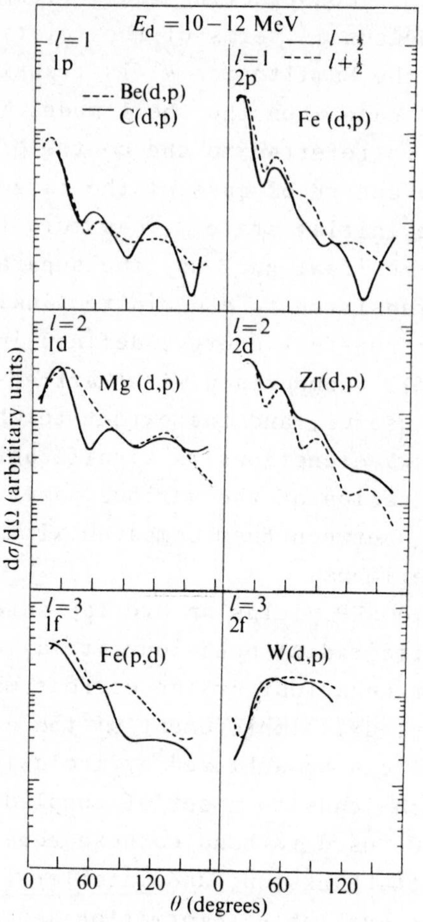

FIG. 8.17. A summary of j-dependent effects in (d,p) reactions. The curves represent the trend of the experimental data for different j-values. (From Schiffer 1968.)

configuration mixing. Angular distributions for (p,d) reactions at 100 MeV (Lee *et al*. 1967), at 155 MeV (Bachelier *et al*. 1969), and at 185 MeV (Sundberg and Källne 1969) show very little structure compared with the data at medium energies. This suggests that the energy region of 100-200 MeV is not particularly useful for a study of the neutron overlap integral, although the lack of sensitivity to the form of the

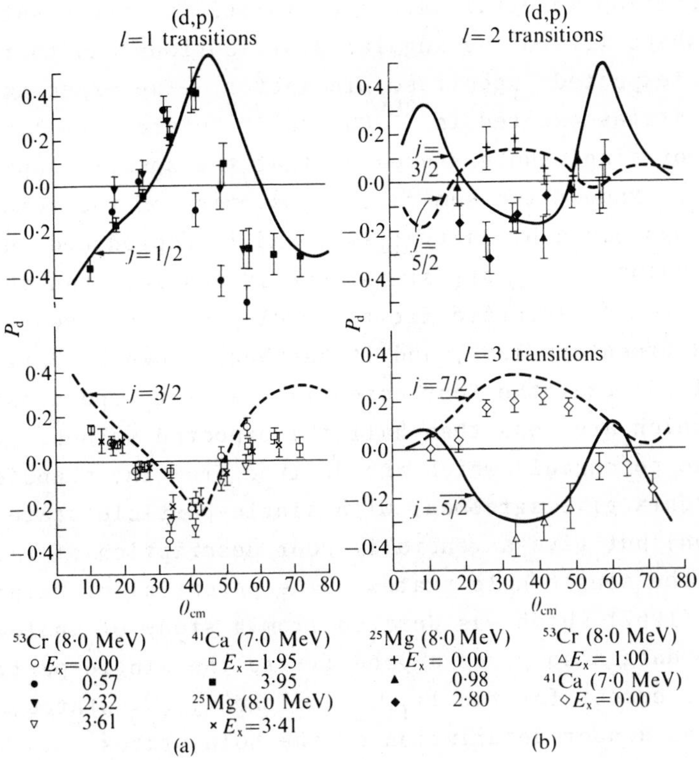

FIG. 8.18. The vector analysing power for various (d,p) reactions. (From Glashauser and Thirion 1969.)

overlap integral would make extraction of the spectroscopic factors more reliable (Towner 1969) if the optical potential parameters were sufficiently well known.

An extensive study of neutron states in ^{208}Pb has been carried out. The neutron levels are shown in Fig. 2.3 where the single-particle states lie above the gap at ~ 5 MeV and are studied by stripping reactions, while the single-hole states lie below the gap and are studied by pick-up reactions. In most cases the radial parameters† of the local single-

---

†The potential strength is varied to give the correct separation energy.

particle potential used to generate the radial part of the neutron overlap integral have been varied to give a satisfactory shape fit to the angular distributions and to reproduce the 'expected' spectroscopic factor. The expectation is that the states excited in ^{209}Pb and ^{207}Pb are pure single-particle or single-hole states so that the spectroscopic factor should exhaust the sum rule. A summary of the results obtained was given by Batty (1970) and is reproduced in Table 8.5. The values of r_0 are significantly smaller than those obtained from an accurate fit to level positions (Rost 1968, Batty and Greenlees 1969) and it has been shown by Parkinson *et al.* (1969) that the Rost potential leads to spectroscopic factors which are less than half the expected values. In contrast, the potentials which are derived from the transfer reaction data give agreement with single-particle states (by definition) but give a generally poor description of the positions of the single-hole states. The potential of Zaidi and Darmodjo (1967) which was derived from a study of analogue resonance data also gives a good fit to the single-particle positions, except for the $1i_{11/2}$ and the $1j_{15/2}$ states, but again gives a poor description of the hole states.†

There is now evidence that the single-particle strength in ^{209}Pb and ^{209}Bi is fragmented in certain cases (Ellegaard, Kantele, and Vedelsby 1967, Bardwick and Tickle 1968, Igo *et al.* 1969) and similar evidence has been presented for the fragmentation of the single-hole strength in ^{207}Pb (Alford and Burke 1969, Moyer, Cohen, and Diehl 1970). Thus, although the so-called single-particle and single-hole states are most strongly excited, the presence of other weakly excited states distributes the strength so that the spectroscopic factors need not be unity. Values obtained for the spectroscopic factors for the pick-up reaction with the neutron potential parameters $r_0 = 1.25$ fm, $a = 0.65$ fm are given in Table 8.6. It appears that a satisfactory description of reaction data and realistic spectroscopic factors can be obtained using a

---

†See also §2.2.1.

TABLE 8.5

Radial parameters for single-particle potentials for ^{208}Pb †

Reference	Method	r_0 (fm)	a (fm)	r_s (fm)	a_{so} (fm)
Neutron levels					
Rost 1968	Fits to levels only	1.35	0.7	1.28	0.7
Batty & Greenlees 1969	Fits to levels only	1.36	0.73	1.26	0.6
Muehllner *et al.* 1967	(d,p)(d,t) 14-25 MeV	1.20	0.65	1.23	0.7
Zaidi and Darmodjo 1967	Analogue resonances	1.19	0.75	1.10	0.75
Dost *et al.* 1967	(d,p) 8 MeV	1.26	0.6	1.26	0.6
Crawley *et al.* 1968	(d,p) 8 MeV	1.23	0.63	1.23	0.6
Igo *et al.* 1969	(t,d) 20 MeV	{1.25	0.65	1.25	0.65
		1.19	0.75	1.19	0.75
Van der Merwe & Heymann 1969	(d,p) 8-24 MeV	1.25	0.65	1.25	0.65
Jeans *et al.* 1969	(d,p) 8-19 MeV	{1.23	0.65	1.23	0.65
		1.19	0.75	1.19	0.75
Parkinson *et al.* 1969	(d,t) 50 MeV	1.25	0.63	1.10	0.50
Alford and Burke 1969	(^{3}He,α) 28 MeV	{1.25	0.65	1.25	0.65
		1.25	0.65	1.10	0.50
Proton levels					
Rost 1968	Fit to levels only	1.28	0.7	0.93	0.7
Batty & Greenlees 1969	Fit to levels only	1.28	0.76	1.09	0.6
Hinds *et al.* 1966	(t,α) 12-14 MeV	{1.30	0.7	1.30	0.7
		1.20	0.7	1.20	0.7
Wildenthal *et al.* 1967	(^{3}He,d) 51 MeV	1.24	0.65	1.24	0.65
Parkinson *et al.* 1969	(d,^{3}He) 50 MeV	1.30	0.75	1.30	0.75
Royer *et al.* 1970	(d,^{3}He) 50, 60 MeV	1.28	0.76	1.09	0.6

† Adapted from Batty (1970).

local potential with $r_0 \sim 1.19$-1.25 fm, but an adequate description of the more tightly bound hole states requires a density-dependent non-local potential such as that introduced by Janiszewski and McCarthy (1972*a*).

The (^{3}He,d) and (d,^{3}He) reactions can be used in a similar way to study proton states in nuclei. Results for the

TABLE 8.6

Spectroscopic strengths, as ratio to the sum rule, for the reaction $^{208}Pb(p,d)$ at 41 MeV†

Hole configuration	Excitation energy (MeV)	Deuteron potential 1		Deuteron potential 4	
		Zero range	Finite range	Zero range	Finite range
$3p_{1/2}$	0	1.2	1.2	1.1	1.1
$2f_{5/2}$	0.57	1.0	1.1	1.0	1.0
$3p_{3/2}$	0.89	1.1	1.1	0.92	0.95
$1i_{13/2}$	1.63	0.56	0.84	0.50	0.61
$2f_{7/2}$	2.34	0.70	0.73	0.63	0.64
$1h_{9/2}$	3.40	0.48	0.68	0.60	0.68

†Results are given for two deuteron optical potentials (Smith *et al.* 1971).

$(d,{}^3He)$ reaction on ^{208}Pb at 50 MeV (Parkinson *et al.* 1969) and at 80 MeV (Royer *et al.* 1970) show quite clearly that the potentials generated by fitting the level positions accurately (Rost 1968, Batty and Greenlees 1969) give very satisfactory agreement with both sets of data. This gives $r_0 \sim 1.25$-1.30 fm for protons, implying a symmetry term in the potential which is surface peaked and of the expected sign or partly of this form and partly of volume form. Further studies of the $(d,{}^3He)$ reaction have been carried out at 34 MeV, 52 MeV, and 80 MeV and results are discussed in §8.4 in relation to results for proton knock-out reactions.

In many cases it has been found necessary to introduce a lower cut-off in the DWBA matrix element in order to improve agreement with the data. The need for this cut-off is removed in (d,p) and (p,d) reactions (Preedom 1972, Harvey and Johnson 1971, McAllen, Pinkston, and Satchler 1971, Cooper, Hornyak, and Roos 1974) when the conventional deuteron optical potentials are replaced by potentials derived from the adiabatic theory of Johnson and Soper (1970). The adiabatic theory takes account of break-up of the deuteron but the structure of the matrix element remains as given in eqn (8.74).

The (d,^{3}He) and (d,t) reactions provide a comparison of proton and neutron overlap integrals. If the overlap integrals and hence the nuclear matrix elements were identical, the ratio of the cross-sections on the same target nuclei would be proportional to the momenta of the outgoing particles, as can be seen from eqn (8.73). Such a comparison has been made for 28 MeV deuterons on nuclei from ^{12}C to ^{40}Ca (Gaillard *et al.* 1968). The data are reasonably well reproduced by DWBA calculations and the ratios of the cross-sections are found to be proportional to the ratios of the squares of the proton and neutron wavefunctions at some radius in the asymptotic region. A similar study at 50 MeV on ^{208}Pb (Parkinson *et al.* 1969) has shown that these reactions are localized in the extreme surface region, and hence the cross-sections are proportional to the squares of the asymptotic tails of the wavefunctions. Comparisons of the (α,t) and (α,^{3}He) reactions have been made at 56 MeV (Gaillard *et al.* 1969) and at 104 MeV (Hauser *et al.* 1972), but in these reactions the situation is more complicated owing to momentum mismatch and differences in the appropriate constant D_0, defined in eqn (8.79).

8.3.4. *Multi-nucleon transfer*

In the case of two-nucleon transfer reactions, such as the (α,d) and (p,^{3}He) reactions, the system x consists of a pair of nucleons. The function ψ_s^μ, defined in equation (8.84), describes the relative motion of the two nucleons inside the nucleus and the function $\psi_j^m(\underset{\sim}{r}_{xC})$ describes the motion of their centre of mass relative to the core C. Thus in a (p,t) reaction, for example, the reduced overlap integral describes the motion in the target nucleus of the centre of mass of a pair of neutrons correlated in the same way as a pair of neutrons in the triton.

The earliest method (Glendenning 1963, 1965) for constructing the overlap integral for two-nucleon transfer is to work within the framework of the shell model, starting from the single-particle states for two nucleons relative to the core. If these are taken to be oscillator functions, it is possible to transform the product of the two single-particle functions into the product of a function of the relative co-ordinate $\mathbf{r}_1 - \mathbf{r}_2$ and a function of the centre-of-mass co-ordinate $\frac{1}{2}(\mathbf{r}_1 + \mathbf{r}_2)$. A more realistic treatment is obtained by matching Hankel functions on to the oscillator functions a a suitable radius.

An improved shell model approach can be obtained by generating each single-particle wavefunction in a realistic one-body potential. This raises the question, however, of what single-particle potentials and what single-particle energies should be used. The overlap integral (8.76) can be written in shell model co-ordinates as

$$\psi(\mathbf{r}_1,\mathbf{r}_2) = \int d^3\xi_C \; \Phi_C^{f*}(\boldsymbol{\xi}_C) \; \Phi_A^{i}(\boldsymbol{\xi}_C,\mathbf{r}_1,\mathbf{r}_2). \qquad (8.97a)$$

The equation for this overlap integral is (Jaffe and Gerace 1969)

$$\{E_A^i - E_C^f - T_1 - T_2 - V(\mathbf{r}_1) - V(\mathbf{r}_2) - W(\mathbf{r}_1,\mathbf{r}_2)\}\,\psi(\mathbf{r}_1,\mathbf{r}_2) = 0 \qquad (8.97b)$$

where $W(\mathbf{r}_1,\mathbf{r}_2)$ is the residual interaction between the two nucleons, $V(\mathbf{r}_1)$ is the potential that binds nucleon 1 to the core, and similarly for $V(\mathbf{r}_2)$. The most common procedure (Rook and Mitra 1964, Bayman and Kallio 1967, Towner and Hardy 1969) is to neglect the residual interaction but bind each nucleon at a separation energy of $\frac{1}{2}\,\varepsilon$, where $\varepsilon = -(E_A^i - E_C^f)$. This implies that the effect of the residual interaction on the energy has been taken into account through the single-particle potentials; the effect of the residual interaction on the wavefunction, i.e. configuration mixing, can be taken into account by using the sum of products of these single-particle wavefunctions. An alternative procedure (Drisko and Rybicki 1966) is to take a single product of single-particle wavefunctions and bind each nucleon at the separation energy ε_n for a nucleus with one nucleon outside the A-2 core; for two neutrons this would imply that $\langle W(\mathbf{r}_1,\mathbf{r}_2)\rangle = \varepsilon - 2\varepsilon_n$ which is the pairing energy. Jaffe and Gerace (1969) include the residual interaction and vary it to obtain the correct separation energy ε.

A substantial amount of data exists for two-nucleon transfer reactions and it has been analysed in detail in spectroscopic terms. However, because of the complexities described above, these data provide rather indirect nuclear size information. This is even more true for transfer of larger numbers of nucleons.

8.4. KNOCK-OUT REACTIONS

In a knock-out reaction the projectile a knocks out a nucleon or group of nucleons x from the target A leaving the residual nucleus C, i.e.

$$a + A \rightarrow a + x + C.$$

A measurement of the final energies of a and x yields information about the separation energies of nucleons or groups of nucleons in the nucleus. In general the energy spectrum will show a set of peaks corresponding to excitation of different states in the residual nucleus, and possibly also to excited states of the system x. The total angular momenta of these states are related by

$$\underset{\sim}{J}_A = \underset{\sim}{j}_x + \underset{\sim}{J}_C$$

As in the case of transfer reactions we seek to identify j_x or l_x, or both, from the shape of the angular distribution and hence to determine the parentage of the nucleus A in terms of the system x + C.

8.4.1. *The matrix element*

A major problem in the description of knock-out reactions arises because the final state consists of a multi-particle system. This has a practical effect on the energy resolution which can be achieved in a coincidence experiment and it has a fundamental effect on the reaction theory. The motion of the outgoing particles a and x is coupled in some way, unless the residual nucleus is infinitely heavy; the form of the coupling depends on the choice of relative co-ordinates and it can appear in the kinetic or potential terms. This aspect of the theory of knock-out reactions has been discussed at length elsewhere (Jackson and Berggren 1965, Jackson 1970*a*) and here we quote the required formulae without derivation.

We take the initial momentum of the projectile in the laboratory frame to be $\underset{\sim}{p}_a$ and take the laboratory momenta of the outgoing particles to be $\underset{\sim}{q}_a$ and $\underset{\sim}{q}_x$, respectively. From momentum conservation we have

$$\underset{\sim}{p}_a - \underset{\sim}{q}_a - \underset{\sim}{q}_x = \underset{\sim}{Q} = -\underset{\sim}{P} \tag{8.98}$$

where Q is the recoil momentum of the residual nucleus C and $\underset{\sim}{P}$ is the momentum of the system x in the target nucleus A. We introduce relative co-ordinates $\underset{\sim}{r}_{aC} = \underset{\sim}{r}_a - \underset{\sim}{R}_C$, etc., and use the symbols $\underset{\sim}{k}_{aC}$, etc., to denote the momenta conjugate to these co-ordinates in the centre-of-mass frame.

With the zero-range interaction (8.79) and the overlap integral defined as in eqns (8.76) and (8.77), the matrix element in DWBA can be written in two alternative forms. These are

$$T_{fi} = V_0 \int \chi_f^{-*}(\underset{\sim}{k}_{aC},\underset{\sim}{r})\, \chi_f^{-*}(\underset{\sim}{k}_{xC},\underset{\sim}{r})\, \psi_x(\underset{\sim}{r})\, \chi_i^{+}(\underset{\sim}{k}_{aA}, g\underset{\sim}{r})\, d^3r \tag{8.99a}$$

$$T_{fi} = V_0 \int \chi_f^{-*}(\underset{\sim}{k}_{bC},\underset{\sim}{r})\, \psi_x(\underset{\sim}{r})\, \chi_i^{+}(\underset{\sim}{k}_{aA}, g\underset{\sim}{r})\, d^3r \tag{8.99b}$$

where $g = M_C/M_A$ and b represents the centre of mass of the particles a and x. Although plane-wave Born approximation is not appropriate for the analysis of data, the structure of the matrix element in this approximation is interesting since it factorizes into two integrals

$$T_{fi}^{PW} = \int V_{ax}(s) \exp\{i(\underset{\sim}{p}_a - \underset{\sim}{q}_a)\cdot\underset{\sim}{s}\}\, d^3s \int \psi_x(\underset{\sim}{r}) \exp(i\underset{\sim}{Q}\cdot\underset{\sim}{r})\, d^3r. \tag{8.100}$$

The first integral is the momentum transform of the two-body interaction and the second integral represents the probability f finding the system x in the nucleus A with momentum $-\underset{\sim}{Q}$. e denote the latter integral for the momentum distribution y $G^{PW}(Q)$. The corresponding quantity in distorted wave heory is called the distorted momentum distribution $G(Q)$.

In the intermediate-energy region it is more appropriate o replace the two-body potential by a t-matrix element. For he (p,2p) reaction we would then require

$$T_{pp} = \langle \bar{\underset{\sim}{p}}_f | \tau_{pp} | \bar{\underset{\sim}{p}}_i \rangle \tag{8.101}$$

here

$$\bar{\underset{\sim}{p}}_i = \tfrac{1}{2}(\underset{\sim}{p}_a + \underset{\sim}{Q}), \qquad \bar{\underset{\sim}{p}}_f = \tfrac{1}{2}(\underset{\sim}{q}_a - \underset{\sim}{q}_x) \tag{8.102}$$

re the momenta of the scattered proton before and after collision in the p-p centre-of-mass system. Denoting the corresponding energies by $\bar{E}_i$ and $\bar{E}_f$ and using the condition for nergy conservation

$$\frac{\hbar^2}{2m} p_a^2 = \frac{\hbar^2}{2m} q_a^2 + \frac{\hbar^2}{2m} q_x^2 + \frac{\hbar^2}{2(A-1)m} Q^2 + E_s \tag{8.103}$$

here E_s is the separation energy, we find

$$\bar{E}_f = \bar{E}_i - \frac{1}{2} E_s - \frac{A}{2(A-1)} \frac{\hbar^2}{2m} Q^2 . \tag{8.104}$$

Thus T_{pp} is an off-shell matrix element and it goes further off shell as the separation energy increases or as the momentum of the struck nucleon in the nucleus increases. For fixed E_s and Q the discrepancy decreases with increasing incident energy. The scattering angle in the p-p system is given by

$$\cos\bar{\theta} = \frac{(\underset{\sim}{p}_a + \underset{\sim}{Q})\cdot(\underset{\sim}{q}_a - \underset{\sim}{q}_x)}{|\underset{\sim}{p}_a + \underset{\sim}{Q}|\,|\underset{\sim}{q}_a - \underset{\sim}{q}_x|} \tag{8.105}$$

In the impulse approximation $|T_{pp}|^2$ is replaced by the free cross-section evaluated at an angle $\bar{\theta}$ and at the laboratory energy corresponding to either $\bar{E}_i$ or $\bar{E}_f$. The difference between these energies gives rise to substantial differences in the (p,2p) cross-section at small angles for incident proton energies below ~160 MeV (Jain 1969), McCarthy and co-workers (Lim and McCarthy 1964, 1966, Deutchman and McCarthy 1968, have used a simple pseudopotential to generate the off-shell matrix elements and have shown that there is disagreement with the impulse approximation for incident energies below 150 MeV. Using a plane wave approximation Redish, Stephenson, and Lerner (1970) have shown that a half off-shell matrix element is required and have evaluated this matrix element for the Reid soft-core potential. This leads to substantial disagreement with PWIA for large values of E_s and incident energies below 150 MeV. Further studies by Stephenson *et al.* (1972) have shown that different nucleon-nucleon potentials yield widely different results but the ratio of the half off-shell scattering amplitude to the on-shell amplitude with the same energy parameter is almost model independent. This ratio can hence be used to correct the observed elastic p-p cross-section.

For knock-out of a bound system of several nucleons, a plausible assumption is that the scattering of the projectile a from the system x inside the nucleus approximates closely to the free scattering of a from x. The matrix element for this model in DWIA is

$$T_{fi} = \langle \bar{\underset{\sim}{p}}_f | \tau_{ax} | \bar{\underset{\sim}{p}}_i \rangle \times \times \int \chi_f^{-*}(\underset{\sim}{k}_{aC}, \underset{\sim}{r})\, \chi_f^{-*}(\underset{\sim}{k}_{xC}, \underset{\sim}{r})\, \psi_x(\underset{\sim}{r})\, \chi_i^{+}(\underset{\sim}{k}_{aA}, g\underset{\sim}{r})\, d^3r \tag{8.106}$$

where

$$\bar{\underset{\sim}{p}}_i = (m_a + m_x)^{-1}(m_x \underset{\sim}{p}_a + m_a \underset{\sim}{Q})$$

$$\bar{\underset{\sim}{p}}_f = (m_a + m_x)^{-1}(m_x \underset{\sim}{q}_a - m_a \underset{\sim}{q}_x) \tag{8.107}$$

As before, the two-body matrix element is to be replaced by the free cross-section evaluated at an appropriate energy and angle. This may be called the *quasi-free* or *peripheral model* for the matrix element, and represents the simplest intuitive picture of the knock-out of a composite particle. This picure is particularly successful for the knock-out of a deuteron or an α-particle from ^{6}Li and for knock-out of an α-particle from ^{9}Be (Roos *et al.* 1968, 1969, Pizzi *et al.* 1969, Pugh *et al.* 1969); in both cases the separation energy is small and there is good evidence for a high degree of clustering in the target nucleus.

If we wish to study the internal structure of the system x, it is more appropriate to use a microscopic model and to express the interaction as the sum of interactions with the individual nucleons in x, so that

$$\tau_{ax} = \sum_j \tau_{aj}.$$

The overlap integral can be written in the form (see §8.4.2)

$$\int \phi_C^{f*}(\underset{\sim}{\xi}_C)\, \phi_A^{i}(\underset{\sim}{\xi}_C, \underset{\sim}{\xi}_x, \underset{\sim}{r}_{xC})\, d^3\xi_C = \sum_{\alpha\beta} C^{\alpha\beta f}\, \phi_x^{\alpha}(\underset{\sim}{\xi}_x)\, \psi_{xC}^{\beta}(\underset{\sim}{r}_{xC}) \tag{8.108}$$

and the DWIA matrix element then factorizes into

$$\begin{aligned} T_{fi} &= \tau_{ax}(q) \sum_{\alpha\beta} C^{\alpha\beta f} \int \chi_f^{-*}(\underset{\sim}{k}_{aC}, \underset{\sim}{r})\, \chi_f^{-*}(\underset{\sim}{k}_{xC}, \underset{\sim}{r})\, \psi_{xC}(\underset{\sim}{r})\, \chi_i^{+}(\underset{\sim}{k}_{aA}, g\underset{\sim}{r})\, d^3r \times \\ &\quad \times \sum_j \int \exp(i\underset{\sim}{q}\cdot\underset{\sim}{\xi}_{x_j})\, \phi_x^{*}(\underset{\sim}{\xi}_x)\, \phi_x(\underset{\sim}{\xi}_x)\, d^3\xi_x \\ &= \tau_{ax}(q) \sum_{\alpha\beta} G_{xC}^{\beta}(Q)\, F_x^{\alpha k}(q) \end{aligned} \tag{8.109}$$

where $\underset{\sim}{q} = \underset{\sim}{p}_a - \underset{\sim}{q}_a = \bar{\underset{\sim}{P}}_i - \bar{\underset{\sim}{P}}_f$ is the momentum transfer and $\underset{\sim}{\xi}_{xj}$ is one of the co-ordinates represented by $\underset{\sim}{\xi}_x$. We have used the symbol k to denote the ground state of x so that elastic scattering of a from x in the same model is given by

$$T_{ax}(q) = \tau_{ax}(q)\, F_x^{kk}(q) \tag{8.110}$$

and hence

$$\begin{aligned} T_{fi} &= \tau_{ax}(q) \sum_{\alpha\beta} \frac{G_{xC}^{\beta}(Q)\, F_x^{\alpha k}(q)}{F_x^{kk}(q)} \\ &= \tau_{ax}(q) \sum_{\beta} \left\{ G_{xC}^{\beta}(Q) + \sum_{\alpha\neq k} \frac{G_{xC}^{\beta}(Q)\, F_x^{\alpha k}(q)}{F_x^{kk}(q)} \right\} \end{aligned} \tag{8.111}$$

Thus, if the term in the expansion with $\alpha = k$ dominates, i.e. if there is a strong overlap between ϕ_x^{α} and ϕ_x^{k}, this matrix element reduces to that given by the quasi-free model. The use of the microscopic model involves certain assumptions about the mechanism of the nuclear reaction, but the use of the quasi-free model evidently involves some additional assumption about the parentage of the target nucleus.

The factorization of the matrix element into a term representing the scattering of particle a from x and a term representing the dissociation of the target nucleus into $A \to x + C$ can conveniently be represented by a diagram, as shown in Fig. 8.19(a). This is the pole graph for the reaction. Fig. 8.19(b) shows the pole graph for heavy-particle knock-out. Thus Fig. 8.19(a) might represent the ^{6}Li(p,pd)α reaction and Fig. 8.19(b) would then represent the ^{6}Li(p,pα)d reaction. Corrections to these graphs are shown in Fig. 8.19(c)-(e). The validity of the pole or peripheral mechanism

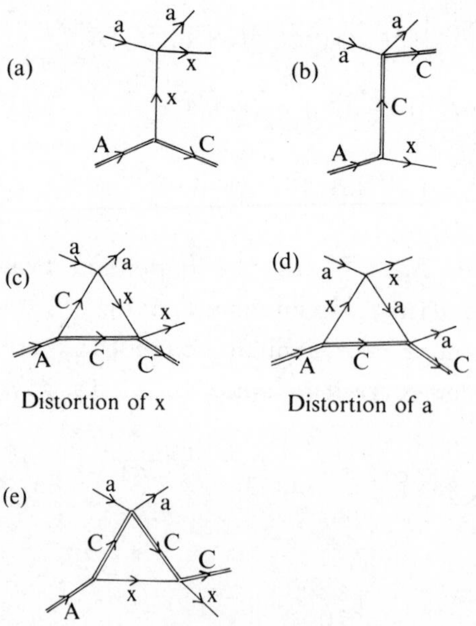

FIG. 8.19. Graphs for the knock-out reaction A(a,ax)C; (a) is the pole graph.

can be examined by applying the Trieman-Yang test. This test requires (Shapiro, Kolybasov, and Augst 1965) that the angular distribution is isotropic with respect to the Trieman-Yang angle, which is the angle between the plane defined by the momenta of particles a and x and the plane defined by the momenta of particles A and C in the anti-laboratory frame, i.e. the frame of reference in which the incident particle is at rest. This implies that the cross-section is invariant under rotation of the plane containing the outgoing particles about the direction of the exchanged particle, provided that particle has spin zero or $\frac{1}{2}$. This test has been applied to data for the ^{6}Li(π^+,2p) reaction (Charpak *et al.* 1967), the ^{6}Li(p,pd) reactions (Liebert, Purser, and Burman 1973), and the ^{12}C(π^-,π^-p) reaction (Aganyants *et al.* 1969); it is generally found that the pole mechanism dominates at low recoil momentum Q although other graphs contribute significantly for higher recoil momenta.

Studies of knock-out reactions initiated by high-energy electrons have been carried out in impulse approximation using the quasi-free model (Jacob and Maris 1962, Ciofi degli Atti 1967, Boffi *et al.* 1968*a*,*b*). In this formalism the cross-section is proportional to the free e-p cross-section (off-shell) and the square of the distorted momentum distribution. Other calculations have been carried out in DWBA (Epp and Griffy 1970, Viollier and Alder 1971, Shanta and Jain 1971, Radhakant 1972*a*,*b*) using the electron-nucleon interaction given by McVoy and Van Hove (1962).

8.4.2. *Overlap integrals*

In the knock-out process a group of nucleons are removed from the target nucleus in the form of a composite particle in state k, leaving the residual nucleus C in a definite final state f. The overlap integral (8.77) can be written as

$$\psi_x^{fi}(\underset{\sim}{\xi}_x, \underset{\sim}{r}_{xC}) = \int \Phi_C^{f*}(\underset{\sim}{\xi}_C)\, \Phi_A^i(\underset{\sim}{\xi}_C, \underset{\sim}{\xi}_x, \underset{\sim}{r}_{xC})\, d^3\xi_C. \tag{8.112}$$

It is also convenient to define a *reduced overlap integral* as (Jackson 1967)

$$\psi_x^{kfi}(\underset{\sim}{r}_{xC}) = \int \phi_x^{k*}(\underset{\sim}{\xi}_x)\, \psi_x^{fi}(\underset{\sim}{\xi}_x, \underset{\sim}{r}_{xC})\, d^3\xi_x \qquad (8.113)$$

where ϕ_x^k is the internal wavefunction for the system x in state k. The wavefunction of the target can be expanded in a complete set of wavefunctions of the residual nucleus to give

$$\Phi_A^i(\underset{\sim}{\xi}_C, \underset{\sim}{\xi}_x, \underset{\sim}{r}_{xC}) = \sum_{\alpha\beta\gamma} C^{\alpha\beta\gamma}\, \phi_x^\alpha(\underset{\sim}{\xi}_x)\, \psi_{xC}^\beta(\underset{\sim}{r}_{xC})\, \Phi_C^\gamma(\underset{\sim}{\xi}_C) \qquad (8.114)$$

where $\alpha\beta\gamma$ represent all necessary quantum numbers and $C^{\alpha\beta\gamma}$ contains the expansion coefficients and coupling constants. This expansion can be regarded as a fractional parentage expansion in the framework of the shell model or as a two-cluster expansion in the framework of the cluster model. The overlap integral is then given by the expression

$$\psi_x^{fi} = \sum_{\alpha\beta} C^{\alpha\beta f}\, \phi_x^\alpha(\underset{\sim}{\xi}_x)\, \psi_{xC}^\beta(\underset{\sim}{r}_{xC}) \qquad (8.115)$$

which we have used in constructing the matrix element in the microscopic model, and the reduced overlap integral is given by

$$\psi^{kfi} = \sum_{\alpha\beta} P_{\alpha k}\, C^{\alpha\beta f}\, \psi_{xC}^\beta(\underset{\sim}{r}_{xC}) \qquad (8.116a)$$

where

$$P_{\alpha k} = \int \phi_x^{k*}(\underset{\sim}{\xi}_x)\, \phi_x^\alpha(\underset{\sim}{\xi}_x)\, d^3\xi_x . \qquad (8.116b)$$

In principle ϕ_x^k, ϕ_x^α are eigenstates of the same Hamiltonian so that $P_{\alpha k} = \delta_{\alpha k}$ and

$$\psi^{kfi} = \sum_\beta C^{k\beta f}\, \psi_{xC}^\beta(\underset{\sim}{r}_{xC}) \qquad (8.117)$$

We have used one term in this expansion in constructing the matrix element in the quasi-free model.

It can be seen from eqn (8.117) that the reduced overlap integral describes the motion in the target nucleus of the centre of mass of a group of nucleons correlated as in the free particle x in state k. If the probability of finding such a system of correlated nucleons in the nucleus is small, this will be reflected by a small value for the coefficient $c^{k\beta f}$. Each transition to a pair of states k and f picks out a particular term in the parentage expansion for the ground state of the target nucleus. If any residual interaction can be neglected, the reduced overlap integral obeys the equation

$$(T_{\mathrm{xC}} + V_{\mathrm{xC}})\,\psi^{kfi} = (E^{i}_{\mathrm{A}} - E^{f}_{\mathrm{C}} - E^{k}_{\mathrm{x}})\,\psi^{kfi} \qquad (8.118)$$

where V_{xC} is the interaction of x with C inside the nucleus. Hence the asymptotic behaviour of the reduced overlap integral is given by

$$\psi^{kfi} \rightarrow e^{-\kappa r}/r, \qquad \kappa^2 = \frac{2\mu_{\mathrm{xC}}(E^{f}_{\mathrm{C}} + E^{k}_{\mathrm{x}} - E^{i}_{\mathrm{A}})}{\hbar^2}\,. \qquad (8.119)$$

Thus the asymptotic behaviour of the reduced overlap integral is determined unambiguously by the separation energy and, if the residual interaction can be neglected, it can be generated from an effective one-body potential. The asymptotic behaviour of ϕ_{x} is determined in a similar way by the separation energy or energies for the system x.

The overlap integral for single-nucleon knock-out, i.e. for reactions such as (p,2p), (p,pn), (e,ep), (α,αp), has the same form as that for single-nucleon transfer reactions. This is discussed in detail in §8.3.3.

The overlap integral for a knock-out process leading to the emission of a bound composite particle is in principle more simple than the overlap integral for the corresponding multi-nucleon transfer reaction, because the internal energy of the composite particle has a definite known value and suitable forms for the wavefunctions of the particle are also known. For processes involving knock-out of α-particles, tritons, or helions the cluster-model formalism has generally been preferred but the shell model has been used, mainly for the (p,pd) reaction.

The overlap integral for the deuteron knock-out has been constructed from shell-model wavefunctions (Sakamoto 1964, Beregi *et al.* 1965, Balashov, Boyarkina, and Rotter 1965) using fractional parentage techniques and transforming the product of two single-particle wavefunctions of oscillator form into the product of a function ϕ of the relative co-ordinate and a function ψ of the centre-of-mass co-ordinate (see also §8.3.4). Both of these transformed functions are also of oscillator form so that neither have the correct asymptotic behaviour. (In principle an overlap integral with the correct asymptotic behaviour can be constructed within the framework of the shell model by using a sufficiently large space of single-particle states, although this has not yet been done for knock-out reactions.) In addition, the function ϕ is not orthogonal to the wavefunction of the free deuteron so that the ratio $F_{\mathrm{d}}^{\alpha k}(q)/F_{\mathrm{d}}^{kk}(q)$ is not unity; consequently the variation of this ratio with momentum transfer must be taken into account and may not be unimportant (Jackson 1965).

In shell-model co-ordinates the matrix element for the (p,pd) reaction in PWBA becomes (Jackson and Elton 1965)

$$T_{\mathrm{fi}} = V_0 \int \phi_{\mathrm{d}}^{*}(\underset{\sim}{r}_1 - \underset{\sim}{r}_2)\, \psi(\underset{\sim}{r}_1,\underset{\sim}{r}_2)\, \exp\{\mathrm{i}(\underset{\sim}{p}_{\mathrm{a}} - \underset{\sim}{q}_{\mathrm{a}} - \underset{\sim}{q}_{\mathrm{d}})\cdot\tfrac{1}{2}(\underset{\sim}{r}_1 + \underset{\sim}{r}_2)\}\, \mathrm{d}^3 r_1\, \mathrm{d}^3 r_2 \tag{8.120a}$$

where the overlap integral is given by

$$\psi(\underset{\sim}{r}_1,\underset{\sim}{r}_2) = \int \Phi_{\mathrm{C}}^{f*}(\underset{\sim}{r}_3,\ldots\underset{\sim}{r}_A)\, \Phi_{\mathrm{A}}^{i}(\underset{\sim}{r}_1,\underset{\sim}{r}_2,\underset{\sim}{r}_3,\ldots\underset{\sim}{r}_A)\, \delta(\underset{\sim}{r}_1 + \underset{\sim}{r}_2 + \underset{\sim}{r}_3 + \ldots + \underset{\sim}{r}_A)\, \times$$
$$\times \exp \frac{-\mathrm{i}\underset{\sim}{Q}\cdot(\underset{\sim}{r}_3 + \ldots + \underset{\sim}{r}_A)}{A-2}\, \mathrm{d}^3 r_3 \ldots \mathrm{d}^3 r_A \tag{8.120b}$$

and the δ-function has been introduced to remove the centre-of-mass motion. These formulae illustrate the remark of Berggren, Brown, and Jacob (1962) that the overlap integral is independent of Q only for the appropriate choice of relative co-ordinates. If this point is ignored, the formalism for the overlap integrals, spectroscopic factors, and sum rules presented in §§8.3 and 8.4 becomes invalid or, at best, subject to corrections of order $1/A$.

The formalism of the cluster model has the considerable advantage that the natural choice of co-ordinates is the set of internal co-ordinates of the clusters and the relative co-ordinates between them (Wildermuth and McClure 1966, Neudatchin and Smirnov 1965) so that, in contrast to the shell model, there is no centre-of-mass motion.

The complete two-cluster expansion is given by eqn (8.144), but the most common procedure is to truncate the expansion and consider only one term in which the two clusters are in specified states, usually their respective ground states, and the relative motion of the two clusters is in the lowest state allowed by the exclusion principle. This wave-function can be written as

$$\Phi_A = \mathscr{A}\{\phi_x(\underset{\sim}{\xi}_x)\ \chi_{xC}(\underset{\sim}{r}_{xC})\ \Phi_C(\underset{\sim}{\xi}_C)\ \chi(ST)\} \tag{8.121}$$

where $\mathscr{A}$ is the antisymmetrization operator and $\chi(ST)$ is a charge-spin function coupled to isospin T and spin S. If the functions ϕ, ψ, Φ are taken to be of oscillator form with identical length parameters, *on antisymmetrization* this cluster-model wavefunction is identical to the shell-model function of oscillator form but with the centre-of-mass motion removed. More complicated cluster model wavefunctions with different length parameters for each function take account of configuration mixing in the shell model sense (Wildermuth 1962, Tang, Wildermuth, and Pearlstein 1962). The parameters of such wavefunctions are generally determined by a variational calculation on the ground-state energy, but this procedure is unfortunately not sensitive to the long-range behaviour of the wavefunction. Fits to data for elastic and inelastic electron scattering provide a more satisfactory method of fixing parameters for subsequent use in reaction calculations (Kudeyarov, Smirnov, and Cherbotarev 1967, 1969, 1971, Neudatchin and Smirnov 1968). The cluster approach can be made much more realistic and more suitable for reaction calculations by requiring that the function for the relative motion has the correct asymptotic behaviour given by (8.120) (Jackson 1967*a*, Watson *et al*. 1971, Jain *et al*. 1969, 1970).

The magnitude of the exchange contribution in cluster knock-out reactions seems to depend very sensitively on the degree of clustering in the target. Exchange between the projectile and the target (projectile exchange) is automatically taken into account if the free two-body interaction is used, but additional exchange contributions to the cross-section can arise as a result of exchange within the target (target exchange or spectator exchange). Jain, Sarma, and Banerjee (1969, 1970) have shown that correct estimation of exchange effects depends on a correct description of the asymptotic behaviour of the reduced overlap integral. In the case of the ^{6}Li(p,pd) reaction, for which the separation energy is very low, their results show that when this asymptotic behaviour is correct the magnitudes of the exchange terms are very much smaller than the magnitudes obtained with a gaussian form for the overlap integral.

The formalism used in this section and in §8.3.3 is based on the assumption that the residual nucleus is stable or long lived. However, in a number of important cases the residual nucleus is formed in a state which is unstable against particle emission. This may occur for removal of one of the least-bound nucleons in certain special cases, for example, $^{6}\text{Li} \rightarrow \text{p} + {}^{5}\text{He}$, but is of more general importance in relation to formation of highly excited states by removal of deeply bound nucleons. An approach to the latter problem has been made by generating the overlap integrals in complex one-body potentials (Herscovitz 1971, Shanta 1973). The physical basis for this procedure is not well established, although Berggren and Ohlén (1973, 1974) have shown that a complex potential can be derived by considering the properties of the unstable residual nucleus. Alternatively, the mechanism of the decay process must be considered and the matrix elements for each separate process must be evaluated; this has been attempted for the ^{12}C(p,2p) reaction (Pittel and Austern 1972, 1974) but clearly involves very formidable calculations if reliable magnitudes are required for the cross-sections.

8.4.3. *Analysis of angular distributions*

Angular distributions for knock-out reactions are obtained by detecting the two outgoing particles in coincidence. The experimental geometry can be classified as follows.

(i) *Symmetric coplanar geometry.* The momenta $\underset{\sim}{p}_a$, $\underset{\sim}{q}_a$, $\underset{\sim}{q}_x$ are in the same plane, $|\underset{\sim}{q}_a| = |\underset{\sim}{q}_x|$, and the two outgoing particles are detected as equal angles on either side of the incident direction. It follows from eqn (8.99) that $\underset{\sim}{Q}$ is parallel or antiparallel to $\underset{\sim}{p}_a$.

(ii) *Symmetric non-coplanar geometry.* The two outgoing particles are detected out of the plane but the other conditions of (i) apply.

(iii) *Non-symmetric coplanar geometry.* The momentum $|\underset{\sim}{q}_a|$ is fixed together with the angle $\hat{\underset{\sim}{p}}_a \cdot \hat{\underset{\sim}{q}}_a$, while $\underset{\sim}{q}_x$ is varied.

(iv) *Unrestricted geometry.* Counter experiments are mainly carried out with geometry (i) or (iii), except for the BOL system, whereas the unrestricted geometry is associated with the use of bubble chambers and spark chambers.

For reactions in which the two outgoing particles are identical geometry (i) has a rather special symmetry because a rotation of $k\pi$, where k is any integer, about the incident direction transforms the system into one which is physically indistinguishable provided that spin directions are not measured. Hence, neglecting spin and characterizing the overlap integral by the quantum numbers n, l, m we have

$$|G(Q)|^2 = \sum_m |g_{nlm}(Q)|^2 \tag{8.122}$$

and the requirement for rotational symmetry about a z-axis taken along the incident direction gives the condition (Jackson and Berggren 1965)

$$g_{nlm} = (-1)^m g_{nlm}$$

i.e.

$$g_{nlm} = 0 \text{ if } m \text{ is odd.} \tag{8.123}$$

It follows from eqn (8.100) that in a plane-wave approximation g_{nlm} is non-zero only when $m = 0$, but in a distorted wave approximation there are contributions from other even-m components. This difference is a consequence of the introduction of additional transverse momentum components by the distortion. In any other geometry the symmetry is lost and $\underset{\sim}{Q}$ is no longer parallel or antiparallel to $\underset{\sim}{p}_a$: in this case there are always transverse components of the recoil momentum and even in a plane wave calculation there will be contributions from all m components. The symmetry is also lost if the outgoing particles are not identical.

Eqn (8.100) also reveals that the cross-section has a maximum at $Q = 0$ if the overlap integral contains s-state components and a minimum at $Q = 0$ if the overlap integral does not contain s-state components. In practice, the minimum is often obscured by the effects of distortion, finite energy and angular resolution, and by the variation of the two-body cross-section.

Experiments restricted to the symmetric coplanar geometry reject a large number of events, but when the geometry is unrestricted the presentation of the data poses greater problems (James *et al.* 1969*a*,*b*, Kullander *et al.* 1971*a*,*b*). Kullander *et al.* (1971*a*,*b*) have presented (p,2p) cross-sections at 600 MeV as functions of the recoil angle θ_Q which is the angle between the incident momentum $\underset{\sim}{p}_a$ and the recoil momentum $\underset{\sim}{Q}$. Their results show considerable dependence on θ_Q, although the dependence of the kinematic factors on θ_Q is small for small values of Q. Using a plane wave approximation it can be shown (Jackson 1967*b*) that, whatever the geometry of the reaction, the quantity $|G(Q)|^2$ is independent of θ_Q. Hence the dependence of the experimental cross-section on θ_Q arises from distortion, and this has been verified by DWIA calculations (Jackson 1971*b*, Gustafsson and Berggren 1971). Thus it appears that these data provide a more stringent test of distorted wave theory but do not provide any additional information about nuclear structure.

The spectroscopic features of knock-out reactions are essentially the same as for pick-up reactions. An important difference between these reactions arises because the knock-out reaction involves two fast outgoing particles, and this gives much greater freedom of choice in what aspects of the reaction are examined. For example, once the incident energy has been chosen for a (p,d) reaction the Q-value essentially determines what momentum components of the transferred neutron can be examined; lower momentum components can be observed by lowering the incident energy, but this makes it more difficult to remove the most tightly bound neutrons. In contrast, in a (p,2p) or (p,pn) reaction very small momentum components are emphasised in the symmetric coplanar geometry, but other momentum components may be studied by changing the geometry of the experiment.

Another characteristic feature of the knock-out reaction is the ability to control the population of the substates of the residual nucleus. This arises from the peculiar symmetry in the symmetric coplanar geometry which, from eqn (8.123), leads to contributions from even-m components only. For example, if a $p_{3/2}$ proton is knocked-out of a 0^+ ground state, the $M_f = \pm\frac{1}{2}$ substates only are populated in the symmetric coplanar experiment, but as soon as the symmetry is relaxed the $M_f = \pm\frac{3}{2}$ substates begin to be populated. Thus the relative population of the substates can be changed by changing the geometry of the experiment.

In the earliest measurements on the (p,2p) reaction in the intermediate-energy region, the energy resolution was of the order of a few MeV. The corresponding angular distributions were analysed in terms of the monopole sum rule or effective occupation number. Subsequently, some individual final states have been resolved and spectroscopic analyses carried out. For 1p shell nuclei satisfactory agreement with the data over the energy range 150 MeV-1 GeV has been achieved using fractional parentage coefficients derived from intermediate coupling calculations and using single-particle wavefunctions which have the correct asymptotic behaviour and fit the elastic electron scattering data (Jain and Jackson 1967, Shanta and Jain 1971). The observed states apparently do not exhaust the sum rule for those nuclei, but there are large uncertainties in some cases. The (p,2p) reaction on 2s1d shell nuclei has also been studied in the intermediate-energy region. The qualitative features of the angular distributions are reproduced using single-particle wavefunctions which fit elastic electron scattering (Jain 1968). Substantial disagreement between results from the (p,2p),(d,^{3}He), and (t,α) reactions was at first observed, particularly for ^{28}Si and ^{24}Mg. This is now understood (Wagner 1969, Arditi *et al.* 1971, Kramer, Mairle, and Kaschl 1971) in terms of the differences in energy resolution and the wide splitting of the 1p strength in 2s1d shell nuclei. Comparison of angular distributions for the (e,ep) and (p,2p) reactions on ^{12}C (Shanta and Jain 1971, Radhakant 1972*a*) and on ^{40}Ca (Campos Venuti *et al.* 1973) show that, within fairly large uncertainties, the same wavefunctions yield agreement with the data. The angular distributions for the ^{12}C(e,ep) reaction at 497 MeV (Bernheim *et al.* 1974*a*,*b*) are compatible with elastic electron scattering. For incident energies in the medium-energy range much better energy resolution has been obtained for the (p,2p) reaction. At these energies the angular distributions are largely determined by the distortion and the variation of the p-p interaction (Pugh *et al.* 1967, Deutchman and McCarthy 1968). Moderate agreement with data for 1p shell nuclei and for ^{40}Ca has been obtained using single-particle wavefunctions generated in a non-local potential (Bray *et al.* 1971, Janiszewski and McCarthy 1972*b*) but it still does not appear possible to obtain a good shape fit in all cases.

Spectroscopic studies of cluster knock-out reactions have so far centred on light nuclei such as ^{6,7}Li, ^{9}Be, ^{12}C, ^{16}O, and ^{28}Si. In most cases the momentum distribution $|G(Q)|^2$ is fitted to or derived from the data, and from this the effective number of clusters

$$N_{\text{eff}} = \int |G(Q)|^2 \, d^3Q \tag{8.124}$$

is obtained. Clearly, N_{eff} is equal to the spectroscopic factor or, in a poor resolution experiment, to the sum rule. Experiments on the ^{9}Be(p,pα) reaction at 57 MeV (Roos *et al.* 1968) and at 160 MeV (Kannenberg 1968) are in good agreement. The quasi-free model appears to work quite well for the (p,pα), (d,2d) and (d,dα) reactions on ^{6}Li, but the (α,2α) reaction is more difficult to analyse owing to off-shell effects.

Results for N_{eff} obtained from these reactions are given in Table 8.7. These show that at comparable energies the values obtained using PWIA are consistent, but omission of distortion

TABLE 8.7

Effective number of clusters for $^{6}Li \rightarrow \alpha + d$

Reference	Reaction	Energy (MeV)	N_{eff} PWIA	N_{eff} DWIA
Devins 1965	^{6}Li(p,pd)α	30	0.04±0.12	
Hendrie *et al.* 1966	^{6}Li(p,pd)α	55	0.15±0.08	
Ruhla *et al.* 1963	^{6}Li(p,pd)α	155	0.31±0.16	
A.K. Jain *et al.* 1973	^{6}Li(d,2d)α	27	0.12-0.18[a]	0.32-0.55
H. Davies *et al.* 1966	^{6}Li(π^{-},2n)α	at rest	0.37±0.10	
M. Jain *et al.* 1970	^{6}Li(p,pα)d	57	0.16±0.08	
Ruhla *et al.* 1963	^{6}Li(p,pα)d	155	0.20±0.10	
A.K. Jain *et al.* 1973	^{6}Li(d,dα)d	27	0.10-0.18[a]	0.34-0.67
A.K. Jain *et al.* 1973	^{6}Li(α,2α)d	43	0.11-0.18[a]	0.30-0.60
Pizzi *et al.* 1969	^{6}Li(α,2α)d	55	0.04±0.11	
A.K. Jain *et al.* 1973	^{6}Li(α,2α)d	55	0.13-0.20[a]	0.32-0.65
M. Jain *et al.* 1970	^{6}Li(α,2α)d	62	0.10±0.06	
Watson *et al.* 1971	^{6}Li(α,2α)d	50-80	0.08±0.04	
A.K. Jain and Sarma 1974	^{6}Li(α,2α)d	70		0.45±0.18

[a]The range of values given in this case arises from different choices for the overlap integral.

introduces an apparent energy dependence into the results. Inclusion of distortion changes the values obtained by factor of ~3 while variations in the overlap integral have a smaller but significant effect. Results obtained for ^{12}C (Jacquot *et al.* 1970) and ^{28}Si (Plieninger, Eichelberger, and Velten 1969) can also be interpreted in the quasi-free model.

The complications arising in the description of the reaction mechanism for knock-out reactions have so far obscured the nuclear structure and nuclear size information which is undoubtedly contained in the data. Consequently, the most meaningful analyses are those which start with overlap integrals and spectroscopic factors consistent with other data.

8.4.4. *Separation energies and energy-weighted sum rules*

Using impulse approximation and the ideas developed in §8.4.1, the summed cross-section for a knock-out reaction can be written in the form

$$\sum_p \frac{d^4\sigma}{d\Omega_1\ dE_1\ d\Omega_2\ dE_2} = \sum_p K\left(\frac{d\sigma}{d\Omega}\right)_{\text{free}} P(Q,\omega)\ \delta(\omega - E_s^{\alpha p}) \tag{8.125}$$

where K is a kinematic factor, ω is the energy loss

$$\omega = E_0 - E_1 - E_2 - E_Q \tag{8.126}$$

and $E_s^{\alpha p}$ is the separation energy for removal of a particle, say a nucleon, from the initial nucleus in state α leaving the nucleons in state p, i.e.

$$E_s^{\alpha p} = E_{A-1}^{p} - E_A^{\alpha}. \tag{8.127}$$

The quantity $P(Q,\omega)$ is also a function of E_0 and E_1/E_2 because of the effect of distortion. Within the framework of the single-particle model the dependence on ω can be represented through the expression

$$P(Q,\omega) = \sum_\gamma S_{\alpha p}(\gamma,\omega)\ |G_\gamma(Q)|^2 \tag{8.128}$$

where γ represents the single-particle labels, usually nlj. Here $S_{\alpha p}(\gamma,\omega)$ is the spectroscopic factor which obeys the sum rule

$$\int S_{\alpha p}(\gamma,\omega)\ d\omega = N_\gamma \tag{8.129}$$

where N_γ is the number of particles in the subshell γ. Assuming that $K\ d\sigma/d\Omega$ can be replaced by average values which may be taken outside the sum, we may also write

$$A^{-}(Q,\omega) = \sum_{p\gamma} S_{\alpha p}(\gamma,\omega)\,|G_{\gamma}(Q)|^{2}\ \delta(\omega - E_{s}^{\alpha p}) \qquad (8.130a)$$

$$= \sum_{p\gamma} |\langle \Phi_{A-1}^{p}|a_{\gamma}(-\underset{\sim}{Q})|\Phi_{A}^{\alpha}\rangle|^{2}\ \delta(\omega - E_{s}^{\alpha p}) \qquad (8.130b)$$

$$= \sum_{\gamma} \langle \Phi_{A}^{\alpha}|a_{\gamma}^{\dagger}(-\underset{\sim}{Q})\ \delta(\omega - H + E_{A}^{\alpha})\ a_{\gamma}(-\underset{\sim}{Q})|\Phi_{A}^{\alpha}\rangle \qquad (8.130c)$$

where $a_{\gamma}(-\underset{\sim}{Q})$ destroys a proton in state γ with momentum $-\underset{\sim}{Q}$. Eqn (8.130*b*) may be linked (Jackson 1971*a*) with the formalism of the response function used in §8.2 to describe summed inelastic scattering, while eqn (8.130*c*) introduces the concept of a spectral function (Gross and Lipperheide 1970) which may be evaluated using the Greens function formalism of many-body theory.

Analysis of the angular distributions $d^{3}\sigma/d\Omega_{1}\,d\Omega_{2}\,dE$ at fixed values of ω provides direct information on $G(Q)$, as discussed in §8.4.3, but for analysis of the energy spectra the expression (8.129) is used. In certain circumstances, valid for the (e,ep) and (p,2p) reactions at high energies, the effect of distortion can be approximately represented by a reduction factor F_{nlj} which is not a rapidly varying function of ω. In this case we have

$$|G_{nlj}(Q)|^{2} = F_{nlj}|G_{nlj}^{\mathrm{PW}}(Q)|^{2} \qquad (8.131)$$

$$P(Q,\omega) = \sum_{nlj} A_{nlj}(\omega)|G_{nlj}^{\mathrm{PW}}(Q)|^{2} \qquad (8.132)$$

$$\int A_{nlj}(\omega)\ d\omega = F_{nlj}\ N_{nlj}. \qquad (8.133)$$

In the earliest (p,2p) experiments, carried out at 185 MeV and 155 MeV with energy resolution $\Delta E \sim$ 4-6 MeV, the energy spectra for 1p-shell nuclei showed two broad peaks for $A \leqslant 12$ and three peaks for $A > 12$. It was reasonable to associate these peaks with knock-out from the $1s_{1/2}$, $1p_{3/2}$ and $1p_{1/2}$ single-particle states. By fitting the shape of each peak an experimental value for the peak position and the width could be determined and, from (8.129), the area under each peak could be associated with the proton occupation number. With better resolution the peaks corresponding to knock-out of the least bound protons are split as the various excited states of the residual nucleus are resolved. This splitting of the single-hole strength is familiar from good resolution pick-up experiments. The relative heights of the peaks can be predicted using spectroscopic factors deduced from appropriate nuclear models (Dietrich 1962, Balashov *et al.* 1965).

The energy spectrum for the $^{12}C(e,ep)$ reaction measured by Amaldi *et al.* (1964) has been examined by several authors. Ciofi degli Atti (1967) used eqn (8.132) with the spectroscopic factors of Balashov *et al.* (1965) and reduction factors calculated from a separate WKB treatment of the distortion, and concluded that the (e,ep) data alone do not serve to distinguish between different single-particle wavefunctions for the knocked-out proton. Other calculations of the same type have been carried out with more complicated wavefunctions (Boffi *et al.* 1968*a*, Boffi, Facati, and Sawicki 1968*b*, Hasan and Naqvi 1974). Radhakant (1972*b*) made a DWBA calculation using eqn (8.128) and found that realistic single-particle wavefunctions gave satisfactory agreement with the data. Both Ciofi degli Atti and Radhakant replaced the energy δ-function by the gaussian function

$$\frac{1}{\sigma}\left(\frac{2}{\pi}\right)^{\frac{1}{2}}\exp\left\{-\left(\frac{\omega-E_s}{\sigma}\right)^2\right\}$$

where

$$\sigma = 0.847(\Delta E^2 + \gamma_h^2)^{1/2}$$

in which ΔE is the experimental energy resolution and γ_h is the width of the hole state. The values of E_s corresponding to the 1s- and 1p-states were fixed by taking the experimental values for the peak positions.

New data for the (e,ep) reaction on light nuclei at 700 MeV (Hiramatsu *et al.* 1973) have been analysed in DWIA using eqn (8.128) with oscillator functions or with realistic single-particle wavefunctions which fit elastic electron cattering. For ^{9}Be and ^{12}C good shape agreement with the nergy spectra and angular distributions is obtained, although he calculated angular distributions for the 1p states must e multiplied by factors of 0.8 and 0.6 respectively. Agree- ent for ^{6}Li and ^{7}Li is not so satisfactory, particularly for he high-energy side of the energy spectrum. Further data for he ^{6}Li(e,ep) reaction at 1200 MeV have been reported by ntoufiev *et al.* (1972), and good shape agreement with the ngular distributions is obtained using PWIA with oscillator avefunctions. Results for the ^{12}C(e,ep) reaction have been btained at 2.5 GeV (Köbberling *et al.* 1974). The angular istribution predicted at this energy using realistic single- article wavefunctions gives good agreement for recoil momenta p to ~1.2 fm^{-1}; beyond this the theory does not provide suf- icient high momentum components and the discrepancy is not esolved by the inclusion of Jastrow correlations. Further esults for the ^{12}C(e,ep) reactions have been obtained at 497 eV (Bernheim *et al.* 1974*a*,*b*), and these also require a reduc- ion in the magnitude of the DWIA calculations for the angular istributions.† The l = 0 component can be identified up to $\omega \sim 50$ MeV.

For medium and heavy nuclei the reduction of the data in a poor resolution experiment is complicated owing to overlap- ping states of the residual nucleus and to the large widths of some of the states reached by knock-out from inner shells. Tyren *et al.* (1966) resolve the energy spectra into symmetric triangular‡ components of different width and assume that the peak positions correspond to well-defined states excited in the reaction. The (e,ep) spectra are analysed by Amaldi and collaborators (Amaldi 1967) with a sum of Maxwellian curves,

---

†This apparent loss of strength in the angular distributions may indicate displacement of the single-hole strength to higher excitation energies, but it is not yet clear whether the effect is due to the choice of parameters in the DWIA calculation.

‡Some aspects of this triangular procedure have been examined by Elton and Sundberg (1972).

whose heights, widths, and positions are free parameters, together with a straight-line background term. The Maxwellian shape is chosen to take account of the radiative tail in the electron-induced process. A new way of analysing the data was developed by James *et al.* (1969*a*,*b*) and applied to their results for the (p,2p) reaction at 385 MeV. Using eqn (8.132) and the formulae for G^{PW}_{nlj} derived from oscillator wavefunctions they analysed the angular distributions at fixed values of ω and at intervals of about 5 MeV, and deduced the reduction factors, listed in Table 8.8, and the strength coefficients, which are displayed in Fig. 8.20. The reduction factors for

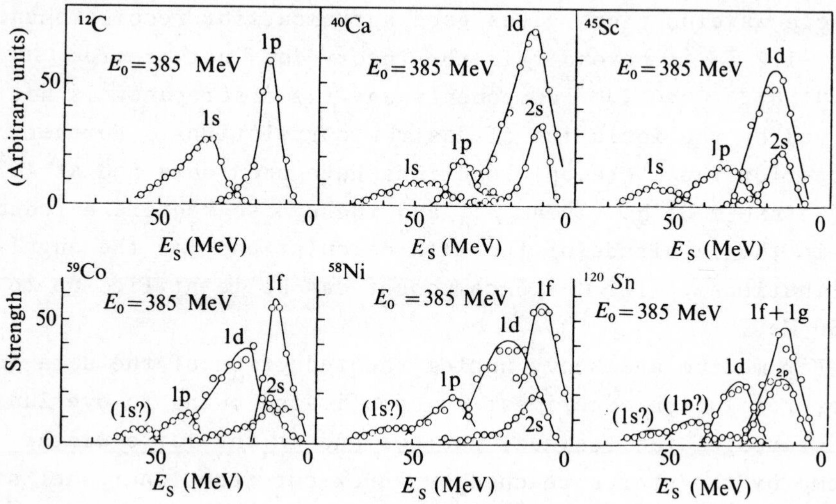

FIG. 8.20. Strength coefficients for various single-particle states seen in the (p,2p) reaction at 385 MeV. (James *et al.* 1969*a*,*b*; this version from Jacob and Maris 1973.)

^{12}C are in very close agreement with the values obtained in a distorted-wave analysis of the same data (McCarthy *et al.* 1971). A similar method has been used to analyse the CERN (p,2p) data at 600 MeV (Landaud *et al.* 1971) and the corresponding reduction factors are listed in Table 8.8. Bernheim *et al.* (1974*a*,*b*) use DWIA to analyse the Saclay (e,ep) data. They use eqn (8.128) for $P(Q,\omega)$, construct $G(Q)$ from realistic single-particle wavefunctions which fit elastic scattering for

TABLE 8.8

Reduction factors for the (p,2p) reaction[†]

	385 MeV				600 MeV		
	^{12}C	^{40}Ca	^{58}Ni	^{120}Sn	^{12}C	^{40}Ca	^{51}V
1s	0.14±0.02	0.10	0.04±0.02	0.03±0.02	0.93±0.015	0.048±0.007	0.034±0.006
1p	0.20±0.02	0.034	0.04	0.017	0.12±0.01		
1d		0.11	0.076	0.030		0.057±0.007	0.0019±0.004
2s		0.15	0.12	0.020			
1f 1g			0.086	0.024			
2p				0.070			

[†] Results at 385 MeV from James *et al.* (1969*a*,*b*) and at 600 MeV from Landaud *et al.* (1971).

momentum transfer $< 2.5\ \text{fm}^{-1}$, and obtain the $S_\gamma(\omega)$ by a fitting procedure. The Tokyo data (Nakamura *et al.* 1974, 1976) are also analysed using DWIA with eqn (8.128) and $G(Q)$ is derived from realistic wavefunctions, but the $S_\gamma(\omega)$ are taken as gaussian functions of ω.

The crucial question in connection with the separation energy data is whether a sufficiently high proportion of the hole strength has been located and identified so that the centroid energy $\Sigma_p \mathscr{I}_p^2(lj)E^p_{A-1}$ is reliably determined. If this energy is known, then the mean separation energy, defined in eqn (2.34) as

$$E_{\mathrm{S}}(lj) = \sum_p \mathscr{I}_p^2(lj)E^p_{\mathrm{A}-1} - E_{\mathrm{A}} \qquad (8.134a)$$

can be deduced. Using eqns (8.85) and (8.128), the mean separation energy can also be expressed as

$$E_{\mathrm{S}}(lj) = \frac{\int P(Q,\omega)\ \omega\ \mathrm{d}\omega}{\int P(Q,\omega)\ \mathrm{d}\omega}\ . \qquad (8.134b)$$

It has been suggested (Brueckner *et al.* 1972) that, if the reaction time is slow, in some sense, rearrangement processes may affect the energies observed and lead to smaller separation energies. However, if the reaction is fast in the sense that formation of the hole state by knock-out and its subsequent decay by particle emission or other means can be regarded as sequential, the rearrangement effects cannot affect the centroid energy (Meldner and Perez 1971). On the other hand, the coupling of the one-hole state to the successive n-hole–(n-1)-particle states spreads the hole strength over several states in the residual nucleus and hence reduces the spectroscopic factor or occupation number associated with the lowest peak (Wille, Gross, and Lipperheide 1971, Wille and Lipperheide 1972, Fritsch, Lipperheide, and Wille 1973). In some cases this fragmentation is more conveniently represented by coupling to collective states, as in the case of fragmentation of neutron states in the lead isotopes discussed in §8.3.3. The coupling to the continuum is also important.

It is evidently essential that experimental measurements

on knock-out reactions should provide energy spectra up to a very high excitation energy in the residual system together with angular distributions, so that the angular momentum character can be identified. Careful distorted wave calculations are then required to determine the distribution of single-particle strength, but few (if any) completely satisfactory spectroscopic studies have yet been carried out. For knock-out from the lower shells such calculations are by no means straightforward, since there may be considerable overlap between states formed by knock-out from two adjacent major shells (Wagner 1969) or between single excitation and double excitation in the shell above (Pittel and Austern 1972, 1974). Further, it is not yet clear whether the broad deep-hole states possess intrinsic structure which has so far been masked by experimental uncertainties.

Values of the mean separation energies obtained in (p,2p) and (e,ep) experiments are listed in Table 8.9. Agreement between the results is satisfactory for the 1s and 1p states in light nuclei, and there is no evidence for any dependence on the incident energy. For the 1s level in ^{40}Ca there is reasonable agreement between the separate (p,2p) experiments giving a value of $E_s \sim 55$ MeV, but the (e,ep) measurement at Frascati gave $E_s \sim 80$ MeV. The analysis by James *et al.* (1969*b*) indicated that the spectral distribution $P(Q,\omega)$ tends to a constant value for $\omega > 60$ MeV giving a uniform background contribution. Consequently, it was assumed that the (e,ep) analysis was suspect owing to the lack of angular distribution measurements, but in a more recent (e,ep) experiment Campos Venuti *et al.* (1973) have shown that the angular distribution for ^{40}Ca shows forward peaking as a function of Q at $\omega = 81$ MeV. Nakamura *et al.* (1974) also observe forward peaking in this energy region for the same reaction and in the ^{12}C(e,ep) reaction up to 66 MeV. Nevertheless, they conclude that the ^{40}Ca spectrum at $\omega > 80$ MeV is explained by the radiative tail and the multiple-collision background. Hiramatsu *et al.* (1973) have made a Monte Carlo calculation of the contribution from multiple collisions. The contribution in ^{12}C is small; it shows a broad peak at about 60 MeV and may be largest at

TABLE 8.9

Mean separation energies in MeV determined from (p,2p) and (e,ep) experiments†

Nucleus	$1s_{1/2}$	$1p_{3/2}$	$1p_{1/2}$	$2s_{1/2}$	$1d_{5/2}$	$1d_{3/2}$	Reaction	Reference
^{6}Li	20.3±1.5	4.5±1.5					(p,2p)	Garron *et al.* (1962)
	22.4±0.7	4.8±0.3					(p,2p)	Tibell *et al.* (1963)
	22.7±0.3	4.9±0.3					(p,2p)	Tyren *et al.* (1966)
	19.0±1.0	3.5±1.0					(e,ep)	Antoufiev *et al.* (1972)
	24.3±0.2	6.2±0.1					(e,ep)	Nakamura *et al.* (1974)
^{7}Li	23.5±0.7	10.0±1.4					(p,2p)	Garron *et al.* (1961)
	24.1±1.5	10.1±1.4					(p,2p)	Pugh & Riley (1961)
	25.8±0.6	11.3±0.5					(p,2p)	Tibell *et al.* (1963)
	23 ±1.5	10.2±1.6					(p,2p)	Garron *et al.* (1962)
	25.5±0.4	11.8±0.3					(p,2p)	Tyren *et al.* (1966)
	26.0±0.2	10.1±0.1					(e,ep)	Nakamura *et al.* (1974)
^{9}Be	27.2±1.9	18.2±1.5					(p,2p)	Pugh & Riley (1961)
	28.7±1.5	18.6±1.0					(p,2p)	Tibell *et al.* (1963)
	25.4±0.5	16.4±0.3					(p,2p)	Tyren *et al.* (1966)
		17.1±0.2					(p,2p)	Roynette *et al.* (1967)
	26.6±0.3‡	18.0±0.2					(e,ep)	Nakamura *et al.* (1974)
^{10}B	31.5±1.5						(p,2p)	Garron *et al.* (1962)
	35 ±2						(p,2p)	Tibell *et al.* (1963)
	30.5±0.6						(p,2p)	Tyren *et al.* (1966)

^{11}B	34 ±3						(p,2p)	Garron *et al.* (1962)
	40 ±5						(p,2p)	Tibell *et al.* (1963)
^{12}C	34 ±3	16 ±3					(p,2p)	Gooding & Pugh (1960)
	34.5±1.5	15.8±1.2					(p,2p)	Garron *et al.* (1962)
	33 ±6	14 ±6					(e,ep)	Amaldi *et al.* (1964)
	34.0±2.0						(p,2p)	Tyren *et al.* (1967)
	28 -46	16 ±0.5					(p,2p)	Yuasa & Hourany (1967)
	38 ±9	17 ±4					(p,2p)	James *et al.* (1969*a*)
	35.5±1.0	15 ±0.5					(p,2p)	Landaud *et al.* (1971)
	38.7±0.8	17.5±0.4					(e,ep)	Bernheim *et al.* (1974*b*)
	37.5±0.3	15.3±0.1					(e,ep)	Nakamura *et al.* (1974)
^{14}N	42 ?		7.5±0.5				(p,2p)	Tyren *et al.* (1966)
^{16}O		18.7±1.4	13.1±1.4				(p,2p)	Pugh & Riley (1961)
	44 ?	19.0±1.0	12.4±1.0				(p,2p)	Tyren *et al.* (1966)
^{23}Na		29.6±2.0?	21.2±1.9?				(p,2p)	Pugh & Riley (1961)
		24-32	22 ±1				(p,2p)	Arditi *et al.* (1967)
^{24}Mg		26.7±0.5	22.2±0.5				(p,2p)	Arditi *et al.* (1967)
^{27}Al				13.4±1.4	19.4±1.4	9.0±1.7	(p,2p)	Pugh & Riley (1961)
	50 ±7	32±3					(e,ep)	Amaldi *et al.* (1964)
				15.6±0.3	20 ?		(p,2p)	Tyren *et al.* (1966)
	56.9±3.2	34.3±0.9		14.3±0.2	14.0±0.6		(e,ep)	§Nakamura *et al.* (1976)

TABLE 8.9 continued

Nucleus	$1s_{1/2}$	$1p_{3/2}$	$1p_{1/2}$	$2s_{1/2}$	$1d_{5/2}$	$1d_{3/2}$	Reaction	Reference
^{28}Si								
				14.3±1.5	19.0±1.4		(p,2p)	Pugh & Riley (1961)
		36 ?	28 ?	13.2±0.3	17 ±0.5		(p,2p)	Tyren *et al*. (1966)
		31.5±1	27 ±1				(p,2p)	Arditi *et al*. (1967)
	47 ±4	27±2		12.0±0.5			(p,2p)	Landaud *et al*. (1971)
	50(51)	37(32)		13.8±0.5	16.2±0.8		(e,ep)	*Bernheim *et al*. (1974*b*)
^{32}S								
				7.2±1.5	12.4±1.7		(p,2p)	Pugh & Riley (1961)
	82 ±9	44±7					(e,ep)	Amaldi *et al*. (1965)
		32.2±1.1?	26.6±1.1?	9.1±0.3			(p,2p)	Tyren *et al*. (1966)
^{40}Ca								
				10.6±0.7	15.0±1.5	8.6±0.8	(p,2p)	Tibell *et al*. (1962)
			24.5±1.5	11.0±0.5	14.9±0.5	8.4±0.5	(p,2p)	Ruhla *et al*. (1964)
	77 ±14	32±4					(e,ep)	Amaldi *et al*. (1966*a*)
				11.6±0.5		8.3±0.6	(p,2p)	Tyren *et al*. (1966)
	48.5±10	35±5		12 ±3	16±5		(p,2p)	James *et al*. (1969*b*)
	51 ±3	30±3		11 ±0.5			(p,2p)	Landaud *et al*. (1971)
	56	42(41)		11.2±0.3	14.9±0.8		(e,ep)	*Bernheim *et al*. (1974*b*)
	58.4±1.1	35.3±0.5		13.6±0.4	18.4±1.6	10.4±1.4	(e,ep)	Nakamura *et al*. (1974)
^{45}Sc								
				12 ±1			(p,2p)	Ruhla *et al*. (1967)
	55 ±10	35±5		13 ±3	17±5		(p,2p)	James *et al*. (1969*b*)

^{51}V			15.7±0.5		(p,2p)	Tyren *et al.* (1966)
			14.1±0.3		(p,2p)	Ruhla *et al.* (1967)
	55 ±5	32±3	14 ±1.5		(p,2p)	Landaud *et al.* (1971)
	59.8±3.2	40.4±0.7	15.1±0.2	19.5±0.2	(e,ep)	§Nakamura *et al.* (1974)
^{58}Ni			10.8±0.3		(p,2p)	Ruhla *et al.* (1967)
	56 ±7	37±5	10 ±3	19±7	(p,2p)	James *et al.* (1969*b*)
	64(62)	44(45)	14.7±0.5	21	(e,ep)	*Bernheim *et al.* (1974*b*)
^{64}Zn			12.4		(p,2p)	Arditi *et al.* (1967)
^{120}Sn	56 ±10	43±8	29 ±5	30±5	(p,2p)	James *et al.* (1969*b*)

†A question mark indicates that the assignment is uncertain.

‡A peak at 37.0 0.4 MeV is also reported.

*The values in brackets arise from a later analysis of the same data by Mougey *et al.* (1976).

§Plane wave analysis.

small *Q*. This broad peak was not confirmed in later calculations by the same group (Nakamura *et al.* 1976), but the forward peaking as a function of *Q* was shown very clearly. Thus, the multiple-collision background may be confused with the 1s peak, particularly at large values of the separation energy where the direct component is small.

It can be seen from Table 8.9 that the mean separation energy for 1s protons increases rapidly in 1p shell nuclei with a slope of about 2 MeV per nucleon. Beyond $A = 16$ the trend indicated by the (p,2p) measurements is a much slower increase of about 0.35 MeV per nucleon with saturation at a constant value of ~ 55 MeV beyond $A = 50$. The 1p separation energy shows somewhat similar behaviour with a rate of increase of about 0.5 MeV per nucleon and less clear evidence of a saturation value of ~ 40 MeV. The 1d and 2s separation energies have even smaller rates of increase with *A*.

There is a substantial body of reliable information on proton hole states at low excitation energies derived from pick-up and knock-out reactions. The data relating to $1d_{5/2}$, $1d_{3/2}$, and $2s_{1/2}$ hole states in fp-shell nuclei and $1p_{3/2}$ and $1p_{1/2}$ states in 2s1d-shell nuclei have been carefully reviewed by Wagner (1973). For the $1d_{3/2}$ and $2s_{1/2}$ states the observed strengths exhaust the sum rule, to a good approximation, but this is not the case for $1d_{5/2}$ transfer. The mean separation energies for the $2s_{1/2}$ and $1d_{3/2}$ states calculated by Wagner (1973) are shown in Fig. 8.21, from which it can be seen that the levels cross over, with the $1d_{3/2}$ protons more strongly bound than the $2s_{1/2}$ at the beginning of the fp-shell but with the positions reversed for Co and Ni. Distribution of strength for 1p states in 2s1d-shell nuclei is shown in Fig. 8.22, which is also due to Wagner. The broken curve shows the calculated position of the mean $1p_{1/2}$ separation energy which increases linearly with *A* in the final nuclei such as ^{15}N, ^{19}F, etc., which are reached by proton removal from $T = 0$ targets. The spin-orbit splitting between the $1p_{3/2}$ and $1p_{1/2}$ centroid energies appears to increase from 6.5 MeV in ^{16}O to about 14 MeV in ^{28}Si (Mairle and Wagner 1974). A survey of the corresponding neutron hole states seen in the (p,d)

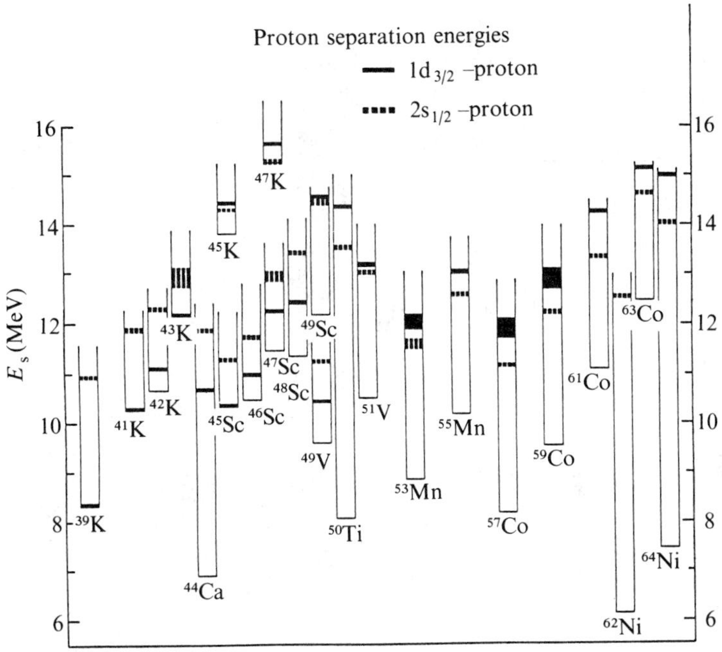

FIG. 8.21. Mean separation energies of $1d_{3/2}$ and $2s_{1/2}$ protons in fp-shell nuclei. The widths of the bars represent estimated uncertainties. (From Wagner 1973.)

reaction has been carried out by Källne (1974); in this case identification of 1p levels proved difficult and it is not clear whether the 1p spin-orbit splitting for neutrons shows a similar increase through the sd-shell.

Information on neutron hole states in nuclei from ^{12}C to ^{56}Fe has been obtained from the (p,d) reaction at 185 MeV (Fägerstrom and Källne 1973, Källne and Fagerström 1974). When adjusted for the energy difference between the least-bound protons and neutrons, the separation energies of the deeply bound neutron states appear to be consistent with the values for protons. Studies of the (p,d) reaction on heavier nuclei such as ^{90}Zr, molybdenum isotopes, tin isotopes, ^{140}Ce, and lead isotopes (Sakai and Kubo 1972, Ishimatsu *et al.* 1972) have revealed evidence for neutron pick-up from the next-to-highest (outer) shell. In general the energy spectra in the

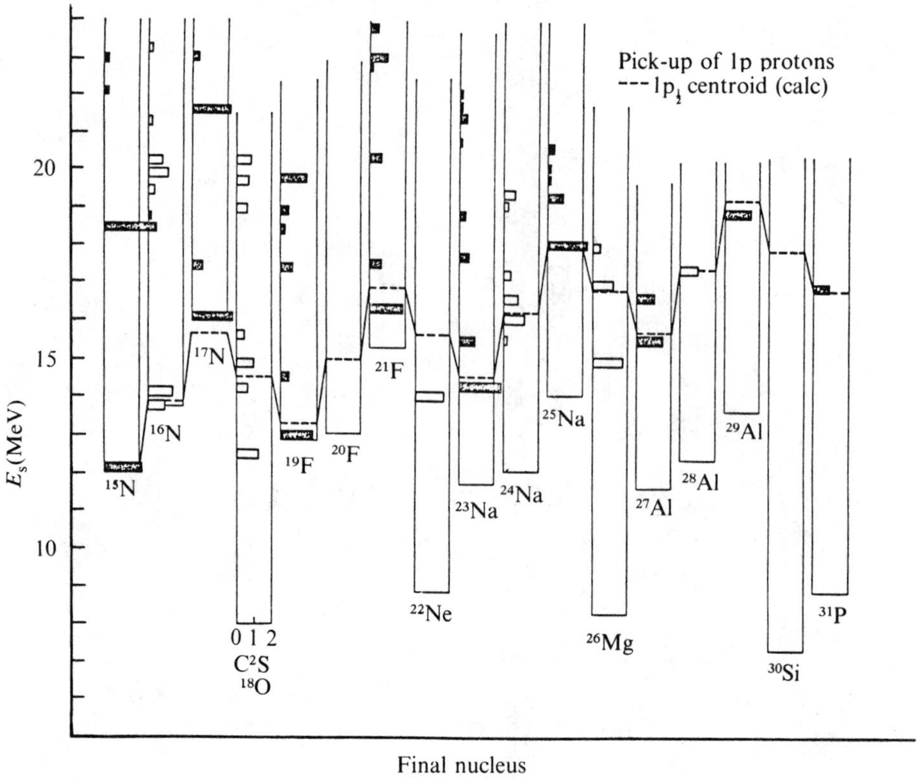

FIG. 8.22. Summary of $\ell = 1$ spectroscopic factors and separation energies observed in (d, ^{3}He) reactions leading to the final nuclei indicated. The broken lines show calculated values of the $1p_{1/2}$ centroid. (From Wagner 1973.)

relevant energy range are continuous and lacking in pronounced structure, but the angular distributions obtained by integrating over parts of the spectrum are strongly forward peaked and in reasonable agreement with summed DWBA distributions. In some cases a broad peak is also observed and is attributed to the summed $g_{9/2}$ hole strength. A similar, very pronounced, broad peak is observed in the (d,t) reaction on tin isotopes (van der Werf *et al.* 1974). The width of this peak has been calculated on the assumption that it results from the splitting of the $g_{9/2}$ strength owing to the interaction of the 1h-0p configuration with 2h-1p and higher configurations, and reasonable agreement with experiment is obtained (Sakai and Kubo

1972).

If reliable data on excitation energies and spectroscopic factors are available, the spectroscopic analysis can be extended though the use of energy-weighted sum rules. For a spin-zero target the monopole sum rule is given by (Bansal and French 1965)

$$E_{\mathrm{s}}^{+}(j)\,\frac{(2j+1-N_j)}{2j+1} - E_{\mathrm{s}}^{-}(j)\,\frac{N_j}{2j+1} = \varepsilon_j + \sum_r N_r(1+\delta_{rj})\,W_{rj} \tag{8.135}$$

where $E_{\mathrm{s}}^{+}(j)$ is the mean excitation energy for particle transfer into state j, $E_{\mathrm{s}}^{-}(j)$ is the mean separation energy for removal of a particle from state j, ε_j is the unperturbed single-particle energy, and W_{rj} is the average two-body interaction between particles in state j and state r. For pick-up or knock-out from a closed shell we have

$$E_{\mathrm{s}}^{-}(j) = -\varepsilon_j - \sum_r N_r(1+\delta_{rj})\,W_{rj} \tag{8.136a}$$

while for transfer into an empty shell we have

$$E_{\mathrm{s}}^{+}(j) = \varepsilon_j + \sum_r N_r(1+\delta_{rj})\,W_{rj}. \tag{8.136b}$$

Clearly the mean energies are equal in magnitude to the unperturbed single-particle energy only when W_{rj} vanishes. These formulae have been used to study the energy systematics of particle states in (d,p) reactions on the calcium and titanium isotopes (Bjerregaard, Hansen, and Sidenius 1965, Barnes *et al.* 1965, Dorenbusch, Belote, and Hansen 1966). The data shown in Figs. 8.21 and 8.22 have been fitted by Wagner (1973) using this method; he has shown that inclusion of the isovector part of the particle-hole interaction reproduced the additional binding seen in $T \neq 0$ targets.

Higher energy moments can also be defined but they contain increasingly large contributions from the highly excited states with small strengths. In addition the moments depend on many-body operators and can be calculated only with an adequate knowledge of particle correlations.

By taking a Hamiltonian which consists of one- and two-body operators only Koltun (1972, 1974) has obtained a sum rule for the total energy of the ground state in the form

$$E_A = \frac{1}{2}\{\langle T\rangle - \sum_i E_s(i)\}$$

where $\langle T\rangle$ represents the expectation value of the kinetic energy operator and $E_s(i)$ is the first energy moment for the state i. Considering only protons, this can be rewritten in terms of the total energy per proton as

$$\frac{E_Z}{Z} = \frac{1}{2}\left(\langle T\rangle_p^{AV} - \langle E_s\rangle_p^{AV}\right) \qquad (8.137a)$$

where $\langle E_s\rangle^{AV}$ is the average value of the mean separation energy. Koltun (1972) compared this sum rule (8.137*a*) with the (p,2p) data of James *et al.* (1969*a*,*b*) and obtained reasonble agreement, but with large uncertainties. The Saclay group (Bernheim *et al.* 1974*b*, Mougey *et al.* 1976) include a correction for the recoil of the centre of mass, so that the sum rule becomes

$$\frac{E_Z}{Z} = \frac{1}{2}\left(\frac{A-2}{A-1}\langle T\rangle_p^{AV} - \langle E_s\rangle_p^{AV}\right). \qquad (8.137b)$$

Their results are given in Table 8.10. These show that, except for ^{40}Ca, the difference Δ between the left-hand and the right-hand sides of eqn (8.137*a*) is significantly different from zero. They estimate that the values of $\langle T\rangle$ and $\langle E_s\rangle$ change by about 10 per cent owing to distortion effects, and attribute the non-zero value of Δ to uncertainties in the treatment of distortion and multiple collisions and to contributions from the unmeasured part of the spectral function at high energy loss.

Faessler *et al.* (1975) have pointed out that the bare nucleon-nucleon interaction is not used in an HF calculation in a restricted model space. Instead a renormalization is carried out which can give rise to a density-dependent effective force or to an effective three-body force as in Skyrme's interaction. In the first case an additional term appears on the right-hand side of eqn (8.137*b*) of the form $\frac{1}{4}\langle \partial G/\partial\rho\rangle^{AV}$

TABLE 8.10

Contributions to Koltun's sum rule[†]

	E_Z/Z	$\langle T\rangle^{AV}$	$\langle E_s\rangle^{AV}$	Δ	$-\frac{1}{4}\langle \partial G/\partial\rho\rangle^{AV}$	$-\frac{1}{2}\langle H_3\rangle^{AV}$	P
^{12}C	-6.93	16.9	23.4	-2.9±0.5	-2.15	-2.93	0.85
^{28}Si	-7.02	17.0	24.0	-3.1±0.6	-2.65	-3.61	0.86
^{40}Ca	-6.73	16.6	27.8	-0.9±0.5	-2.81	-3.81	0.97
^{58}Ni	-7.11	18.8	25.0	-3.6±0.7	-2.97	-4.06	0.83

[†]The experimental results given in the first four columns are taken from Mougey *et al.* (1975) and the calculated values are taken from Faessler *et al.* (1975). All energies are in MeV.

where G is the effective interaction and $\partial G/\partial\rho$ is its derivative with respect to density, and hence the difference Δ is equal to minus this quantity. Results obtained using a version of the modified delta interaction of Ehlers and Moszkowski (1972) are given in Table 8.10. In the second case the difference Δ is equal to $-\frac{1}{2}\langle H_3\rangle^{AV}$ where H_3 is the effective three-body force. Results obtained for the Skyrme force are also given in Table 8.10. Alternatively, in renormalized BHF theory[†] depletion of the deeply bound states is taken into account, and with an average occupation probability $P = 1 - d$ the sum rule becomes (Faessler *et al.* 1975)

$$\frac{E_Z}{Z(1+d)} = \frac{1}{2}\left(\frac{1-d}{1+d}\,\frac{A-2}{A-1}\,\langle T\rangle_p^{AV} = \langle E_s\rangle_p^{AV}\right). \tag{8.137c}$$

The values of P deduced from this sum rule are given in Table 8.10, and are consistent with those deduced by Becker (1970) using eqn (2.29).

It can be seen that each of these methods of handling the renormalization yields a satisfactory explanation for the data

[†]Details of the HF and RBHF theories and the interactions used in the calculations are given in §§2.2.2 and 2.2.3.

except for ^{40}Ca, so that there is no need to introduce additional 'bare' three-body forces or to seek missing regions of the spectral function. However, the RBHF approach does imply that spectroscopic strength is missing from the low-lying single-particle states so that

$$\int P(Q,\omega)\, d\omega\, d^3Q < Z$$

if the sum over γ in eqn (8.128), which defines $P(Q,\omega)$, is restricted to the filled states of the simplest shell model.

8.5. PHOTONUCLEAR PROCESSES

The interaction of photons with nuclei has been the subject of extensive theoretical and experimental work. For low photon energies interest is centred on the excitation of nuclear states in the giant resonance region and the subsequent decay of these states by emission of a nucleon. The nuclear structure information so obtained is connected very closely with that obtained by other methods of nuclear excitation such as inelastic electron scattering. The use of sum rules is important and nuclear size information tends to relate to properties of the nucleus as a whole.

At photon energies above about 50 MeV the process of photo-excitation gives rise to very short-lived states so that emission of a nucleon can be regarded as a direct process, and as the photon wavelength decreases the photon may be considered to interact with one or a limited number of nucleons in the nucleus. In this case comparison with knock-out and transfer reactions is useful. At high photon energies photon production of mesons is the dominant process and the hadronic components of the photon propagator become important.

8.5.1. *Direct emission of one or two nucleons*

The electromagnetic interaction can be expressed as a sum of one-body operators. Consequently, the simplest picture of a direct (γ,p) reaction is that a photon of energy E interacts with the charge and magnetic moment of the proton, and the proton is ejected from the nucleus with a certain kinetic energy T. The equation for conservation of energy is

$$E_\gamma = T + E_Q + E_s$$

where E_Q is the recoil energy of the residual nucleus and E_s is the proton separation energy. Combining this equation with the condition for momentum conservation we have

$$E_\gamma\{1 + a\cos\theta(T^2 + 2mc^2T)^{1/2} - a(T + E_s)\}$$
$$= aAmc^2T + E_s - aE_s(T + \tfrac{1}{2}E_s) \qquad (8.138a)$$

where m is the proton mass, θ is the angle of emission of the proton, and $a = \{(A-1)mc^2\}^{-1}$. For small values of E_s this reduces to the expression usually quoted (Manuzio *et al.* 1969):

$$E_\gamma = (aAmc^2T + E_s)\{1 + a\cos\theta(T^2 + 2mc^2T)^{1/2} - aT\}^{-1}. \qquad (8.138b)$$

This gives the values of E_γ at which kinematic resonances occur in the (γ,p) cross-section for fixed values of T. Some results for the $^7Li(\gamma,p)$ reaction (Sanzone *et al.* 1970) with $E_\gamma < 84$ MeV and $\Delta T = 2.5$ MeV are shown in Fig. 8.23; the resonance energy agrees with the prediction of eqn (8.138*b*) using $E_s \sim 10$ MeV which is consistent with the 1p separation energy for ^7Li measured in knock-out reactions. Similar results have been obtained for ^{12}C (Manuzio *et al.* 1969) giving a 1p separation energy of ~ 16 MeV. Further photoproton spectra have been obtained for ^6Li and ^7Li at $E_\gamma = 60$ MeV (Gardiner, Matthews, and Owens); these spectra are qualitatively in agreement with those seen in the (p,2p) reaction and include a peak at high excitation energy which is consistent with proton removal from the 1s shell.

These similarities with knock-out reactions suggests that the (γ,p) reaction may be described by a single-particle model. There is, however, an importance difference. The knock-out reaction, in the most frequently used coplanar geometry, is sensitive to low momentum components of the

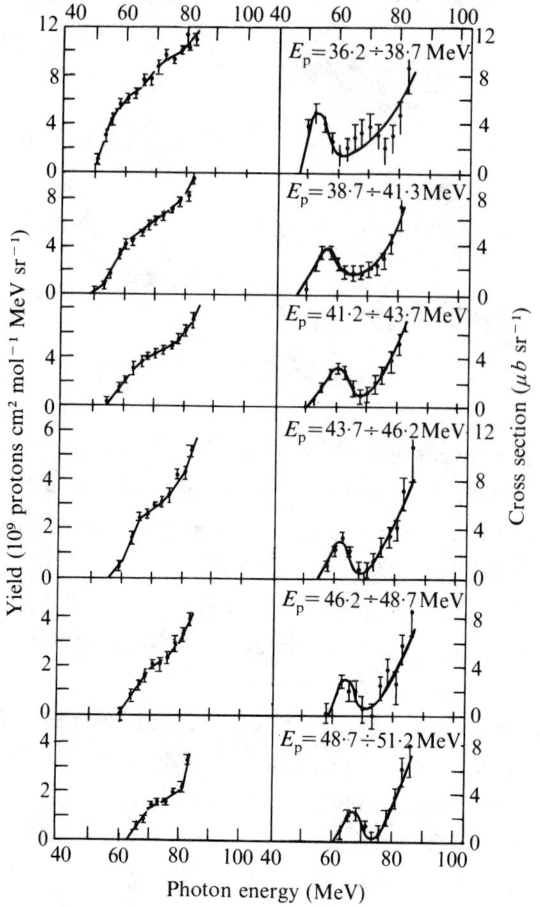

FIG. 8.23. Proton yields and cross-sections for the ^{7}Li(γ,p) reaction obtained using a proton energy interval of 2.5 MeV. (From Sanzone *et al.* 1970.)

nucleon motion in the nucleus (see §8.4.3). In contrast, a photon carries relatively little momentum. Hence, a 100 MeV photon can cause emission of an 80 MeV proton with momentum of 400 MeV/*c* only if such high momentum components are present in the nucleus.

A formalism based on the single-particle mechanism was proposed by Shklyarevskii (1959) and developed by Matthews *et al.* (1968). The outgoing proton is described by a plane wave but with the local wave number k inside the nucleus. All multipoles in the electromagnetic interaction are retained for the interaction with the charge and magnetic moment. The differential cross-section for emission of a nucleon from angular momentum state l in the nucleus is then given by

$$\frac{d\sigma_l}{d\Omega} = \frac{e^2}{4\pi mc^2}\frac{k^2 k_p}{E_\gamma}\left[\sin^2\theta\left|\frac{Z_1}{A}\Phi_l(P_1) - \frac{Z_2}{A-1}\Phi_l(P_2)\right|^2 + \right.$$
$$\left. + \frac{k_\gamma^{\,2}}{2k^2}\left|g_1\Phi_l(P_1) + g_2\Phi_l(P_2)\right|^2\right] \qquad (8.139)$$

where $k = E_\gamma/\hbar c$ is the photon momentum, k_p is the asymptotic momentum of the proton in the centre-of-mass system, Z_1 and g_1 are the charge and magnetic moment of the emitted proton, and Z_2 and g_2 are the charge and magnetic moment of the residual nucleus. The momenta absorbed by the nucleon and the residual nucleus, respectively, are

$$\underset{\sim}{P}_1 = \underset{\sim}{k} - (A-1)\underset{\sim}{k}_\gamma/A, \qquad \underset{\sim}{P}_2 = \underset{\sim}{k}_\gamma + \underset{\sim}{k}/A \qquad (8.140)$$

and Φ_l is the momentum distribution

$$\Phi_l(q) = \int \exp(-i\underset{\sim}{q}.\underset{\sim}{r})\ \psi_l(\underset{\sim}{r})\ d^3r \qquad (8.141)$$

where ψ_l is the single-particle wavefunction for the proton. This is the same quantity as the $G^{PW}(Q)$ which appears in the plane wave theory for knock-out reactions (see eqn 8.100). Matthews *et al.* (1968) found that this simple model fails to fit the data on the ^{6}Li(γ,p) reaction at 100 MeV by at least an order of magnitude unless additional high momentum components are introduced into the single-particle wavefunction, but Sanzone *et al.* (1970) found that a simple oscillator function for ^{7}Li leads to only slight under-estimation of the cross-section for E_γ = 50-75 MeV. Further results for ^{6}Li and ^{7}Li at 60 MeV (Gardiner 1971) suggest that this simple model describes emission of 1p protons fairly well but fails by an order of magnitude to describe removal from the 1s shell. Shklyarevskii (1959) obtained reasonable agreement with early data on ^{12}C (Whitehead *et al.* 1958) by varying the constant potential in which the local wavenumber of the outgoing proton is calculated. More recently, Manuzio *et al.* (1969) found quite satisfactory agreement for $^{12}C(\gamma,p)$ in the region E_γ = 50-80 MeV with the realistic single-particle wavefunctions used by Boffi *et al.* (1968*b*) to fit electron scattering. However, it is doubtful if comparison of these plane wave calculations with the data is meaningful. More recent work (Fink, Hebach, and Kummel 1972, Findlay *et al.* 1974) has demonstrated the importance of distortion of the outgoing proton.

Fujii (1963) has presented a different approach to the single-particle mechanism. He neglects higher multipoles and the recoil of the residual nucleus but he includes the final-state interactions (distortion) of the outgoing proton using a real energy-dependent square-well potential. This leads to good shape agreement with the data of Whitehead *et al.* (1958).

The difficulties arising with the simple single-particle model have led to various attempts to introduce high momentum components into the matrix element. These enter in the final state through distortion, but most authors have been concerned with modification of the shell model wavefunction Ψ_{sm}, using the Jastrow model, with a correlated wavefunction Ψ_{c} of the form

$$\Psi_{\mathrm{c}}(1,2,\ldots,A) = \prod_{i>j}^{A} f(r_{ij})\, \Psi_{\mathrm{sm}}(1,2,,\ldots,A) \qquad (8.142)$$

where $\underset{\sim}{r}_{ij} = \underset{\sim}{r}_i - \underset{\sim}{r}_j$. The correlation function obeys the conditions

$$f(r_{ij}) = 0,\ r_{ij} \to 0;\ f(r_{ij}) = 1,\ r_{ij} \to \infty. \qquad (8.143)$$

Shlyarevskii (1962) used the function $1 - \exp(-\lambda r_{ij}^2)$ and found some improvement in the fit to the (γ,p) data with $\lambda \sim 0.5\text{-}0.6\ \mathrm{fm}^{-2}$ Weise and Huber (1971) write the two-nucleon correlation function in the form

$$\Psi_c(1,2) = n_c\{\Psi_{sm}(1,2) - \frac{1}{4\pi}\int dq'\,\omega(q')\int d\Omega\,\exp(i\underset{\sim}{q}'.\underset{\sim}{r}_{12})\,\Psi_{sm}(1,2)\}$$

where n_c is a normalization constant, and observe that $f(r_{12})$ simulates scattering or exchange of momentum between particles 1 and 2. They arbitrarily choose $\omega(q') = \delta(q - q')$ so that

$$f(r_{12}) = n_c\{1 - j_0(qr_{12})\}.$$

This correlation function is plotted in Fig. 8.24. When the

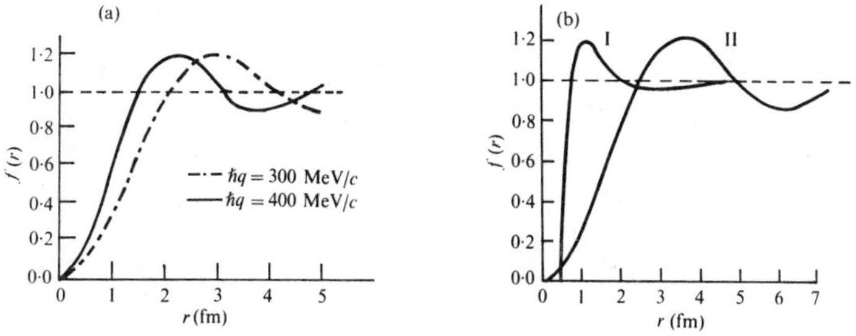

FIG. 8.24. (a) The correlation function $f(r) = n_c\{1 - j_0(qr)\}$ for different values of $\hbar q$ (Weise and Huber 1971). (b) Curve II is the correlation function calculated for ^{16}O from the Bethe-Goldstone equation while curve I is the function $1 - j_0(qr)$ with $\hbar q$ = 250 MeV/c. (After Fink *et al.* 1972.)

single-particle wavefunctions for ^{16}O are generated in a Saxon-Woods potential, the inclusion of the correlation function increases the (γ,p) cross-section by at least an order of magnitude. The shape of the cross-section as a function of proton energy is quite sensitive to the value of the parameter q, and best agreement with the somewhat meagre data for $^{16}O(\gamma,p)$ at E_γ = 40-70 MeV is obtained with $\hbar q$ = 200 MeV/c ($\sim$1 fm^{-1}). Further comparisons by Gardiner *et al.* (1973) and

Findlay *et al*. (1974) of their data for $^{6,7}Li$ and ^{12}C with this model using $\hbar q$ = 300 MeV/c gave good agreement at E_γ = 60 MeV but poorer results at 80 and 100 MeV.

Malecki and Picchi (1973) introduce a correlation function generated by means of a unitary operator. They conclude that the effects of short-range correlations may be seen for $E_\gamma > 100$ MeV but for lower photon energies the correlations play a minor role. The contradiction between these various results may be understood from the work of Fink *et al*. (1972) who take a correlation function for ^{16}O calculated from the Bethe-Goldstone equation. This is compared in Fig. 8.24 with the form given by eqn (8.144). The matrix elements depend critically on the positions of the zeros of $j_0(qr_{12})$ and, if $\hbar q$ is increased to 500 MeV/c to give a function which more closely resembles the Bethe-Goldstone prediction, the (γ,p) cross-section falls by two orders of magnitude. It can also be seen that for $\hbar q$ = 200 MeV/c the wavefunction is modified at quite large values of r_{12}. As Ripka and Gillespie (1970) have pointed out, the use of the Jastrow model with correlation functions containing significant low momentum components is equivalent to a change in the basic single-particle wavefunctions, so that orthogonality may be destroyed and comparisons of results obtained with shell model and correlation wavefunctions become less meaningful. Fink *et al*. (1972) have also emphasized the importance of orthogonality between the bound-state and continuum wavefunctions.

In a single-particle model, and neglecting recoil, the photon interacts with a neutron only through its magnetic moment. Since this interaction is small at the photon energies in question, the (γ,n) cross-section could be expected to be negligible in comparison to the (γ,p) cross-section. Experimentally, the cross-sections have been found to be comparable in magnitude. It is also found (Miller, Buss, and Rawlins 1971) that eqn (8.138) adequately describes the kinematic conditions with values of E_s consistent with other measurements. Emission of a neutron can come about, however, through the ground state correlations, as shown in Fig. 8.25. Thus a study of the (γ,n) reaction might yield more direct

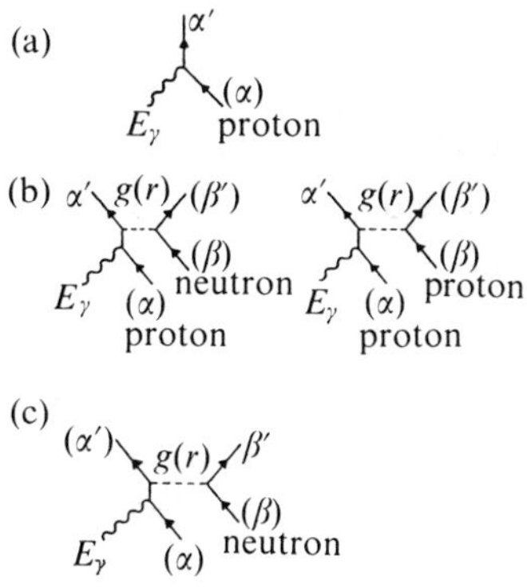

FIG. 8.25. Diagrams for the (γ,p) and (γ,n) reactions. Bound states for the nucleons are indicated by (α) (α') (β) and (β'), while continuum states are indicated by α' and β'. The function $g(r)$ is $1-f(r)$, where $f(r)$ is the correlation function defined in the text. (From Weise and Huber 1971.)

information about correlations. Weise and Huber (1971) found good agreement with the data for $^{16}O(\gamma,n)$ using the expression (8.145) with $\hbar q$ = 200 MeV/c. In contrast, Fink *et al.* (1972) predicted cross-sections two orders of magnitude smaller than the experimental data and concluded that short-range correlations do not contribute significantly to (γ,n) reactions for E_γ = 40-100 MeV. The discrepancy between these results arises from the different treatment of the correlation function.

In an alternative approach to the (γ,n) reaction it is assumed that the photon interacts with a preformed cluster c. The interaction is then enhanced owing to the electric multipole interaction of the photon with the charge of the cluster. This model yields a (γ,n) cross-section of the form

$$\frac{d\sigma}{d\Omega} \propto F(P) \frac{d\sigma_c}{d\Omega} \qquad (8.145)$$

where P is the momentum of the cluster c in the nucleus, $F(P)$ is the square of the momentum distribution of the cluster c, and $d\sigma_c/d\Omega$ is the (γ,n) cross-section for the free nucleus c. Thus this approach is essentially the same as the quasi-free model of knock-out reactions, and is based on the assumption that some of the nucleons in the nucleus are correlated as in the cluster c. Results for the $^{12}C(\gamma,n)$ reaction indicate that the quasi-alpha model is most successful in describing the data up to $E_\gamma \sim 100$ MeV (Miller *et al.* 1971, Rawlins *et al.* 1968). In this model the basic process is assumed to be absorption of the photon by ^4He leaving a bound ^3He nucleus (Mamasakhlisov and Dzhibuti 1962). When an oscillator form is used for the initial nuclear wavefunction it is necessary, as E_γ increases, to modify the length parameters for the internal motion of the α-particle and for its relative motion in the nucleus. This indicates a deficiency in the oscillator wavefunctions.

The quasi-deuteron model (Levinger 1951, Dedrick 1955, Gottfried 1958) has been studied in much greater detail than any other quasi-cluster model because of its importance in the description of the (γ,np) reaction. Gottfried developed Levinger's model, assuming that the photon is absorbed by an np pair in a relative $^3S_1$ state and that at short separation distances the relative wavefunction is the same as that of the free deuteron. (The probability of interaction with pp pairs is small if the electric dipole interaction predominates.) This gives a cross-section for the (γ,np) reaction on a nucleus proportional to the cross-section for the photo-disintegration of the deuteron multiplied by the function $F(P)$ previously defined. Garvey *et al.* (1965) have measured the angular distribution of protons detected in coincidence with neutrons from the $^{16}O(\gamma,np)$ reaction for $E_\gamma \sim 250$ MeV, and have compared the data with the quasi-deuteron predictions using oscillator functions to calculate $F(P)$. Not surprisingly the agreement is unsatisfactory, although shape fits can be obtained with modifications of the parameters. Further measurements on the $^{16}O(\gamma,np)$ reaction have been made by Hartmann *et al.* (1973). Using the oscillator form for $F(P)$ they find the same discrepancy between experiment and theory, namely that the measured angular distribution is narrower than the predicted one. However, good fits can be obtained by omitting the s-shell contribution to $F(P)$, which presumably simulates in a rough way the effect of distortion in the final state. The energy dependence of the cross-section divided by $F(P)$ is shown in Fig. 8.26, where it is compared with the cross-section for photodisintegration of the deuteron. This comparison lends support to the quasi-deuteron model.

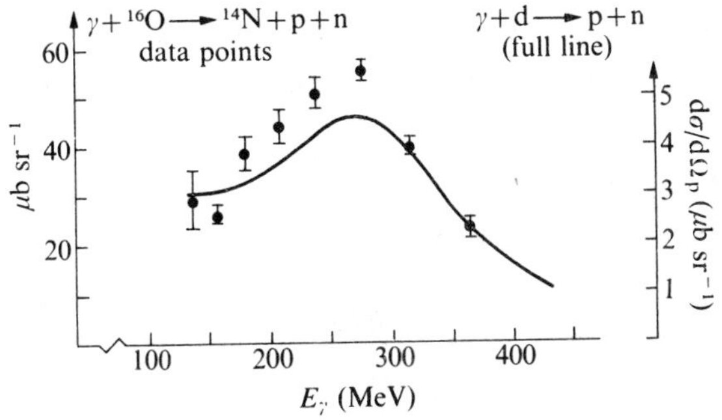

FIG. 8.26. Cross-section for the $^{16}O(\gamma,pn)$ process divided by $F(P)$ (data points) compared with the cross-section for photodisintegration of the deuteron. (After Hartmann *et al.* 1973.)

The quasi-deuteron model can also be used to describe (γ,p) and (γ,n) reactions if the theoretical cross-section for (γ,np) is integrated over the energy and angle of the unobserved nucleon. Matthews *et al.* (1968) found that this model gave good agreement with their data for $^{6}Li(\gamma,p)$ and $^{12}C(\gamma,p)$ for photon energies up to ~100 MeV. Similar good agreement was obtained by Schier and Schoch (1974) with their data on $^{12}C(\gamma,n)$ and $^{16}O(\gamma,n)$ for E_γ = 60-150 MeV. However, Rawlins *et al.* (1968) and Miller *et al.* (1971) found that the quasi-alpha model is preferred for $E_\gamma \gtrsim 30$ MeV.

Some authors have developed a microscopic model for the (γ,np) reaction. The cross-section is essentially of the form

$$d^4\sigma = \frac{2\pi}{\hbar c}\frac{d\underset{\sim}{k}_1}{(2\pi)^3}\frac{d\underset{\sim}{k}_2}{(2\pi)^3}\sum_\beta |\langle \Phi^\beta_{A-2}\,\chi^-(1,2)|\mathcal{O}(1,2)|\Phi^\alpha_A\rangle|^2 \times$$

$$\times\ \delta(E_1 + E_2 + E_s^{\alpha\beta} - E_\gamma + E_Q) \qquad (8.148)$$

where E_Q is the recoil energy of the A-2 system, $\mathcal{O}(1,2)$ is the operator for photon absorption and $\chi^-(1,2)$ is the wavefunction of the outgoing proton and neutron. Kopaleishvili and Jibuti (1963) took an oscillator form for Φ_A in order to use the Brody-Moshinsky transformation and obtain functions of

$$\underset{\sim}{R} = \tfrac{1}{2}(\underset{\sim}{r}_1 + \underset{\sim}{r}_2), \qquad \underset{\sim}{r} = \underset{\sim}{r}_1 - \underset{\sim}{r}_2.$$

The function of the separation distance r was then multiplied by a correlation function of the form due to Dabrowski (1958)

$$f(r) = 1 - \exp\{-\beta\left(\frac{r-a}{a}\right)^2\}, \qquad r > a$$

$$= 0 \qquad\qquad , \qquad r \leqslant a \qquad (8.147)$$

but the final-state interaction was neglected. Reitan (1962) and Ostgaard (1965) have included the final-state interaction between the np pair, but not distortion of the motion of their centre-of-mass by the residual nucleus. The interaction in the final state appears to affect the cross-section significantly but the initial state correlation has much less effect. All these calculations over-estimate the integrated cross-section for ^{16}O by a substantial amount.

Similar calculations have been carried out by Bramanis (1971) but he includes a very approximate treatment of distortion by the residual nucleus. The calculations are compared with data on $^{32}S(\gamma,np)$ and $^{40}Ca(\gamma,np)$; reasonable agreement is obtained and it appears the final-state interactions between the neutron and the proton and between the pair and the residual nucleus have a very important effect on the cross-section.

It is evident from these studies of photonuclear reactions leading to emission of one or two nucleons that the reactions can in principle give information about separation energies and momentum components in the nucleus. In practice, neither the data nor the theoretical models are yet sufficiently precise to yield detailed information. In particular, the treatment of correlations is very confused. It is far from clear whether short-range correlations are needed; if not, the medium-range effects of configuration mixing (or clustering) and the Pauli principle must be considered carefully, together with the final-state interactions.

8.5.2. *Excitation of the giant dipole resonance*

The giant dipole resonance (GDR) is seen as a broad peak in the photonuclear cross-section at an excitation energy of 15-25 MeV, depending on the mass number. Because of the strength of the resonance, the initial explanation was given in terms of a collective model of the excitation (Goldhaber and Teller 1948, Steinwedel, Jensen, and Jensen 1950). It was postulated that the protons and neutrons in the nucleus behave as two interpenetrating incompressible fluids, so that absorption of the photon leads to opposing oscillations of the two fluids and the restoring force is provided by the symmetry term which is proportional to $K(\rho_p - \rho_n)^2/(\rho_p + \rho_n)$. Thus the parameters of this model are the coefficient K of the nuclear symmetry energy, a nuclear radius defining the rigid surface, and the width of the Lorentzian line shape. The model does not allow calculation of the width which is associated with a frictional damping. An improved collective model has been developed in which the restruction to a rigid nuclear surface is removed, and recent results have been reviewed by several authors (Spicer 1969, Bramblett, Fultz, and Berman 1973, Danos 1973).

An alternative description of the excitation of the GDR has been attempted in terms of the shell model. Since the photonuclear interaction is represented by a one-body operator, the shell model description involves the excitation of one-particle—one-hole states. The residual interaction mixes these states and raises the energy of the coherent dipole state. This was first demonstrated in the schematic model (Brown and

Bolsterli 1959). More accurate particle-hole calculations have been carried out by many authors and have been reviewed by Spicer (1969) and R.F. Barrett *et al.* (1973). Important features of these calculations are the correct description of the single-particle wavefunctions for continuum states and the correct location of the single-particle and single-hole states. Thus there is a significant link with other topics discussed in this chapter which should provide the information required as input for realistic particle-hole calculations.

8.5.3. *Sum rules*

It is convenient to introduce sum rules to eliminate the explicit dependence of the theory on the detailed behaviour of the excited states of the target nucleus. The energy-weighted sum rules are defined by the expression

$$\sigma_n = \int \sigma(E)\ E^n\ \mathrm{d}E = \sum_f E^n |\langle f|\mathcal{O}|i\rangle|^2 \tag{8.148}$$

where $\mathcal{O}$ is the relevant operator and $E = E_\gamma$. Using standard closure techniques the sum rule for $n = 0$ is found to be (Levinger 1960)

$$\sigma_0 = 2\pi^2 \frac{e^2\hbar}{mc} \frac{NZ}{A} \simeq 60 \frac{NZ}{A} \text{ MeV mb}. \tag{8.149}$$

This is the Thomas-Reiche-Kuhn (TRK) formula and is subject to small corrections for relativistic effects and also for the Fermi momentum of the nucleons in the nucleus (O'Connell 1973). It is a model-dependent formula because the derivation depends on the assumption that the nucleon-nucleon force depends only on the relative separation, and consequently the result is modified by the presence of exchange and velocity-dependent forces.

The sum σ_{-1} is sometimes called the bremsstrahlung-weighted cross-section and is given by (Levinger 1960)

$$\sigma_{-1} = \frac{4\pi^2}{3} \frac{e^2}{\hbar c} \frac{NZ}{A-1} \langle r^2 \rangle_m \tag{8.150}$$

The result for $n = -2$ depends on the nuclear polarizability which may be calculated semiclassically in the two-fluid model to give (Levinger 1960)

$$\sigma_{-2} = \frac{\pi^2}{20}\frac{e^2}{\hbar c}\frac{A\langle r^2\rangle_m}{K} \tag{8.151}$$

where K is the coefficient of the nuclear symmetry energy.

The experimental results for these quantities have been reviewed by O'Connell (1973) and Fuller (1973). For $A < 40$ the experimental value for σ_0 is twice σ_0^{TRK} while for heavier nuclei it is $(1.1\text{-}1.4)\sigma_0^{\mathrm{TRK}}$. This discrepancy can be removed theoretically by introducing a particle-hole interaction with a suitable exchange mixture, but difficulties then arise with σ_{-1}. O'Connell (1973) suggests that σ_{-1} is sensitive to long-range correlations which are isospin dependent. Further calculations (Lane and Mekjian 1973, Weng *et al.* 1974) confirm this suggestion but indicate that σ_{-1} is insensitive to short-range correlations.

It is now known that for a target nucleus with $T \neq 0$ the GDR has components with $T_> = T + 1$ and $T_< = T$, and the splitting of these states has been estimated as $\Delta E = V(T + 1)/A$ with $V = 55 \pm 15$ MeV (Akyuz and Fallieros 1971, Paul, Amann, and Snover 1971). This is consistent with $2V = U_1$ where U_1 is the coefficient in the Lane potential defined by eqn (10.8). The bremsstrahlung-weighted cross-sections for E1 excitation of final states with these isospins are given by (O'Connell 1973)

$$\sigma_{-1}(T) = \frac{4\pi^2}{3(T+1)}\frac{e^2}{\hbar c}\left\{\frac{1}{3}T\langle r_{\mathrm{S}}^2\rangle + T\langle r_{\mathrm{V}}^2\rangle + \frac{1}{6}T(2T-1)(2T+3)\langle r_{\mathrm{T}}^2\rangle\right\} \tag{8.152a}$$

$$\sigma_{-1}(T+1) = \frac{4\pi^2}{3(T+1)}\frac{e^2}{\hbar c}\left\{\frac{1}{3}\langle r_{\mathrm{S}}^2\rangle - T\langle r_{\mathrm{V}}^2\rangle - \frac{1}{6}T(2T-1)\langle r_{\mathrm{T}}^2\rangle\right\} \tag{8.152b}$$

$$\sigma_{-1} = \frac{4\pi^2}{3(T+1)}\frac{e^2}{\hbar c}\left\{\frac{1}{3}(T+1)\langle r_{\mathrm{S}}^2\rangle + \frac{1}{3}T(2T-1)(T+1)\langle r_{\mathrm{T}}^2\rangle\right\} \tag{8.152c}$$

where the mean-square isoscalar, isovector, and isotensor radii are defined as

$$\langle r_S^2 \rangle = \langle TT_3 | \sum_{ji} (\underset{\sim}{r}_i \cdot \underset{\sim}{r}_j)(\underset{\sim}{t}_i \cdot \underset{\sim}{t}_j) | TT_3 \rangle \quad (8.153a)$$

$$\langle r_V^2 \rangle = \frac{\langle TT_3 | \sum_i r_i^2 \, t_{iz} | TT_3 \rangle}{2T_3} = \frac{N\langle r^2 \rangle_n - Z\langle r^2 \rangle_p}{4T_3} \quad (8.153b)$$

$$\langle r_T^2 \rangle = \langle TT_3 | \sum_{ij} (\underset{\sim}{r}_i \cdot \underset{\sim}{r}_j)(3t_{iz}t_{jz} - \underset{\sim}{t}_i \cdot \underset{\sim}{t}_j) | TT_3 \rangle \{3T_3^2 - T(T+1)\}^{-1}. \quad (8.153c)$$

Leonardi (1972, 1973) has pointed out that the data for σ_{-1} can be used to obtain information on the difference between the proton and neutron distributions. He concludes that for $A \sim 90$ the data require $\langle r^2 \rangle_n^{1/2} - \langle r^2 \rangle_p^{1/2} \lesssim 0$, while for ^{208}Pb he finds $\langle r^2 \rangle_n^{1/2} - \langle r^2 \rangle_p^{1/2} \sim -0.15$ fm. The latter result is not in agreement with the Hartree-Fock and other calculations discussed in Chapter 2, but does agree with some of the other experimental studies of the neutron distribution discussed in Chapter 10.

The notation used in eqn (8.148) implies that the upper limit for the integration of the experimental data is infinity. Since the main contribution to the photo-absorption cross-section at high energies comes from meson production, it is more usual to set the upper limit at the energy $\mu = m_\pi c^2$. Using dispersion relations Gell-Mann, Goldberger, and Thirring (1954) obtained a sum rule for $n = 0$ which makes explicit reference to the mesonic contributions. They find

$$\sigma_{int} = \int_0^\mu \sigma_{\gamma A}(E) \, dE$$

$$= 2\pi^2 \frac{e^2\hbar}{mc} \frac{NZ}{A} + \int_\mu^\infty \{A\sigma_{\gamma N}(E) - \sigma_{\gamma A}(E)\} \, dE \quad (8.154)$$

where $\sigma_{\gamma N}$ is defined as

$$A\sigma_{\gamma N}(E) = Z\sigma_{\gamma p}(E) + N\sigma_{\gamma n}(E) \quad (8.155)$$

and $\sigma_{\gamma p}$, $\sigma_{\gamma n}$ are the elementary photon-nucleon cross-sections.

The second term in eqn (8.154) arises from the difference in the photoproduction on bound and free nucleons due to the exclusion principle and other effects, and has been estimated by Gell-Mann *et al.* (1954) to be ~ 0.4 σ_0^{TRK} for all A.

New measurements of the total cross-section for photo-nuclear absorption in the energy range 2-20 GeV indicate a correction to the TRK sum rule of (0.4-1.2) σ_0^{TRK} depending on the mass number (Weise 1973*a*,*b*). The essential assumptions in the derivation of the GGT sum rule are that

$$\left.\begin{aligned} \sigma_{\gamma A}(E) &\xrightarrow[E\to\infty]{} A\sigma_{\gamma N}(E) \\ \mathrm{Re}\, F_{\gamma A}(E) &\xrightarrow[E\to\infty]{} A\, \mathrm{Re}\, F_{\gamma N}(E) \end{aligned}\right\} \tag{8.156}$$

where $F(E)$ denotes the forward-scattering amplitude for photon-nucleus or photo-nucleon scattering. Weise (1973*a*,*b*) has derived a new sum rule which yields the GGT result in the same limiting situation. This sum rule provides a link between the low-energy region $E_\gamma < 140$ MeV and the high-energy region $E_\gamma > 1$ GeV. In this intermediate region there are at present very few data.

8.5.4. *High-energy processes*

The total cross-section for photon absorption by nuclei in the energy region 2-20 GeV can be written as

$$\sigma_{\gamma A}(E) = A_{eff}(E)\, \sigma_{\gamma N}(E) \tag{8.157}$$

where $A_{eff}(E) \simeq A^{0.91}$. Experimental values of $A_{eff}(E)/A$ are shown in Fig. 8.27.

The limit of $A_{eff} = A$ in eqn (8.157) would imply that the scattering from the nucleus could be represented as the sum of single scatterings from individual nucleons (see also §8.2). For hadron-nucleus interactions this relation is not expected; for example neutron-nucleus total cross-sections in the same energy region behave approximately as $A^{0.82}$ (Franco 1972), while the pion-nucleus reaction cross-sections at 1-2 GeV behave very accurately as $A^{0.69}$ (Allardyce *et al.* 1973).

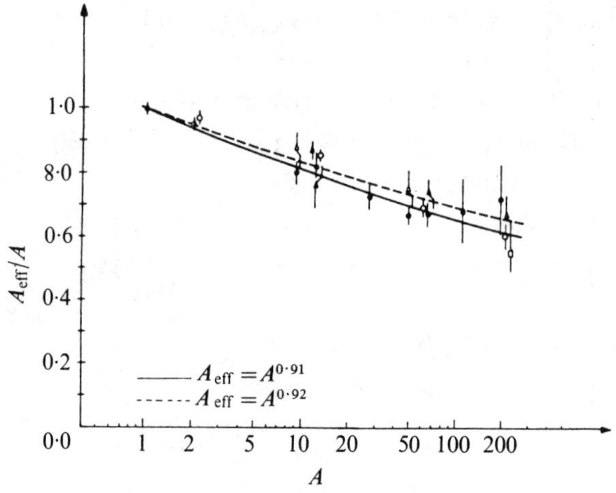

FIG. 8.27. Values of $A_{\text{eff}}(E)/A$ as determined by various measurements of the photo-absorption cross-section. (After Weise 1973*a*.)

Proportionality of the cross-sections to $A^{2/3}$ is taken to be a signature of absorption by a 'black' nucleus with probable localization of the interaction in the nuclear surface (see §10.4.1). The mean free path of photons in the nuclear medium is very large compared with the nuclear size and a simple argument would predict $\sigma_{\gamma A} = A\sigma_{\gamma N}$. The observed departure from this expectation is taken as an indication of hadronic components in the photon propagator.

In the vector-dominance model (Bell 1968, Stodolsky 1967) it is assumed that the photons may be converted to vector mesons, i.e. the ρ, ω, and ϕ, during propagation through the nucleus. The vector mesons interact strongly with the nucleons in the nucleus giving rise to absorption or 'shadowing' of the interior nucleons by those on the surface. The scattering of a photon thus involves a direct or one-step contribution, in which the photon interacts directly with a nucleon, and a two-step process, in which the photoproduced vector meson interacts with a nucleon. The amplitudes for these processes interfere, but at low energies the two-step process is unimportant so that $A_{eff} \sim A$.

Calculations of the hadronic part of the photo-absorption cross-section on nuclei in this model involve the photon-meson coupling constant and the appropriate hadron-nucleus optical potential. The latter is usually taken to be of a very simple form; for example the rho-nucleus potential is written as

$$V_{\rho A}(r) \propto \sigma_{\rho N} \, A\rho_m(r) \tag{8.158}$$

where $\sigma_{\rho N}$ is the total cross-section for rho-nucleon scattering, and the real part of the rho-nucleon scattering amplitude has been neglected. Brodsky and Pumplin (1969) have taken $\sigma_{\rho N}$ to be energy independent and equal to 30 mb, while Gottfried and Yennie (1969) deduce an energy dependence for $\sigma_{\rho N}$. It is convenient to study the ratio $\sigma_{\gamma A}/A\sigma_{\gamma N} = A_{eff}/A$ as this eliminates the $\gamma\rho$ coupling constant. Theoretical predictions and experimental data (Brookes *et al.* 1973) are compared in Fig. 8.28; these indicate that shadowing increases with energy over the range 1-4 GeV but becomes energy independent at higher energies.

Brodsky and Pumplin (1969) have generalized the model to include other intermediate hadrons in addition to the rho. This leads to a substantial reduction in the shadowing effect, i.e. an increase in A_{eff}.

Photoproduction of ρ mesons has been studied extensively over the energy range 2-9 GeV by detecting the $\pi^{\pm}$ pair produced by decay of the rho:

$$\gamma + A \rightarrow \rho + A \rightarrow \pi^{+} + \pi^{-} + A.$$

The differential cross-section for coherent production is given by (Kolbig and Margolis 1968).

$$\frac{d\sigma}{dt}(\gamma A \rightarrow \rho A) = \frac{d\sigma}{dt}(\gamma N \rightarrow \rho N) \; F(\sigma_{\rho N}, t) \tag{8.159}$$

where t is the 4-momentum transfer,

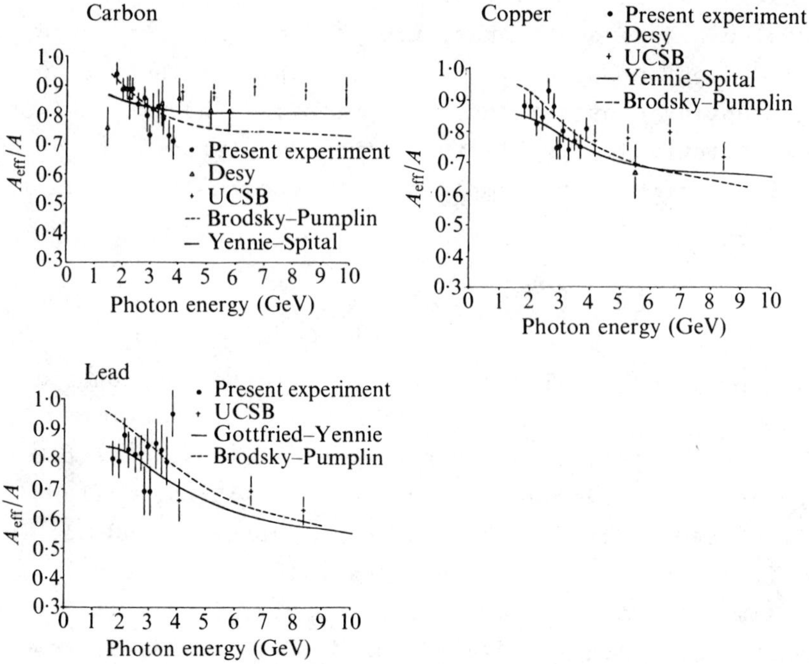

FIG. 8.28. The energy dependence of $A_{eff}(E)/A$ compared with theoretical predictions. (From Brookes *et al.* 1973.)

$$F(\sigma_{\rho N},t) = \left|A\int J_0(q_T b)\ \exp(iq_L z)\rho_m(b,z)\exp\{-\tfrac{1}{2}\sigma_{\rho N}\int_z^{\infty} A\rho_m(b,z')dz'\}d^2b\ dz\right|^2 \tag{8.160}$$

q_L and q_T are the longitudinal and transverse components of the momentum transfer given by

$$q_T \simeq k\theta, \qquad q_L \simeq m_\rho^2/2k \simeq \sqrt{-t_{min}} \tag{8.161}$$

and t_{min} is the minimum value of t. Measurement of the production cross-section at $\theta = 0$, $t = t_{min}$ for a range of nuclei allows the determination of $\sigma_{\rho N}$ and, if vector dominance is assumed in order to obtain an expression for $(d\sigma/dt)(\gamma N \to \rho N)$, of the coupling constant $\gamma_\rho^2/4\pi$. Summaries of the results up to 1969 (Leith 1969, Gilman 1969) show that there is disagreement between values of $\gamma_\rho^2/4\pi \sim 0.5$, $\sigma_{\rho N} \sim 26$ mb, which support the vector-dominance model, and values of $\gamma_\rho^2/4\pi \sim 1$, $\sigma_{\rho N} \sim 35$ mb. More recent results (Alvensleben *et al.* 1970*a*,

Coddington *et al.* 1974) support the lower value for the coupling constant.

The dependence of the photoproduction cross-section on the nuclear matter distribution $\rho_m(r)$ suggests that, if the other parameters can be determined, the photoproduction data for a range of nuclei could be used to obtain information on the matter distribution. This has been attempted by Alvensleben *et al.* (1970*a*,*b*), who use an optical potential of the form

$$V_{\rho A}(r) \propto \sigma'_{\rho N}(1 - i\beta)\, A\rho_m(r) \qquad (8.162)$$

where β is the ratio of the real to imaginary part of the forward-scattering amplitude and σ' is a rho-nucleon cross-section modified to take account of the double-scattering correction to the cross-section (Moniz and Dixon 1969). This is a forward-scattering or large-A approximation to the first-order potential (see §7.2.2), and consequently the size parameters of ρ_m must incorporate the range of the ρN interaction. Alvensleben *et al.* take a Fermi distribution for all nuclei with diffuseness a = 0.545 fm and halfway radius $r_0A^{1/3}$. They vary r_0 in order to fit the data on 13 nuclei between Be and U and find r_0 = 1.12 ± 0.02 fm. It is noted that a variation of 10 per cent in the diffuseness changes the results by less than 2 per cent and that a 5 per cent change in $r_0A^{1/3}$ changes the value of $\sigma_{\rho N}$ by 1 mb. Coddington *et al.* (1974) use the same parameters for the matter distribution but deduce a slightly lower value for $\sigma_{\rho N}$.

The nuclear r.m.s. radii obtained by Alvensleben *et al.* (1970*a*,*b*) exceed the values obtained by other methods by ~ 0.35 fm in C and 0.15-0.25 fm in Pb. Part of this discrepancy is probably due to the mistaken assumption that the halfway radius varies as $A^{1/3}$ (see §1.2.3), but in view of the many simplifying features of the theory it would seem to be more profitable to insert known nuclear matter distributions to obtain a more reliable estimate for $\sigma_{\rho N}$.

Incoherent photoproduction of single pions has also been studied. The production of charged pions has been measured at

2.85 GeV (Abrahamian *et al.* 1972) and at 16 GeV (Boyarski *et al.* 1969). Examination of Z_{eff} for π^+ production suggests that there is little energy dependence, in contradiction to the vector-dominance model, and this is supported by measurements of A_{eff} for neutral pion production in the same energy region (Meyer *et al.* 1972). Comparison of Z_{eff} for π^+ production and N_{eff} for π^- production allows a comparison of proton and neutron distributions in the target nuclei. These data are interpreted in terms of the Fermi distribution to suggest that $c_{\text{p}} > c_{\text{n}}$ or $a_{\text{p}} > a_{\text{n}}$.

We now consider in more detail the derivation of the formula (8.159) for the cross-section, which was developed by Kölbig and Margolis (1968) using the Glauber formalism.

Coherent production is defined as the process

$$\text{a} + \text{A} \rightarrow \text{c} + \text{A}$$

where the target nucleus A is left in its ground state. The projectile a may be any particle and is not restricted to being a photon. The scattering amplitude in DWIA is given by eqns (8.3) and (8.4) with $\rho_{if}(\underset{\sim}{r})$ replaced by $A\rho_{\text{m}}(\underset{\sim}{r})$ since $f = i$ in this case. We denote the scattering amplitude for the corresponding production process on a single nucleon by M_{ac} and the scattering amplitude for scattering from a single nucleon by M_{a} or M_{c}, where each M is defined in the appropriate two-body centre-of-mass system. Hence the required DWIA amplitude is

$$F(q) = R_{\text{a}}\, M_{\text{ac}}(q) \int \chi_{\text{c}}^{-*}(\underset{\sim}{k}', \underset{\sim}{r})\, A\rho_{\text{m}}(\underset{\sim}{r})\, \chi_{\text{a}}^{+}(\underset{\sim}{k}, \underset{\sim}{r})\, \text{d}^3 r \quad (8.163)$$

where R_{a} is the ratio of the reduced mass of a in the a-nucleus system to the reduced mass in the a-nucleon system, and $\underset{\sim}{q} = \underset{\sim}{k} - \underset{\sim}{k}'$ is the momentum transfer. If we use the semi-classical expression (7.28) for the distorted wavefunctions and (7.52) for the optical potential, we have

$$\chi_{\text{a}}^{+}(\underset{\sim}{k}, \underset{\sim}{r}) = \exp\{\text{i}\underset{\sim}{k}\cdot\underset{\sim}{r} + \text{i}\,\frac{2\pi}{k}\, R_{\text{a}}\, M_{\text{a}}(0) \int_{-\infty}^{z} A\rho_{\text{m}}(\underset{\sim}{r}')\, \text{d}z'\} \quad (8.164a)$$

$$\chi_c^{-*}(\underset{\sim}{k}'\underset{\sim}{r}) = \exp\{-i\underset{\sim}{k}'\cdot\underset{\sim}{r} + i\,\frac{2\pi}{k}\,R_c\,M_c(0)\int_z^{\infty} A\rho_m(\underset{\sim}{r}'')\,dz''\}. \tag{8.164b}$$

For convenience in notation we now absorb the ratio R into the scattering amplitude M, but it should be noted that our definitions of the scattering matrix M in Chapters 7 and 8 require that this ratio is taken into account. On the other hand, Glauber defines a two-body scattering amplitude in the particle-nucleus system so that the ratio of reduced masses does not appear (see Jackson 1969*a*, appendix 3).

If we now neglect any component of q along the forward direction z (i.e. set $q_L = 0$) and introduce the impact parameter b, we find

$$F(q) = M_{ac}(q)\int \exp(i\underset{\sim}{q}\cdot\underset{\sim}{b})\,A\rho_m(b,z)\,\exp\{i\,\frac{2\pi}{k}\,M_a(0)\int_{-\infty}^{z} A\rho_m(b,z')\,dz'\}\,\times$$
$$\times\,\exp\{i\,\frac{2\pi}{k}\,M_c(0)\int_z^{\infty} A\rho_m(b,z'')\,dz''\}\,d^2b\,dz \tag{8.165}$$

and integrating over z yields

$$F(q) = \frac{ik}{2\pi}\,\frac{M_{ac}(q)}{M_a(0) - M_c(0)}\int \exp(i\underset{\sim}{q}\cdot\underset{\sim}{b})\Big[\exp\{i\,\frac{2\pi}{k}\,M_a(0)\,T(b)\} -$$
$$-\,\exp\{i\,\frac{2\pi}{k}\,M_c(0)\,T(b)\}\Big]\,d^2b \tag{8.166}$$

where

$$T(b) = A\int_{-\infty}^{\infty} \rho_m(b,z)\,dz. \tag{8.167}$$

The two-body forward amplitudes may be written as

$$M(0) = \frac{ik}{4\pi}\,\sigma(1 - i\beta) \tag{8.168}$$

so that the coherent cross-section becomes

$$\frac{d\sigma^C}{d\Omega}(aA \rightarrow cA) = |M_{ac}(q)|^2\,\frac{4}{(\sigma_a - \sigma_c)^2 + (\beta_a\sigma_a - \beta_c\sigma_c)^2}$$
$$\times\,\Big|\int J_0(qb)\Big[\exp\{-\tfrac{1}{2}\sigma_a(1 - \beta_a)T(b)\} - \exp\{-\tfrac{1}{2}\sigma_c(1 - i\beta_c)T(b)\}\Big]\,d^2b\Big|^2 \tag{8.169}$$

In the limit $q = 0$ the cross-section can be written in the form

$$\frac{d\sigma^C}{d\Omega}(aA \to cA) = |M_{ac}(0)|^2 \, N^2(A, \tfrac{1}{2}\sigma_a, \tfrac{1}{2}\sigma_c) \qquad (8.170)$$

where the real parts of the amplitudes have been omitted to give

$$N(A, \tfrac{1}{2}\sigma_a, \tfrac{1}{2}\sigma_c) = \frac{2}{\sigma_a - \sigma_c} \int \left[\exp\{-\tfrac{1}{2}\sigma_a \, T(b)\} - \exp\{-\tfrac{1}{2}\sigma_c \, T(b)\} \right] d^2 b. \qquad (8.171)$$

For finite q_L eqn (8.165) can be used to give the cross-section

$$\frac{d\sigma^C}{dt}(aA \to cA) = \frac{d\sigma}{dt}(aN \to cN) \left| A \int J_0(q_T b) \exp(iq_L z) \rho_m(b,z) \exp\{-\tfrac{1}{2}\sigma_a(1-i\beta_a)T(b)\} \times \right.$$
$$\left. \times \exp\left[-\tfrac{1}{2}\{\sigma_a(1-i\beta_a) - \sigma_c(1-i\beta_c)\} \int_z^\infty A\rho_m(b,z')dz' \right] d^2 b \, dz \right|^2 \qquad (8.172)$$

where $t = -q^2$. Eqns (8.159) and (8.160) are obtained by setting $\sigma_a = 0$ for photoproduction and $\beta_c = 0$.

Incoherent production is defined to include excitation of all possible final nuclear states. Using the same formalism together with closure yields (Kölbig and Margolis 1968)

$$\frac{d\sigma^I}{d\Omega}(aA \to cA) = |M_{ac}(q)|^2 \, N(A, \sigma_a, \sigma_c) + \text{higher terms in } q \qquad (8.173)$$

where

$$N(A, \sigma_a, \sigma_c) = \frac{1}{\sigma_a - \sigma_c} \int \left[\exp\{-\sigma_a \, T(b)\} - \exp\{-\sigma_c \, T(b)\} \right] d^2 b. \qquad (8.174)$$

For $\sigma_a = \sigma_c$ this reduces to the semiclassical result for the $N(A)$ defined in eqn (8.62*a*,*b*) for summed inelastic scattering.

Auger and Lombard 1973*b*, 1974 have plotted

$$(\sigma_a - \sigma_c)^{-1} \left[\exp\{-\sigma_a \, T(b)\} - \exp\{-\sigma_c \, T(b)\} \right]$$

i.e. the integrand for $N(A, \sigma_a, \sigma_c)$ for ^{27}Al and ^{208}Pb. Their results are shown in Fig. 8.29, together with the nuclear matter distributions used in the calculation which was Beiner's

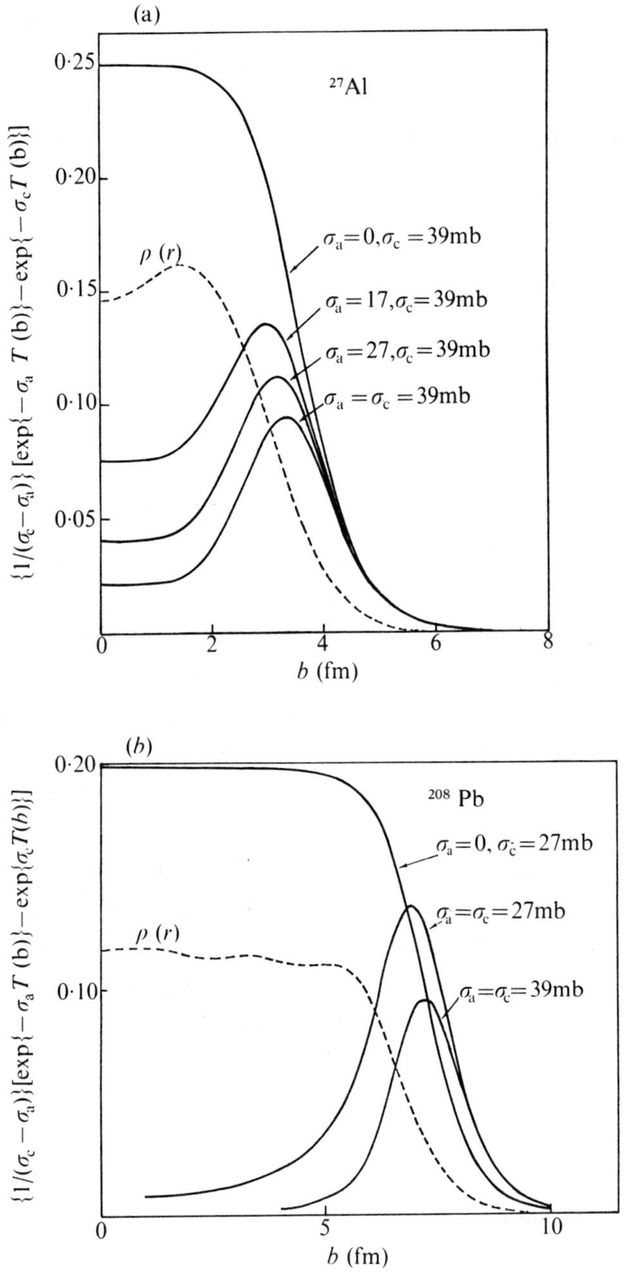

FIG. 8.29. Behaviour of the integrand which gives the effective nucleon number $N(A, \sigma_a, \sigma_c)$ for various values of σ_a and σ_c. The broken curves show the nuclear matter distributions for ^{27}Al and ^{208}Pb, respectively, used in the calculation. (After Auger and Lombard 1974.)

phenomenological distribution generated in a Saxon-Woods type of potential (see §2.2.1). These calculations show that in the case of particle production by photons, i.e. $\sigma_a = 0$, the process probes essentially the whole of the nucleus. The corresponding values of $N(A,\sigma_a,\sigma_c)$ are listed in Table 8.11† for several nuclei, for the phenomenological density and for a Thomas-Fermi distribution which falls much more rapidly in the region $\rho < 0.1\ \rho_0$. It is evident that both the coherent and incoherent photoproduction processes are very insensitive to differences between the densities.

8.6. ALPHA-DECAY

The emission of α-particles by radioactive nuclei was one of the first topics in nuclear physics to be elucidated by the use of quantum mechanics. Gamow (1928) and Condon and Gurney (1928, 1920) derived a relation between the kinetic energy of the α-particle and the radioactive decay constant in terms of the tunnel effect through a potential barrier. When combined with the knowledge obtained from α-particle scattering, these results indicated the behaviour of the nuclear and Coulomb potentials and provided estimates of nuclear sizes (Rutherford 1929).

The essential assumption of the elementary theory of α-decay is that the process may be separated into two parts – the formation of an α-particle at the nuclear surface and the penetration of the α-particle through the potential barrier. The decay constant λ can then be written as

$$\lambda = fP \tag{8.175}$$

where P is the probability of barrier penetration. The factor f contains the nuclear structure aspects of the process. It depends on the overlap of the initial and final nuclear wave-functions, and hence determines the relative decay rates for transitions to different final states.

When a simple WKB approximation is used to evaluate the barrier penetrability, the result can be expressed in the form

† See page 417.

$$P \sim e^{-G} \tag{8.176}$$

where G is given by

$$G = 2\left(\frac{2m}{\hbar^2}\right)^{1/2} \int \{U(r) - T\}^{1/2}\, dr\ , \tag{8.177}$$

T is the kinetic energy of the emitted α-particle, and the integration goes from one side of the barrier to the other. If the interaction between the α-particle and the residual nucleus is taken to be

$$\begin{aligned} V &= -V_0 & r &< R \\ &= 2Ze^2/R, & r &\geqslant R \end{aligned}$$

and it is assumed that the barrier is thick, the exponent can be written as

$$G \simeq A - BR^{1/2} \tag{8.178}$$

where A and B are constants independent of the nuclear size. Kaplan (1951) determined R for a range of heavy even-even nuclei by fitting the disintegration constants using the measured α-particle energies and assuming that the factor f in eqn (8.175) is unity. He found, as an average value,

$$r_0 = RA^{-1/3} = 1.57 \pm 0.015 \text{ fm}$$

where A is the mass number of the residual nucleus. The results showed some structure dependence. For example, for ^{214,216,218}Po he found $r_0 = 1.56$ fm, but for ^{212}Po which decays to ^{208}Pb he found $r_0 = 1.53$ fm. This radius R is a parameter of the α-nucleus interaction, and consequently is only indirectly connected with the nuclear matter distribution.

More sophisticated methods have been developed for the calculation of barrier penetrabilities to take account of the diffuse surface of the nuclear interaction and of the possibility of an asymmetric barrier in deformed nuclei. Some of these methods are reviewed by Preston (1962) and Davidson (1968). In early calculations the optical potential parameters obtained by Igo (1959) for 40 MeV α-particle scattering from ^{208}Pb were often used, although this led to some uncertainty over the ambiguities in the optical potential. Bencze and Sandulescu (1966) calculated a microscopic potential using Watanabe's method (see §7.4.1) and obtained a real depth of 231 MeV. For the transition $^{238}\text{Pu} \rightarrow {}^{234}\text{U}$ this increased the penetrability by a factor of 5 compared with results obtained by Poggenburg (1965) using a standard Saxon-Woods potential with a depth of 74 MeV. They also showed that the WKB method underestimates the penetrability by a factor of 2 to 5 compared with numerical integration of Schrödinger's equation. Other studies (Chaudhury 1963, Bencze 1967) suggest that the non-locality of the α-nucleus interaction can also increase the penetrability, although the non-locality parameters introduced in these calculations may not be realistic.

Studies of α-particles reduced widths have been mainly based on formulae derived from R-matrix theory. This yields the relations (Mang 1964)

$$\lambda = \Gamma/\hbar \tag{8.179a}$$

$$\Gamma = 2\gamma_\alpha^2\, P \tag{8.179b}$$

$$\mathrm{P} = \left\{\frac{kr}{F^2(r) + G^2(r)}\right\}_{r=R_0} \tag{8.179c}$$

where γ_α^2 is the α-particle reduced width and F,G are the solutions of Schrödinger's equation with the α-nucleus potential whose asymptotic behaviour corresponds to the regular and irregular Coulomb functions, respectively. The radius R_0 is a parameter of R-matrix theory. It should be taken sufficiently far from the centre of the residual nucleus so that there is no coupling between different final channels in the exterior region. In practice it is found that the calculated width Γ depends rather sensitively on R_0, and for calculations in heavy nuclei a value of ~ 8 fm has been preferred, even though Igo's potential gives a barrier beginning at 9-9.5 fm (Mang and Rasmussen 1962, Rasmussen 1963), and the use of even smaller radii has been proposed (Rauscher, Rasmussen, and Haroda 1967). A detailed study of the dependence of the α-particle widths for light nuclei on the channel radius in R-matrix theory has been carried out by Arima and Yoshida (1974). They find that the width defined by eqn (8.180*b*) decreases with increasing channel radius and converges on a constant value, provided that the resonance is narrow. For a broad resonance the formulae (8.179*b*,*c*) must be modified to take account of the energy shift of the resonance and the energy dependence of the resonance energy and width.

Many calculations have been carried out within the framework of R-matrix theory with special emphasis on the determination of the reduced width γ. Mang (1960) used the shell model for spherical nuclei and showed that the behaviour of the decay rates for the polonium isotopes could be associated with shell closure at ^{208}Pb. Harada (1961, 1962) studied the same nuclei and found that the inclusion of configuration mixing increases the reduced widths very substantially; this result has been confirmed by other calculations (Rasmussen 1963, Chang 1966). The calculation of γ has also been carried out for collective transitions in vibrational and rotational nuclei (Sandulescu 1962, 1963, Mang and Rasmussen 1962, Sandulescu and Dumitrescu 1965, Dumitrescu and Sandulescu 1967, Rauscher *et al.* 1967).

Harada (1961, 1962) found that the overlap of the wave-function of the initial nucleus with that of the residual nucleus and of the α-particle, which is a function of the distance between the centres of mass of the α-particle and the residual nucleus, peaks very sharply at a distance near to or about 0.5 fm less than the nuclear quadratic radius

$$Q = (5\langle r^2\rangle/3)^{1/2}$$

at least for heavy nuclei. Consequently, this overlap is evaluated at the channel radius in many calculations. Nevertheless, in the formalism of Mang and of Harada the evaluation of the reduced width still involves full integration over the relative co-ordinates of the nucleons in the α-particle, and for this reason a gaussian wavefunction for the α-particle and oscillator functions for the single-particle states in the initial nucleus are usually preferred. Rasmussen (1963) simplified the expression for the reduced width by introducing δ-functions in these relative co-ordinates. Unfortunately, this approximation gives a consistent over-estimation of the contribution of high-j orbitals in shell model theory.

The difficulties associated with the channel radius have led to attempts to reformulate the theory of α-decay in a manner which does not require an arbitrary division of configuration space into inner and outer regions. Harada and Rauscher (1968) have followed Feshbach's approach to low-energy nuclear reactions and obtain a prescription for the width Γ as the matrix element of the residual interaction between the initial wavefunction and the final wavefunction for the channel containing an α-particle moving in an optical potential. They carry out calculations for a point α-particle and meet difficulties owing to the ambiugities in the phenomenological potential. Kadmenskii and Kalechits (1971) have obtained an expression for the width in the diagonal case using the formalism for decaying states developed by Goldberger and Watson (1964). In this treatment the width Γ is given by the matrix element of the full nuclear interaction between the residual nucleus and the α-particle and the wavefunction for the emitted α-particle is generated from a point Coulomb potential, but it is still necessary to interpret the system in terms of two spatially separated regions. A review of these and other methods has recently been given and a new method introduced which avoids the use of arbitrary radii and ambiguous phenomenological potentials (Jackson and Rhoades-Brown 1976*b*). This method, which is based on a microscopic treatment of the alpha-nucleus interaction similar to that described in §7.4.4, leads to spectroscopic factors for the polonium isotopes 10^2-10^3 times larger than the predictions of the shell model, and these magnitudes are supported by another study (DeVries *et al.* 1976) which is based on R-matrix theory but removes much of the ambiguity in the potential. These results suggest that for these heavy nuclei the shell model, even with substantial configuration mixing, is far from adequate and attempts to compensate in *R*-matrix theory lead to the abnormally low values of the channel radii, discussed above.

It is evident that data on α-decay can provide some detailed information about the structure of the extreme surface region in nuclei but a unified microscopic theory of α-decay must be used. Such studies of α-decay combined with studies of α-particle scattering at low energies give very precise information about the barrier radius and the behaviour of the alpha-nucleus interaction in the barrier region.

TABLE 8.11

Values of $N(A,\sigma_a,\sigma_c)$ calculated for Thomas-Fermi distributions (TF) and for Beiner's phenomenological matter distribution (SW).†

σ_a [mb]		$\sigma_a = 0$			$\sigma_a = 17$		$\sigma_a = 27$
σ_c [mb]		17	27	39	27	39	39
^{27}Al	TF	16.5	13.3	10.7	7.76	6.17	4.85
	SW	16.7	13.6	11.1	8.26	6.73	5.46
^{90}Zr	TF	43.7	32.6	24.9	13.8	10.4	7.56
	SW	44.0	33.1	25.5	14.6	11.3	8.47
^{140}Ce	TF	43.7	32.6	24.9	13.8	10.4	7.56
	SW	62.3	45.8	34.8	17.8	13.5	10.0
^{208}Pb	TF	84.2	60.3	44.8	19.7	14.4	10.0
	SW	85.0	61.4	46.2	21.3	16.1	11.8

†Auger and Lombard 1973.

9
HADRONIC ATOMS

9.1. PIONIC ATOMS

Negatively charged pions, and other negatively charged hadrons such as the negative kaon and the antiproton, can be captured to form a hadronic atom. In many respects these hadronic atoms resemble the muonic atoms discussed in Chapter 4. The essential difference arises from the strong interaction between the hadron and the nucleus, which leads to an additional energy shift in the atomic energy levels for the hadron and to broadening of these levels associated with nuclear absorption of the hadron. The full interpretation of these effects in terms of nuclear sizes and nuclear structure clearly requires a considerable understanding of the hadron-nucleus interaction, and much of the initial work on the properties of hadronic atoms has been devoted to a study of this interaction.

9.1.1. *Basic properties of pionic atoms*

The initial capture of a pion into an atomic orbit is a complicated process dominated by the electromagnetic interaction. According to the simple Bohr theory, the binding energies, and radii are given by

$$B_n = m_\pi c^2 \frac{(Z\alpha)^2}{2n^2} \simeq 3.7\ Z^2\ \text{keV} \tag{9.1}$$

$$r_n = \frac{\hbar}{m_\pi c}\,\frac{n^2}{Z\alpha} \simeq 200\ Z^{-1}\ \text{fm} \tag{9.2}$$

where m_π is the pion mass, α is the fine-structure constant, and n is the principal quantum number. The total energy is $E = m_\pi c^2 - B$, as in eqn (4.4). The value of n for which the radius of the pion orbit is approximately equal to that of the K-shell electrons is $n \sim (m_\pi/m)^{1/2} \sim 16$. Hence, for pion orbits with $n \ll 16$ the pion is well inside the electron cloud so that there is no shielding of the nucleus by the electrons.

The pion falls from this highly excited state by non-radiative transitions (Auger emission) and by radiative X-ray transitions.

A more accurate expression for the energy may be obtained by including relativistic effects. Since the pion is a boson, this means solving the Klein-Gordon equation, which for a particle of mass m moving in an electrostatic potential V has the form

$$\{\hbar^2 c^2 \nabla^2 + (E - V)^2 - m^2 c^4\}\psi = 0. \tag{9.3}$$

For a pion moving in the field V_{pc} of a point nucleus of mass M the binding energies (neglecting relativistic recoil and other radiative corrections) are obtained by using the reduced mass $\mu_\pi = m_\pi/(1 + m_\pi/M)$ in the Klein-Gordon equation. This gives

$$B_{nl} = \mu_\pi c^2 \left\{1 - \left(1 + \frac{Z^2\alpha^2}{n'^2}\right)^{-1/2}\right\} \tag{9.4a}$$

$$\simeq \mu_\pi c^2 \left\{\frac{Z^2\alpha^2}{2n^2} + \frac{Z^4\alpha^4}{2n^4}\left(\frac{n}{l+\frac{1}{2}} - \frac{3}{4}\right) + \ldots\right\} \tag{9.4b}$$

where $n' = n-l-\frac{1}{2}+\{(l+\frac{1}{2})^2-Z^2\alpha^2\}^{1/2}$. The expressions (9.4$a$) and (9.4$b$) are not valid for s-states if $Z > 68$, in which case the Klein-Gordon equation has solutions with infinite binding energy. If V is the potential V_{C} for a nucleus of finite size this divergence problem disappears and the equation can be solved numerically. For small Z or large l the effect of the finite extension of the nuclear charge may be estimated using perturbation theory, which gives the energy shift as

$$\Delta B_{nl} = \int |\psi_{nl}|^2 (V_{\text{C}} - V_{\text{pC}})\, d^3 r \tag{9.5}$$

but in other cases the effect of the finite size of the nucleus cannot be regarded as a small perturbation.

The inclusion of the strong-interaction potential V_{N} is not straightforward, but it is customary to add it to V_{C} and to approximate its effect by neglecting the V_N^2 term and replacing $(E-V_C)V_N$ by $\mu_\pi c^2 V_N$, which gives

$$\{\hbar^2 c^2 \nabla^2 + (E - V_C)^2 - \mu_\pi^2 c^4\}\psi = 2\mu_\pi c^2 V_N \psi. \tag{9.6}$$

For pions this procedure is justified in terms of the relative weakness of the pion-nucleon force; it is less obvious that this justification can be made for kaons (Deloff and Law 1974*a*) but the same equation is used.

As in the case of muonic atoms the vacuum polarization correction is important since it leads to a change in the electrostatic potential due to the nucleus at distances of the order of the reduced Compton wavelength $\lambda\!\!\!^{-}_e$ of the electron, which is comparable with the radii of the pion orbits. To lowest (second) order the additional potential is given by (see §4.7)

$$V_{vp}(\mathbf{r}) = -\frac{2\hbar c}{3\pi} Z\alpha^2 \int \rho_{ch}(\mathbf{r}') \frac{K(|\mathbf{r} - \mathbf{r}'|)}{|\mathbf{r} - \mathbf{r}'|} d^3r' \tag{9.7}$$

where, for $r \ll \lambda_e$,

$$K(|\mathbf{r} - \mathbf{r}'|) = \frac{\log \lambda_e}{C|\mathbf{r}-\mathbf{r}'|} - \frac{5}{6} \tag{9.8}$$

with $C = 1.781$ (Uehling 1935, Schwinger 1949). The condition for the validity of eqn (9.8) is $n^2/Z \ll \frac{1}{2}$ so that it is not useful for very light elements or large n. The contribution to the energy is given approximately by

$$\varepsilon_{vac}(n) = \int |\psi_n|^2 V_{vp}(\mathbf{r}) d^3r \tag{9.9}$$

where ψ_n is the normalized pionic wavefunction in the presence of the strong interaction, i.e. the solution of eqn (9.6).

Krell and Ericson (1969) have evaluated the vacuum polarization corrections using eqns (9.7)-(9.9). The shifts are quite large and the calculations are time consuming. Fortunately, the contribution to ε_{vac} comes from regions outside the strong-interaction region, so that the effect of the strong interaction on the vacuum polarization correction is nearly model independent. They find that the change in ε_{vac} due to strong interaction effects is typically 1-2 per cent of the level shifts due to the strong interaction. Some values obtained for these shifts are given in Table 9.1. The effect of the higher-order diagrams for vacuum polarization yields a level shift of

$$\Delta B_n \simeq \alpha^3 Z^2 B_n (0.04188 + 0.01424\alpha^2 Z^2 F_0) \qquad (9.10)$$

where $1 < F_0 < 1.5$ for $Z < 100$ (Wichmann and Kroll 1956). This expression applies to point nuclei and is inaccurate for s-states, particularly for large Z.

Schafer (1967) estimated the energy shifts due to vacuum polarization for the 3d and 4f states[†] in calcium and titanium using the Klein-Gordon equation and the non-relativistic Schrödinger equation. The results given in Table 9.2 show that the differences between the two calculations are negligible. The magnitude in eV in Table 9.2 may be compared with the magnitude of other corrections to the 4f-3d transition energies calculated by Schafer and given in Table 9.3. The Lamb shift (second-order self-energy correction) is small because of the greater mass of the pion compared with the electron. The same situation arises in muonic atoms, as discussed in Chapter 4. A re-evaluation of the higher-order vacuum polarization corrections has been carried out (Backenstoss *et al.* 1973, Tauscher 1973) for the CERN measurements of the 5g-4f transition in iodine and the 6h-5g transition in thallium, and the results are given in Table 9.4. This has led to a fresh determination of the pion mass, which is consistent with the revised Berkeley value (Schafer 1973).

9.1.2 *X-ray transitions*

The energy levels of pionic atoms cannot be determined directly but the energy differences may be deduced from measurements of X-ray transition energies. Early measurements of pionic X-rays

---

[†]As in Chapter 4 we use atomic notation for n and l when referring to states of the hadronic atom.

TABLE 9.1

Strong vacuum polarization shifts (in keV)†

Element	$\delta\varepsilon_{vac}(1s)$	$\delta\varepsilon_{vac}(2p)$	$\delta\varepsilon_{vac}(3d)$
^{6}Li	-0.004	0	0
^{9}Be	-0.017	0	0
^{10}B	-0.032	0	0
^{11}B	-0.040	0	0
^{12}C	-0.064	0	0
^{14}N	-0.114	0	0
^{16}O	-0.184	0	0
^{18}O	-0.231	0	0
^{19}F	-0.309	0	0
^{23}Na	-0.590	0	0
^{32}S	-1.703	0.009	0
^{40}Ca	-2.993	0.027	0
^{48}Ti		0.036	0
^{51}V		0.042	0
^{52}Cr		0.054	0
^{55}Mn		0.060	0
^{56}Fe		0.075	0
^{59}Co		0.068	0
^{88}Sr		0.050	0.008
^{115}In		-0.627	0.049
^{181}Ta		-6.304	0.395

†From Krell and Ericson 1969. Parameters for strong interaction are those given in Table 9.9.

TABLE 9.2

Values for 4f–3d transitions in pionic calcium and titanium obtained using a reduced mass derived from $m_\pi c^2$ = 139.58 MeV†

Energy level	Vacuum polarization shift (ev)	
	Relativistic	Non-relativistic
Calcium 3d	-316.4	-315.1
Calcium 4f	-87.4	-87.2
Titanium 3d	-420.1	-418.1
Titanium 4f	-120.4	-120.0

†From Schafer 1967.

TABLE 9.3

Calculations of the 4f–3d pionic calcium and titanium transition energies with $m_\pi c^2$ = 139.580 MeV†

Effect	Transition energy (keV)	
	Calcium	Titanium
Klein-Gordon equation	72.388±0.001	87.622±0.001
Reduced mass	-0.270±0.001	-0.273±0.001
Vacuum polarization (second-order)	+0.230±0.002	+0.301±0.002
Vacuum polarization (fourth-order)	+0.002±0.002	+0.002±0.002
Strong-interaction shift	+0.002±0.001	+0.004±0.002
Orbital electron screening	-0.001±0.001	-0.001±0.001
Electromagnetic form factors	negligible	negligible
Lamb shift	negligible	negligible
Pionic atom recoil	negligible	negligible
Hyperfine effects	negligible	negligible
Calculated transition energy	72.351±0.003	87.655±0.004

†From Schafer 1967.

TABLE 9.4

Experimental energies and calculated corrections for transition in pionic iodine and thallium used for π^- mass determination[†]

	Iodine 5g-4f	Thallium 6h-5g
Measured energy (keV)	237.136±0.017	301.733±0.015
Fermi parameters c (fm)	5.86	6.45
t (fm)	2.00	2.50
Strong-interaction shift (keV)	0.025±0.002	0.008±0.000
Vacuum polarization total (keV)	1.037±0.010	1.324±0.010
Higher-order vacuum polarization and other corrections (keV)	-0.001	-0.021
Electron screening (keV)	-0.014±0.005	-0.075±0.005
Resulting π^- mass	139.563±0.012	139.572±0.009

[†] From Backenstoss *et al*. 1973.

have been summarized by Stearns (1957) and West (1958). Intensive accurate measurements have been made at Berkeley (Jenkins *et al.* 1966, Jenkins and Kunselmann 1966, Jenkins and Crowe 1966), at SREL (Harris *et al.* 1968, Miller *et al.* 1968, Eckhause *et al.* 1972, Sapp *et al.* 1972), and at CERN (Backenstoss *et al.* 1967*a*, 1976, Poelz *et al.* 1968, Schmitt *et al.* 1968, Koch *et al.* 1968, 1969, von Egidy and Povel 1974).

As the pion cascades to lower orbits the energy is released by X-ray emission or electron emission (Auger effect). The transition probabilities for Auger emission and for electric dipole radiative transitions are respectively (Burbidge 1953, Backenstoss 1970)

$$W_{\mathrm{A}} = \frac{2\pi}{\hbar} |\langle \psi_{\pi}^{\mathrm{f}}\phi_{\mathrm{e}}^{\mathrm{f}}| r_{\mathrm{e}\pi} |\psi^{\mathrm{i}}\phi_{\mathrm{e}}^{\mathrm{i}}\rangle|^2 \tag{9.11}$$

$$W_{\mathrm{X}} = \frac{4e^2(\Delta E)^3}{3\hbar^4 c^3} |\langle \psi_{\pi}^{\mathrm{f}}| r |\psi_{\pi}^{\mathrm{i}}\rangle|^2 \tag{9.12}$$

where $\psi_{\pi}, \phi_{\mathrm{e}}$ are the pion and electron wavefunctions in the initial state i or the final state f, $r_{\mathrm{e}\pi}$ is the distance between the electron and the pion, and ΔE is the transition energy. In both cases the selection rule is $\Delta l = \pm 1$, with $\Delta l = -1$ favoured. The probability for Auger emission is greatest when there is a large overlap between the electron and pion wavefunctions, i.e. for large n and small binding energies, and is most important at the upper end of the cascade process.

Ericson (1969) argues on dimensional grounds that

$$W_{\mathrm{X}} \propto r_n^2(\Delta E)^3 \propto Z^4 \tag{9.13}$$

since $B_n \propto Z^2$ and $r_n \propto Z^{-1}$. If it is assumed that the strong interaction can be represented by a complex square well and that perturbation theory applies at least approximately, the corresponding energy shift and line width are given by

$$\varepsilon_{nl} - \mathrm{i}\Gamma_{nl} \propto \int_0^R |\psi_{nl}|^2 \, r^2 \mathrm{d}r. \tag{9.14}$$

On dimensional grounds Ericson (1969) takes the pion wavefunction to be

$$\psi_{nl} \propto r_n^{-3/2}(r/r_n)^l$$

at small distances, which gives

$$\varepsilon_{nl} \propto Z^{4(2l+3)/3}, \qquad \Gamma_{nl} \propto Z^{4(2l+3)/3}. \tag{9.15}$$

This very strong dependence on Z in the higher states implies that, for these states, the energy shifts and level widths are difficult to measure for low Z. The experimentally observed Z dependence is illustrated in Figs. 9.1 and 9.2.

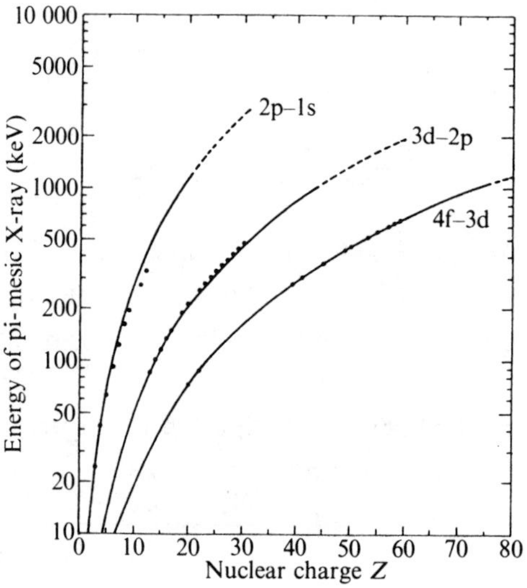

FIG. 9.1. X-ray transition energies in pionic atoms. The dots represent the experimental points and the curves represent the predicted values for a point nucleus. (From Lock and Measday 1970).

Hyperfine splitting of the X-ray lines of pionic tantalum (Carrigan *et al.* 1967) and of pionic holmium (Ebersold *et al.* 1974) has been observed. The hyperfine structure was assumed to be due to interaction with the nuclear quadrupole moment. There is also an energy shift due to static and dynamic interactions with the nuclear moments. The formalism for this is essentially the same as that for muonic atoms, which is

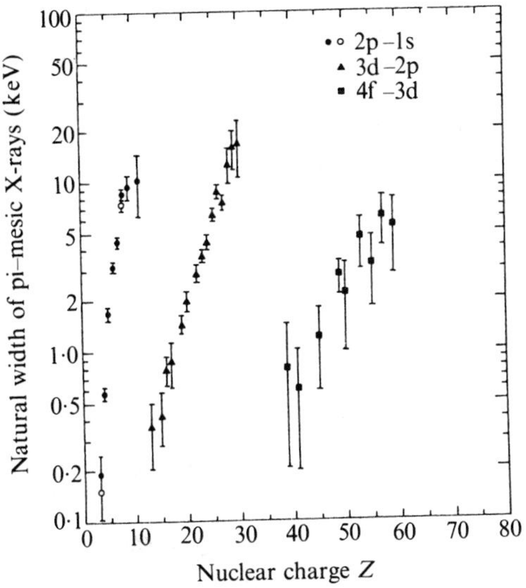

FIG. 9.2. The widths of X-ray lines in pionic atoms. For the 2p-1s transitions, the open circles correspond to a lighter isotope. (From Lock and Measday 1970.)

discussed in Chapter 4.

9.1.3. *Strong-interaction effects*

The strong interaction between the pion and the nucleus causes the pionic levels to be shifted in energy and broadened. The change in energy and line broadening of a transition $n'l'$-nl is dominated by the energy shift ε_{nl} and width Γ_{nl} of the lower level, although the most recent work (Koch 1973) has taken into account the width of both the lower and the upper level. The broadening of the levels is associated, in lowest order, with the removal of the pion by absorption on one or, more probably, two nucleons through the process

$$\pi + N + N \rightarrow N + N. \quad (9.16)$$

This reduces the lifetime of the pion in the atomic orbit to the point where the widths of the levels may be as much as the order of keV.

Because pions are removed from higher levels by nuclear absorption the yield for radiative transitions to lower levels is correspondingly decreased. The rate of nuclear absorption in the upper level can be deduced by measuring the X-ray yield Y for transitions from this level. For the transition $n'l'$-nl this is given by

$$Y_{n'l',nl} = P_{n'l'} \frac{\Gamma^{\mathrm{rad}}_{n'l',nl}}{\Gamma_{n'l'} + \Gamma^{\mathrm{abs}}_{n'l'}} \tag{9.17}$$

where $\Gamma^{\mathrm{rad}}_{n'l',nl}$ is the partial width for radiative emission, $\Gamma_{n'l'}$ is the total width for electromagnetic transitions including Auger emission, and $\Gamma^{\mathrm{abs}}_{n'l'}$ is the width for nuclear absorption. The population probability $P(n'l')$ must be obtained from cascade calculations or observations on other X-ray transitions. The widths corresponding to electromagnetic transitions can be calculated from the transition rates given in eqns (9.11) and (9.12), and it follows that the radiative widths are proportional to Z^4. Eqn (9.15) gives an estimate of the Z dependence for the strong-interaction width. Comparison of these estimates indicates that the low l transitions are increasingly difficult to observe as Z increases, and experimental confirmation of this is evident in Fig. 9.2.

An estimate of the strong-interaction shift can be obtained by considering pion-nucleus scattering at zero energy (Ericson 1969). This gives

$$\varepsilon_{nl} - i\Gamma_{nl} \propto Z^{2l+3} \lim_{k\to 0} (\delta^l/k^{2l+1}). \tag{9.18}$$

The experimental results indicate that ε_{nl} is negative for $l = 0$ (i.e. the binding energy is decreased) and positive for $l > 0$. Hence the pion-nucleus interaction is repulsive for $l = 0$ leading to a negative phase shift, while for $l > 0$ the potential is attractive and $\delta_l > 0$.

An extensive analysis of the experimental data obtained up to 1969 has been carried out by Krell and Ericson (1969) using eqn (9.6) with a non-local potential of the form

(Ericson and Ericson 1966)

$$2\mu_\pi c^2 V_N(r) = q(r) - \underset{\sim}{\nabla}\alpha(r)\underset{\sim}{\nabla} \tag{9.19}$$

$$q(r) = -4\pi\hbar^2 c^2 [p_1 b_0 A\rho_m(r) + p_1 b_1\{N\rho_n(r) - Z\rho_p(r)\} + i\, p_2 \text{Im } B_0 A^2\, \rho_m^2(r)] \tag{9.20}$$

$$\alpha(r) = \alpha_0(r)\{1 - \tfrac{1}{3}\,\xi\alpha_0(r)\}^{-1} \tag{9.21}$$

$$\alpha_0(r) = -4\pi\hbar^2 c^2\{p_1^{-1} c_0 A\rho_m(r) + i\, p_2^{-1}\, \text{Im}C_0 A^2\rho_m^2(r)\} \tag{9.22}$$

where b_0, b_1, Im B_0, Im C_0 and c_0 are constants associated with the pion-nucleon interaction, and p_1, p_2 are kinematical factors. The correlation parameter ξ is derived from a consideration of the polarization of the nuclear medium (Lorentz-Lorenz effect). The imaginary part of the potential is derived from the two-nucleon absorption process (eqn 9.16) with the approximation that the two-nucleon distribution function is uncorrelated. The results obtained using a two-parameter Fermi distribution for $\rho_m(r)$ and taking $\rho_p = \rho_n$ are given in Tables 9.5-9.8.

The theoretical results obtained by Krell and Ericson for the shifts of the 2p-1s transitions are compared in Fig. 9.3 with the experimental values. The theoretical results are changed very little when the gradient part of the interaction and the absorption terms are omitted. This indicates that the 1s shift is essentially determined by the local part of the interaction which is repulsive and is derived in this model from the s-wave part of the pion-nucleon interaction. The 1s widths are also independent of the gradient term. In Fig. 9.4 the same results are displayed in a modified form to reduce the strong dependence on Z; this figure illustrates the importance of the isovector term proportional to b_1. As the neutron excess increases the shifts become more repulsive. The 2p shifts are plotted on the reduced scale in Fig. 9.5. The attractive gradient term, derived from the p-wave part of the pion-nucleon interaction, is essential to describe

TABLE 9.5

Shifts and widths of the 1s states in pionic atoms[†]

Element	$(E^{(2p)}_{point} - E^{(1s)}_{point})$ ‡	$\langle r^2\rangle^{1/2}$ (fm)	t (fm)	$\varepsilon^{(1s)}_{ch}$	$\varepsilon^{(1s)}_{th}$ §	$\varepsilon^{(1s)}_{exp}$	$\Gamma^{(1s)}_{th}$	$\Gamma^{(1s)}_{exp}$
$^{6}_{3}Li$	24.56	2.43	1.00	-0.05	-0.51	-0.38±0.06[a]	0.14	0.15±0.05[a]
$^{7}_{3}Li$	24.65	2.37	1.00	-0.05	-0.74	-0.64±0.06[a]	0.17	0.19±0.05[a]
$^{9}_{4}Be$	44.09	2.40	1.00	-0.16	-2.00	-1.77±0.05[a]	0.56	0.58±0.05[a]
$^{10}_{5}B$	69.12	2.45	1.00	-0.40	-3.51	-3.18±0.18[b]	1.30	1.27±0.25[b]
$^{11}_{5}B$	69.21	2.42	1.00	-.039	-4.30	-4.23±0.18[b]	1.38	1.87±0.25[b]
$^{12}_{6}C$	99.93	2.42	1.00	-0.81	-6.80	-6.99±0.15[b]	2.89	2.96±0.25[b]
$^{14}_{7}N$	136.48	2.48	1.77	-1.58	-11.77	-11.74±0.15[b]	5.19	4.48±0.30[b]
$^{16}_{8}O$	178.80	2.71	1.83	-3.16	-19.43	-18.85±0.25[b]	6.80	7.56±0.50[b]
$^{18}_{8}O$	178.98	2.77	1.87	-3.30	-23.85	-23.97±0.25[b]	6.27	8.67±0.70[b]
$^{19}_{9}F$	227.02	2.85	2.00	-5.51	-32.58	-31.12±0.5 [b]	9.21	9.4 ±1.5 [b]

$^{23}_{11}Na$	340.94	2.94	2.00	-12.62	-63.37	-64.74±1.0 [b]	17.9	10.3 ±4.0 [b]
$^{32}_{16}S$	731.66	3.12	2.50	-57.61	-203.5		61.1	
$^{40}_{20}Ca$	1158.97	3.41	2.70	-149.1	-419.2		99.1	

†From Krell and Ericson 1969. Notation: energies in keV; ε_{ch} is the expected electromagnetic shift in the absence of strong interactions; ε_{th} and Γ_{th} are the calculated full shift (including finite-size effect) and width according to the parameters of Table 9.9; ε_{exp} and Γ_{exp} are corresponding experimental values; $\langle r^2 \rangle$ means $\langle r^2 \rangle$ charge.

‡Includes vacuum polarization.

§Includes the modification of the vacuum polarization by strong interaction (see Table 9.1).

[a] Harris *et al.* (1968).

[b] Backenstoss *et al.* (1967*a*).

TABLE 9.6

Shifts and widths of the 2p states in pionic atoms[†]

Element	$(E^{(3d)}_{point} - E^{(2d)}_{point})$[‡]	$\langle r^2 \rangle^{1/2}$ (fm)	t (fm)	$\varepsilon^{(2p)}_{ch}$	$\varepsilon^{(2p)}_{th}$[§]	$\varepsilon^{(2p)}_{exp}$	$\Gamma^{(2p)}_{th}$	$\Gamma^{(2p)}_{exp}$
$^{14}_{7}N$	25.064	2.48	1.77	0	0.006		4.0 eV	2.1±0.3 eV[a]
$^{16}_{8}O$	32.797	2.71	1.83	0	0.012		7.5 eV	4.7±0.8 eV[a]
$^{18}_{8}O$	32.831	2.77	1.87	0	0.012		8.2 eV	3.8±0.7 eV[a]
$^{19}_{9}F$	41.504	2.85	2.00	0	0.023		16 eV	11.2±1.9 eV[a]
$^{23}_{11}Na$	62.295	2.94	2.00	0	0.069		56 eV	34.6±7.6 eV[a]
$^{27}_{13}Al$	87.20	2.91	2.28	-0.01	0.17	0.20 0.10	0.18	0.36±0.15 [b]
$^{31}_{15}P$	116.33	3.07	2.45	-0.02	0.36	0.45 0.10	0.41	0.43±0.15 [b]
$^{32}_{16}S$	132.46	3.12	2.50	-0.02	0.52	0.60 0.10	0.59	0.79±0.15 [b]
$^{40}_{20}Ca$	207.72	3.41	2.70	-0.11	1.52	1.94 0.18	2.03	2.00±0.25 [b]
$^{48}_{22}Ti$	251.83	3.56	2.20	-0.20	1.94	2.15 0.20	3.37	2.89±0.25 [b]

$^{51}_{23}V$	275.50	3.58	2.10	-0.26	2.26	2.35±0.20	4.43	3.66±0.25[b]
$^{52}_{24}Cr$	300.21	3.66	2.30	-0.36	2.74	2.55±0.25	5.41	4.46±0.35[b]
$^{55}_{25}Mn$	326.04	3.72	2.20	-0.48	2.98	3.08±0.25	6.69	6.38±0.40[b]
$^{56}_{26}Fe$	352.92	3.75	2.50	-0.63	3.65	3.51±0.30	8.34	8.65±0.60[b]
$^{59}_{27}Co$	380.93	3.83	2.50	-0.84	3.72	3.81±0.35	9.91	7.37±0.70[b]
$^{88}_{38}Sr$	762.49	4.14	2.30	-7.12	-3.68		55.2	
$^{115}_{49}In$	1283.87	4.49	2.30	-35.2	-70.8		148.5	
$^{181}_{73}Ta$	2956.21	5.50	2.80	-404.0	-885.0		353.0	

†From Krell and Ericson 1969. Notation as in Table 9.5.

‡Includes vacuum polarization.

§Includes the modification of the vacuum polarization by strong interaction.

[a]Koch *et al*. (1968, 1969).

[b]Poelz *et al*. (1968).

TABLE 9.7

Shifts and widths of the 3d states in pionic atoms[†]

Element	$(E^{(4f)}_{\text{point}} - E^{(3d)}_{\text{point}})$‡	$\langle r^2 \rangle^{1/2}$ (fm)	t (fm)	$\varepsilon^{(3d)}_{\text{ch}}$	$\varepsilon^{(3d)}_{\text{th}}$	$\varepsilon^{(3d)}_{\text{exp}}$	$\Gamma^{(3d)}_{\text{th}}$	$\Gamma^{(3d)}_{\text{exp}}$
$^{115}_{49}$In	440.31	4.49	2.30	-0.08	2.1	2.6±0.5	2.5	2.6±0.6[a]
$^{122}_{51}$Sb	477.52	4.67	2.40	-0.14	2.7		3.3	
$^{127}_{53}$I	516.28	4.74	2.00	-0.19	3.5	4.5±0.8	4.7	4.4±1.5[a]
$^{133}_{55}$Cs	556.62	4.80	2.50	-0.28	4.3	5.4±1.5	5.9	3.3±1.5[a]
$^{139}_{57}$La	598.54	4.86	2.50	-0.39	5.2	6.4±2.0	7.7	6.2±2.0[a]
$^{141}_{59}$Pr	642.04	4.88	2.50	-0.51	6.6	6.1±2.0	10.1	5.4±2.5[a]

† From Krell and Ericson 1969. Notation as in Table 9.5.

‡ Includes vacuum polarization.

§ Includes the modification of the vacuum polarization by strong interaction.

[a] Schmitt *et al*. (1968).

TABLE 9.8

Shifts and widths of the 4f states in pionic atoms

Element	$(E^{(5g)}_{point} - E^{(4f)}_{point})$‡	$\langle r^2\rangle^{1/2}$ (fm)	t (fm)	$\varepsilon^{(4f)}_{ch}$	$\varepsilon^{(4f)}_{th}$ §	$\varepsilon^{(4f)}_{exp}$	$\Gamma^{(4f)}_{th}$	$\Gamma^{(4f)}_{exp}$
$^{181}_{73}Ta$	453.03	5.50¶	2.80	-0.01	0.48	0.87±0.2	0.34	0.5±0.2[a]
$^{197}_{79}Au$	531.87	5.32	2.32	-0.02	0.90	1.29±0.2	0.85	1.0±0.3[a]
$^{209}_{83}Bi$	588.11	5.50	2.10	-0.04	1.45	1.95±0.3	1.46	1.7±0.5[a]
$^{232}_{90}Th$	693.68	5.80¶	2.80	-0.13	2.89	4.32±0.6	3.01	6.0±0.9[b]
$^{232}_{90}Th$	693.68	5.80¶	2.00	-0.11	3.01	4.32±0.6	3.36	6.0±0.9[b]
$^{232}_{90}Th$	693.68	5.70¶	2.00	-0.10	2.93	4.32±0.6	3.39	6.0±0.9[b]
$^{238}_{92}U$	725.54	5.87¶	2.80	-0.17	3.38	5.96±1.1	3.74	6.1±1.0[b]
$^{239}_{94}Pu$	758.14	5.85¶	2.80	-0.20	4.15	8.06±1.6	4.61	9.1±2.5[b]

† From Krell and Ericson 1969. Notation as in Table 9.5.

‡ Includes vacuum polarization.

§ Includes the modification of the vacuum polarization by strong interaction

¶ Strongly deformed nuclei; r.m.s. radii and diffuseness t chosen arbitrarily.

[a] Schmitt *et al.* (1968). [b] Jenkins and Kunselmann (1966).

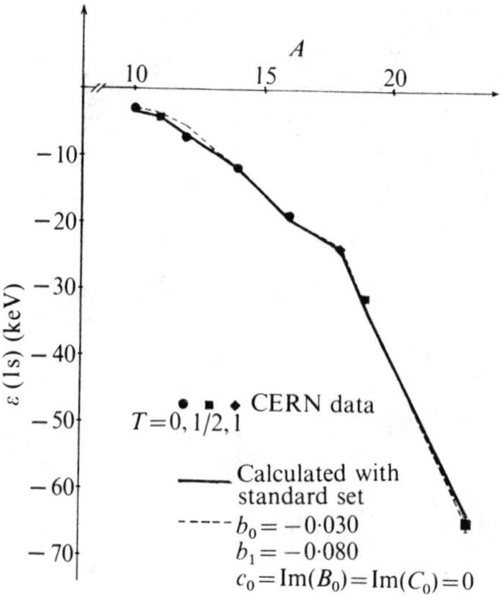

FIG. 9.3. The 1s strong-interaction shift in pionic atoms. The solid curve corresponds to calculated values using the potential (9.19) while the broken curve corresponds to calculated values obtained when the absorptive parts of the potential are omitted, i.e. Im B_0 = Im C_0 = 0. (From Krell and Ericson 1969.)

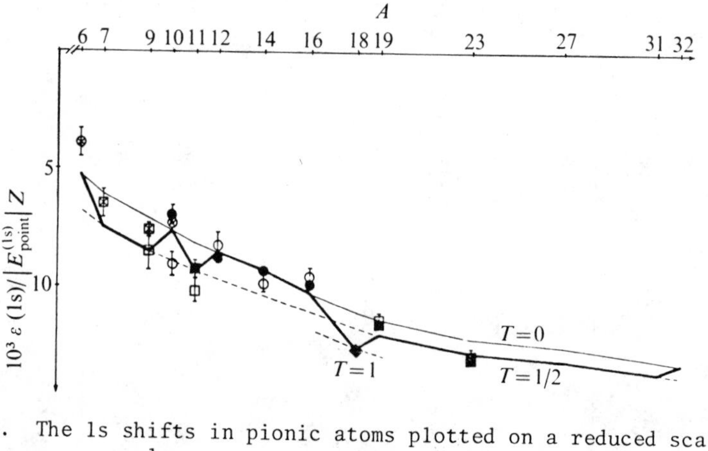

FIG. 9.4. The 1s shifts in pionic atoms plotted on a reduced scale. The lines marked T = 0, $\frac{1}{2}$, 1, respectively, connect nuclei of the same isospin. The solid heavy line is calculated with the full potential (9.19) while the lighter line corresponds to calculated values obtained by omitting the isovector term, i.e. with b_1 = 0. (After Krell and Ericson 1969.)

these attractive shifts. However, the calculations indicate that the 2p shift may also become repulsive for a value of $Z \sim 36$, when the repulsive local term becomes important. It is argued (Ericson, Ericson, and Krell 1969) that this is essentially a nuclear size effect. For the higher states the predicted shifts are somewhat low for the heavier nuclei. Krell and Ericson (1969) suggest that this could be due to neglect of the isospin dependence of the gradient term in the pion-nucleus interaction (Ericson and Ericson 1966). If so, more accurate data should be sensitive to differences between the proton and neutron distributions.

Eqn (9.17) has been used by Sapp *et al.* (1972) to obtain absorption widths for the upper level in 2p-1s transitions in 1p-shell nuclei. The population probabilities and capture widths range from 68 per cent and 0.015 ± 0.004 eV in ^{6}Li to 57 per cent and 12 ± 4 eV in ^{16}O. The data can be satisfactorily described with the optical potential (9.19) and standard parameters. Egidy and Povel (1974) obtain a 2p width for ^{16}O of 11 ± 6 eV, in agreement with Sapp *et al.*, and a strong-interaction shift of 4.1 ± 2.3 eV.

Anderson, Jenkins, and Powers (1970) used the same form for the potential but allowed the r.m.s. radii of the proton and neutron distributions to differ. By varying the parameters of the potential to obtain the best fit to data for a wide range of nuclei they obtained the result

$$\langle r^2\rangle_n^{1/2} - \langle r^2\rangle_p^{1/2} = -0.01 \pm 0.16 \text{ fm.} \tag{9.23}$$

Thus, in terms of the average behaviour of a range of nuclei, they find no significant difference between the r.m.s. radii. Kunselmann and Grin (1970) have examined data for the calcium isotopes and, on the assumption of equal r.m.s. radii in ^{40}Ca, they obtained for ^{44}Ca

$$\langle r^2\rangle_p^{1/2} = 3.55 \text{ fm}, \qquad \langle r^2\rangle_n^{1/2} = 3.9 \pm 0.15 \text{ fm.}$$

This difference in r.m.s. radii is somewhat in excess of that predicted by the Hartree-Fock calculations discussed in Chapter

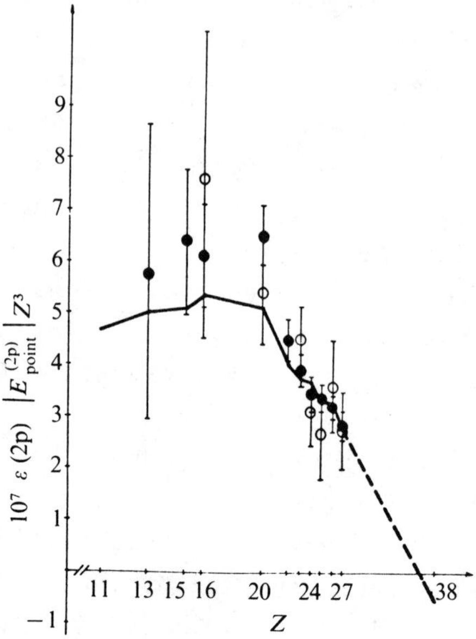

FIG. 9.5. The 2p strong-interaction shift in pionic atoms. The solid line corresponds to calculations with the potential (9.19). (After Krell and Ericson 1969.)

2. There has not been a study of precisely what moments of the nuclear matter distribution contribute to the various transitions in pionic atoms. Because of the modification of the pion wavefunctions by the strong interaction (Krell and Ericson 1969) there is no reason to suppose that the results obtained for muonic atoms are relevant to the pionic case.

9.1.4. *The pion-nucleus interaction at low energies*

It is evident that interpretation of data for pionic atoms is intimately connected with the study of the pion-nucleus interaction at low energies, and that a reliable knowledge of the interaction is essential for the extraction of nuclear size information. In the low-energy region the pion-nucleon scattering amplitude can be written as

$$f(\mathbf{k},\mathbf{k}') = b_0 + b_1 \mathbf{t}_\pi\cdot\boldsymbol{\tau} + [c_0 + c_1\, \mathbf{t}_\pi\cdot\boldsymbol{\tau}]\mathbf{k}\cdot\mathbf{k}' + i f_3\, \boldsymbol{\sigma}\cdot\mathbf{k}'\wedge\mathbf{k} \tag{9.24}$$

where the coefficients can be expressed in terms of the s-wave scattering lengths α_{2T} and the p-wave scattering lengths $\alpha_{2T,2J}$ as

$$b_0 = \frac{1}{3}(\alpha_1 + 2\alpha_3), \qquad b_1 = -\frac{1}{3}(\alpha_1 - \alpha_3) \tag{9.25}$$

$$c_0 = \frac{1}{3}(\alpha_{11} + 2\alpha_{13} + 2\alpha_{31} + 4\alpha_{33}) \tag{9.26}$$

$$c_1 = -\frac{1}{3}(\alpha_{11} + 2\alpha_{13} - \alpha_{31} - 2\alpha_{33}). \tag{9.27}$$

Neglecting spin, this leads to a low-energy optical potential of the form (Kisslinger 1955)

$$\frac{2\mu}{\hbar^2} V(r) = -4\pi\left[b_0 A\rho_{\text{m}}(r) + b_1\{N\rho_{\text{n}}(r) - Z\rho_{\text{p}}(r)\}\right] + + 4\pi\, \underset{\sim}{\nabla}\left[c_0 A\rho_{\text{m}}(r) + c_1\{N\rho_{\text{n}}(r) - Z\rho_{\text{p}}(r)\}\right]\underset{\sim}{\nabla}. \tag{9.28}$$

A number of corrections, mentioned in the previous section, lead to the potential due to Ericson and Ericson given in eqns (9.19)-(9.22). An alternative approach leads to the local Laplacian potential discussed in §10.4.1.

Jenkins and Kunselmann (1966) have analysed their level-shift data using the potential (9.28), and Koltun (1969) has compared these parameters with those obtained from pion-nucleon scattering. Both sets of values are given in Table 9.9 together with the parameters obtained by Krell and Ericson using the potential (9.19). The isoscalar coefficient b_0 must be increased to fit the data for pionic atoms; this effect appears to arise from the effective field correction in multiple scattering (Ericson 1969, Koltun 1969). Estimates of the parameters Im B_0 and Im C_0 depend on experimental information on the process (9.16) and its inverse. Krell and Ericson (1969) derive predicted values of

$$\text{Im}\, B_0 = 0.0174 \quad \mu^{-4}, \qquad \text{Im}\, C_0 = 0.073 \quad \mu^{-6}$$

which are a factor of 2 lower than the phenomenological values given in Table 9.9.

TABLE 9.9

Pion-nucleon scattering parameters†

	π-N scattering	Pionic atoms (Kisslinger potential)	Pionic atoms (Ericsons potential)
b_0 (μ^{-1})	-0.012 ± 0.004	-0.0197 ± 0.0004	-0.030
b_1 (μ^{-1})	-0.097 ± 0.007	-0.064 ± 0.013	-0.080
c_0 (μ^{-3})	0.21 ± 0.01	0.131 ± 0.011	0.24
c_1 (μ^{-3})	0.18 ± 0.01	-0.018 ± 0.090	—
Im B_0 (μ^{-4})	—	—	0.040
Im C_0 (μ^{-6})	—	—	0.14

†In units of μ^{-1} = 1.415 fm.

The formulae (9.19) and (9.28) omit any explicit consideration of the finite size of the pion. Iachello and Lande (1971) have examined the contribution of the size effect in pionic atoms using perturbation theory and find level shifts of 1-4 per cent of the strong-interaction shifts for the 2p-1s transition. For Z = 8 the effect of the pion finite size is of the same order of magnitude as the lowest-order correction for vacuum polarization, and could therefore have an effect on the extraction of parameters of the pion-nucleus potential.

9.2. KAONIC ATOMS

9.2.1. *Basic properties of kaonic atoms*

In kaonic atoms, as in other mesic atoms, the meson is captured into a highly excited atom with a principal quantum number n ~ 60-100. As the kaons cascade down the Auger process initially dominates. It is assumed that the electrons are stripped from the atom and not replaced, so that all the electrons have been removed by the time the kaon orbit has the same radius as the K-shell electrons. The kaonic atom then de-excites by electric dipole transitions until the strong interaction with the nucleus becomes effective and nuclear absorption occurs. The selection rules favour population of circular orbits with $l = n - 1$, and it is usually assumed that by the time nuclear absorption takes place approximately 90 per cent of the kaons are in circular orbits. This is the theory of Wightman (Bethe and de Hoffmann 1955) on which most calculations are based, but the assumptions have been challenged (Rook 1963, 1971, Wilkinson 1968) and are discussed further in §9.2.3.

The position and widths of the energy levels in kaonic atoms are determined in a manner similar to that for pionic atoms and the calculations are subject to the same corrections. However, there are a number of essential differences between the pionic and kaonic case. The ratios of the pion and kaon masses to the electron mass are, respectively, 273 and 968 and hence, using the simple Bohr formula (9.2), the kaonic radii for a given n are substantially smaller. The value of n at which the radius of the kaon orbit is comparable with that of the K-shell electrons is $n \sim 30$. Further, the kaon-nucleon interaction is stronger than the pion-nucleon interaction. These two effects imply that the presence of nuclear absorption should be evident at larger values of n than is the case for pionic atoms, and the strong-interaction shifts and widths should be larger.

As in the pionic case the radiative probability depends on Z^4, while the probability for nuclear absorption depends more strongly on Z. Typical transition probabilities are plotted in Fig. 9.6. It can be seen that for some value of Z, say Z_c, the transition probabilities for a given orbit, say n_0, are comparable; these are the points marked on the figure. It follows that for $Z > Z_c$ the X-ray transition $n_0 \rightarrow n_0 - 1$ is heavily suppressed and may even not be observed. Experimental confirmation of this effect is given in Table 9.10, which contains results from the earliest observations of kaonic X-rays from light nuclei (Burleson *et al.* 1965, Wiegand and Mack 1967). The yield Y gives the number of transitions per kaon captured. It can be seen from the values of Y that in ^{12}C about 29 per cent of all the captured kaons reach the 3d state but the 3d-2p X-ray transition from this state is too weak to be observed, implying that nuclear absorption is

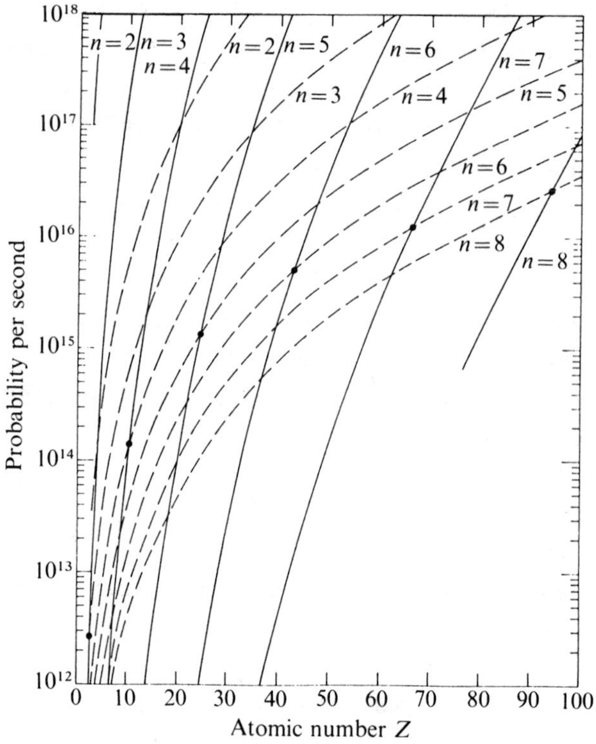

FIG. 9.6. X-ray transition probabilities for circular orbits $(n,n-1)$ in kaonic atoms (broken lines) and probabilities for nuclear absorption from the same orbits (solid lines). The dots indicate the values of Z at which both processes are equally probable for a given n. (After Wilkinson 1968.)

entirely dominant for the 3d state. For boron just over a quarter of the kaons which reach the 3d state undergo a radiative transition to the 2p state, and as Z decreases the yield increases.

9.2.2. *The kaon-nucleon and kaon-nucleus interactions*
The mass, isospin, and strangeness of the kaon states are listed in Table 9.11. Since the baryon number is zero, the hypercharge is equal to the strangeness, and because of the difference in the hypercharge for positive and negative kaons

TABLE 9.10

Yields of K-mesonic X-rays

Isotope	Transition	Estimated energy (keV) (Klein-Gordon)	Yield	Ref.
^{4}He	3d-2p	6.7 ± 0.2	$0.1 < Y_L < 1.0$	a
^{4}He	2p-1s	34.7 ± 0.3	$Y_K = 0.2\ Y_L$	a
^{7}Li	3d-2p	15.3	0.21 ± 0.07	b
^{9}Be	3d-2p	27.6	0.15 ± 0.06	b
B	4f-3d	15.2	0.3 ± 0.15	b
B	3d-2p	43.6	0.07 ± 0.03	b
^{12}C	6f-3d	37.8	0.04 ± 0.02	b
^{12}C	5f-3d	32.2	0.04 ± 0.02	b
^{12}C	4f-3d	22.0	0.21 ± 0.05	b
^{12}C	3d-2p	63.0	< 0.04	b

[a] Burleson *et al.* 1965.
[b] Wiegand and Mack 1967.

TABLE 9.11

The K-meson states

Particle	t	t_3	S	Mass (MeV)
K^+	$\frac{1}{2}$	$\frac{1}{2}$	+1	493.9
K^0	$\frac{1}{2}$	$-\frac{1}{2}$	+1	497.8
$\bar{K}^0$	$\frac{1}{2}$	$+\frac{1}{2}$	-1	497.8
K^-	$\frac{1}{2}$	$-\frac{1}{2}$	-1	493.9

the K^+N and K^-N systems have different properties. Thus the K^+p and K^+n total cross-sections are featureless but the K^-N cross-sections exhibit resonant behaviour (Bransden 1969). The K^-p system forms two resonant states: the $Y_0^*(1405)$ is an iso-singlet $S_{1/2}$ state with binding energy of ~27 MeV compared with the K^-p threshold at 1432 MeV, and the Y_1^* (1385) is an iso-triplet $P_{3/2}$ state with binding energy of ~47 MeV. Both resonances decay by the strong interaction, the former into $\Sigma\pi$ with $T = 0$ and the latter predominantly into $\Lambda\pi$. The widths of the states are 35 MeV and 50 MeV respectively. From isospin considerations the Y_0^* cannot appear in the K^-n channel.

In the low-energy KN scattering there is no evidence of a p-wave contribution for centre-of-mass energies below 38 MeV. Kim (1965, 1967) and Martin and Sakitt (1969) have analysed the data using the multi-channel reaction matrix method of Dalitz and Tuan (1960). Martin and Sakitt use a zero-range method, while Kim uses an effective-range method which yields the scattering lengths A_T

$$A_0 = -1.674 + i\,0.722 \text{ fm} \qquad (9.29a)$$

$$A_1 = -0.003 + i\,0.688 \text{ fm}. \qquad (9.29b)$$

Kim (1967) also suggests that the Y_1^* is due to final-state interactions and should not be regarded as a bound state.

The dominant mode of nuclear absorption of kaons involves a single nucleon through inelastic processes of the form

$$K^- + N \rightarrow Y + \pi \qquad (9.30)$$

where Y stands for a Σ or Λ hyperon. These processes are listed in full in Table 10.7. They release sufficient energy to overcome the binding of the kaon and nucleon. In the nucleus there is also a small probability of absorption on two nucleons; this process is discussed in §9.2.4.

In view of the dominance of absorption on a single nucleon, it is reasonable to construct a kaon-nucleus optical potential which is proportional to the nuclear density in both real and imaginary parts. In his early calculations, Rook (1962, 1963) took the imaginary part of the kaon-nucleus potential to be of the form

$$|W(r)| = \frac{1}{2}\hbar\, v\, \sigma_{KN}\, A\, \rho_m(r) \tag{9.31}$$

where v is the relative kaon-nucleus velocity and σ_{KN} is the averaged total cross-section. This formula can be extended to allow for differences in the proton and neutron distributions to give (Burhop 1967, Rook 1968*b*)

$$|W(r)| = \frac{1}{2}\hbar v\{\sigma_{Kp}\, Z\rho_p(r) + \sigma_{Kn}\, N\rho_n(r)\} \tag{9.32}$$

$$= 2\pi\, \hbar^2\, m_K^{-1}\{\frac{1}{2}\, \mathrm{Im}(A_0 + A_1)\, Z\rho_p(r) + \mathrm{Im}\, A_1\, N\rho_n(r)\} \tag{9.33}$$

where we have used the optical theorem to obtain the second line. Burhop has taken the relative velocity for a kaon at rest and a nucleon of momentum† 160 MeV/c and the corresponding cross-sections are

$$\sigma_{Kp} = 338 \text{ mb}, \qquad \sigma_{Kn} = 155 \text{ mb}. \tag{9.34}$$

By analogy with the s-wave part of the pion-nucleus potential (9.28) Ericson and Scheck (1970) take the potential

$$2\mu_K\, c^2\, V_N = -4\pi\, \hbar^2 c^2\, p_1\{\frac{1}{2}(A_0 + A_1)\, Z\rho_p(r) + A_1\, N\rho_n(r)\} \tag{9.35}$$

where μ_K is the kaon reduced mass and $p_1 = (1 + m_K/m_N)$ is a kinematic factor. With the parameters (9.29) this potential has a repulsive real part. Ericson and Scheck (1970) and

---

†It has been pointed out (Jackson 1968, Rook 1968*b*) that this nucleon monentum is rather high compared with the value for nucleons in the nuclear surface obtained from the (p,2p) reaction (see Chapter 8) or predicted from the single-particle model. This implies (Bethe and Siemens 1970) that the estimated proton cross-section at threshold is too high.

Krell (1971) have used p-wave potentials similar to the gradient form for pions, but find that this additional term gives a negligible correction to the level shifts and widths.

When a nucleon is bound in a nucleus, the formation of the Y_0^* becomes energetically possible in the K^-p channel. Taking the K^-p system at rest as the zero of energy, the energy of relative motion is given by (Burhop 1971)

$$E_{\text{rel}} = -\varepsilon_{\text{N}} - \tfrac{1}{2}P^2/(m_{\text{K}} + m_{\text{N}})$$

where ε_{N} is the nucleon binding energy and P is the sum of the kaon and nucleon momenta which, in an impulse approximation, is equal to the sum of the $\Sigma\pi$ momenta and can be determined by measuring these momenta and the opening angle (Bruxelles Collaboration 1963). Estimates of E_{rel} for a range of nuclei (Burhop 1971) yield a value close to -27 MeV, which is the position of the Y_0^* resonance. Bardeen and Torigoe (1971) obtained similar estimates of -10 to -30 MeV.

Formation of the Y_0^* may be expected to have a profound influence on the kaon-nucleon interaction and to dominate over elastic scattering (Uretsky 1967, Bloom, Johnson, and Teller 1969). This is evident from the plot of K^-p and K^-n scattering amplitudes given in Fig. 9.7. It can be seen that in both the Kim and Martin-Sakitt solutions the K^-p amplitude is varying rapidly below threshold.

Bethe and Siemens (1970) take the effect of the resonance into account by replacing the threshold value of A_0, given by (9.29*a*), by a Breit-Wigner form

$$A_0 \sim \frac{\tfrac{1}{4}\Gamma^2}{\{E - M(Y_0^*)\}^2 + \tfrac{1}{4}\Gamma^2}.$$

Bardeen and Torigoe (1971, 1972) derive energy-averaged† K^-N cross-sections using the formula

$$\bar{f}_{\text{KN}} = \int f_{\text{KN}}(W)\ P(W)\ \mathrm{d}W \qquad (9.36)$$

---

†This procedure is essentially the same as the Fermi-averaging procedure adopted in the calculation of pion optical potentials (see §10.4).

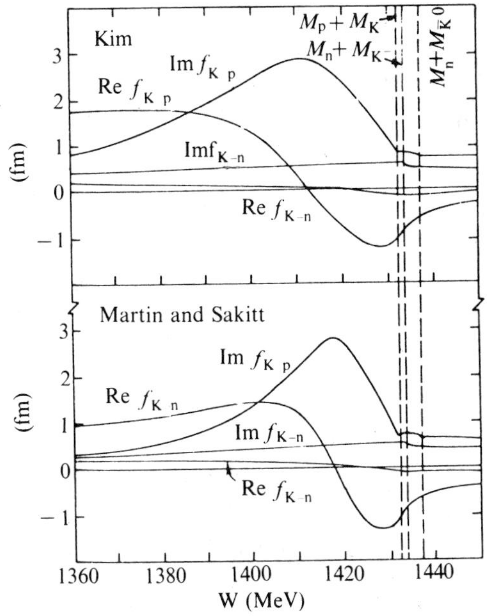

FIG. 9.7. The real and imaginary parts of the kaon-nucleon scattering amplitudes plotted against centre-of-mass energy W for the Kim and Martin-Sakitt solutions. (From Bardeen and Torigoe 1971.)

where $P(W)$ is the probability of finding the kaon-nucleon system with a total centre-of-mass energy given by

$$W = m_K + m_N + E_{rel}.$$

This integral is evaluated using a single-particle model of the nucleus. The values obtained are slowly varying with the range of nuclei studied (^{16}O, ^{40}Ca, ^{96}Mo), and consequently these energy-averaged cross-sections have been used to determine constant effective potential strengths for use in further calculations. Their potential then consists of two terms proportional to the proton and neutron densities respectively, as in eqn (9.35), but with modified coefficients.

Chattarji and Ghosh (1973) have attempted to take the effects of the Y_0^* into account by making an analytic continuation of the amplitudes from free KN scattering and then averaging these amplitudes over the nucleon momentum distribution which is taken to be of gaussian form. This also yields a potential essentially of the form (9.35) but with modified coefficients.

Wycech (1971) has extended the method of Dalitz and Tuan (1960) to go off the energy shell and to describe the kaon-nucleon interaction inside the nucleus. In general the off-shell effects are not very important because the range of the KN interaction is ~ 0.25 fm which is much less than the average internucleon distance. However, the propagation of the Y_0^* in the nuclear medium has a significant effect, and even at low density the width of the resonance is increased as a result of collisions with other nucleons. There is a marked difference between the modified scattering amplitude for K^-p scattering and the free amplitude. The results indicate that the K^-p interaction will dominate over the K^-n interaction if the proton and neutron densities are equal, but, because of the exclusion principle, collisions between the Y_0^* and another proton are less probable than collisions with a neutron, so that if there is a neutron excess the decay of the Y_0^* is enhanced and the probability of capture on protons is reduced. The kaon-nucleus optical potential becomes a non-linear function of the density owing to the effect of the Y_0^* and the real part is essentially zero in the capture region.

The potential of Wycech can be approximated by the formula (Rook and Wycech 1972)

$$A\ V(r) = -2(U_p + iW_p)\ Z\ f(r) - 2(U_n + iW_n)(A - Z)\ f(r) \tag{9.37a}$$

where $f(r)$ is a Saxon-Woods radial function with $R = 1.07\ A^{1/3}$, $a = 0.54$ fm, and

$$
\begin{aligned}
U_p &= 35 \text{ MeV} && \text{for } f(r) > 0.09 \\
&\quad -50 \text{ MeV} && \text{for } f(r) < 0.01 \\
&\quad 161 + 1060\, f(r) \text{ MeV} && \text{otherwise} \\
W_p &= 110 \text{ MeV} && \text{for } f(r) > 0.2 \qquad (9.37b) \\
&\quad 12.6 + 47.4\, f(r) \text{ MeV} && \text{otherwise} \\
U_n &= 7 \text{ MeV} \\
W_n &= 27 \text{ MeV.}
\end{aligned}
$$

A modified set of parameters is

$$
\begin{aligned}
U_p &= 35 \text{ MeV} && \text{for } f(r) > 0.2 \\
&\quad -50 \text{ Mev} && \text{for } f(r) < 0.05 \qquad (9.37c) \\
&\quad -78 + 566\, f(r) && \text{otherwise}
\end{aligned}
$$

Alberg, Henley, and Wilets (1973) have derived a non-local kaon-nucleus optical potential using separable two-body potentials of Yamaguchi form, whose parameters are determined by comparison with Kim's analysis of KN scattering, and an independent pair model for the nucleus. They define an equivalent local potential and also derive a local potential using a local density approximation. In the interior region the real parts of both potentials are attractive; in the capture region the LDA potential remains attractive while the equialent local potential is slightly repulsive.

Deloff and Law (1974*a*) argue that the optical model approach is inappropriate for the description of bound states in a kaonic atom. They construct a kaon-nucleus potential by folding a finite-range complex kaon-nucleon potential into the nuclear density distribution. They fix the strength of the gaussian KN potential from the scattering lengths and determine the range by fitting K-^{4}He scattering. The folded potential for heavier nuclei then has no free parameters if the nuclear distributions are assumed to be known, but the shape of the potential is not the same as the density distribution.

9.2.3. *X-ray transitions and strong-interaction effects*
Theoretical interpretation of the measured X-ray yields for given transitions involves evaluation of the transition rates for X-ray and Auger emission and for nuclear absorption. The formulae for the first two are given by eqns (9.12) and (9.11) respectively, while the transition rate for nuclear absorption is usually taken to be

$$W_{\mathrm{N}} = \frac{2W}{\hbar} \int |\psi_{\mathrm{K}}(\underset{\sim}{r})|^2 \rho_{\mathrm{m}}(r)\, \mathrm{d}^3 r \qquad (9.38)$$

where W is the strength of the imaginary part of the kaon-nucleus potential. Using the expression (9.31) for the potential, the transition rate becomes

$$W_{\mathrm{N}} = \sigma_{\mathrm{KN}}\, v\, A \int |\psi_{\mathrm{K}}(\underset{\sim}{r})|^2 \rho_{\mathrm{m}}(r)\, \mathrm{d}^3 r. \qquad (9.39)$$

Bethe and Siemens (1970) also write

$$W_{\mathrm{N}} = B\, A \int |\psi_{\mathrm{K}}(\underset{\sim}{r})|^2 \rho_{\mathrm{m}}(r)\, \mathrm{d}^3 r \qquad (9.40)$$

which defines B in relation to the other coefficients as

$$B = 2W/\hbar A. \qquad (9.41)$$

The values obtained for B using the various methods discussed in the preceding section are listed in Table 10.9.

There are a number of approximations involved in the derivation of eqn (9.38) which have been examined by Rook (1968*a*). In perturbation theory the transition rate is given by

$$W_{\mathrm{N}} = \frac{2\pi}{\hbar} |M|^2 \rho(E)$$

where M is the matrix element for the transition (9.30) and $\rho(E)$ is the density of states for the $\mathrm{Y}\pi$ system at energy E. The matrix element can be written as

$$M = \langle \phi_{\pi}\, \phi_{\mathrm{Y}}\, \Phi_{\mathrm{f}} | g | \Phi_{\mathrm{i}}\, \psi_{\mathrm{K}} \rangle \qquad (9.42)$$

where g is the interaction operator and Φ_i, Φ_f are the initial and final nuclear wavefunctions. With the assumption that g is a point interaction depending only on channel spin, taking plane waves for the hyperon and pion, and taking a single-particle model for the nucleus, Rook obtains the transition rate

$$W_N = \sum_{\substack{\text{all}\\ \text{nucleons}}} \sigma_{KN}\, C \int \sum_l (l_K 0\; l_N 0 | l 0)^2 |I(P)|^2 \left(\frac{2\varepsilon M}{\hbar^2} - P^2\right) P^2 \mathrm{d}P \qquad (9.43a)$$

where

$$I(P) = \int j_l(Pr)\, \psi_{l_K}(r)\, u_{jl_N}(r)\, r^2 \mathrm{d}r\,, \qquad (9.43b)$$

C is a kinematic factor, P is a mean value for the sum of the hyperon and pion momenta, M is the sum of the hyperon and pion masses, and ε is the difference between the Q-value of the reaction and the recoil energy of the residual nucleus. The factor g^2 has been replaced by the averaged cross-section σ_{KN} and u_{jl_N} is the radial part of the single-particle wavefunction for the bound nucleon. If it is now assumed that the Q-value is large and the final states form a complete set, eqn (9.43) becomes

$$W_N \simeq \sum_{\substack{\text{all}\\ \text{nucleons}}} \frac{\sigma_{KN} v}{4\pi} \left(\frac{Q - E_{jl_N}}{Q}\right)^{1/2} \int |\psi_{l_K}(r)|^2\, u^2_{jl_N}(r)\, r^2 \mathrm{d}r \qquad (9.44)$$

$$\simeq Z\, \sigma_{Kp}\, v \int |\psi_{l_K}(r)|^2\, \rho_p(r)\, \mathrm{d}^3 r + N\, \sigma_{Kn}\, v \int |\psi_{l_K}(r)|^2\, \rho_n(r)\, \mathrm{d}^3 r \qquad (9.45)$$

where E_{jl_N} is the binding energy for a nucleon in state jl_N.

Rook (1968*a*) has compared the transition rates obtained using eqns (9.43*a*,*b*) and (9.44) for selected nuclei. The results are given in Table 9.12 together with the parameters of the single-particle potentials used to generate the u_{jl_N}. The square root in eqn (9.44) has been replaced by unity, and this probably accounts for the increased discrepancy between the two calculations for the larger values of Z. The magnitude

TABLE 9.12

Results of comparison of eqns (9.43) and (9.44)[†]

Nucleus		Well depths (MeV)		Meson state		Transition probabilities (s^{-1})		Ratio
A	Z	protons	neutrons	n_K	l_K	Eqn (9.43)	Eqn (9.44)	
				2	1	2.83×10^{-4}	3.43×10^{-4}	0.82
12	6	55	53	3	2	7.44×10^{-8}	9.06×10^{-8}	0.82
				4	3	8.89×10^{-12}	1.08×10^{-11}	0.82
				4	3	1.21×10^{-2}	1.61×10^{-2}	0.75
				5	4	0.99×10^{-4}	1.34×10^{-4}	0.74
107	47	65	57	6	5	3.95×10^{-7}	5.42×10^{-7}	0.73
				7	6	8.87×10^{-10}	1.23×10^{-9}	0.72

[†] From Rook 1968*a*.

of the discrepancy is not very sensitive to l_K. If the closure approximation is not valid, eqn (9.43) can be re-written in the form

$$W_N \propto \int \psi_{l_K}(r)\, u_{jl_N}(r)\, W(r,r')\, \psi_{l_K}(r')\, u_{jl_N}(r')\, r^2 dr\, r'^2 dr' \qquad (9.46)$$

which is equivalent to introducing a non-local optical potential. Rook (1962) represented the non-locality by the expression

$$u_{jl_N}(r)\, W(r,r')\, \psi_{l_K}(r') = \int \frac{\exp(-\mu|\underset{\sim}{r}-\underset{\sim}{r}'|)}{\mu|\underset{\sim}{r}-\underset{\sim}{r}'|}\, u_{jl_N}(r')\, \psi_{l_K}(r)\, d\hat{\underset{\sim}{r}}\, d\hat{\underset{\sim}{r}}' \qquad (9.47)$$

The results obtained for emulsion nuclei (Z = 41, A = 93) suggest that this effect is not of great importance. However, a more recent study of $W(r,r')$ (Rook 1968*a*) indicates that it is an oscillatory function and not sharply localized, so that the choice (9.47) may not be realistic.

Formulae such as (9.44) can be used to examine the

relative importance of absorption on nucleons in different subshells. This has been done by Aslam and Rook (1970*b*) for ^{208}Pb using the parameters (9.34), and their results are given in Table 9.13. It can be seen that the difference in the proton and neutron binding energies is relatively unimportant compared with the ratio of $\sigma_{Kp}/\sigma_{Kn} = 2.15$. Although the subshells with the small values of binding energy tend to dominate, the degree depends on $2j+1$, i.e. the number of nucleons in the subshell, and there are important contributions from all the subshells with binding energies in the range 7-15 MeV. Thus, estimates of absorption rates derived from approximate nuclear wavefunctions or density distributions which are fitted, in the surface region, to the exponential fall for the least-bound nucleon may be in error but not seriously so.

As noted in §9.2.1 X-ray transitions lead to population of circular orbits because the dipole selection rule for a transition from state $n_1 l_1$ to $n_2 l_2$ requires $l_2 = l_1 - 1$. The dipole matrix element favours transitions with $n_2 = n_1 - 1$ but the presence of the factor $(\Delta E)^3 = \{E(n_1) - E(n_2)\}^3$ favours the largest possible change in n. The Auger effect, on the other hand, favours small changes in n. For the emulsion nucleus $Z = 41$, $A = 93$ Rook (1963) has calculated the X-ray, Auger, and absorption rates for a meson starting in any one of the angular momentum states corresponding to $n = 15$. Table 9.14 lists the quantum numbers of the state reached when the nuclear absorption dominates. Thus if the meson was already in a circular orbit $n = 15$, $l = 14$ the cascade process would continue through circular orbits until nuclear absorption took place and the last observed X-ray would be for the transition 6h-5g. Alternatively, if the low angular momentum states were strongly populated at $n = 15$ the nuclear absorption would take place in an s-state and no circular orbits would have been populated. An intermediate situation would arise if the population at the state $n = 15$ were statistical, i.e. the population $P(l)$ were proportional to $2l + 1$, and in this case the nuclear absorption is largest from the circular orbit. The radial dependence of the absorption

TABLE 9.13

Relative absorption of a K^- meson in the 7i state by different bound states of protons and neutrons in ^{208}Pb†

Nuclear state			Neutron		Proton		
n	l	j	Binding energy (MeV)	Rate ($\times 10^{14}\ s^{-1}$)	Binding energy (MeV)	Rate ($\times 10^{14}\ s^{-1}$)	Ratio p/n
3	1	$\frac{1}{2}$	7.4	65.2			
2	3	$\frac{5}{2}$	8.9	111.7			
3	1	$\frac{3}{2}$	9.7	98.2			
1	6	$\frac{13}{2}$	11.6	156.9			
2	3	$\frac{7}{2}$	13.7	109.4			
1	5	$\frac{9}{2}$	14.7	66.2			
3	0	$\frac{1}{2}$	21.4	17.7	8.5	41.3	2.3
2	2	$\frac{3}{2}$	22.6	28.0	9.9	64.8	2.3
1	5	$\frac{11}{2}$	23.4	82.8	12.7	180.7	2.0
2	2	$\frac{5}{2}$	26.0	40.4	14.2	89.3	2.2
1	4	$\frac{7}{2}$	28.2	29.6	18.0	66.9	2.3
1	4	$\frac{9}{2}$	32.6	46.2	24.4	98.4	2.1
2	1	$\frac{1}{2}$	35.0	7.4	27.6	15.1	2.0
2	1	$\frac{3}{2}$	36.8	15.3	30.3	30.5	2.0
1	3	$\frac{5}{2}$	40.1	13.9	33.6	29.9	2.1
1	3	$\frac{7}{2}$	43.3	23.1	37.7	48.7	2.1
2	0	$\frac{1}{2}$	47.4	4.2	40.9	8.8	2.1
1	2	$\frac{3}{2}$	50.6	5.6	44.8	12.8	2.3
1	2	$\frac{5}{2}$	55.6	9.7	48.0	22.8	2.4
1	1	$\frac{1}{2}$	63.2	1.5	57.1	3.8	2.5
1	1	$\frac{3}{2}$	64.7	3.4	58.6	8.4	2.5
1	0	$\frac{1}{2}$	71.0	0.7	64.5	2.1	3.0

† From Aslam and Rook 1970*b*.

TABLE 9.14

Kaonic states (n'l') at which nuclear absorption occurs for a cascade starting in the state (15,l)[†]

n = 15, l	0	2	4	6	8	9	10	11	12	13	14
Capture n'	12	12	11	10	11	10	9	8	7	6	5
State l'	0	0	0	1	3	4	4	4	4	4	4

[†]From Rook 1963.

probability, the integrand of eqn (9.38), is plotted in Fig. 9.8 for the three cases discussed above.

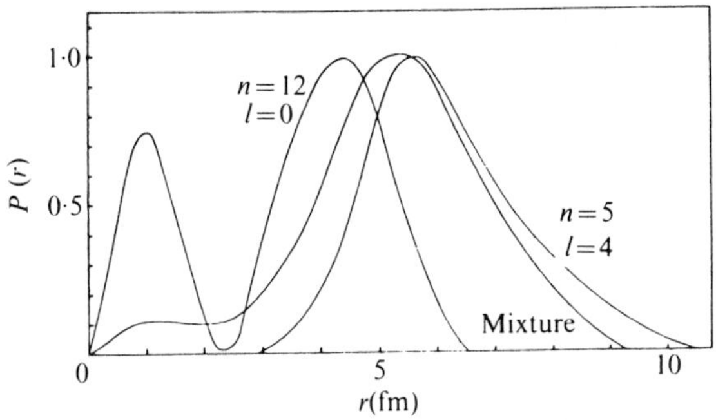

FIG. 9.8. The radial dependence of the nuclear absorption probability corresponding to different assumptions about the population $P(l)$ of the states with n = 15. (From Rook 1963.)

More elaborate cascade calculations have been carried out (Eisenberg and Kessler 1963, Martin 1963, Rook 1970) usually starting at $n \sim 100$. Rook assumes a statistical distribution and for Z in the range 17-92 obtains values for percentage population of circular orbits between 25 and 69 per cent. More recently, Rook (1971) has challenged the statistical assumption and has proposed an empirical distribution of the form $(2l + 1)\exp(-al)$ where a is a positive constant.

The yields for light nuclei have been studied by Rook

(1968*b*) using eqn (9.45) with Burhop's parameters (9.34). The theoretical results obtained assuming that all the electrons have been removed from the electron shells are given in column A of Table 9.15, and are to be compared with the experimental results given in Table 9.10. If it is assumed that

TABLE 9.15

Theoretical kaonic X-ray yield per meson in light nuclei†

Nucleus	Transition	Yield per meson		
		A	B	C
^{7}Li	3d-2p	0.47	0.40	0.17
^{9}Be	3d-2p	0.26	0.25	0.13
B	4f-3d	1.00	0.19	0.20
B	3d-2p	0.16	0.16	0.09
^{12}C	4f-3d	1.00	0.27	0.20
^{12}C	3d-2p	0.08	0.08	0.05

†From Rook 1968*b*.

the electron shells are completely refilled so that Auger transitions compete with X-ray transitions, the results of column B are obtained and are in substantially better agreement with the data. Column C is obtained by neglecting the Auger transitions, assuming that only 20 per cent of the mesons decay from circular orbits, and reducing the nuclear absorption rate by a factor of 5. The remaining discrepancy for ^{12}C can be removed by allowing this nucleus to be deformed. Intermediate situations between the two extremes represented by columns B and C are possible, although it is difficult to justify a reduction in the nuclear absorption rate in view of the small values of B^{p} and B^{n} derived from Burhop's parameters compared with other calcualtions. However, these results do lend support to the argument that 100 per cent decay by circular orbits is not feasible. For medium and heavy nuclei Auger transitions do not compete with X-ray transitions from the lowest states because the Auger rate depends only weakly on Z while the X-ray rate goes as Z^4.

Wiegand (1969) has measured X-ray yields from a range of medium and heavy nuclei. He observes transitions with $n_2 = n_1 + 1$ and assumes that they correspond to transitions between circular orbits. He also determines the critical or cut-off value $Z_c(n_0)$ at which the transition $n_0 \rightarrow n_0 - 1$ is suppressed. His calculated values of the nuclear absorption rate are too small to reproduce these values correctly, but this is due to the choice of very small values for B^p and B^n. Aslam and Rook (1970*b*) have calculated the absorption rates in a similar manner but using a single-particle model and the parameters (9.34). Except for ^{208}Pb, the theoretical results given in Table 9.16 are generally satisfactory compared with Wiegand's data given in Table 9.17 and can be improved slightly by inclusion of nuclear deformation. Bardeen and

TABLE 9.16

Theoretical kaonic X-ray yields per meson†

Nucleus	Transition	Theoretical X-ray yields without deformation	Theoretical X-ray yields with deformation	Deformation parameter β
^{35}Cl	4f-3d	0.12	0.12	0
^{58}Ni	5g-4f	0.58	0.55	0.187
^{98}Mo	6h-5g	0.73	0.70	0.168
^{158}Gd	7i-6h	0.71	0.59	0.358
^{208}Pb	8j-7i	0.91	0.91	0
^{238}U	8j-7i	0.63	0.53	0.292

†From Aslam and Rook 1970*b*.

Torigoe (1971) have also made calculations using their effective potential strengths and Fermi distributions; their results are displayed in Fig. 9.9 and are again generally

TABLE 9.17

Experimental and theoretical yields (columns (4) and (5)) for the transition $(n+1,\ n) \to (n,n-1)$ and capture and radiation widths for $(n,n-1)$ (columns (6) and (7)) in keV†

Z (1)	Element (2)	Transition $(n+1,\ l+1) \to (n,l)$ (3)	Y_{exp} (4)	Y_{th} (5)	Γ_{nl}^{abs} (6)	Γ_{nl}^{rad} (7)	Vacuum polarization (8)	E_{K-G} + vacuum polarization (9)
17	Cl	5g-4f			7.60×10^{-3}	7.23×10^{-4}	0.39	84.68
		4f-3d	0.083 ± 0.039	0.09	5.09	3.40×10^{-3}	1.16	183.37
20	Ca	6h-5g			5.52×10^{-5}	4.28×10^{-4}	0.24	63.72
		5g-4f	0.57 ± 0.20	0.89	4.43×10^{-2}	1.39×10^{-3}	0.60	117.51
28	Ni	7i-6h			4.08×10^{-6}	6.38×10^{-4}	0.286	75.62
		6h-5g	0.67 ± 0.40	0.99	2.79×10^{-3}	1.65×10^{-3}	0.61	125.60
		5g-4f	0.50 ± 0.35	0.37	1.11 ×	5.37×10^{-3}	1.43	231.69
30	Zn	7i-6h			1.26×10^{-5}	8.42×10^{-4}	0.35	86.94
		6h-5g	1.20 ± 0.25	0.99	7.57×10^{-3}	2.18×10^{-3}	0.73	144.40
34	Se	7i-6h			8.80×10^{-5}	1.39×10^{-3}	0.49	111.88
		6h-5g	1.0 ± 0.7	0.94	3.91×10^{-2}	3.61×10^{-3}	1.02	185.85
42	Mo	9k-8j			6.24×10^{-9}	7.35×10^{-4}	0.25	75.96

		8j-7i			3.90×10^{-6}	1.46×10^{-3}	0.46	110.92
		7i-6h			1.69×10^{-3}	3.25×10^{-3}	0.86	171.13
		6h-5g	0.50 ± 0.24	0.66	4.54×10^{-1}	8.42×10^{-3}	1.74	284.35
56	Ba	9k-8j			1.58×10^{-6}	2.33×10^{-3}	0.57	135.48
		8j-7i			5.16×10^{-4}	4.63×10^{-3}	0.99	197.87
		7i-6h	1.0 ± 0.47	0.90	1.11×10^{-1}	1.03×10^{-2}	1.81	305.40
64	Cd	10ℓ-9k			3.80×10^{-8}	2.17×10^{-3}	0.50	126.59
		9k-8j			1.37×10^{-5}	3.98×10^{-3}	0.82	177.18
		8j-7i	1.00 ± 0.35		3.48×10^{-3}	7.92×10^{-3}	1.41	258.82
		7i-6h	0.50 ± 0.16	0.70	5.67×10^{-1}	1.76×10^{-2}	2.54	399.57
74	W	10ℓ-9k			7.77×10^{-7}	3.89×10^{-3}	0.74	169.50
		9k-8j			2.12×10^{-4}	7.12×10^{-3}	1.21	237.28
		8j-7i	1.0 ± 0.9	0.97	3.89×10^{-2}	1.42×10^{-2}	2.04	346.68
80	Hg	10ℓ-9k			5.03×10^{-6}	5.31×10^{-3}	0.92	198.27
		9k-8j			1.06×10^{-3}	9.73×10^{-3}	1.48	277.58
		8j-7i	0.67 ± 0.20	0.90	1.51×10^{-1}	1.94×10^{-2}	2.49	405.63
82	Pb	12n-11m			7.13×10^{-11}	2.10×10^{-3}	0.41	117.01
		11m-10ℓ			2.49×10^{-8}	3.42×10^{-3}	0.63	153.99
		10 -9k			6.78×10^{-6}	5.87×10^{-3}	0.98	208.37

Table 9.17 continued

Z (1)	Element (2)	Transition $(n+1,\ l+1) \to (n,l)$ (3)	Y_{exp} (4)	Y_{th} (5)	Γ^{abs}_{nl} (6)	Γ^{rad}_{nl} (7)	Vacuum polarization (8)	E_{K-G} + vacuum polarization (9)
		10ℓ-9k			6.78×10^{-6}	5.87×10^{-3}	0.98	208.37
		9k-8j			1.36×10^{-3}	1.07×10^{-2}	1.58	291.72
		8j-7i	0.25 ± 0.14	0.89	1.85×10^{-1}	2.14×10^{-2}	2.65	426.32
92	U	12n-11m			2.38×10^{-9}	3.33×10^{-3}	0.57	147.45
		11m-10ℓ			5.66×10^{-7}	5.42×10^{-3}	0.86	194.07
		10ℓ-9k			1.05×10^{-4}	9.31×10^{-3}	1.33	262.64
		0k-8j			1.43×10^{-2}	1.71×10^{-2}	2.12	367.78
		8j-7i	0.40 ± 0.20	0.55	1.30	3.40×10^{-2}	3.53	537.62

†Column (8) shows the vacuum polarization correction to the transition energies; column (9) gives the full energy difference $E_{n+1,n} - E_{n,n-1}$ in keV for a point-like nucleus and including the vacuum polarization.

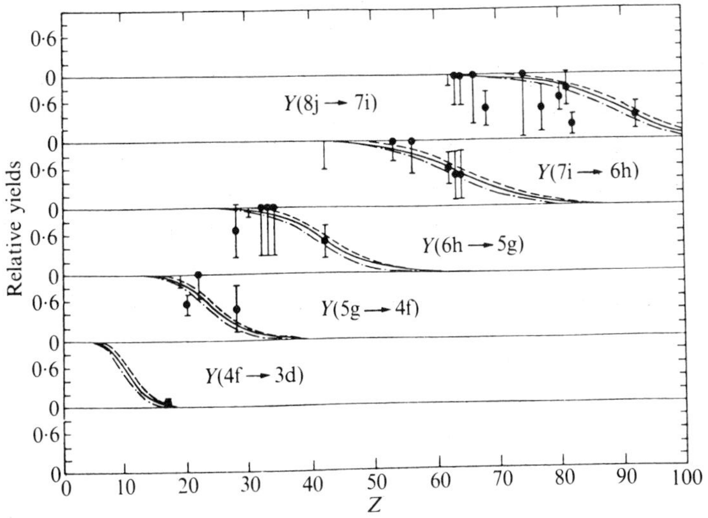

FIG. 9.9. The observed relative X-ray yields compared with the calculated values for the Martin-Sakitt solution (broken line), and the Kim solution (solid line). The dot-dashed line corresponds to a calculation with Kim's solution but with the neutron distribution in the extreme surface region increased by a factor of 3. (After Bardeen and Torigoe 1971.)

satisfactory except for ^{208}Pb. Ericson and Scheck (1970) have used the optical potential (9.35) with the threshold parameters (9.29) and Fermi distributions to calculate the theoretical yields, widths, and energy differences including vacuum polarization corrections. Some of these results are given in Table 9.17. The theoretical yield is given by

$$Y_{\text{th}}(n+1, l+1 \rightarrow nl) = \frac{\Gamma^{\text{rad}}_{n+1,l+1}}{\Gamma^{\text{rad}}_{n+1,l+1} + \Gamma^{\text{abs}}_{n+1,l+1}} \tag{9.48}$$

Thus, each value of Y_{th} in the table is calculated from the values of Γ^{rad}, Γ^{abs} given in the line above. As in the other calculations, there is a disagreement for the 8j-7i transition in ^{208}Pb.

The theory of Wycech, discussed in the previous section, suggests that the real part of the optical potential may be near to zero in the capture region. Ericson and Scheck have investigated this using the potential (9.35) but varying Re $\bar{A}$, where $\bar{A}$ is the averaged coefficient when $\rho_p = \rho_n$. The effect on the yield is not very pronounced but the absorption width of the lower level is more strongly affected. This is illustrated in Fig. 9.10. The measurement by Backenstoss *et al.* (1970*a*) of the strong-interaction shift for the 4f-3d transition in ^{32}S allows further investigation of this point, and the theoretical and experimental results are given in Table 9.18. It is evident that the free parameters do not describe the data.

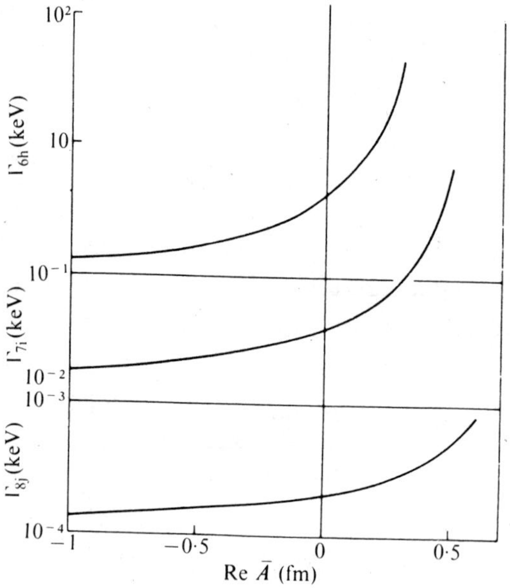

FIG. 9.10. Variation of the strong-interaction widths with the strength of the real part of the potential (9.35) for Z = 74. (After Ericson and Scheck 1970.)

More complete measurements on strong-interaction shifts and widths have since been published (Backenstoss *et al.* 1972*a*). Analysis of these data shows that use of the free (threshold) KN parameters leads to predictions for the widths which are two to three times too small and confirms the con-

TABLE 9.18

Linewidth and strong-interaction shift for the 4f-3d transition in sulphur[†]

	Re $\bar{A}$ (fm)	Im $\bar{A}$ (fm)	Γ_{3d} (keV)	ΔE (keV)
Experiment	—	—	2.2 ± 0.6	-0.36 ± 0.25
Free parameters	-0.84	0.7	1.04	-0.76
Zero real potential	0	0.7	1.54	-0.50
Best fit	+0.42	0.7	2.2	-0.37

[†]From Backenstoss *et al*. 1970*a*.

clusion obtained from the earlier result for sulphur. Results from various calculations are collected together in Table 9.19. The theoretical predictions are obtained by solving the Klein-Gordon equation to obtain the complex eigenvalue. The theoretical shift is then given by the difference between the experimental energy and the theoretical energy including the lowest-order vacuum polarization correction, and the theoretical width is twice the imaginary part of the eigenvalue. General agreement with the data is obtained and little more can be achieved until the experimental uncertainties are reduced.

Further experimental results for energy shifts and widths arising from the strong interaction are given in Table 9.20. The value of the width for S is in reasonable agreement with the value given in Table 9.19 but there is some disagreement for the energy shift. The disagreement between widths measured for Ni is significant since calculations based on a kaon-nucleus optical potential with modified parameters do not reproduce the lower value of Barnes *et al*. (1974).

9.2.4. *K absorption in the nuclear surface*

In the earliest work on K^- interactions with emulsion nuclei (Chadwick *et al*. 1958, Jones 1958) it was shown that if the nuclear capture takes place from circular orbits the capture

TABLE 9.19

Energy shifts and widths of light kaonic atoms

Shifts in keV

Transition	Experiment[a]	Theory					
		b	c	d	e	f	g
^{10}B(3d-2p)	-0.208 ± 0.035	—	—	-0.25	-0.238	-0.210	-0.234
^{11}B(3d-2p)	-0.167 ± 0.035	—	—	-0.26	-0.247	-0.168	-0.245
^{12}C(3d-2p)	-0.59 ± 0.08	—	—	-0.67	-0.619	-0.540	-0.723
^{31}P(4f-3d)	-0.33 ± 0.08	-0.36	-0.46	-0.36	-0.384	-0.314	-0.266
^{32}S(4f-3d)	-0.55 ± 0.66	-0.61	-0.73	-0.61	-0.620	-0.487	-0.475
^{35}Cl(4f-3d)	-0.77 ± 0.40	-1.00	-1.16	-1.05	-1.040	-0.801	-0.884

Widths in keV

Transition	Experiment[a]	Theory					
		b	c	d	e	f	g
^{10}B(3d-2p)	0.81 ± 0.10	—	—	0.72	0.532	0.799	0.963
^{11}B(3d-2p)	0.70 ± 0.08	—	—	0.70	0.554	0.700	0.948
^{12}C(3d-2p)	1.73 ± 0.15	—	—	1.58	1.270	1.712	2.088
^{31}P(4f-3d)	1.44 ± 0.12	2.01	1.78	1.68	0.958	1.399	1.895
^{32}S(4f-3d)	2.33 ± 0.06	3.06	2.72	2.78	1.496	2.200	3.067
^{35}Cl(4f-3d)	3.8 ± 1.0	4.68	4.14	4.28	2.252	3.80	4.441

[a] Backenstoss *et al.* 1972*a*.
[b] Rook and Wyčech 1972, parameters (9.37*b*).
[c] Rook and Wycech 1972, parameters (9.37*c*).
[d] Alberg *et al.* 1973.
[e] Bardeen and Torigoe 1972.
[f] Chattarji and Ghosh 1973.
[g] Deloff and Law 1974*a*. Corrected values provided by the authors.

TABLE 9.20

Experimental energy shifts and widths in kaonic atoms

Transition	Width (keV)	Shift (keV)	Reference
Al(4f-3d)	0.49 ± 0.16	-0.13 ± 0.05	a
Si(4f-3d)	0.81 ± 0.12	-0.24 ± 0.05	a
S(4f-3d)	3.11 ± 0.60	-0.228 ± 0.160	b
Co(5g-4f)	0.677 ± 0.293	-0.463 ± 0.104	b
Ni(5g-4f)	1.24 ± 0.14	-0.250 ± 0.10	b
Ni(5g-4f)	0.59 ± 0.21	-0.180 ± 0.07	a
Cu(5g-4f)	1.65 ± 0.72	-0.240 ± 0.22	a
Ag(6h-5g)	2.45 ± 0.54	-0.535 ± 0.180	b
Cd(6h-5g)	3.01 ± 0.48	-0.210 ± 0.140	b

[a]Barnes *et al.* (1974).
[b]Batty *et al.* (1975).

process is localized in the extreme surface region of the nucleus. This is evident from Fig. 9.8 since the halfway radius of the matter distribution for emulsion nuclei is about 5 fm. This figure also indicates the importance of the assumption of circular orbits. In general, it appears that the radial absorption probability has its maximum at about 1 fm outside the halfway radius in light- and medium-mass nuclei and about 2 fm outside in heavy nuclei (Wilkinson 1968, Bethe and Siemens 1970). For a Fermi distribution with diffuseness parameter a = 0.55 fm these positions correspond to densities of 1/7 and 1/40, respectively, of the central value. For single-particle distributions this ratio depends on the separation energies of the least-bound nucleons and may be quite different for protons or neutrons, as can be seen from Fig. 2.2. These results have led to the view that the study of kaonic atoms provides a method of studying the nuclear matter distribution in the extreme surface region (Wilkinson 1961, 1968).

Ericson and Scheck (1970) have studied the sensitivity of the capture width $\Gamma_{n,l}^{\text{abs}}$ to variations in the parameters c,t

of a nuclear matter distribution of Fermi shape. They find the empirical relation

$$\Gamma^{abs}_{n,l}(c,t) \propto \exp\{-\alpha(c-c_0) - \beta(t-t_0)\} \tag{9.49}$$

where c_0, t_0 are some fixed values taken from fits to other data and α,β are positive constants which vary with Z and n,l. Since α and β are very similar in magnitude it is not possible to determine c or t independently.

Both Ericson and Scheck (1970) and Bardeen and Torigoe (1971) conclude that the X-ray data of Wiegand (1969) can be reproduced, within the large experimental errors, without requiring any difference between the parameters of Fermi distributions for the protons and neutrons. Fig. 9.9 shows that the effect of enhancement of the neutron distribution in the periphery is not significant at present. Similar conclusions were reached by Rook (1962). The use of single-particle models for the nucleus does, however, lead to differences between the proton and neutron distributions in nuclei. Aslam and Rook (1970*b*) have investigated the effect of this by calculating the ratio

$$X = \frac{N \int |\psi_K|^2 \rho_n(r)\, d^3r}{Z \int |\psi_K|^2 \rho_p(r)\, d^3r} \tag{9.50}$$

using single-particle distributions. The results are given in Table 9.21. These indicate that the capture rate on neutrons is enhanced, although the model gives differences between the r.m.s. radii larger than are now considered acceptable (see Chapters 2 and 10). This enhancement is significant in relation to the relative capture rate $R^{L/H}_{pn}$ discussed in §10.7.2; it is not so significant in the calculation of X-ray yields because these depend on a nuclear capture rate given by eqn (9.45) which involves the coefficients $v\,\sigma_{KN}$ or B^N and, as we have seen, B^p is substantially larger than B^n. Thus the X-ray yields are at present a highly insensitive probe of neutron distributions, because the nuclear capture is dominated by the

TABLE 9.21

Ratio of K^- absorption on neutrons and protons. Absorption occurs in the two states $n\ l$ and the contributions are weighted according to the experimental yields[†]

Nucleus	nl	X	N/Z	$\langle r^2\rangle_n^{1/2}$	$\langle r^2\rangle_p^{1/2}$
^{35}Cl	4f,3d	1.03	1.05	3.30	3.32
^{58}Ni	5g,4f	1.17	1.07	3.58	3.56
^{98}Mo	6h,5g	2.32	1.33	4.56	4.34
^{158}Gd	7i,6h	3.03	1.47	5.23	4.95
^{208}Pb	8j,7i	3.68	1.54	5.67	5.38
^{238}U	8j,7i	3.21	1.59	5.99	5.72

[†]From Aslam and Rook 1970*b*.

K^-p interaction. If the yields could be measured to a degree of accuracy which gave sensitivity to the neutron distributions, an extremely accurate and reliable theory of the K^-p interaction, including the role of the Y_0^*, would be required in order to interpret the data.

The capture of a kaon may also proceed through interaction with two nucleons

$$K^- + N + N \rightarrow Y + N$$

which yields a fast hyperon and no pion. The ratio of this process to capture on a single nucleon is 20 per cent in emulsion nuclei, 15 per cent in helium and 1 per cent in deuterium. Wilkinson (1961) has proposed that these results could be understood if the nucleons in the surface of medium and heavy nuclei are correlated into α-particle clusters. This proposition has been examined by several authors (Rook 1962, Wycech 1967, Bethe and Siemens 1970). Aslam and Rook (1970*a*) have made a careful comparison of the calculations by Rook (1962) which implied that some strong correlation, such as might occur in α-particle clustering, is indeed necessary to explain the results in emulsion nuclei and the calculation of Wycech (1967) which reproduced the experimental results without recourse to clustering. Their conclusion is that the discrepancy between the theoretical and experimental results for emulsion nuclei without clustering is not as large as had at first appeared and that there is no evidence at present for clusters in the nuclear surface.

More information on the nature of the absorption process may be obtained from detection of the nuclear γ-rays following kaon capture, since a measurement of the γ-ray energies provides a means of identifying the residual nucleus from which they have been emitted and determining the excitation energies. Such γ-rays have been observed following kaon capture in S and C (Wiegand, Gallup, and Godfrey 1972) in ^{27}Al, ^{28}Si, Ni, and Cu (Barnes *et al.* 1972*a*), and in ^{27}Al (Batty *et al.* 1975). Theoretical predictions for the γ-ray intensity following kaon capture in ^{16}O and ^{208}Pb have been given by Bloom, Weiss, and Shakin (1972) on the assumption of single-nucleon removal.

For the lighter nuclei removal of a single proton appears to be the dominant process, while single neutron removal gave a very low cross-section or in some cases was not observed at all. This low contribution from single neutron removal is consistent with the dominance of the K^-p interaction compared with the K^-n interaction. Even in ^{27}Al a relatively large yield of γ-rays corresponding to α-particle removal was seen. In Ni some evidence for single proton removal remains, but removal of one or more α-particles is an important mode. In Cu the favoured mode is removal of a triton or a triton plus one or more α-particles. However, it should be noted that examination of the γ-ray spectrum does not determine whether the residual nucleus was reached by a single direct process with emission of a 'real' α-particle or triton, or whether a sequence of interactions occurred starting with the recoil of a single nucleon.

Further indication of the interaction mechanism for kaon capture may perhaps be obtained by comparing the results with those observed for γ-ray emission following nuclear excitation by pions and by nucleons. Capture of fast pions (Jackson *et al.* 1973, Lind *et al.* 1974) also leads to emission of γ-rays from residual nuclei corresponding to removal of one or more α-particles. For medium-mass nuclei there is little evidence for single nucleon knock-out, while for a target of ^{28}Si γ-rays from the residual nucleus ^{27}Al are seen but not from ^{27}Si. The same effect with a target of ^{28}Si is seen following bombardment with low-energy pions (Ashery *et al.* 1974), but with a target of ^{27}Al the γ-rays resulting from both proton and neutron removal are seen. Absorption of pions at rest (Plendl 1973, Ullrich *et al.* 1974) gives similar results but the dominance of α-particle removal is less pronounced. (Energy and momentum conservation requires that absorption at rest involves at least two correlated nucleons).

The γ-rays emitted following the interaction of 100 MeV protons with ^{56}Fe and ^{58}Ni have been studied by Chang, Wall, and Fraenkel (1974). Strong transitions to residual nuclei equivalent to the target nucleus minus one, two, or three α-particles were noted. For the ^{58}Ni target the transitions involving emission of a single nucleon were weak. Chang *et al.* explain their results by assuming that in the first stage of the reaction (pre-equilibrium state) a few nucleon-nucleon collisions lead to an average excitation energy of ~50 MeV. They take this situation as input for an evaporation calculation and obtain quite good agreement with the relative yields of γ-rays, following α-particle or single nucleon emission, but the importance of single-nucleon removal is over-estimated. A further study of interactions of 200 and 100 MeV pions and 200 MeV protons with ^{58}Ni and ^{60}Ni has been made by Jackson *et al.* (1975). The average number of nucleons removed from the target by pions appears to be independent of the bombarding energy and of the target nucleus, but there are other features which appear to depend on the target nucleus and on pion charge. The absorption of 140 MeV of mass energy in the pion interaction is also a significant feature.

Much more experimental work will be required before the

mechanism of kaon and pion capture in nuclei is fully understood, and detection of the emitted charged particles will be particularly important. Preliminary results obtained by Amann *et al.* (1975) for the spectra of particles emitted following absorption of 235 MeV pions suggest that complex particle emission is qualitatively explained by a pick-up mechanism rather than direct removal. Cascade calculations apparently over-estimate the number of high-energy protons emitted by a large factor.

9.3. HYPERONIC AND ANTIPROTONIC ATOMS

Observations of X-ray transitions in Σ^- atoms of S, Cℓ, and Zn have been reported (Backenstoss *et al.* 1970*a*, 1972*a*,*b*). In S and Cℓ the lowest observed transitions were 6-5 transitions, while in Zn the lowest observed transition was 8-7. The measured energies were reproduced closely by a calculation with only electromagnetic interactions including vacuum polarization. Strong-interaction shifts and widths should be evident for the lower transitions but these also have low yields.

X-ray transitions in antiprotonic atoms were first observed in Tℓ (Bamberger *et al.* 1970). Since then measurements of strong-interaction shifts and widths for the 4f-3d transition in ^{14}N and ^{16}O have been reported (Barnes *et al.* 1972*b*). A simple analysis of the latter data in terms of an optical potential of the form (9.35) yields an effective scattering length $\bar{A} = (2.9 \pm 2.0) + i(1.5 \pm 1.1)$ fm. Bugg *et al.* (1973) have reported further studies of the interaction of stopping antiprotons with nuclei.

Measurements of X-ray transitions have also been reported (Backenstoss *et al.* 1972*b*) for P, S, Cℓ, K, for Sn, I, Pr, and for W. In the group of light elements the last observable transition is the 5-4 transition, while for the next group it is 8-7, and in W the 9-8 transition is not observed but an upper limit of 25 per cent can be given for the yield. Values for the yield of the last observable transition are used to determine the absorption width of the upper level through eqn (9.48). Theoretical calculations using perturbation theory for the absorption rate, eqn (9.38), with the imaginary potential (9.33) yield reasonable agreement with the data. The $\bar{p}N$ scattering lengths of Puget (1970), which have somewhat larger imaginary parts than those of Bryan and Phillips (1968), seem to be favoured. In sulphur, the natural linewidth of the 5-4 transition has been determined and this gives the width of the lower level as 310 ± 180 eV. The strong-interaction energy shift is found to be -80 ± 40 eV. The latter value and the yield can be reproduced with

$$A_0 = A_1 = 0.20 + i\,1.01$$

which implies a weakly attractive real part for the optical potential, whereas the free parameters give a repulsive potential.

Leon and Seki (1974) have also used a potential of the form (9.35); they find that the interpretation of the $\bar{p}$ annihilation data of Bugg *et al.* depends rather sensitively on the coefficient $\bar{A}$ determined from the X-ray measurments, but conclude that the annihilation data are consistent with other information on the proton and neutron distributions in the extreme surface. Deloff and Law (1974*b*) construct a $\bar{p}$-nucleus potential by folding a gaussian $\bar{p}N$ potential into the nuclear matter distribution. The strength of the $\bar{p}N$ potential is determined by fitting the scattering lengths of Bryan and Phillips and, with a suitable choice for the range of the two-body potential, qualitative agreement with the data of Backenstoss *et al.* (1972*b*) is obtained.

Wycech (1973) has estimated that nuclear absorption in hyperonic and antiprotonic atoms should occur in the 1 per cent density region, while Leon and Seki's calculation emphasizes the 10 per cent region. In either case, analyses such as those described above, which use Fermi distributions, are totally unrealistic and the conclusions must be subject to considerable doubt.

NEUTRON DISTRIBUTIONS

10.1. INTRODUCTION

Theoretical predictions for proton and neutron distributions have already been discussed in detail in Chapter 2, and certain models for the distribution of excess neutrons were also examined. It was evident that the differences between proton and neutron distributions, particularly in the surface region, arise through the effects of the symmetry potential and the Coulomb potentials. The behaviour of the symmetry term in the single-particle potential was discussed in Chapter 2 and the symmetry term in the optical potential was discussed in Chapter 7.

A considerable amount of information can be obtained from high-energy nucleon scattering and from α-particle scattering, as explained in Chapter 7, but the analyses of these data are concerned essentially with the matter distributions so that the behaviour of the neutron distribution is inferred indirectly. The same is true for the analysis of summed inelastic scattering described in Chapter 8. Further information can be obtained from inelastic scattering to a definite final state, also described in Chapter 8, but in this case the quantity studied is the transition density which in general does not depend exclusively on the distribution of the valence neutrons. Certain nuclear reactions described in Chapter 8, such as neutron pick-up reactions and the photonuclear (γ,n) process, also provide useful information, but in these cases the information relates to the single-neutron states rather than to the neutron distribution as a whole.

In this chapter we examine a number of different processes which have been used to obtain rather more direct information about the neutron distribution or the excess-neutron distribution.

10.2. INTERPRETATION OF COULOMB ENERGY DIFFERENCES

The Coulomb energy difference, or Coulomb displacement energy, is the energy difference E_d between the parent state in the nucleus (Z,N) and the isobaric analogue state in the nucleus $(Z+1, N-1)$. This is illustrated in Fig. 10.1. The study of

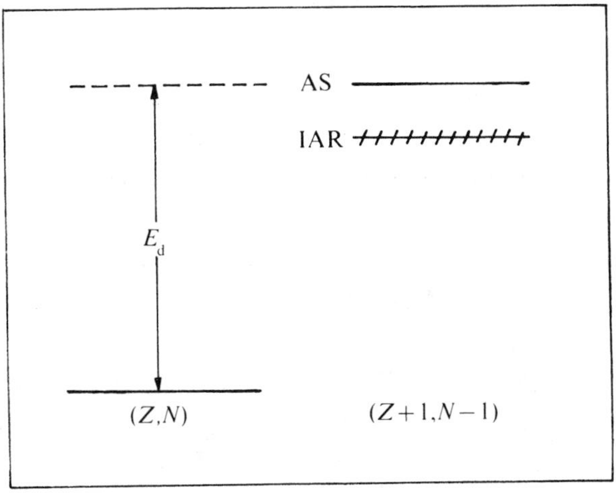

FIG. 10.1. Location of the analogue state (AS) and the isobaric analogue resonance (IAR), and definition of the Coulomb energy difference E_d. (From Jackson 1974.)

these energies has long been regarded as a source of nuclear size information, both for the average properties of a wide range of nuclei and for the individual properties of an isobaric pair (Jänecke 1969, Nolen and Schiffer 1969).

In the earliest calculations the nucleus was regarded as a uniformly charged sphere of radius $r_0A^{1/3}$ so that the Coulomb energy difference becomes

$$E_d = \frac{6\ e^2\ \bar{Z}}{5\ r_0\ A^{1/3}} \tag{10.1}$$

where $\bar{Z}e$ is the average of the charge of the isobaric states. An exchange term must be subtracted from this simple expression, and a correction for the deformation of the nuclear charge distribution can also be included. A much more important correction arises from the diffuseness of the nuclear surface. This may be taken into account phenomenologically using a charge distribution of Fermi shape. Lindner (1968) has given an expansion in powers of the ratio of the diffuseness parameter to the halfway radius which has as leading term the expression given in eqn (10.1). The nuclear charge distributions may also be constructed using a single-particle model. The earliest calculations were made with oscillator functions (Carlson and Talmi 1954, Sengupta 1960), but later calculations with single-particles wavefunctions generated in Saxon-Woods potentials (Sood and Green 1957, Wilkinson and Mafethe 1966, Wilkinson and Hay 1966) have shown the importance of fitting correctly the single-particle binding energies for the least-bound protons.

The basic assumption in the analyses of the energy differences is that the distribution of the extra proton in the isobaric state is the same as the distribution ρ_{ne} of the excess N-Z neutrons in the parent state which is given by

$$\rho_{\mathrm{ne}}(\underset{\sim}{r}) = N\ \rho_{\mathrm{n}}(\underset{\sim}{r}) - Z\ \rho_{\mathrm{nc}}(\underset{\sim}{r}) \tag{10.2}$$

where ρ_{nc} is the distribution of the core of Z neutrons. The principal contribution to the energy E_{d} is then given by the interaction of this proton with the Coulomb potential due to the core of Z protons. This may be written as

$$\Delta E_{\mathrm{d}} = \frac{1}{N-Z}\int V_{\mathrm{c}}(\underset{\sim}{r})\ \tilde{\rho}_{\mathrm{ne}}(\underset{\sim}{r})\ \mathrm{d}^3r' \tag{10.3}$$

In this expression $\tilde{\rho}_{\mathrm{ne}}$ includes the finite size of the proton, i.e.

$$\tilde{\rho}_{\mathrm{ne}}(\underset{\sim}{r}) = \int \rho_{\mathrm{ne}}(\underset{\sim}{r}')\ \rho_{\mathrm{d}}(|\underset{\sim}{r}-\underset{\sim}{r}'|)\ \mathrm{d}^3r \tag{10.4}$$

where ρ_{d} represents the charge distribution of a single proton, as in eqn (1.5). There are a number of additional corrections to the energy which arise from exchange, the electromagnetic spin-orbit interaction, the n-p mass difference, and the Thomas-Ehrman shift; these quantities have been discussed extensively in the literature and estimates given (Nolen and Schiffer 1969, N. Auerbach *et al.* 1969, Friedman and Mandelbaum 1969).

In most calculations it is assumed that $\rho_{nc} = \rho_p$. It is then possible to construct $\tilde{\rho}_{ne}$ from a suitable model, either microscopic or macroscopic, and using eqn (10.3) plus the correction terms the predicted energy difference can be compared with the experimental value. Nolen, Schiffer, and Williams (1968) used a Fermi charge distribution for ^{208}Pb which reproduced electron scattering data to construct the Coulomb potential and then varied the parameters of a Fermi distribution for $\tilde{\rho}_{ne}$ until the experimental value of the Coulomb energy difference was obtained. Subsequently, Nolen and Schiffer (1969) generated ρ_{ne} for a range of nuclei in a Saxon-Woods single-particle potential. They found that the results are rather insensitive to the detailed shape of ρ_{ch} and $\tilde{\rho}_{ne}$ but are very sensitive to the r.m.s. radius of $\tilde{\rho}_{ne}$. They also found that the experimental Coulomb energy data could not be reproduced unless the r.m.s. radius of the neutron excess distribution was only slightly greater than that of the distribution of the core. In the single-particle model this implies that the radii of Saxon-Woods potentials which yield the required ρ_{ne} to fit the Coulomb energy differences for a range of nuclei from ^{13}C to ^{208}Pb are 10-20 per cent smaller than those required to give proton distributions in agreement with electron scattering data. Alternatively, if the same Saxon-Woods potentials are used to generate both ρ_p and ρ_{ne}, the calculated Coulomb energy differences are ~7 per cent too small.

The consequence of Nolen and Schiffer's result is that the difference between the r.m.s. radii of the total neutron distribution and the proton distribution becomes quite small, for example ~ 0.07 fm for ^{208}Pb. Another analysis of the data for ^{208}Pb by Bethe and Siemens (1968) using the macroscopic Thomas-Fermi model gave a neutron halfway radius slightly smaller than the proton radius. Results obtained by Nolen and Schiffer (1969) and Friedman and Mandelbaum (1969) for the r.m.s. radius of the *charge* distribution ρ_{ch}^{ex} of the proton in the analogue state compared with the r.m.s. radius of the charge distribution of the core are given in Table 10.1. The finite electromagnetic size of the proton must be

TABLE 10.1

Root-mean-square radii of the charge distribution ρ^{ex}_{ch} *derived from Coulomb energy differences using a given value of* ρ_{ch}

	$\langle r^2 \rangle^{1/2}_{ch}$	$\langle r^2 \rangle^{ex\ 1/2}_{ch}$
Nolen and Schiffer 1969: single-particle model		
^{48}Ca	3.48	3.69
^{62}Ni	3.87	4.07
^{120}Sn	4.64	4.87
^{208}Pb	5.51	5.95
Friedman and Mandelbaum 1969: single-particle model		
^{90}Zr	4.30	4.38
^{120}Sn	4.66	4.84
^{208}Pb	5.50	5.76
Friedman 1971a: hydrodynamical model and isospin correction		
^{48}Ca	3.48	3.53
^{60}Ni	3.86	3.99
^{120}Sn	4.64	4.71
^{208}Pb	5.50	5.57
Friedman 1971b: single-particle model and isospin correction		
^{48}Ca	3.48	4.12
^{208}Pb	5.50	6.03

subtracted in order to compare with values for ρ_{ne} and ρ_p.

Several explanations of the apparent anomaly in the results for Coulomb energy differences have been suggested (E.H. Auerbach, Kahana, and Weneser 1969, N. Auerbach *et al.* 1969, 1972, Damgaard *et al.* 1970, Wong 1970). It has been recognized that isospin impurity in the core implies that $\rho_{nc} \neq \rho_p$, so that there is an additional contribution to the energy arising from the interaction of the extra proton with isospin impurity of the form (E.H. Auerbach *et al.* 1969)

$$\delta E_{\mathrm{d}} = \frac{4\alpha}{2(N-Z)} \int \frac{\tilde{\rho}_{\mathrm{ne}}(r)}{\tilde{\rho}_0(r)} \, [Z\,\tilde{\rho}_{\mathrm{p}}(\underset{\sim}{r}) - Z\,\tilde{\rho}_{\mathrm{nc}}(\underset{\sim}{r})] \, \mathrm{d}^3 r \qquad (10.5)$$

where

$$\tilde{\rho}_0(\underset{\sim}{r}) = Z\,\tilde{\rho}_{\mathrm{p}}(\underset{\sim}{r}) + N\,\tilde{\rho}_{\mathrm{n}}(\underset{\sim}{r}), \qquad (10.6)$$

$\tilde{\rho}_{\mathrm{ne}}$ is defined by eqn (10.4) and $\tilde{\rho}_{\mathrm{p}}$ and $\tilde{\rho}_{\mathrm{n}}$ are defined by a similar equation so that $\tilde{\rho}_{\mathrm{p}} \equiv \rho_{\mathrm{ch}}$. The quantity α is the coefficient of the symmetry term in the single-particle potential, as defined in eqns (2.10) and (7.61), so that $4\alpha \simeq 100$ MeV. This correction has been evaluated in the hydrodynamical model (Damgaard *et al.* 1970, Friedman 1971*a*) and in the single-particle model (Friedman 1971*b*). It appears that the correction due to isospin asymmetry is sufficient to remove the discrepancy between the calculated and experimental values of the Coulomb energy differences and to remove the discrepancy between the radii of the single-particle potentials needed to generate ρ_{ch} and ρ_{ne}. There remains, however, a large discrepancy between the value of the r.m.s. radius of $\rho_{\mathrm{ch}}^{\mathrm{ex}}$ given by the hydrodynamical model and the value given by the single-particle models, which may be due to shell effects. It can be seen from Table 10.1 that this amounts to 0.46 fm for ^{208}Pb. The larger value for ^{208}Pb predicted by the single-particle model has been confirmed by measurements on single neutron transfer reactions (Korner and Schiffer 1971, Friedman *et al.* 1972, Schiffer and Korner 1973). Similarly, values for the r.m.s. radii of the neutron excess in the calcium isotopes obtained from heavy ion transfer reactions (Jones *et al.* 1974) exceed the values deduced from Coulomb energy differences by Nolen and Schiffer (1969) by ~0.7 fm.

The formulation of the correction due to the isospin impurity described above has been criticized (Negele 1971) on the grounds that it gives an unreasonably large enhancement of the symmetry interaction in the nuclear surface and the use of the local density in the denominator of eqn (10.5) is difficult to justify. Coulomb energy differences have been studied in the Hartree-Fock approximation by Van Giai *et al.* (1971) using the Skyrme interaction; their method takes into account both isospin impurities in the core and polarization of the core by the extra-core nucleon. They conclude that polarization of the core is of negligible importance and that the correction due to isospin impurity is not large enough to remove the anomaly noted by Nolen and Schiffer. A similar result was obtained by Negele (1971) and by Auerbach (1974).

A more detailed study of the nuclear structure of the isobaric states suggests a number of additional corrections which might contribute to the Coulomb energy difference such as configuration mixing and deformation of the core. At the present time there is little agreement on the magnitude of these corrections or, in some cases, even on their sign. There is also the possibility of corrections due to any breakdown in charge symmetry in the nuclear force (Negele 1971). However, Shlomo (1972) has calculated the contribution to the Coulomb energy difference for mirror nuclei due to symmetry-breaking potentials proposed by various authors. He concludes that, because of the short-range character of these potentials, their inclusion may resolve the discrepancy for $A = 3$ but not for $A = 41$. Brown, Horsfjord, and Liu (1973) support this view and argue that approximately half the discrepancy for $A = 41$ can be attributed to correlation in the nuclear surface. Such correlations or clustering can occur in the low-density region where the effect of the exclusion principle is reduced and are not adequately represented in G-matrix calculations.

10.3. CHARGE-EXCHANGE REACTIONS AND ISOBARIC ANALOGUE RESONANCES

The contribution to the optical potential from the symmetry term can be written in the form suggested by Lane (1962), i.e.

$$V(r) = V_0(r) + V_1(r)\, \underset{\sim}{t}\cdot\underset{\sim}{T}/A \tag{10.7}$$

where $\underset{\sim}{t}$, $\underset{\sim}{T}$ are the isospin operators for the projectile and target nucleus, respectively. The diagonal matrix element of this potential between states with $T_3 = \frac{1}{2}(N-Z)$ and $t = \mp \frac{1}{2}$ is $V_0(r) \mp V_1(r)(N-Z)/4A$. Hence, if V_0 and V_1 have the same radial behaviour and strengths $-U_0$ and U_1 respectively, we have

$$V(r) = -\{U_0 \pm U_1(N-Z)/4A\}\ f(r) \qquad (10.8)$$

so that $U_1 = 4\alpha$, where α is the coefficient defined in eqn (7.60). The matrix element of the off-diagonal part of the potential (10.7) containing t_+T_- is $(2T)^{1/2}\ V_1(r)/2A$. This represents a nuclear process in which the T_- operator acting on the ground state of the target nucleus (Z,N) has produced the isobaric analogue state in the nucleus $(Z+1,\ N-1)$. Such states may be directly excited in a charge-exchange reaction, such as (p,n) or (^{3}He,t), and in such a reaction the transition arises solely from the isovector term in the potential, whereas the same term gives only a small correction to elastic scattering.

Extensive studies of the (p,n) reaction have been made in the medium energy region and a few studies of the (p,n) and (n,p) reactions have been made in the intermediate-energy region. Various methods of solving the Schrödinger equation with the potential (10.7) in order to obtain the charge-exchange cross-sections have been described by Satchler (1969) who also discussed fits to the data. In general, it may be said that there is a preference for surface-peaked behaviour for the real part of the symmetry term and some uncertainty about an imaginary part. Carlson, Lind, and Zafiratos (1973) have recently examined the (p,n) reaction on several nuclei at a proton energy of 23 MeV. They fixed the parameters of the proton optical potential to be those determined by Becchetti and Greenlees (1969) to give a good fit to proton elastic scattering near $E_p = 23$ MeV and modified the parameters of $V_1(r)$ to give a good fit to the (p,n) reaction. Good fits to the (p,n) reaction were obtained by slightly reducing the strengths of the real and imaginary parts of V_1 and increasing the diffuseness.† Eqn (10.7) was then used to predict an optical potential for neutron elastic scattering which gave good agreement with the data. This result confirms the validity of Lane's formula which evidently gives a self-consistent description of proton and neutron elastic scattering and charge-exchange reactions, at least in the low-energy region.

Charge-exchange reactions may also be described by a microscopic model. A number of calculations have been carried out at intermediate energies using impulse approximation and at medium energies using an effective interaction (Madsen 1966, Wong *et al.* 1967, J.D. Anderson *et al.* 1969, Clough *et al.* 1969, Mahalanabis 1969). In this work the main interest was in the microscopic structure of the nuclear states and the parameters of the effective interaction. However, it is also possible to calculate the symmetry term using eqn (7.93) and this involves the distribution of excess neutrons. Such a calculation has been carried out by Batty, Friedman, and Greenlees (1969) using a single Yukawa potential for the nucleon-nucleon interaction and generating the excess-neutron distribution in a Saxon-Woods well. Plots of $\rho_{ne}(r)$ are shown in Fig. 10.2 and these show a change from a predominantly surface shape to a volume shape with increasing mass number, and also show the effect of shell structure. (Similar effects are seen in the calculations of N. Auerbach *et al.* (1972).) The resulting potentials yield reasonably good agreement with (p,n) data at 30 and 50 MeV, and in some cases significantly better agreement than the phenomenological potentials.

Microscopic calculations have also been carried out for the (^{3}He,t) reaction (Wesolowski *et al.* 1968, Kossanyi-Demay *et al.* 1970). Again, the interest is centred on the microscopic structure of the nuclear states and the parameters of the effective interaction, and reasonable agreement is achieved for the analogue state transitions.

---

†Woods, Whitten, and Igo (1972) have since suggested that this change is necessary owing to neglect of the correction for non-locality.

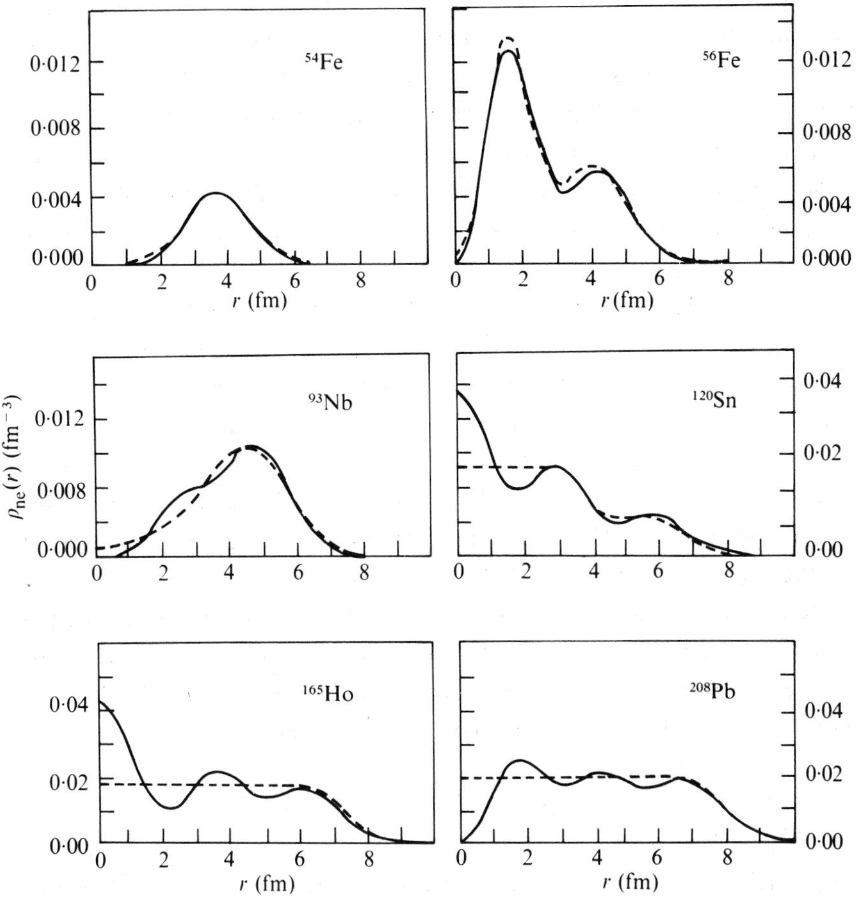

FIG. 10.2. Distributions for excess neutrons calculated from the Batty-Greenlees single-particle potentials (full line) compared with parametrizations based on Saxon-Woods shapes. (From Batty *et al.* 1969.)

At low proton energies the neutron remains bound in the nucleus; the excitation of isobaric analogue states then gives rise to resonances in proton scattering. The position of the isobaric analogue resonance is shifted from the position of the analogue state owing to the coupling of the analogue state to the compound and continuum states. This is illustrated in Fig. 10.1.

The theoretical description of isobaric analogue resonances involves the calculation of the position of a resonance, the total width, and the partial widths for emission of protons from various configurations (N. Auerbach *et al.* 1972). This requires a real potential in which the neutron form factor can be generated and a complex optical potential for the scattered proton. Detailed studies in the lead region (Zaidi and Darmodjo 1967, Bondorf and Bund 1969, Bund and Blair 1970) and in the $A = 140$ region (Harney, Wiedner, and Wurm 1968, Harney 1968) indicate that a surface-peaked isovector term is needed, or equivalently that the halfway radius of the neutron potential must be less than that of the proton potential. The neutron potential derived by Rost (1968) for ^{208}Pb leads to partial widths which are too large by a factor of 2 (Bund and Blair 1970) and so does use of a volume form for the isovector term (Bondorf and Bund 1969). This result is consistent with the results for neutron potentials in the lead region noted in §§2.2.1 and 10.2.

Because of the sensitivity of the spectroscopic factors and single-particle widths to the parameters of the neutron potential, Clarkson, von Brentano, and Harney (1971) have introduced a reduced single-particle width

$$G_c = \kappa^3 \, \Gamma_c / N^2$$

where κ and N are the quantities defined in eqns (8.94) and (8.95). The measured partial width is related to the calculated single-particle width Γ_c by the relation

$$\Gamma = S(c)\Gamma_c$$

so that

$$\Gamma = \Lambda_c G_c$$

where $S(c)$ is the spectroscopic factor for the channel c and Λ_c is the reduced normalization defined in eqn (8.96). Thus G_c is now largely insensitive to variations in the parameters of the neutron potential. There remains sensitivity to the radius of the nuclear part of the proton potential and to the

Coulomb radius, but the latter can be fixed by requiring agreement with the Coulomb displacement energy (Clarkson *et al.* 1971, Kent *et al.* 1972). Satisfactory agreement with the data for ^{208}Pb is obtained with $R_c = R_p = R_n = 1.20A^{1/3}$ fm.

10.4. SCATTERING OF MESONS AND ANTIPARTICLES

10.4.1. *Elastic scattering, total, and reaction cross-sections*

Pion scattering from nuclei has been studied over a wide energy range from ~20 MeV to 60 GeV. Phenomenological analyses of the data are relatively unusual, and in most analyses the pion-nucleus interaction is connected with the pion-nucleon data by a variety of methods essentially similar to those described in Chapter 7 for other projectiles. We first review these methods and their success in fitting the data, and then describe those analyses of pion-nucleus scattering which are directly concerned with the study of the neutron distribution.

The pion differs from the nucleon in a number of important respects. For example, even at relatively low kinetic energies the pion is a relativistic particle. At the very least this requires the use of relativistic kinematics, and it also requires some consideration of which wave equation should be solved to determine the scattering wavefunction and phase shifts. A number of prescriptions exist for using the Klein-Gordon equation or a modified Schrödinger equation; these have been reviewed by Allardyce *et al.* (1973). Dedonder (1972) has also pointed out the need for a relativistic multiple-scattering theory. It should also be noted that use of the multiple-scattering theory developed for nucleon scattering implies that the basic two-body process is $\pi + N \to \pi + N$, but pions may also be removed from the elastic beam by the process $\pi + N + N \to N + N$, i.e. by absorption on two nucleons. For convenience we refer to the latter process as annihilation to distinguish it from absorption in the optical model sense. Because the annihilation process requires two nucleons with high relative momentum it is usually argued that this process depends on $\rho^2(r)$ rather than $\rho(r)$ (Koltun 1969) although some authors include the effect of this process in the term dependent on $\rho(r)$ (see, for example, Johnston and Watson 1961). An additional characteristic of the pion which is of considerable significance is the existence of the pion-nucleon resonances, and the dominance of the p-wave part of the interaction in the region of 100-200 MeV is due to the $\Delta(1236)$ resonance with $T = \frac{3}{2}$, $J = \frac{3}{2}$, i.e. the (3,3) resonance.

The construction of the pion-nucleus optical potential in non-relativistic theory has been reviewed by Koltun (1969). Use of the first term in a multiple-scattering expansion and the impulse approximation leads to an optical potential of the form given by eqn (7.49), or in the forward-scattering approximation by (7.53). However, the pion-nucleus interaction differs from the nucleon-nucleon interaction in that it is rather weak and is dominated by s- and p-waves up to a few hundred MeV. This means that the scattering amplitude (7.7) can be written in the form

$$M = A + B\, \underset{\sim}{k}\cdot\underset{\sim}{k}' + C\, \underset{\sim}{\sigma}\cdot\underset{\sim}{k} \wedge \underset{\sim}{k}' \tag{10.9}$$

where $\underset{\sim}{\sigma}$ is the spin operator for the nucleon, and $A = b_0 + b_1\, \underset{\sim}{t}_\pi\cdot\underset{\sim}{t}$, etc. This leads to Kisslinger's potential (Kisslinger 1955) which in configuration space has the form

$$2E_\pi\, U(r) = -4\pi\, \hbar^2 c^2\, A\{b_0\, \rho_m(r) - c_0\, \underset{\sim}{\nabla} \cdot \rho_m(r)\, \underset{\sim}{\nabla}\} \tag{10.10}$$

for a nucleus with spin and isospin zero, where the coefficients b_0 and c_0 are derived from the s- and p-wave parts of the pion-nucleon forward-scattering amplitude, as discussed in §9.1.4, and E_π is the pion total energy in the pion-nucleus centre-of-mass system. A number of corrections to this potential have been developed, particularly for low-energy pions (Ericson and Ericson 1966). An alternative formulation leads to a potential of the form (Lee and McManus 1971, Kisslinger *et al.* 1972)

$$2E_\pi\, U(r) = -4\pi\, \hbar^2 c^2\, A\{(b_0 + \bar{c}_0)\, \rho_m(r) + \tfrac{1}{2}\, c_0\, \nabla^2\, \rho(r)\} \tag{10.11}$$

where $\bar{c}_0 = k^2 c_0$. Because of the different character of the s- and p-wave pion-nucleon interaction the first term in eqn (10.10) gives a repulsive real part for the potential while the first term in eqn (10.11) gives an attractive real part. The difference between the two expressions for $U(r)$ arises from the different assumptions concerning the off-shell behaviour of the pion-nucleon interaction in the two cases (Landau, Phatak, and Tabakin 1972).

A number of analyses of differential, reaction, and total cross-sections in the energy region 20-280 MeV have been carried out using an optical potential and solving a suitable wave equation either exactly or approximately. The Kisslinger potential has been used with reasonable success to fit data on light and medium nuclei (E.H. Auerbach *et al.* 1967, Krell and Barmo 1970, Sternheim and Auerbach 1970, Dedonder 1971) using parameters derived from pion-nucleon phase shifts and a variety of models for the nuclear matter distribution. Lee and McManus (1971) have used impulse approximation to construct an optical potential for ^{12}C which, in the limit of s- and p-waves only, reduces to the form (10.11). They show that this potential leads to reasonable agreement with the data when used in a relativistic optical model code or in a semiclassical calculation. The real part of the potential becomes repulsive just beyond 180 MeV, and both the real and imaginary parts of the potential are substantially different from the conventional shape owing to the contribution from the second term in eqn (10.11). The r.m.s. radii of the calculated potentials at various energies are given in Table 10.2, and it can be seen that, as the energy increases, these values converge towards the value of 2.45 fm assumed for the r.m.s. radius of the matter distribution.

Kujawski and Aitken (1974) have developed a model for the pion optical potential using the approach previously adopted by Lerner and Redish (1972) for proton scattering; this model takes into account the Fermi motion of the target nucleons and utilizes fully off-shell t-matrices calculated from a pion-nucleon potential. Initial calculations give quite good agreement with data on ^{12}C at 180 MeV but poor agreement at 120 MeV.

TABLE 10.2

Root-mean-square radii of pion potentials†

Pion energy (MeV) / R.M.S. radius (fm)	120	150	180	200	230	260	280
Real potential	3.2	3.05	—	2.64	2.66	2.6	2.56
Imaginary potential	3.1	2.96	2.88	2.84	2.87	2.72	2.7

†From Lee and McManus 1971.

Total cross-sections for π^+ and π^- scattering from ^{12}C have been measured for incident pion energies in the range 90-850 MeV (Clough *et al.* 1973), and differences of a few per cent were noted. The 'total' cross-sections in this context are obtained by subtracting estimates for the Coulomb contribution and the Coulomb-nuclear interference term. Since all other evidence suggests that the difference between the proton and neutron distributions in ^{12}C are negligible, it is important to establish that the difference between the measured cross-sections arises from the different effect of the repulsive or attractive Coulomb potential on the wavefunctions for pion scattering from the nuclear potential. Fig. 10.3 shows the experimental results for the cross-section differences together with theoretical predictions from a semiclassical model (Fäldt and Pilkuhn 1972) and from optical model calculations with the Kisslinger potential (10.10) and the Laplacian potential (10.11). It is evident that the calculations give a satisfactory explanation of the source of the difference in the cross-sections. The Laplacian potential gives the best agreement with absolute magnitudes of the cross-section.

There have been several studies of pion-nucleus scattering from ^{12}C using Glauber's multiple-scattering expansion with a gaussian form for the pion-nucleon scattering amplitude:

$$f^{\pm}(q) = f^{\pm}(0)\ \exp(-\tfrac{1}{2}\beta_{\pm}^2 q^2) \qquad (10.12)$$

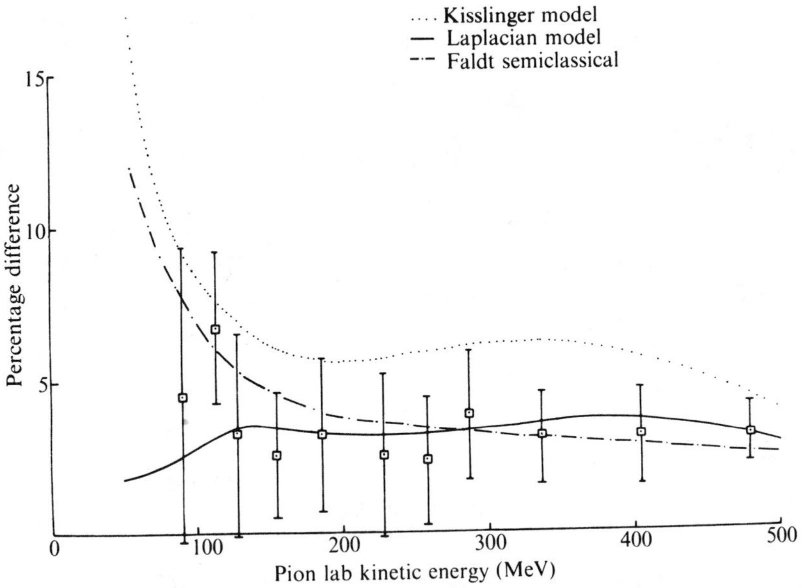

FIG. 10.3. The percentage difference in the total cross-sections for π^+ and π^- scattering from ^{12}C compared with predictions obtained using the Kisslinger optical potential, the Laplacian optical potential, and a semiclassical method. (From Clough *et al.* 1973.)

where $\pi^{\pm}(q)$ are the amplitudes for $\pi^{\pm}$-p (or equivalently for $\pi^{\mp}$-n) scattering (Bjornenak *et al.* 1970, Schmit 1970, Wilkin 1970, Kohmura 1971). Wilkin obtains a real parameter β^2 averaged over the pion-nucleon values by fitting the cross-section data, while Kohmura uses complex parameters obtained from the pion-nucleon phase shifts. There is also a difference in the treatment of the nuclear form factor and Kohmura argues that correct choice of the r.m.s. radius is more important than a detailed description of the density distribution. Fermi motion is not taken into account in either calculation. Fig. 10.4 shows the fit to the total cross-section as a function of energy.

There was initially some surprise that use of both the Kisslinger potential and the Glauber multiple-scattering approach led to reasonable agreement with the data in the vicinity of the (3,3) resonance. This has been studied by Ericson and Hüfner (1970), who conclude that pion-nucleus

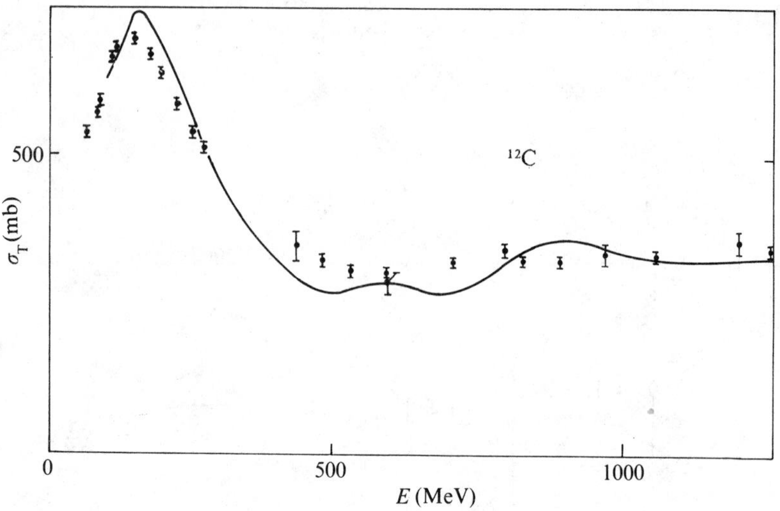

FIG. 10.4. Total cross-sections for pion scattering from ^{12}C calculated using a Glauber model compared with experimental data. (From Kohmura 1971.)

scattering in this region is dominated by strong absorption arising from the resonance character of the pion-nucleon amplitude and that the scattering is insensitive to the finer details of the pion-nucleus interaction. Silbar and Sternheim (1972) have used the same model but have shown that the data are sensitive to the off-shell behaviour of the pion-nucleon amplitude, and this conclusion is supported by a rather different type of calculation by Dedonder (1972). Silbar and Sternheim also find an energy-dependent effective radius consistent with the results of Lee and McManus.

For pion-nucleon scattering, as for nucleon-nucleon scattering, the coefficients of the scattering amplitude are isospin dependent and can be written as in eqn (7.44). The first-order optical potential for π^+ or π^- scattering then becomes

$$U^{\pm}(r) \propto \int \{Z\, f^{\pm}(q)\, F_{\rm p}(q) + N\, f^{\mp}(q)\, F_{\rm n}(q)\} \exp(-i\underset{\sim}{q}.\underset{\sim}{r})\, d^3q \qquad (10.13)$$

or, if we use the forward-scattering approximation and the optical theorem

$$U^{\pm}(r) \propto \frac{1}{2}(\sigma^{+} + \sigma^{-})\, A\, \rho_{m}(r) \pm \frac{1}{2}(\sigma^{-} - \sigma^{+})\, \{N\, \rho_{n}(r) - Z\, \rho_{p}(r)\} \tag{10.14}$$

where $\sigma^{\pm}$ are the $\pi^{\pm}$-p total cross-sections. Thus, a suitable comparison of π^{+} and π^{-} scattering should yield information on the difference between the proton and neutron distributions. If, for example, at a certain energy $\sigma^{-} \gg \sigma^{+}$ it follows that a π^{+} beam will mainly scatter from neutrons while a π^{-} beam will mainly scatter from portons, or if $\sigma^{+} \gg \sigma^{-}$ the reverse is the case. Kisslinger *et al.* (1972) have studied the validity of this argument with respect to the differential cross-sections for $\pi^{\pm}$ scattering at 50-300 MeV since, in this region, the dominance of the (3,3) resonance means that σ^{-} is negligible in comparison with σ^{+}. They find that the depth and separation of the minima in the differential cross-sections for $\pi^{\pm}$ scattering from ^{12}C, ^{40}Ca, and ^{208}Pb are sensitive to changes in the parameters of the distributions. In these calculations it is again important to include the Coulomb scattering accurately since this affects the π^{+} and π^{-} scattering differently. It is also important to take account of the Fermi motion of the nucleons in the nucleus, and a plausible method of doing this is to average the pion-nucleon total cross-sections over the nucleon momentum distribution (Crozon *et al.* 1965). In calculations on ^{12}C and ^{40}Ca at 1 GeV Kembhavi (1971) found that the position and depth of the minima are also very sensitive to the magnitude and sign of the real part of the pion-nucleon scattering and also the parameter β defined in eqn (10.12), and that lack of precise knowledge of these quantities was likely to inhibit extraction of nuclear size information from differential cross-sections.

Uncertainties in the pion-nucleon data and more fundamental uncertainties arising from approximations made in derivation of the potential may be reduced in importance by studying a ratio of cross-sections rather than absolute values. This procedure also minimizes the effect of experimental uncertainties. The classic example of this method is the study of the ratio of the reaction cross-sections $\sigma_R^{\pm}$ for π^+ and π^- scattering from the same target nucleus at the same energy. The original experiment of this type was carried out by Abashian *et al.* (1956) at incident pion momenta of 0.84 GeV/*c* and 1.24 GeV/*c* on Pb. It can be seen from the plot of $\sigma^{\pm}$ against pion momentum in Fig. 10.5 that there is a substantial difference in the pion-nucleon total cross-sections

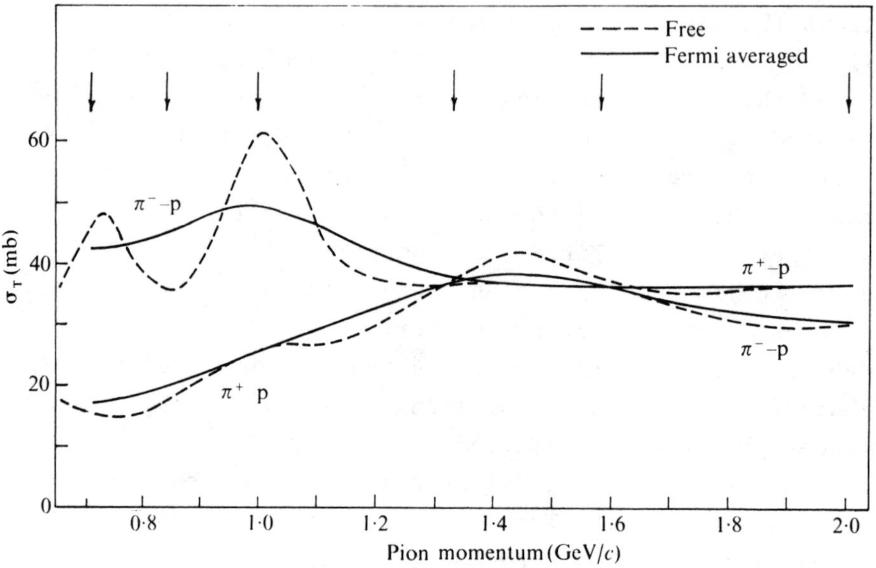

FIG. 10.5. The free and Fermi-averaged total cross-sections for pion-nucleon scattering as a function of incident pion momentum. (From Allardyce *et al.* 1973.)

at 840 MeV/*c*, so that a measurement at this point should give information on the difference between the proton and neutron distributions in Pb. Calculations using a semiclassical method (Elton 1961*a*) and solving a Klein-Gordon equation (E.H. Auerbach, Qureshi, and Sternheim 1968) indicated that only a very small difference was compatible with the data and that there was a preference for $\langle r^2\rangle_n^{1/2} - \langle r^2\rangle_p^{1/2}$ to be small and negative.

The error on the experimental ratio σ_R^-/σ_R^+ obtained by Abashian *et al.* was 10 per cent. Measurements of these ratios for a range of nuclei have recently been made by Allardyce *et al.* (1972, 1973) with an accuracy of ~1 per cent, which makes possible a much more precise comparison of proton and neutron distributions. The nuclei studied were C, Al, Ca, Ni, Sn, ^{120}Sn, Ho, Pb, and ^{208}Pb. The pion momenta at which the measurements were made are indicated by arrows on Fig. 10.5. At 0.71, 0.84, and 1.00 GeV/*c* the differences between the pion-nucleon total cross-sections are large, so that there should be strong dependence of the data on the proton and neutron distributions, and the measurement at 0.84 GeV/*c* can be compared with the earlier result. At 1.36 and 1.58 GeV/*c* the Fermi-averaged values of $\sigma^{\pm}$ are identical, and it follows from eqns (10.13) or (10.14) that, even for a target nucleus with $N \neq Z$ and $\rho_n \neq \rho_p$, the nuclear potentials for $\pi^{\pm}$ scatter- are identical, and within the framework of this theory the only effect causing the ratios to depart from unity is the Coulomb scattering. These measurements therefore provide important checks of the validity of the theoretical approach and the accuracy of the calculations. An additional measurement has been included at 2.00 GeV/*c* where the difference between $\sigma^{\pm}$ is rather small but the pion-nucleon cross-sections are rather flat so that resonance effects are less important.

A detailed description of the method of calculation is given by Allardyce *et al.* (1973). It appears that an accurate treatment of the Coulomb scattering and use of Fermi-averaged pion-nucleon cross-sections is essential, as was previously noted by E.H. Auerbach *et al.* (1968). (The Fermi-averaged cross-sections have been deduced using a variety of realistic momentum distributions (Murugesu 1971), but fortunately a unique set of averaged values is obtained and these are plotted in Fig. 10.5.) The parameters $\beta_{\pm}^2$ have been obtained by fitting experimental data for pion-nucleon cross-sections, and it is found that use of the gaussian form (10.12) for the scattering amplitude instead of the forward-scattering approximation has a significant effect on the absolute values but a negligible effect on the ratios. Similarly, estimates of the contribution from the second-order term in the multiple-scattering expansion for the potential (Foldy and Walecka 1969) indicate that this has a small effect on the absolute values but a negligible effect on the ratios.

A comparison of the experimental and theoretical values of the ratios of pion-nucleus reaction cross-sections at 1.36 GeV/*c* is given in Table 10.3. The agreement is excellent.

TABLE 10.3

Experimental and theoretical values for the ratio σ_R^-/σ_R^+ *at an incident pion momentum of 1.36 GeV/c* †

Target nucleus	C	Al	Ca	Ni
Experimental ratio	1.007 ± 0.007	1.014 ± 0.009	1.015 ± 0.007	1.027 ± 0.005
Theoretical ratio	1.006	1.012	1.016	1.021

Target nucleus	Sn	Ho	Pb
Experimental ratio	1.030 ± 0.005	1.041 ± 0.009	1.043 ± 0.007
Theoretical ratio	1.032	1.038	1.043

† From Allardyce *et al.* 1973.

Provided that the momentum dependence of the pion-nucleon amplitudes is taken into account through eqn (10.12), the agreement between the experimental and theoretical values for the absolute values of $\sigma_R^{\pm}$ at the same pion momentum is better than 5 per cent for all nuclei and is better than 1 per cent for Pb. The calculations have been carried out with many of the density distributions discussed in Chapter 2, for example the Elton-Swift distributions for ^{12}C and ^{40}Ca, the Batty-Greenlees (BG) distribution for ^{40}Ca and ^{208}Pb, the Zaidi-Darmodjo (ZD) and Negele (NEG) distributions for ^{208}Pb, the hydrodynamical model (HYD) for Ni, Sn, and ^{208}Pb, and with identical Fermi distributions (F). The agreement with experiment confirms that these models give a generally correct description of the nuclear matter distributions. The differences between the models become evident at the lower pion momenta studied, and Allardyce *et al.* conclude, from all nuclei studied, that those distributions in which the neutron r.m.s. radius is the same as the proton r.m.s. radius to within ±0.1 fm are consistent with the data, while those which predict a large difference in the r.m.s. radii, such as the Batty-Greenlees and Negele distributions for ^{208}Pb, are not consistent with the data. The magnitude of the discrepancy between theory and experiment for these distributions is shown in Fig. 10.6. For light and medium nuclei this conclusion is

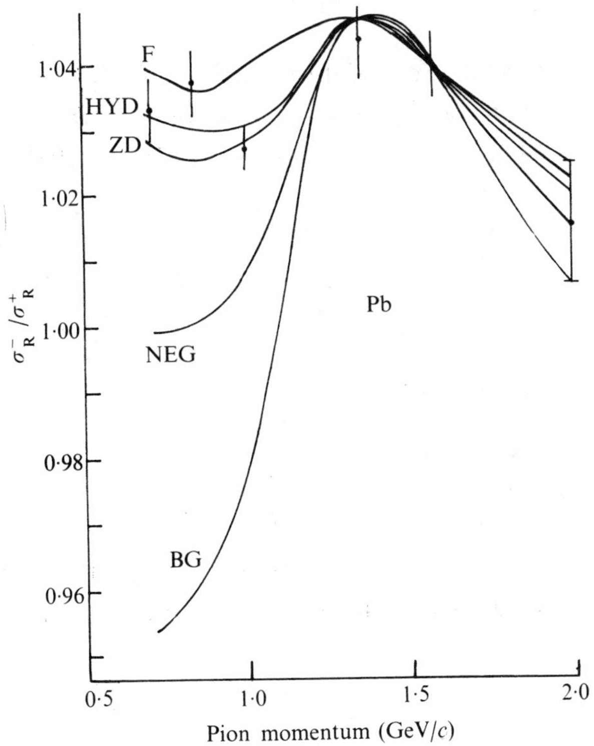

FIG. 10.6. The ratio of reactions for π^- and π^+ scattering from Pb calculated for various matter distributions (see text) and compared with the experimental data. (From Allardyce *et al.* 1973.)

consistent with the calculations and other experiments discussed in previous chapters, while for the heaviest nuclei there is disagreement with the Hartree-Fock and Thomas-Fermi calculations and with the conclusions drawn from some of the other experiments.

We have endeavoured to stress that conclusions about the relative merits of various distributions should be accompanied by some discussion of what region of the nucleus is being probed. In the present context this can conveniently be done using the semiclassical approximation which gives

$$\sigma_R = 2\pi \int_0^\infty \{1 - |\eta(b)|^2\}\, b\, db \tag{10.15}$$

$$\sigma_T = 4\pi \int_0^\infty \{1 - \mathrm{Re}\, \eta(b)\}\, b\, db. \tag{10.16}$$

In the semiclassical approximation $\eta(b) = \exp\{i\chi(b)\}$, where $\chi(b)$ is given by eqn (7.31), and substituting into this equation the potential (10.14) with the correct kinematic factors we obtain

$$\eta(b) = \exp\{-\tfrac{1}{2}\sigma_1 T_1(b) \mp \tfrac{1}{2}\sigma_2 T_2(b)\} \tag{10.17}$$

where

$$\sigma_1 = \tfrac{1}{2}(\sigma^+ + \sigma^-), \qquad \sigma_2 = \tfrac{1}{2}(\sigma^- - \sigma^+) \tag{10.18}$$

$$T_1(b) = \int_{-\infty}^{\infty} A\,\rho_m(b,z)\, dz, \qquad T_2(b) = \int_{-\infty}^{\infty} \{N\rho_n(b,z) - Z\rho_p(b,z)\} dz. \tag{10.19}$$

Glauber (1967) calls $T_1(b)$ the thickness function. As in Chapter 8, we also defined the effective nucleon number

$$N^{\pm}(A,\sigma_1,\sigma_2) = 2\pi \int_0^\infty T_1(b) \exp\{-\sigma_1 T_1(b) \mp \sigma_2 T(b)\}\, b db. \tag{10.20}$$

Substituting (10.17) into (10.15), and then differentiating with respect to σ_1, it is easy to show that

$$\frac{\partial\sigma_R^{\pm}}{\partial\sigma_1} = N^{\pm}(A,\sigma_1,\sigma_2) \tag{10.21}$$

and hence

$$\sigma_R^{\pm} = \int_0^{\sigma_1} N^{\pm}(A,\sigma_1',\sigma_2')\ d\sigma_1'. \tag{10.22}$$

Similarly, it may be shown that

$$\sigma_T^{\pm} = \int_0^{\sigma_1} N^{\pm}(A,\tfrac{1}{2}\sigma_1',\tfrac{1}{2}\sigma_2')\ d\sigma_1'. \tag{10.23}$$

Thus, in order to examine the localization of the absorption of the projectile by the nucleus we may examine the integrands in eqns (10.15), (10.16), or (10.20) as functions of impact parameter b. Allardyce *et al.* (1973) have examined the localization of pion absorption for the momentum range of their experiment and show that the principal contribution comes from the transition region of the nucleus, i.e. the 90-10 per cent region for a Fermi matter distribution. The same conclusion was reached by Franco (1972) from his study of total cross-sections for 1 GeV neutrons on nuclei. It appears that the integrands fall exponentially at large distances owing to the fall in the density while the factor b eliminates the contributions from the inner region. These conclusions encourage the view that these measurements can be used to study neutron distributions since they are sensitive to the region of the nucleus where the proton distributions are best known from electromagnetic methods.

Reaction cross-sections for π^- scattering from a range of nuclei have been measured by Allaby *et al.* (1971) for pion momenta in the range 20-60 GeV/c. They also measured reaction cross-sections for negative kaons and antiprotons at momenta in the range 20-40 GeV/c. These data have been analysed by Abul-Magd, Alberi, and Bertocchi (1969) and by Batty and Friedman (1972) using the semiclassical approximation. In the latter analysis an optical potential of the form given by eqn (10.13) is used, but since the pion-nucleon amplitudes are fairly constant with pion momentum it is not necessary to consider resonance effects or to carry out the Fermi-averaging procedure. Good agreement with the absolute values for C, Al, Sn, and Pb is obtained with identical proton and neutron distributions and, as at lower pion energies, the Batty-Greenlees and Negele distributions do not yield agreement with data. The results for K^- and $\bar{p}$ scattering are consistent with the π^- data.

10.4.2. *Charge-exchange reactions*

The presence of the $\underset{\sim}{t}_\pi \cdot \underset{\sim}{T}$ operator in the pion optical potential allows the possibility of single charge-exchange reactions completely analogous to the nucleon charge-exchange reactions. Because the pion has isospin unity, the possibility also arises of double charge-exchange reactions which can occur through the $(\underset{\sim}{t}_\pi \cdot \underset{\sim}{T})$ acting in second order or through an additional isotensor term $(\underset{\sim}{t}_\pi \cdot \underset{\sim}{T})^2/A^2$. Theoretical and experimental work on pion charge-exchange reactions up to 1968 has been reviewed by Koltun (1969) and further theoretical studies in the 20-120 MeV region have been carried out by Koren (1969) using the Kisslinger potential. In general, the cross-sections are disappointintly small, except for low energies near threshold.

A theoretical study of pion charge-exchange reactions in terms of nuclear size effects has been carried out by Rost and Edwards (1971) in the energy range 100-300 MeV. They show that the total cross-section for charge-exchange has a deep minimum at ~140 MeV owing to absorption associated with the (3,3) resonance, but that the cross-section rises at lower energies and higher energies. The variation of the total cross-section for ^{93}Nb at 200 MeV with the halfway radius of the neutron distribution is given in Table 10.4. Experimental studies (Zaider *et al.* 1973) show that the excitation curve for $^{13}C(\pi^+,\pi^0)$ is relatively flat. The Laplacian potential (10.11) gives good agreement with the limited data but the Kisslinger potential does not. The excitation function is very very sensitive to the isovector term in the potential arising from the excess neutron distribution.

Single charge-exchange reactions for high-energy pions and kaons have been studied by Kofoed-Hansen and Margolis (1969). They argue that the yield in such reactions as (π^+,π^0), (K^+,K^0) is proportional to an effective neutron num-

TABLE 10.4

Total cross-sections for pion charge-exchange reactions on ^{98}Nb at 200 MeV†

$a_p = a_n$ (fm)	c_p (fm)	c_n (fm)	$\sigma(\pi^+,\pi^0)$ (μb)	$\sigma(\pi^+,\pi^-)$ (nb)
0.52	1.08 $A^{1/3}$	1.02 $A^{1/3}$	11	9
0.52	1.08 $A^{1/3}$	1.08 $A^{1/3}$	53	26
0.52	1.08 $A^{1/3}$	1.14 $A^{1/3}$	190	190

†From Rost and Edwards 1971.

ber, while in the reactions (π^-,π^0), (K^-,K^0) the yield is proportional to an effective proton number. They then construct certain ratios of these cross-sections which should be particularly sensitive to differences in the proton and neutron distributions in nuclei. Unfortunately, these experiments appear extremely difficult at the present time.

10.5. PION PRODUCTION

The production of pions in nucleon scattering from nuclei has been studied in experiments with intermediate-energy protons by Heer *et al.* (1966) at 600 MeV and by Haddock, Zeller, and Crowe (1964) at 725 MeV. The energies of the outgoing pions are in the range 100-400 MeV. These results have been interpreted in terms of neutron and proton distributions using a model originally due to Margolis (1968).

In the model of Margolis it is assumed that the pion production process proceeds through formation of the Δ(1236) isobar which subsequently decays into a pion and a nucleon. Positive pions can then be formed in the reactions

$$p + p \rightarrow \Delta^{++} + n \rightarrow \pi^+ + p + n \tag{10.24}$$

$$\rightarrow \Delta^{+} + p \rightarrow \pi^+ + n + p \tag{10.25}$$

$$p + n \rightarrow \Delta^{+} + n \rightarrow \pi^+ + n + n \tag{10.26}$$

and negative pions are formed in the reaction

$$p + n \rightarrow \Delta^0 + p \rightarrow \pi^- + p + p. \qquad (10.27)$$

The relative strengths of the reactions (10.24)-(10.27) are given by the appropriate Clebsch-Gordan coefficients and are such that, if $\rho_p = \rho_n$, the ratio of production of π^+ to π^- is $(10Z + N)/N$. This simple argument appears to give a qualitative explanation of the observed ratios.

This model may be made more quantitative (Glauber 1967, Margolis 1968, Hirt 1969) by using the semiclassical approximation with an optical potential of the form (7.6). The summed cross-section for pion production on nuclei then becomes

$$\frac{d\sigma}{d\Omega}(\pi^\pm) = \frac{d\sigma}{d\Omega}(\Delta)\ 2\pi \int_0^\infty T^\pm(b)\ \exp\{-\sigma_1\ T_1(b)\}\ b\ db \qquad (10.28)$$

where we have omitted terms containing the two-nucleon correlation function; $(d\sigma/d\Omega)(\Delta)$ is the differential cross-section for isobar production in nucleon-nucleon scattering,

$$10\ T^+(b) = \int_{-\infty}^\infty [10Z\ \rho_p(b,z) + N\ \rho_n(b,z)]\ dz \qquad (10.29)$$

$$10\ T^-(b) = \int_{-\infty}^\infty N\ \rho_n(b,z)\ dz \qquad (10.30)$$

and σ_1, $T_1(b)$ are as defined in eqns (10.18) and (10.19). Integrating eqn (10.28) over the solid angle, we obtain

$$\sigma(\pi^\pm) = \sigma(\Delta)\ N^\pm(A,\sigma_1) \qquad (10.31)$$

where $\sigma(\Delta)$ is the total cross-section for isobar production in nucleon-nucleon scattering. The formulae (10.28) and (10.31) have been used to interpret the data with a variety of assumptions about the nuclear distributions. It should be noted that in deriving these equations we have assumed the validity of the closure approximation, have set $\sigma_2 = 0$, where σ_2 is defined by eqn (10.18), and have replaced the distortion of the outgoing pion by distortion of a nucleon. Any modification of σ_1 and $\sigma(\Delta)$ due to Fermi motion of the nucleons in the nucleus has also been neglected.

The usual procedure is to extract $N^{\pm}$ from the data and to compare it, as a function of A, with calculated values. Margolis (1968) takes $\rho_p = \rho_n$ and uses gaussian and Fermi shapes with parameters derived from charge distributions. He obtains reasonable agreement with the data using a Fermi distribution. Hirt (1969) has used Fermi distributions and has examined the sensitivity of $N^{\pm}(A,\sigma_1)$ to differences in the neutron and proton distributions. He concludes that the data are consistent with $\langle r^2\rangle_n^{1/2} - \langle r^2\rangle_p^{1/2} \sim 0.6$ fm in ^{208}Pb. Lombard, Auger, and Basile (1971) have made similar calculations using distributions derived from the single-particle potential of Beiner and from the energy-density formalism (see §2.3) and find that both distributions give an equally acceptable fit. They conclude that the r.m.s. radius difference in ^{208}Pb is substantially less than the value obtained by Hirt. It appears that this method does not as yet provide precise information on the difference between proton and neutron distributions although it is in principle capable of doing so.

A number of very interesting studies of pion production have been made at lower proton energies. These are concerned with the nature of the production process at threshold or the mechanism of excitation of specific nuclear states. They do not directly provide information on nuclear distributions.

10.6. BETA-DECAY

For β-decay transitions which satisfy the Fermi selection rules, the ft-value is given by (Blin-Stoyle 1969a)

$$ft(1 + \delta_R) = 2\pi^3 \log 2/G_V^2|M_V|^2 \qquad (10.32)$$

where δ_R is the electromagnetic radiation correction, G_V is the polar vector coupling constant, and M_V is the nuclear matrix element for allowed transitions between an initial state a and a final state b

$$M_{\mathrm{V}} = \langle b|T_{\pm}|a\rangle = \langle b|\sum_{j} \tau_{\pm}^{(j)}|a\rangle. \tag{10.33}$$

For superallowed $0^{+} \rightarrow 0^{+}$ transitions the nuclear matrix element can be written as

$$|M_{\mathrm{V}}|^{2} = 2(1 - \delta_{\mathrm{C}}) \tag{10.34}$$

where δ_{C} represents the correction due to charge-dependent effects in the nucleus.

The calculation of f usually involves radial wavefunctions for the electron which are solutions of the Dirac radial equation with a potential due to a uniform nuclear charge distribution of radius R and a screening potential due to the electron cloud. These wavefunctions are evaluated at the radius $r = R$, although they are functions of the radial coordinate r and should contribute to the integrand of the nuclear matrix element. Behrens and Bühring (1967) have shown that a suitable modification of eqn (10.32) is

$$\tilde{f}t(1 + \delta_{\mathrm{R}})\,\overline{C(W)} = 2\pi^{3} \log 2/G_{\mathrm{V}}^{2}|M_{\mathrm{V}}|^{2} \tag{10.35}$$

where $\tilde{f}$ is calculated from electron wavefunctions evaluated at $r = 0$, and $\overline{C(W)}$ is a correction term given by

$$\overline{C(W)} = 1 + \frac{A_{1}\,\langle b|\sum_{j} \tau_{\pm}^{(j)}\, r_{j}^{2}|a\rangle}{R^{2}\,\langle b|\sum_{j} \tau_{\pm}^{(j)}|a\rangle}\,. \tag{10.36}$$

The total energy of the electron is $W(mc^{2})$, A_{1} is a numerical coefficient, and $\langle b|\sum \tau_{\pm}^{(j)}\, r_{j}^{2}|a\rangle$ is the nuclear matrix element for second forbidden transitions. Values of $\tilde{f}t$ for superallowed $0^{+} \rightarrow 0^{+}$ transitions calculated without screening for various values of $R = r_{0}\, A^{1/3}$ are given in Table 10.5.

Blin-Stoyle (1969*b*) has shown that for superallowed $0^{+} \rightarrow 0+$ transitions the quantity $\overline{C(W)}$ can be expressed in terms of the r.m.s. radii of the proton and neutron distributions in the neutron-rich nucleus, i.e.

TABLE 10.5

Dependence of $\tilde{\text{f}}\text{t}$ *values (without screening) on the nuclear radius* $R = r_0 A^{1/3}$†

	r_0 (fm)			
Decay	1.2	1.3	1.4	1.5
$^{14}O \rightarrow {}^{14}N$	3075	3074	3074	3074
$^{26}Al^m \rightarrow {}^{26}Mg$	3038	3037	3036	3035
$^{34}Cl \rightarrow {}^{34}S$	3096	3094	3092	3091
$^{42}Sc \rightarrow {}^{42}Ca$	3077	3074	3072	3069
$^{46}V \rightarrow {}^{46}Ti$	3087	3083	3081	3078
$^{50}Mn \rightarrow {}^{50}Cr$	3077	3073	3070	3067
$^{54}Co \rightarrow {}^{54}Fe$	3082	3078	3074	3071

†From Behrens and Bühring 1967.

$$\overline{C(W)} = 1 + 0.3\, A_1 \left[\frac{\langle r^2\rangle_n + \langle r^2\rangle_p}{\langle r^2\rangle_p} + \frac{A}{2}\,\frac{\langle r^2\rangle_n - \langle r^2\rangle_p}{\langle r^2\rangle_p} \right]. \tag{10.37}$$

Thus, the quantity $\overline{C(W)}$ is sensitive to the difference $\langle r^2\rangle_n - \langle r^2\rangle_p$ particularly for large A. Using the values of $\tilde{f}t$ and A_1, given by Behrens and Bühring, Blin-Stoyle obtained the results given in Table 10.6. Since the ft-values are expected to increase with increasing Z these results suggest that $\langle r^2\rangle_n - \langle r^2\rangle_p \lesssim 0$. More precise conclusions about the differences in r.m.s. radii require more accurate data. Reliable estimates of δ_c are also required and these depend on similar effects to those discussed in §10.2, for example isospin impurity in the core, and also depend on the nuclear model (Blin-Stoyle 1969*a*).

Behrens and Bühring (1972) have examined the dependence

TABLE 10.6

ft *values for superallowed transitions*[†]

Decay	$ft(1+\delta_R)\overline{C(W)}$		
	$\langle r^2\rangle_n = \langle r^2\rangle_p$	$\langle r^2\rangle_n - \langle r^2\rangle_p = 0.2$ fm	$\langle r^2\rangle_n - \langle r^2\rangle_p = -0.2$ fm
$^{14}O \rightarrow {}^{14}N$	3143 ± 11	3144 ± 11	3142 ± 12
$^{26}Al^m \rightarrow {}^{26}Mg$	3096 ± 9	3092 ± 9	3100 ± 9
$^{34}Cl \rightarrow {}^{34}S$	3149 ± 19	3141 ± 19	3158 ± 19
$^{42}Sc \rightarrow {}^{42}Ca$	3122 ± 9	3107 ± 9	3136 ± 9
$^{46}V \rightarrow {}^{46}Ti$	3132 ± 8	3113 ± 8	3146 ± 8
$^{50}Mn \rightarrow {}^{50}Cr$	3116 ± 9	3093 ± 9	3132 ± 9
$^{54}Co \rightarrow {}^{54}Fe$	3118 ± 18	3090 ± 18	3135 ± 18

[†]From Blin-Stoyle 1969*b*.

of $\tilde{f}t$ and $C(W)$ on the shape of the nuclear charge distribution by using several distributions with the same r.m.s. radius and showed that the effect is negligible. The same conclusion was reached by Asai and Ogata (1974). However, even for a uniform distribution there is sensitivity to $R = r_0A^{1/3}$ for medium-mass nuclei. Wilkinson (1973) has examined the latter point more quantitatively and concluded that it is necessary to use r_0 values accurate to 2 per cent to give *ft* values reliable to 1 per cent for medium-mass nuclei.

The difference in the *ft* values of *GT* decays involving mirror nuclei is found to be particularly sensitive to the difference between the binding energies of the proton giving β^+-decay and the neutron giving β^--decay (Wilkinson 1970, 1971). In a further study Towner (1973) has shown that there is considerable sensitivity to the single-particle potentials in which the wavefunctions of the decaying proton or neutron are generated and that there is considerable uncertainty in defining these potentials.

10.7. KAON-NUCLEUS INTERACTIONS

10.7.1. *Regenerative scattering*

The neutral kaon K^0 and its antiparticle $\bar{K}^0$ are distinct and are not separately eigenstates of the charge conjugation operator. It is convenient, therefore, to represent a neutral kaon beam as a mixture of a long-lived component

$$K_L^0 = \frac{1}{\sqrt{2}} (|K^0\rangle + |\bar{K}^0\rangle)$$

and a short-lived component

$$K_S^0 = \frac{1}{\sqrt{2}} (|K^0\rangle - |\bar{K}^0\rangle).$$

After a time of the order of a nanosecond the kaon beam contains only the long-lived component. The scattering of such a beam may be described in terms of the scattering amplitudes f_{K^0N} and $\bar{f}_{\bar{K}^0N}$ for the K^0 and $\bar{K}^0$ components, respectively. It follows that if $f \neq \bar{f}$ the composition of the state after scattering will be different from that before scattering and the final state may be resolved into K_L^0 and K_S^0 components. This process is known as regenerative scattering, and may be represented by the equation

$$K_L^0 + N \rightarrow K_S^0 + N. \tag{10.38}$$

The presence of the K_S^0 component can be detected by looking for decays of the type $K_S^0 \rightarrow \pi^+ + \pi^-$.

Foeth *et al.* (1970) have measured the elastic differential cross-section for regenerative scattering from Cu and Pb for incident kaon momenta in the range 2.5-6.5 GeV/*c* and for momentum transfer q up to 0.17 GeV/*c*. Results are shown in Fig. 10.7.

The kaon-nucleus scattering amplitudes $f(q^2)$ and $\bar{f}(q^2)$ have been calculated using eqns (7.31) and (7.33). The optical potentials for kaon-nucleus scattering have been constructed using the forward-scattering approximation and the

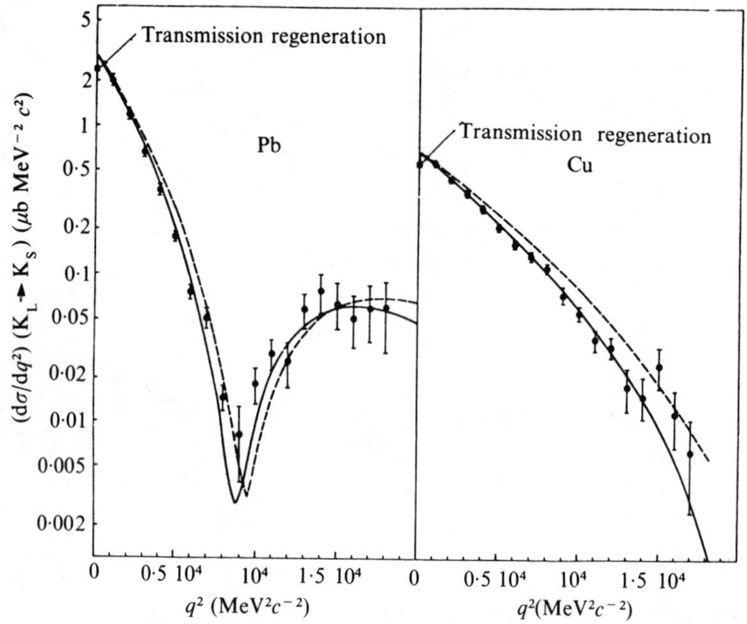

FIG. 10.7. Angular distributions for kaon regenerative scattering from Cu and Pb. The solid line shows the best fit to the data and the broken line shows results obtained with $\rho_n = \rho_p$. (From Foeth *et al.* 1970).

single-scattering approximation to give

$$U(r) \propto f_{K^0p}(0)\ Z\ \rho_p\ r) + f_{K^0n}(0)\ N\ \rho_n(r) \tag{10.39a}$$

$$\bar{U}(r) \propto \bar{f}_{\bar{K}^0p}(0)\ Z\ \rho_p(r) + \bar{f}_{\bar{K}^0n}(0)\ N\ \rho_n(r). \tag{10.39b}$$

The imaginary parts of the forward-scattering amplitudes can be obtained from the optical theorem and isospin symmetry with expressions of the form

$$\frac{4\pi}{k}\ \mathrm{Im}\ f_{K^0p}(0) = \sigma_T(K^0p) = \sigma_T(K^+n). \tag{10.40}$$

The real parts of f and $\bar{f}$ have been obtained from dispersion relations (Boehm *et al.* 1968) and are found to be small. At 4 GeV/c, $\mathrm{Im}\ f_{K^0p}(0) \simeq \mathrm{Im}\ f_{K^0n}(0)$ and $\mathrm{Im}\ \bar{f}_{K^0p} \simeq 0.8\ \mathrm{Im}\ \bar{f}_{K^0n}(0)$.

Thus, regenerative scattering is sensitive to the difference between the neutron and proton distribution although not nearly as sensitive as the comparison of π^+ and π^- scattering.

The calculations of Foeth *et al.* (1970) indicate that the regenerative scattering occurs predominantly in the nuclear transition region. The broken curves in Fig. 10.7 were obtained using equal parameters for the proton and neutron distributions, which were taken to be of Fermi shape, while the full curves were obtained with ~10 per cent increase in the neutron halfway radius or ~30 per cent increase in the neutron-diffuseness parameter.

This method appears to offer scope for comparison of neutron and proton distributions, but at this stage it is difficult to assess the reliability of the results since we do not know whether other, more realistic, distributions with equal r.m.s. radii will fit the data. Nor do we know the magnitude of corrections to the kaon-nucleus scattering due to the nucleon Fermi momentum, multiple scattering, etc., although such corrections can in principle be evaluated.

10.7.2. *Kaon absorption*

As we have seen in Chapter 9 the negative kaon can form a K-mesic atom. The kaon makes transitions through the atomic states, accompanied by the emission of Auger electrons and X-rays, until it is sufficiently close to the nucleus for nuclear capture through the strong interaction to dominate over the atomic transitions. The competition between the nuclear capture and the atomic cascade processes leads to a cut-off in the X-ray yield at some lower limit n_0 in the principal quantum number of the atomic states. Since the nuclear capture probability depends on the nuclear density distribution the X-ray measurements lead to nuclear size information, as described in Chapter 9.

There is a further method of obtaining nuclear size information from K-mesic atoms which yields more direct evidence on the relative proportions of protons and neutrons in the surface region. This arises from examination of the particular final states arising from the interaction of a kaon with a nucleon. The possible processes which lead to one pion are listed in Table 10.7. Neglecting the kaon energies, which are of the order of a few hundred keV, and taking the nucleon to be at rest, the total available energy is 1432 MeV in the K^-+p system and 1433 MeV in the K^-+n system. This means that the processes listed in Table 10.7 are energetically possible and there is an energy release in each case. It can be seen

TABLE 10.7

Kaon-nucleon interactions leading to one pion in the final state

Initial state	Final state	Isospin T	Available energy (MeV)
$K^- + p$	$\Sigma^- + \pi^+$	0,1	96.6
	$\Sigma^+ + \pi^-$	0,1	103.1
	$\Sigma^0 + \pi^0$	0,1	105.6
	$\Lambda^0 + \pi^0$	1	181.8
$K^- + n$	$\Sigma^- + \pi^0$	1	102.4
	$\Sigma^0 + \pi^-$	1	102.3
	$\Lambda^0 + \pi^-$	1	178.5

from the table that the emission of a $(\Sigma^\pm\ \pi^\mp)$ pair can be taken as a signature for kaon capture by a proton, while the emission of a Σ^- without an accompanying π^+ is a signature for capture by a neutron, so that the ratio

$$R_{\mathrm{pn}} = \frac{\text{number of events } (\Sigma^+\ \pi^- + \Sigma^-\ \pi^+)}{\text{number of events } (\Sigma^-\ \text{no}\ \pi^+)} \tag{10.41}$$

may be connected with the relative proportions of protons and neutrons encountered. Following Bethe and Siemens (1970) we use b^T to denote the transition rate for the process K^- + nucleon → hyperon + π in the isospin channel T. The ratio (10.41) can then be written as

$$R_{\mathrm{pn}} = \frac{\text{number of protons encountered}}{\text{number of neutrons encountered}} \times \frac{b^1(\mathrm{p}\Sigma) + \frac{2}{3}\, b^0(\mathrm{p}\Sigma)}{b^1(\mathrm{n}\Sigma)} \tag{10.42}$$

where the coefficient $\frac{2}{3}$ arises from the Clebsch-Gordan coefficients for isospin coupling.

It has been noted by Davis *et al.* (1967) that the presence of an Auger electron track may be taken as the signature for a K^--capture on a heavy nucleus, whereas the observation of a recoil track may be taken as an indication of capture on a light nucleus. This leads to the experimental ratios for R_{pn} given in Table 10.8. The important result here is that

TABLE 10.8

Ratio R_{pn} *for kaon absorption*

Nuclei	Experiment Burhop (1967)	Theory		
		Bethe & Siemens (1970)	Aslam & Rook (1970*b*)	Bardeen & Torigoe (1971)†
Hydrogen	7.1 ± 1.5			
Deuterium	10.2 ± 2.0			
Helium	11.0 ± 1.8			
Light nuclei (C,N,O)	22.4 ± 4.4	3.04		12.9
Heavy nuclei (Ag,Br)	4.5 ± 0.6	0.73		7.2
$R_{pn}^{L/H}$ light/heavy	$5.0^{+1.2}_{-0.8}$	4.16	2.32	1.8

†Kim solution.

capture on neutrons is apparently five times more probable in heavy nuclei than in light nuclei, or alternatively that capture by protons in heavy nuclei is suppressed.

The analysis of the ratio R_{pn} has proceeded along the lines of the analysis of the K-mesic X-ray data and in many cases a satisfactory description of the combined data has been sought. The nuclear capture rate from an atomic orbit with quantum numbers n,l may be written as

$$\Gamma_i(n,l) = B^i \int |\phi_K^{nl}(\underset{\sim}{r})|^2 \rho_i(\underset{\sim}{r})\, d^3r \qquad i = n,p \quad (10.43)$$

where, as in Chapter 9, we are using first-order perturbation theory and closure together with a zero-range K-N interaction. The coefficients B are given by

$$B^{\mathrm{p}} = \frac{1}{2}\, b^1(\mathrm{p}) + \frac{1}{2}\, b^0(\mathrm{p}) \qquad (10.44a)$$

$$B^{\mathrm{n}} = b^1(\mathrm{n}). \qquad (10.44b)$$

For a free nucleon, the relation

$$b^{\mathrm{T}} = v\, \sigma^{\mathrm{T}} \qquad (10.45)$$

applies, where v is the relative velocity in the KN system. Burhop (1967) has evaluated the coefficients using (10.44) and (10.45) taking the K^- to be at rest, the nucleon to have a mean Fermi momentum of 160 MeV/c,† and taking the average reaction cross-sections to be σ_{p} = 338 mb and σ_{n} = 155 mb. These values are given in Table 10.9. It can be seen that $B^{\mathrm{p}}/B^{\mathrm{n}} \sim 2$ but that the ratio $(b^1 + \frac{2}{3}\, b^0)/b^1$ which appears in R_{pn} is ~3.

TABLE 10.9

The square of the strong-interaction amplitudes for the K^-N system system in units of fm^3 s^{-1}

	B^{p}	B^{n}	$B^{\mathrm{p}}/B^{\mathrm{n}}$
Burhop (1967)	3.27×10^{23}	1.50×10^{23}	2.18
Bloom *et al.* (1969)	5.36×10^{24}	1.10×10^{24}	4.88
Bethe and Siemens (1970)	3.07×10^{24}	1.49×10^{24}	2.06
Bardeen and Torigoe (1971) (Kim solution)	5.22×10^{24}	1.08×10^{24}	4.85
Bardeen and Torigoe (1971) (Martin-Sakitt solution)	3.60×10^{24}	9.25×10^{23}	3.90

†For nucleons in the nuclear surface, this value is rather high. See footnote on p. 445, Chapter 9.

The ratio $(\Gamma_p/\Gamma_n)_{light}/(\Gamma_p/\Gamma_n)_{heavy}$ is independent of the B coefficients and should be equal to the experimental value of ~5 for the ratio $R_{pn}^{L/H}$ of R_{pn} in light nuclei to R_{pn} in heavy nuclei. Burhop (1967) obtained a ratio of 1.2 using Fermi density distributions with equal parameters for the protons and neutrons, and further showed that reasonable agreement with the experimental value is obtained for a 50 per cent increase in the diffuseness parameter of the neutron distribution or an increase of 0.8 fm in the neutron halfway radius. Rook (1968*a*) has shown that the approximations involved in deriving the expression (10.43) do not signifianctly affect this conclusion. However, the simple calculation may be challenged on two counts: (i) the Fermi distribution may well not give an adequate description of the nucleon distribution in the region $\rho(r) < 0.2\ \rho_0$ where it is believed that the kaon capture takes place (see Chapter 9), and (ii) the relation (10.45) is inappropriate for the $T = 0$ channel which is dominated by the presence of the Y_0^* resonance (see also Chapter 9).

Aslam and Rook (1970*b*) have generated the proton and neutron density distribution in a single-particle potential with the same radial parameters for protons and neutrons, and with the remaining potential parameters chosen to give the correct separation energy for the last particle and the accepted r.m.s. radius for the proton distribution. Assuming that their results for ^{98}Mo should be comparable to the Ag,Br mixture, their value for $R_{pn}^{L/H}$ is 2.32. Their neutron distributions have not been checked against other data. Bethe and Siemens (1970) have used a density distribution of the form

$$\rho_i(r) = \rho_0 \min[1, \tfrac{1}{2} \exp\{-(r-R)\gamma_i\}], \qquad i = \text{n,p} \qquad (10.46)$$

where

$$\gamma_n^2 = +\ 8\ M\ E_n/\hbar^2, \qquad \gamma_p^2 = 8\ M(V+E_p)/\hbar^2$$

$$R = 1.1\ A^{1/3}\text{fm}, \qquad V = Ze^2/R_{av}$$

and E_i is the separation energy of the least-bound proton or neutron. This leads to a ratio $R_{\mathrm{pn}}^{\mathrm{L/H}}$ of 4.16; this result is consistent with the experimental result and the improvement has arisen from the faster fall in the tail of the proton distribution due to the Coulomb barrier.

Bloom *et al.* (1969) have shown that the ratio of 5 for R_{pn} in light to heavy nuclei can be explained by taking special account of the formation of the Y_0^* resonance in the $T = 0$ channel. This has also been taken into account by Bethe and Siemens (1970) who derive the quantities b^T using an effective-range approximation with a Breit-Wigner function for the resonance. The amplitudes obtained are comparable with those of Bloom *et al.* but are much larger than the values used by other authors, as can be seen from Table 10.9. The ratio of $B^{\mathrm{p}}/B^{\mathrm{n}}$ is, however, very similar to that obtained by Burhop, so that the difference in the values of $R_{\mathrm{pn}}^{\mathrm{L/H}}$ arises from the different treatment of the proton and neutron distributions. Bardeen and Torigoe (1971) obtained a ratio of ~1.8 for both the Kim and Martin-Sakitt KN parameters. Wycech (1971) has extended the effective-range theory of Dalitz and Tuan (1960) which includes Y_0^*, as described in §9.2.2, and concludes that the experimental value of $R_{\mathrm{pn}}^{\mathrm{L/H}}$ can be reproduced if there are about three times as many neutrons as protons in the region of nuclear density for which $\rho(r) < 0.2\rho_0$. This is consistent with the ratios predicted by Aslam and Rook (1970*b*) and by Bethe and Siemens (1970).

As has been pointed out by Nolen and Schiffer (1969) the evidence for a neutron excess in the extreme surface is not inconsistent with approximately equal r.m.s. radii for the proton and neutron distributions. This point is illustrated in Fig. 10.8 where we have plotted the ratio $N\rho_{\mathrm{n}}/Z\rho_{\mathrm{p}}$ for the ZD neutron and the BG proton distributions in $^{120}\mathrm{Sn}$. These distributions are plotted separately in Fig. 2.2 and have equal r.m.s. radii. Without any artificial assumptions, these distributions yield a substantial neutron excess beyond 6 fm. The regions of the nucleus probed by various processes are indicated on the figure.

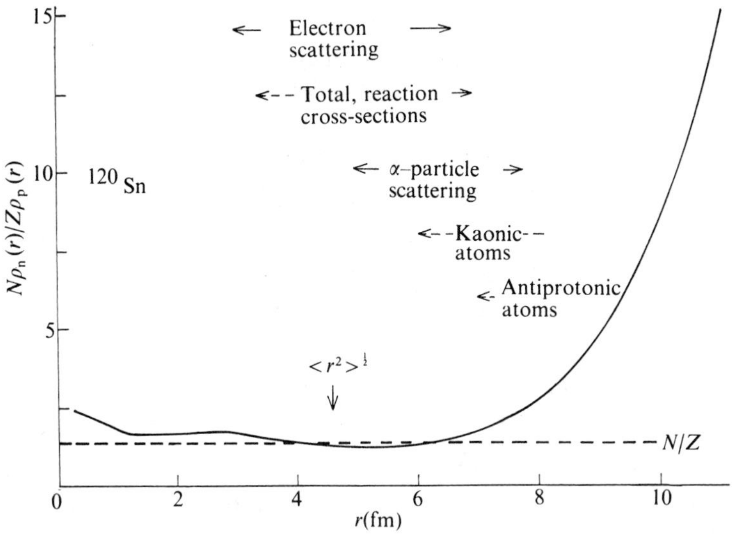

FIG. 10.8. The ratio of $N\rho_n(r)/Z\rho_p(r)$ for the BG (proton) and ZD (neutron) distributions for ^{120}Sn shown in Fig. 2.2, compared with the ratio N/Z. The horizontal lines show the region of the nucleus probed by various processes; where these lines are broken the extent of the region is uncertain. (From Jackson 1975.)

11
CONCLUSIONS

In recent years, and indeed during the time we have been compiling this book, there has been an explosion in the amount of information relating to nuclear sizes and a dramatic change in the experiments and analyses which yield this information. At the same time, Hartree-Fock and other theoretical methods are beginning to produce meaningful ground state distributions which are in reasonable agreement with experiment.

Most of the attention has been focused on a few selected nuclei, especially closed-shell nuclei. The charge densities of these nuclei are now being given with errors which are sensible and significant, and although interpretations of strong-interaction processes are still open to debate, some statements can be made with reasonable confidence about the comparison of proton and neutron distributions in these nuclei.

Wherever possible we have tried to emphasize the importance of deducing precisely which properties of the nucleus are determined by each type of measurement and of relating these properties to certain spatial regions of the nucleus or to particular integral properties of the charge or matter distribution. Our present conclusions on this aspect of the nuclear size problem are summarized in Figs. 10.8 and 11.1.

In strong-interaction processes, particularly elastic scattering and particle production, many analyses suffer deficiencies arising from over-confidence in simple theories. An increased awareness of the importance of exchange processes, energy dependence of basic interactions, and, in some cases, of the role of multiple scattering is needed if the increasing amount of data is to yield reliable nuclear size information. For those processes which are known to probe low density regions of the nucleus, for example α-particle and heavy-ion scattering, and transitions in hadronic atoms, it is essential to use representations of the nuclear matter distri-

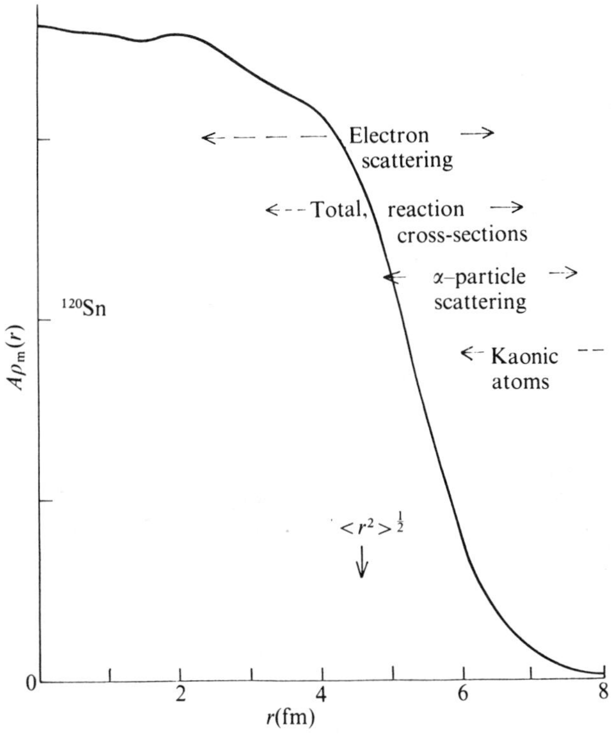

FIG. 11.1. The nuclear matter distribution $A\rho_m(r)$ for ^{120}Sn given by the BG (proton) and ZD (neutron) distributions shown in Fig. 2.2. The horizontal lines show the region of the nucleus probed by various processes; where the lines are broken the extent of the region is still uncertain. (From Jackson 1975.)

butions that are physically meaningful in these regions. Model-independent analyses of the most accurate strong-interaction data are quite urgently needed.

In the case of deformed nuclei and magnetic moment distributions the amount and accuracy of the experimental information is increasing rapidly, although there has not yet been the sort of break-through in the analysis which occurred for spherical nuclei. The problem is much more difficult and it is unlikely that model independent forms of dipole, quadrupole, and higher multipole shapes will ever be obtained. These quantities will come from theoretical calculations, however, and with improved experiments the constraints on these calculations will become so great that we shall be able to give the radial shapes with reasonable confidence.

In the future we expect to see experiments extended to a much larger range of nuclei, but perhaps the most interesting and challenging results will come from very accurate electron scattering measurements over a wide range of momentum transfer for a few selected nuclei. The analysis will then provide a very stringent test of Hartree-Fock calculations and it may be possible to investigate meson-exchange and nucleon-isobar effects.† As indicated in §4.9 a very accurate determination of the ^4He mean-square-radius by electron scattering could provide evidence for the existence of a scalar boson ϕ, which slightly changes the muon-nucleon interaction.

Above all, we have to stress that the study of nuclear sizes intrudes into a very large part of nuclear physics and involves some of the most exciting developments in the subject. Many of the problems are all the more challenging because a combination of atomic, nuclear, and elementary particle physics is needed to study them.

---

†Vector meson contributions have been studied by Blankenbecler and Gunion (1971) and by Chemtob, Moniz, and Rho (1974). The contribution of the N*(1238) has been investigated by Kallio *et al.* (1974) and by Arenhövel and Miller (1974).

REFERENCES

ABASHIAN, A., COOL, R., and CRONIN, J.W., 1956. *Phys. Rev.* **104**, 855.

ABRAHAMIAN, L.O., AGANIANTS, A.O., ADAMIAN, F.V., DANELIAN, V.D., DEMIOKHINA, N.A., GEVORKIAN, S.R., ISRAELIAN, M.KH., KAZARIAN, G.KH., KHUDAVERDIAN, A.G., KHURSHOODIAN, L.S., LEBEDEV, A.N., MANUKIAN, J.V., MURADIAN, E.G., SIRUNIAN, A.M., VARTAPETIAN, H.A., and VATIAN, A.L., 1972. *Phys. Lett.* **38B**, 544.

ABUL-MAGD, A.Y., 1968. *Nucl. Phys.* **B8**, 638.

——— ALBERI, G., and BERTOCCHI, L., 1969. *Phys. Lett.* **30B**, 182.

——— and EL-NADI, M., 1966. *Prog. theor. Phys.* **35**, 798.

ACKER, H.L., BACKENSTOSS, G., DAUM, C., SENS, J.C. and DE WIT, S.A., 1966. *Nucl. Phys.* **87**, 1.

ADLER, S.L. 1974. Preprint *FNAL Publ. 74/63 -THY.*

AGANYANTS, A.O., BAYUKOV, Y.D., DEZA, V.N., DONSKOV, S.V., KOLYBASOV, V.M., LEKSIN, G.A., STOLIN, V.L., and VOROBYEV, L.S., 1969. *Nucl. Phys.* **B11**, 79.

AGASSI, D. and WALL, N.S. 1972. *Phys. Rev.* **C7**, 1368.

AKHIEZER, A.I. and BERESTETSKII, V.B., 1965. *Quantum electrodynamics*, transl. G.M. Volkoff (New York: Interscience), p. 703.

AKIMOV, Yu.K., ANDERT, K., KAZARINOV, Yu.M., KALININ, A.I., KISELEV, V.S., LAPIDUS, L.I., OSIPENKO, B.P., PETROV, M.M., SHURAVIN, V.N., ARVANOV, A.N., BADALYAN, G.U., BEGLARYAN, Dzh.M., KOVALENKO, V.I., MARKARYAN, A.A., MELIKOV, G.I., PETROSYAN, Zh.V., POGOSOV, V.S., CHATRCHYAN, A.M., BORCHEA, K., BUCE, A., DORCHOMAN, D., and PETRASCU, M., 1972. *Sov. Phys. JETP* **35**, 651.

AKYÜZ, R.O. and FALLIEROS, S., 1971. *Phys. Rev. Lett.* **27**, 1016.

ALBERG, M., HENLEY, E.M., and WILETS, L., 1973. *Phys. Rev. Lett.* **30**, 255.

ALDER, K., BOHR, A., HUUS, T., MOTTELSON, T., and WINTHER, A., 1956. *Rev. mod. Phys.* 28, 432.

ALFORD, W.P. and BURKE, D.G., 1969. *Phys. Rev.* **185**, 1560.

ALKHAZOV, G.D., AMALSKY, G.M., BELOSTOTSKY, S.L., VOROBYOV, A.A., DOMCHENKOV, O.A., DOTSENKO, Yu.V., and STARODUBSKY, V.E., 1972. *Phys. Lett.* **42B**, 121.

——— BELOSTOTSKY, S.L., VOROBYOV, A.A., DOTSENKO, Yu.V., DOMCHENKOV, O.A., STARODUBSKY, V.E., and SHUVAEV, M.A., 1973. *Proc. Fifth Int. Conf. on High Energy Physics and Nuclear Structure*, ed. G. Tibell (Amsterdam: North-Holland), p. 176.

——— ——— DOMCHENKOV, O.A., DOTSENKO, Yu.V., KUROPATKIN, N.P., SCHUVAEV, M.A. and VOROBYOV, A.A., 1975. *Phys. Lett.* **57B**, 47.

——— 1976. Preprint *LINP*, 218.

ALLABY, J.C., BUSHNIN, Yu.B., GORIN, Yu.P., DENISOV, P., GIACOMELLI, G., DIDDENS, A.N., ROBINSON, R.W., DENISKOV, S.V., KLOVNIG, A., PETRUKHIN, A.I., PROKOSHIN, Yu.D., STAHL-BRANDT, C.A., STOYONOVA, D.A., and SHUVALOV, R.S., 1971. *Sov. J. nucl. Phys.* **13**, 295.

ALLARDYCE, B.W., BATTY, C.J., BAUGH, D.J., FRIEDMAN, E., HEYMANN, G., CAGE, M.E., PYLE, G.J., SQUIER, G.T.A., CLOUGH, A.S., JACKSON D.F., MURUGESU, S., and RAJARATNAM, V.H., 1973. *Nucl. Phys.* **A209**, 1.

——— ——— ——— ——— ——— WEIL, J.L., CAGE, M.E., PYLE, G.J., SQUIER, G.T.A., CLOUGH, A.S., COX, J., JACKSON, D.F., MURUGESU, S., and RAJARATNAM, V.H., 1972. *Phys. Lett.* **41B**, 577.

ALSTER, J. and CONZETT, H.E., 1965. *Phys. Rev.* **139**, B50.

——— SHREVE, D., and PETERSON, R.J., 1966. *Phys. Rev.* **144**, 999.

ALVENSLEBEN, H., BECKER, U., BERTRAM, W.K., CHEN, M., COHEN, K.J., KNASEL, T.M., MARSHALL, R., QUINN, D.J., ROHDE, M., SANDERS, G.H., SCHUBEL, H., and TING, S.C.C., 1970*a*. *Phys. Rev. Lett.* **24**, 786.

——— 1970*b*. *Phys. Rev. Lett.* **24**, 792.

AMALDI, E., FIDECARO, G., and MARIANI, J., 1950. *Nuovo Cim.* **7**, 553.

AMALDI, U., 1967. *Proc. Int. School of Physics 'E Fermi'* Course No. 38, ed. T. Ericson (New York, London: Academic Press), p. 284.

——— CAMPOS VENUTI, G., CORTELLESSA, G., FRONTEROTTA, G., REALE, A., and SALVADORI, P., 1965. *Accad. naz. Lincei* **39**, 470.

——— ——— ——— ——— ——— ——— and HILLMANN, P., 1964. *Phys. Rev. Lett.* **13**, 341.

——— ——— ——— de SANCTIS, E., FRULLANI, S., LOMBARD, R., and SALVADORI, P., 1966*a*. *Phys. Lett.* **22**, 593.

——— 1966*b*. *Accad. naz. Lincei* **41**, 494.

AMANN, J.F., BARNES, P.D., DOSS, M., DYTMAN, S.A., EISENSTEIN, R.A., PENKROT, J., and THOMPSON, A.C. 1975. *Phys. Rev. Lett.* **35**, 1066.

AMOS, K., MADSEN, V.A., and McCARTHY, I.E., 1967. *Nucl. Phys.* **A94**, 103.

ANDERSON, B.L., BACK, B.B., and BANG, J.M., 1970. *Nucl. Phys.* **A147**, 33.

ANDERSON, D.K., JENKINS, D.A., and POWERS, R.J., 1970. *Phys. Rev. Lett.* **24**, 71.

ANDERSON, H.L., HARGROVE, C.K., HINCKS, E.P., McANDREW, J.D., McKEE, R.J., BARTON, R.D., and KESSLER, D., 1969. *Phys. Rev.* **187**, 1565.

ANDERSON, J.D., BLOOM, S.D., WONG, C., HORNYAK, W.F., and MADSEN, V.A., 1969. *Phys. Rev.* **177**, 1416.

ANNI, R. and TAFFARA, L., 1970. *Riv. nuovo Cim.* **2**, 1.

ANTOUFIEV, Yu.P., AGRANOVICH, V.L., KUZMENKO, V.S., and SOROKIN, P.V., 1972. *Phys. Lett.* **42B**, 347.

ARDITI, M., BIMBOT, L., DOUBRE, H., FRASCARIA, N., GARRON, J.P., RIOU, M., and ROYER, D., 1971. *Nucl. Phys.* **A165**, 129.

——— DOUBRE, H., RIOU, M., ROYER, D., and RUHLA, C., 1967. *Nucl. Phys.* **A103**, 319.

ARENHÖVEL, H. and MILLER, H.G., 1974. *Z. Phys.* **266**, 13.

ARIMA, A. and YOSHIDA, S., 1974. *Nucl. Phys.* **A219**, 475.

ASAI, J. and OGATA, H., 1974. *Nucl. Phys.* **A233**, 55.

ASHERY, D., ZAIDER, M., SHAMAI, Y., COCHAVI, S., MOINESTER, M.A., YAVIN, A.I., and ALSTER, J., 1974. *Phys. Rev. Lett.* **32**, 943.

ASLAM, K. and ROOK, J.R., 1970*a*. *Nucl. Phys.* **B20**, 159.

——— ——— 1970*b*. *Nucl. Phys.* **B20**, 397.

ATKINSON, J. and MADSEN, V.A., 1970. *Phys. Rev.* **C1**, 1377.

ATKINSON, J.H., HESS, W.N., PEREZ-MENDEZ, V., and WALLACE, R.W., 1959. *Phys. Rev. Lett.* **2**, 168.

AUERBACH, E.H., FLEMING, D.M., and STERNHEIM, M.M., 1967. *Phys. Rev.* **162**, 1683.

——— KAHANA, S., and WENESER, J., 1969. *Phys. Rev. Lett.* **23**, 1253.

——— QURESHI, H.M., and STERNHEIM, M.M., 1968. *Phys. Rev. Lett.* **21**, 162.

AUERBACH, N., 1974. *Nucl. Phys.* **A229**, 447.

——— HÜFNER, J., KERMAN, A.K., and SHAKIN, C.M., 1969. *Phys. Rev. Lett.* **23**, 484.

——— 1972. *Rev. Mod. Phys.* **44**, 48.

AUFMUTH, P., BEHRENS, H.O., HEILIG, K., and WILDE, D., 1971. Private communication referred to by Heilig and Steudel (1974).

AUGER, J.P. and Lombard, R.J., 1973*a*. *Phys. Lett.* **45B**, 115.

——— ——— 1973*b*. *Phys. Lett.* **45B**, 487.

——— ——— 1974. *Nuovo Cim.* **21A**, 529.

AUSTERN, N., 1961. *Ann. Phys.* **15**, 299.

——— 1970. *Direct reaction theories* (New York: John Wiley).

——— and BLAIR, J.S., 1965. *Ann. Phys.* **33**, 15.

——— DRISKO, R.M., HALBERT, E., and SATCHLER, G.R., 1964. *Phys. Rev.* **133**, 83.

AVERDUNG, H., 1974. *Internal Rep. No. KPH 3/74 Mainz* (unpublished).

BABAEV, A., BRACHMAN, E., ELISEEV, G., ERMILOV, A., GALAKTIONOV, Yu., KAMISHKOV, Yu., LUBIMOV, V., LUGETSKY, N., NAGOVITZIN, V., NOVIK, V. V., SHEVCHENKO, V., SHUMILOV, E., ZELDOVICH, O., ZVETKOVA, T., BALAMATOV, N., GORYACHEV, B., LEIKIN, E., SIROTKIN, A., TITOV, V., and TURIN, V., 1974. *Phys. Lett.* **51B**, 501.

BACHELIER, D., BERNAS, M., BRISSAUD, I., DETRAZ, C., and RADVANYI, P., 1969. *Nucl. Phys.* **A126**, 60.

BACKE, H., ENGFER, R., JAHNKE, U., KANKELEIT, E., PEARCE, R.M., PETITJEAN, C., SCHELLENBERG, L., SCHNEUWLY, H., SCHRÖDER, W.U., WALTER, H.K., and ZEHNDER, A., 1972. *Nucl. Phys.* **A189**, 472.

BACKENSTOSS, G., 1970. *A. Rev. nucl. Sci.* **20**, 467.

——— BAMBURGER, A., BERGSTRÖM, I., BOUNIN, P., BUNACIU, T., EGGER, J., HULTBERG, S., KOCH, H., KRELL, M., LYNEN, U., RITTER, H.G., SCHWITTER, A., and STEARNS, R., 1972*a*. *Phys. Lett.* **38B**, 181.

——— ——— BUNACIU, T., EGGER, J., KOCH, H., LYNEN, U., RITTER, H.G.,

SCHMITT, H.A., and SCHWITTER, A., 1972*b*. *Phys. Lett.* **41B**, 552.

—— —— EGGER, J., HAMILTON, W.D., KOCH, H., LYNEN, U., RITTER, H.G., and SCHMITT, H., 1970*a*. *Phys. Lett.* **32B**, 399.

—— BUNACIU, T., CHARALAMBUS, S., EGGER, J., KOCH, H., BAMBURGER, A., LYNEN, U., RITTER, H.G., and SCHMITT, H., 1970*b*. *Phys. Lett.* **33B**, 230.

—— CHARALAMBUS, S., DANIEL, H., KOCH, H., POELZ, G., SCHMITT, H., and TAUSCHER, L., 1967*a*. *Phys. Lett.* **25B**, 365.

—— 1967*b*. *Phys. Lett.* **25B**, 547.

—— DANIEL, H., JENTZSCH, K., KOCH, H., POVEL, H.P., SCMEISSNER, F., SPRINGER, K., and STEARNS, R.L., 1971. *Phys. Lett.* **36B**, 422.

—— —— KOCH, H., MALSBURG, Ch. von der, POELZ, G., POVEL, H.P., SCHMITT, H., and TAUSCHER, L., 1973. *Phys. Lett.* **43B**, 539.

—— EGGER, J., EGIDY, T. von, HAGELBERG, R., HERRLANDER, C.J., KOCH, H., POVEL, H.P., SCHWITTER, A., and TAUSCHER, L., 1974. *Nucl. Phys.* **A232**, 519.

—— GOEBEL, K., STADLER, B., HEGEL, U., and QUITMANN, D., 1965. *Nucl. Phys.* **62**, 449.

BAKER, S.D. and McINTYRE, J., 1967. *Phys. Rev.* **161**, 1200.

BALASHOV, V.V., BOYARKINA, A.N., and ROTTER, I., 1965. *Nucl. Phys.* **59**, 417.

BALL, J.B., FULMER, C.B., GROSS, E.E., HALBERT, M.L., HENSLEY, D.C., LUDEMANN, C.A., SALTMARSH, M.J., and SATCHLER, G.R., 1975. *Nucl. Phys.* **A252**, 208.

BAMBERGER, A., LYNEN, U., PIEKARZ, H., PIEKARZ, J., POVH, B., RITTER, H.G., BACKENSTOSS, G., BUNACIU, T., EGGER, J., HAMILTON, W.D., and KOCH, M., 1970. *Phys. Lett.* **33B**, 233.

BANSAL, R.K. and FRENCH, J.B., 1965. *Phys. Lett.* **19**, 223.

BARANGER, M., 1967. *Proc. Gatlinburg Int. Conf. on Nuclear Physics*, eds. R.L. Becker *et al.* (New York, London: Academic Press), p. 659.

—— 1970. *Nucl. Phys.* **A149**, 225.

BARBIERI, R., 1975. *Phys. Lett.* **56B**, 266.

BARDEEN, W.A. and TORIGOE, E.W., 1971. *Phys. Rev.* **C3**, 1785.

—— 1972. *Phys. Lett.* **38B**, 135.

BARDIN, T.T., BARRETT, R.C., COHEN, R.C., DEVONS, S., HITLIN, D., MACAGNO, E., NISSIM-SABAT, C., RAINWATER, J., RUNGE, K., and WU, C.S., 1965. *Columbia Univ. Rep. No. GEN-72, ONR 266(72).*

—— COHEN, R.C., DEVONS, S., HITLIN, D., MACAGNO, E., RAINWATER, J., RUNGE, C., WU, C.S., and BARRETT, R.C., 1967. *Phys. Rev.* **160**, 1043.

BARDWICK, J. and TICKLE, R., 1968. *Phys. Rev.* **171**, 1305.

BARNES, P.D., BOCKELMAN, C.K., HANSEN, O., and SPERDUTO, A., 1965. *Phys. Rev.* **140**, B42.

—— DYTMAN, S., EISENSTEIN, R.A., LAM, W.C., MILLER, J., SUTTON, R.B., POWERS, R.J., ECKHAUSE, M., KANE, J.R., ROBERTS, B.L., WELSH, R.E., KUNSELMAN, A.R., REDWINE, R.P., and SEGEL, R.E., 1972*b*. *Proc. Conf. on Few Nucleon Problems in the Nuclear Interaction*, eds. I. Slaus,

S.A. Moszkowski, R.P. Haddock, and W.H.T. van Oers (Amsterdam:North-Holland/American Elsevier), p. 944.

——— EISENSTEIN, R.A., LAM, W.C., MILLER, J., SUTTON, R.B., ECKHAUSE, M., KANE, J.R., WELSH, R.E., JENKINS, D.A., POWERS, R.J., KUNSELMAN, A.R., REDWINE, R.P., and SEGEL, R.E., 1974. *Nucl. Phys.* **A231**, 477.

——— ——— ——— ——— ——— ——— ——— ——— ——— ——— ——— ———

——— and SCHIFFER, J.P., 1972*a*. *Phys. Lett.* **29**, 230.

BARNETT, A.R. and LILLEY J.S., 1974. *Phys. Rev.* **C9**, 2010.

BARRETT, R.C., 1966. *Nucl. Phys.* **88**, 128.

——— 1968. *Phys. Lett.* **28B**, 93.

——— 1970. *Phys. Lett.* **33B**, 388.

——— 1974. *Rep. Prog. Phys.* **37**, 1.

——— 1976. *Muon Physics*, vol. I, eds. V.W. Hughes and C.S. Wu (New York: Academic Press).

——— 1975*a*. *Proc. Saclay Conf. on Electron Scatt. at Intermediate Energies*, eds. J.B. Bellicard, M. Bernheim, N. de Bottan, A. Bussière, J. Duclos, B. Frois, and C. Schuhl, p. 413.

——— BRODSKY, S.J., ERICKSON, G.W., and GOLDHABER, M.H., 1968. *Phys. Rev.* **166**, 1589.

——— OWEN, D.A., CALMET, J., and GROTCH, H., 1973. *Phys. Lett.* **47B**, 297.

BARRETT, R.F., BIEDENHARN, L.C., DANOS, M., DELSANTO, P.P., GREINER, W., and WAHSWEILER, H.G., 1973. *Rev. mod. Phys.* **45**, 44.

BASSEL, R.H., DRISKO, R.M., and ROOS, P.G., 1968. *Suppl. J. phys. Soc. Japan* **24**, 347.

——— ——— and SATCHLER, G.R., 1962. *Oak Ridge Nat. Lab. Rep. No. 3240.*

BASSICHIS, W.H., KERMAN, A.K., and SVENNE, J.P., 1967. *Phys. Rev.* **160**, 746.

BATTY, C.J., 1961. *Nucl. Phys.* **23**, 562.

——— 1970. *Phys. Lett.* **31B**, 496.

——— 1971. *Nucl. Phys.* **A178**, 17.

——— BIAGI, S.F., RIDDLE, R.A.J., ROBERTS, A., ROBERTS, B.L., WORLEDGE, B.H., BEROVIC, N., PYLE, G.J., SQUIER, G.T.A., CLOUGH, A.S., CODDINGTON, P., and HAWKINS, R.E., 1975. *Proc. Sixth Int. Conf. on High Energy Physics and Nuclear Structure*, eds. D.E. Nagle *et al.* (New York: American Institute of Physics).

——— and FRIEDMAN, E., 1971. *Phys. Lett.* **34B**, 7.

——— 1972. *Nucl. Phys.* **A179**, 701.

——— FRIEDMAN, E., and GREENLEES, G.W., 1969. *Nucl. Phys.* **A127**, 368.

——— ——— and JACKSON, D.F., 1971. *Nucl. Phys.* **A175**, 1.

——— and GREENLEES, G.W., 1969. *Nucl. Phys.* **A133**, 673.

BAUCHE, J., 1969. *Ph.D. Thesis* Orsay.

——— 1974. *J. Phys. (Paris)*, 19.

BAUGH, D.J., 1969. *Nucl. Phys.* **A131**, 417.

BAYMAN, B.F. and KALLIO, A., 1967. *Phys. Rev.* **156**, 1121.

BECHETTI, F.G., 1968. *M.S. Thesis* University of Minnesota.

—— and GREENLEES, G.W., 1969. *Phys. Rev.* **182**, 1190.

BECKER, R.L., 1970. *Phys. Rev. Lett.* **24**, 400; *Phys. Lett.* **32B**, 263.

—— and DAVIES, K.T.R., 1969. *Proc. Int. Conf. on Properties of Nuclear States* eds. M. Harvey *et al.* (Montreal: University of Montreal Press), p. 164.

—— and PATTERSON, M.R., 1971. *Nucl. Phys.* **A178**, 88.

BEHRENS, H. and BÜHRING, W., 1967. *Nucl. Phys.* **A106**, 433.

—— —— 1972. *Nucl. Phys.* **A179**, 297.

BEINER, M., 1970. *Trieste Lectures 1969* (Vienna: IAEA).

—— and LOMBARD, R.J., 1973. *Phys. Lett.* **47B**, 399.

BELL, J.S., 1968. *CERN Report No. TH877.*

BELLETTINI, G., COCCONI, G., DIDDENS, A.N., LILLETHUN, E., MATTHIAE, G., SCANLON, J.P., and WETHERALL, A.M., 1966. *Nucl. Phys.* **A79**, 609.

BELLICARD, J., 1975. *Proc. Saclay Conf. on Electron. Scatt. at Intermediate Energies* eds. J.B. Bellicard, M. Bernham, N. de Botton, A. Bussière, J. Duclos, B. Frois, and C. Schuhl, p. 271.

—— BARREAU, P., and BLUM, D., 1964. *Nucl. Phys.* **60**, 319.

—— BOUNIN, P., FROSCH, R.F., HOFSTADTER, R., McCARTHY, J.S., URHANE, F.J., YEARIAN, M.R., CLARK, B.C., HERMAN, R., and RAVENHALL, D.G., 1967. *Phys. Rev. Lett.* **19**, 527.

BENCZE, G., 1967. *Phys. Lett.* **23**, 713.

—— and SANDULESCU, A., 1966. *Phys. Lett.* **22**, 473.

BERARD, R.W., BUSKIRK, F.R., DALLY, E.B., DYER, J.N., MARUYAMA, X.K., TOPPING, R.L., and TRAVERSO, T.J., 1973. *Phys. Lett.* **47B**, 355.

BEREGI, P., ZELENSKAYA, N.S., NEUDATCHIN, V.G., and SMIRNOV, Yu.F., 1965. *Nucl. Phys.* **66**, 513.

BERG, R.A. and WILETS, L., 1956. *Phys. Rev.* **101**, 201.

BERGGREN, T., 1965. *Nucl. Phys.* **72**, 337.

—— BROWN, G.E., and JACOB, J., 1962. *Phys. Lett.* **1**, 88.

—— and OHLÉN, G., 1973. *Lett. nuovo Cim.* **7**, 701.

—— —— 1974. *Proc. Int. Conf. on Nuclear Structure and Spectroscopy* eds. H.P. Blok and A.E.L. Dieperink (Amsterdam: Scholars Press), p. 173.

BERNABÉU, J. and JARLSKOG, C., 1972. *Nucl. Phys.* **B75**, 59.

—— —— 1976. *Phys. Lett.* **60B**, 197.

—— and ROS, J., 1974. *Nucl. Phys.* **A220**, 1.

BERNHEIM, M., BUSSIÈRE, A., GILLEBERT, A., MOUGEY, J., PHAN XUAN HO, PRIOU, M., ROYER, D., SICK, I., and WAGNER, G.J., 1974*a*. *Phys. Rev. Lett.* **32**, 898.

—— —— —— —— PRIOU, M., ROYER, D., SICK, I., and WAGNER, G.J., 1974*b*. *Proc. Int. Conf. on Nuclear Structure and Spectroscopy* eds. H.P. Blok and A.E.L. Dieperink (Amsterdam: Scholars Press), p. 412.

BERNOW, S., DEVONS, S., DUERDOTH, I., HITLIN, D., KAST, J.W., MACAGNO, E.R., RAINWATER, J., RUNGE, K., and WU, C.S., 1967. *Phys. Rev. Lett.* **18**, 787.

BERNSTEIN, A.M., 1969. *Adv. Nucl. Phys.* **3**, 325.

——— and SEIDLER, W.A., 1971. *Phys. Lett.* **34B**, 569.

——— ——— 1972. *Phys. Lett.* **39B**, 583.

BERTHOT, J., BERTIN, P.Y., GIRARDEAU-MONTAUT, J.P., and ISABELLE, D.B., 1969. *Proc. Int. Conf. on Properties of Nuclear States* eds. M. Harvey *et al.* (Montreal: University of Montreal Press), p. 699.

BERTIN, A., CARBONI, G., DUCLOS, J., GASTALDI, U., GORINI, G., NERI, G., PICARD, J., PITZURRA, O., PLACCI, A., POLACCO, E., TORELLI, G., VITALI, A., and ZAVATTINI, E., 1975. *Phys. Lett.* **55B**, 411.

BERTIN, M.C., TABOR, S.L., WATSON, B.A., EISEN, Y., and GOLDRING, G., 1971, *Nucl. Phys.* **A167**, 216.

BERTINI, R., BEURTEY, R., BROCHARD, F., BRUGE, G., CATZ, H., CHAUMEAUX, A., DURAND, J.M., FAIVRE, J.C., FONTAINE, J.M., GARRETA, D., GUSTAFSSON, C., HENDRIE, D.L., HIBOU, F., LEGRAND, D., SAUDINOS, J., and THIRION, J., 1973. *Phys. Lett.* **45B**, 119.

BERTOZZI, W., COOPER, T., ENSSLIN, N., HEISENBERG, J., KOWALSKI, S., MILLS, M., TURCHINETZ, W., WILLIAMSON, C., FIVOZINSKY, S.P., LIGHTBODY, J.W., and PENNER, S., 1972*a*. *Phys. Rev. Lett.* **28**, 1711.

——— FRIAR, J., HEISENBERG, J., and NEGELE, J.W., 1972*b*. *Phys. Lett.* **41B**, 408.

BERTSCH, G.F. and KUO, T.T.S., 1968. *Nucl. Phys.* **A112**, 204.

BETHE, H.A., 1968. *Phys. Rev.* **167**, 879.

——— 1971. *Ann. Rev. nucl. Sci.* **21**, 93.

——— and de HOFFMANN, F., 1955. *Mesons and fields* vol. II (Evanston, Ill.: Row, Peterson).

——— and ELTON, L.R.B., 1968. *Phys. Rev. Lett.* **20**, 745.

——— and MOLINARI, A., 1971. *Ann. Phys.* **63**, 393.

——— and NEGELE, J.W., 1968. *Nucl. Phys.* **117A**, 575.

——— and SIEMENS, P.J., 1968. *Phys. Rev.* **27B**, 549.

——— 1970. *Nucl. Phys.* **B21**, 589.

BIMBOT, L., TATISCHEFF, B., BRISSAUD, I., LE BORNEC, Y., FRASCARIA, N., and WILLIS, A., 1973. *Nucl. Phys.* **A210**, 397.

BINGHAM, C.R. and HALBERT, M.L., 1968. *Phys. Rev.* **169**, 933.

BISHOP, G.R., 1965. *Nuclear structure and electromagnetic interactions* ed. N. MacDonald (New York: Plenum Press), p. 211.

——— ISABELLE, D.B., and BÉTOURNÉ, C., 1964. *Nucl. Phys.* **54**, 97.

BJERREGAARD, J.H., HANSEN, O., and SIDENIUS, G., 1965. *Phys. Rev.* **138**, B1097.

BJORKEN, J.D. and DRELL, S.D., 1964. *Relativistic quantum mechanics* (New York: McGraw-Hill).

BJORKLAND, J.A., RABOY, A., TRAIL, C.C., EHRLICH, R.D., and POWERS, R.J., 1965. *Nucl. Phys.* **69**, 161.

BJORNENAK, K., FINFJORD, J., OSLAND, P., and REITAN, A., 1970. *Nucl. Phys.* **B22**, 179.

BLAIR, J.S., 1954. *Phys. Rev.* **95**, 1218.

—— 1966. *Lectures in theoretical physics* vol. VIIIc (Boulder: University of Colorado Press), p. 343.

—— 1970. *Nuclear reactions induced by heavy ions* eds. R. Bock and W.R. Hering (Amsterdam: North-Holland), p. 1.

—— NAIR, K.G., POLTER, D., and VANDENBOSCH, R., 1972. *Univ. Washington Ann. Rep.* p. 116.

BLAISE, J., 1958. *Annls. Phys.* **3**, 1019.

BLANKENBECLER, R. and GUNION, J.F., 1971. *Phys. Rev.* **D4**, 718.

BLATT, J.M. and WEISSKOPF, V.F., 1952. *Theoretical nuclear physics* (New York: John Wiley).

BLIN-STOYLE, R.J., 1955. *Phil. Mag.* **46**, 973.

—— 1969*a*. *Isospin in nuclear physics* ed. D.H. Wilkinson (Amsterdam: North-Holland), chap. 4.

—— 1969*b*. *Phys. Lett.* **29B** 12.

BLOMQVIST, J., 1972. *Nucl. Phys.* **B48**, 95.

BLOOM, S.D., JOHNSON, M.H., and TELLER, E., 1969. *Phys. Rev. Lett.* **23**, 28.

—— WEISS, M.S., and SHAKIN, C.M., 1972. *Phys. Rev.* **C5**, 238.

BOCHMANN, G.V., KOFOED-HANSEN, O., and MARGOLIS, B., 1970. *Phys. Lett.* **33B**, 222.

BODMER, A.R., 1953. *Proc. phys. Soc.* **A66**, 1041.

BOEHM, A., DARRIULAT, P., GROSSO, G., KAFTANOV, V., KLEINKNECHT, K., LYNCH, H.L., RUBBIA, C., TICHO, H., and TITTEL, K., 1968. *Phys. Lett.* **27B**, 594.

BOEHM, F., 1974. *California Inst. Technol. Rep. No. CALT-63-200.*

—— and LEE, P.L., 1974. *Atom. Data Nucl. Data Tables* **14**, 605.

BOFFI, S., BOUTEN, M., CIOFI DEGLI ATTI, C., and SAWICKI, J., 1968*a*. *Nucl. Phys.* A120, 135.

—— FACATI, F.D., and SAWICKI, J., 1968*b*. *Nuovo Cim.* **57B**, 103.

BOHR, A. and MOTTELSON, B.R., 1953. *Mat.-fys. Meddr.* **27**, No. 16.

—— and WEISSKOPF, V.W., 1950. *Phys. Rev.* **77**, 94.

BOHR, N., 1922. *Nature (Lond.)* **109**, 746.

BONDORF, J.P. and BUND, G.W., 1969. *Nucl. Phys.* **A127**, 481.

BONN, J., HUBER, G., KLUGE, H.-J., KÖPF, U., KUGLER, L., OLTEN, E.-W., and RODRIGUEZ, J., 1973*b*. *Atomic physics 3* (New York, London: Plenum Press), p. 471.

BORKOWSKI, F., PEUSER, P., SIMON, G.G., WALTHER, V.H., and WENDLING, R.D. 1974. *Nucl. Phys.* **A222**, 269.

BORKOWSKI, F., SIMON, G.G., WALTHER, V.H., and WNDLING, R.D., 1975. *Nucl. Phys.* **B93**, 461.

BORYSOWICZ, J. and HETHERINGTON, J.H., 1973. *Phys. Rev.* **C7**, 2293.

BOWEN, P.H., SCANLON, J.P., STAFFORD, G.H., THRESHER, J.J., and HODGSON, P.E., 1961. *Nucl. Phys.* **22**, 640.

BOYARSKI, A.M., DIEBOLD, R., ECKLUND, S.D., FISCHER, G.E., MURATA, Y., RICHTER, B., and SANDS, M., 1969. *Phys. Rev. Lett.* **23**, 1343.

BOYD, R.N., FENTON, J., WILLIAMS, M., KRUSE, T., and SAVIN, W., 1971. *Nucl. Phys.* **A162**, 497.

——— and GREENLEES, G.W., 1968. *Phys. Rev.* **176**, 1394.

BRAIN, S., 1974. *Ph.D. Thesis* University of Glasgow.

BRAMANIS, E., 1971. *Nucl. Phys.* **A175**, 17.

BRAMBLETT, R.L., FULTZ, S.C., and BERMAN, B.L., 1973. *Proc. Asilomar Conf. on Photonuclear Reactions and Applications* ed. B.L. Berman (Springfield: NTIS), p. 13.

BRANDOW, B.M., 1966. *Phys. Rev.* **152**, 863.

——— 1967. *Rev. mod. Phys.* **39**, 771.

BRANSDEN, B.H., 1969. *High energy physics* vol. III, ed. E.H.S. Burhop (New York: Academic Press), p. 1.

BRAY, K.H., BUNKER, S.N., JAIN, M., JAYARANA, K.S., MILLER, C.A., NELSON, J.M., VAN OERS, W.H.T., WELLS, D.O., JANISZEWSKI, J., and McCARTHY, I.E., 1971. *Phys. Lett.* **35B** 41.

BREIT, G., 1958. *Rev. mod. Phys.* **30**, 507.

BRINK, D.M. and BOEKER, E., 1967. *Nucl. Phys.* **A91**, 1.

——— FRIEDRICH, H., WEIGUNY, A., and WONG, C.W., 1970. *Phys. Lett.* **33B**, 143.

——— and ROWLEY, N., 1974. *Nucl. Phys.* **A219**, 79.

——— and N. TAKIGAWA, 1976. Preprint.

——— and VAUTHERIN, D., 1969. *Proc. Int. Conf. on Properties of Nuclear States* eds. M. Harvey *et al.* (Montreal: University of Montreal Press), p. 232.

BRISSAUD, I., BIMBOT, L., le BORNEC, Y., TATISCHEFF, B., and WILLIS, N., 1974. *Phys. Lett.* **48B**, 319.

——— le BORNEC, Y., TATISCHEFF, B., BIMBOT, L., BRUSSEL, M.K., and DUHAMEL, G., 1972*a*. *Nucl. Phys.* **A191**, 145.

——— and BRUSSEL, M.K., 1976*a*. Preprint.

——— ——— 1976*b*. Preprint.

——— TATISCHEFF, B., BIMBOT, L., BRUSSEL, M.K. COMPARAT, V., and WILLIS A., 1972*b*. *Phys. Rev.* **C6**, 595.

BRIX, P. and KOPFERMANN, H., 1949. *Z. Phys.* **126**, 344.

——— 1958. *Rev. mod. Phys.* **30**, 517.

BRODSKY, S.J. and PUMPLIN, J., 1969. *Phys. Rev.* **182**, 1794.

BROOKES, G.R., CLOUGH, A.S., FREELAND, J.H., GALBRAITH, W., KING, A.F.,

ARMSTRONG, T.H., HOGG, W.R., LEWIS, G.M., ROBERTSON, A.W., RAWLINSON, W.R., TAIT, N.R.S., THOMPSON, J.C., and TOLFREE, D.W.F., 1973. *Phys. Rev.* **8D**, 136.

BROWN, G.E. and BOLSTERLI, M., 1959. *Phys. Rev. Lett.* **3**, 472.

——— and ELTON, L.R.B., 1955. *Phil. Mag.* **46**, 164.

——— GUNN, J.H., and GOULD, P., 1963. *Nucl. Phys.* **46**, 598.

——— HORSFJORD, V., and LIU, K.F., 1973. *Nucl. Phys.* **A205**, 73.

BRUCH, R., HEILIG, K., KALETTA, D., STEUDEL, A., and WENDLANDT, D., 1969. *J. Phys. (Paris)* **30** Suppl. C1-51.

BRUECKNER, K.A., BUCHLER, J.R., JORNA, S., and LOMBARD, R.J., 1968. *Phys. Rev.* **171**, 1188.

——— and GOLDMAN, D.T., 1959. *Phys. Rev.* **116**, 424.

——— LOCKETT, A.M., and ROTENBERG, M., 1961. *Phys. Rev.* **121**, 255.

——— MELDNER, H.W., and PEREZ, J.D., 1972. *Phys. Rev.* **C6**, 773.

BRUGE, G., FAIVRE, J.C., FARAGGI, H., and BUSSIÈRE, 1970. *Nucl. Phys.* **A146**, 597.

BRUXELLES, OXFORD, and UNIVERSITY COLLEGE, LONDON, COLLABORATION 1963. *Proc. Siena Int. Conf. on High Energy Physics*, p. 199.

BRYAN, R.A. and PHILLIPS, R.J.N., 1968. *Nucl. Phys.* **B5**, 201.

——— and SCOTT, B.L., 1964. *Phys. Rev.* **135**, B434.

BUCK, B. and PEREY, F.G., 1962. *Phys. Rev. Lett.* **8**, 444.

BUDINI, P. and FURLAN, G., 1959. *Nuovo Cim.* **13**, 790.

BUDZANOWSKI, A., DUDEK, A., GROTOWSKI, K., and STRZALKOWSKI, A., 1970. *Phys. Lett.* **32B**, 431.

——— ——— ——— MAJKA, Z., and STRZALKOWSKI, A., 1974. *Particles and Nuclei* **6**, 97.

BUGG, W.M., CONDO, G.T., HURT, E.L., COHN, H.O., and McCULLOCH, R.D., 1973. *Phys. Rev. Lett.* **31**, 475.

BUMILLER, F.A., BUSKIRK, F.R., DYER, J.N., and MONSON, W.A., 1972. *Phys. Rev.* **C5**, 391.

BUND, G.W. and BLAIR, J.S., 1970. *Nucl. Phys.* **A144**, 384.

BURBIDGE, G.R., 1953. *Phys. Rev.* **89**, 189.

BURGE, E.J., CALDERBANK, M., LEWIS, V.E., RUSH, A.A., and THOMAS, G.L., 1967. *Proc. Tokyo Conf. on Nuclear Structure* (Tokyo: Institute for Nuclear Study, University of Tokyo), p. 247.

BURHOP, E.H.S., 1967. *Nucl. Phys.* **B1**, 438.

——— 1971. RHEL meeting on stopping kaons and nuclear structure. *Rep. No. RHEL/M/H9* Rutherford High Energy Laboratory, Chilton, Oxon.

BURLESON, G.R., COHEN, D., LAMB, R.C., MICHAEL, D.N., SCHLUTER, R.A., and WHITE, T.O., 1965. *Phys. Rev. Lett.* **15**, 70.

BUTTLE, P.J.A. and GOLDFARB, L.J.B., 1964. *Proc. phys. Soc.* **83**, 701.

——— ——— 1966. *Nucl. Phys.* **A78**, 409.

——— ——— 1971. *Nucl. Phys.* **A176**, 299.

CAGE, M.E., CLOUGH, D.L., COLE, D.J., ENGLAND, J.B.A., PYLE, G.J., ROLPH, P.M., WATSON, L.H., and WORLEDGE, D.H., 1972. *Nucl. Phys.* **A183**, 449.

——— COLE, A.J., and PYLE, G.J., 1973. *Nucl. Phys.* **A201**, 418.

CALDERBANK, M., BURGE, E.J., LEWIS, V.E., SMITH, D.A., and GANGULY, N.K., 1967. *Nucl. Phys.* **A105**, 601.

CAMPI, X, 1975. Communication to the Trieste conference on Hartree-Fock and self-consistent field theories in nuclei *Rep. No. IPNO/TH 75-14*.

——— and SPRUNG, D.W.L., 1972. *Nucl. Phys.* **A194**, 401.

——— ——— 1973, *Phys. Lett.* **B46**, 291.

——— ——— and MARTORELL, J., 1974. *Nucl. Phys.* **A223**, 541.

CAMPOS VENUTI, G., CORTELLESSA, G., FARCHI, G., FRULLANI, S., GIORDANO, R., SALVADORI, P., TAKAMATSU, K., CAPITANI, G.P., and De SANCTIS, E., 1973. *Nucl. Phys.* **A205**, 628.

CARLSON, B.C. and TALMI, I., 1954. *Phys. Rev.* **96**, 436.

CARLSON, J.D., LIND, D.A., and ZAFIRATOS, C.D., 1973. *Phys. Rev. Lett.* **30** 99.

CARRIGAN, R.A. Jr., GUPTA, P.D., SUTTON, R.B., SUZUKI, M.N., THOMPSON, A.C., COTÉ, R.E., PRESTWICH, W.V., GAIGALAS, A.K., and RABOY, S., 1968. *Phys. Lett.* **27B**, 622.

——— SAPP, W.W., SCHULER, W.B., SIEGAL, R.T., and WELSH, R.E., 1967. *Phys. Lett.* **25B**, 193.

CHADWICK, G.B., DURANI, S.A., JONES, P.B., WIGNELL, J.W.G., and WILKINSON, D.H., 1958. *Phil. Mag.* **3**, 1193.

CHALMERS, J. and SAPERSTEIN, A., 1967. *Phys. Rev.* **156**, 1099.

——— ——— 1968. *Phys. Rev.* **168**, 1145.

CHALOUPKA, V., BRICMAN, C., BARBARO-GALTIERI, A., CHEW, D.M., KELLY, R.L., LASINSKI, T.A., RITTENBERG, A., ROSENFELD, A.H., TRIPPE, T.G., UCHIYAMA, F., BARASH-SCHMIDT, N., SÖDING, P., and ROOS, M., 1974. *Phys. Lett.* **50B**, 1.

CHAMPEAU, R.-J., MICHEL, J.J., and WALTHER, H., 1973. Private communication referred to by Heilig and Steudel 1974.

——— and MILADI, M., 1974. *J. Phys. (Paris)* **35**, 105.

CHANG, C.C., WALL, N.S., and FRAENKEL, Z., 1974. *Phys. Rev. Lett.* **33**, 1493.

CHANG, F.C., 1966. *Phys. Rev.* **141**, 1136.

CHARPAK, G., FAVIER, J., MASSONET, L., and ZUPANCIC, C., 1967. *Proc. Gatlinburg Int. Conf. on Nuclear Physics* eds. R.L. Becker *et al.* (New York: Academic Press), p. 465.

CHASE, D.M., WILETS, L., and EDMONDS, A.R., 1958. *Phys. Rev.* **110**, 1080.

CHASMAN, C., RISTINIEN, R.A., COHEN, R.C., DEVONS, S., and NISSIM-SABAT, C., 1965. *Phys. Rev. Lett.* **14**, 181.

CHATTARJI, D. and GHOSH, P., 1973. *Phys. Rev.* **C8**, 2115.

CHAUDHURY, M.L., 1963. *Phys. Rev.* **130**, 2339.

CHAUMEAUX, A., LALY, V., and SCHAEFFER, R., 1976. To be published.

CHEMTOB, M., MONIZ, E.J., and RHO, M., 1974. *Phys. Rev.* **C10**, 344.

CHEN, M.Y., 1968. *Ph.D. Thesis* Princeton University.

—— 1970*a*. *Phys. Rev.* **C1**, 1167.

—— 1970*b*. *Phys. Rev.* **C1**, 1176.

—— 1975. *Phys. Rev. Lett.* **34**, 341.

CHENG, S.C., CHEN, M.Y., KAST, J.W., LEE, W.Y., MACAGNO, E.R., RUSHTON, A.M., and WU, C.S., 1971. *Phys. Lett.* **34B**, 615.

CHERTOK, B.T., JONES, E.C., BENDEL, W.L., and FAGG, L.W., 1969. *Phys. Rev. Lett.* **23**, 34.

CHESLER, R.B., 1967. *Ph.D. Thesis* California Institute of Technology.

—— and BOEHM, F., 1968. *Phys. Rev.* **166**, 1206.

CIOFI DEGLI ATTI, C., 1967. *Nucl. Phys.* **A106**, 215.

CLARK, B.C., MERCER, R.L., RAVENHALL, D.G., and SAPERSTEIN, A.M., 1973. *Phys. Rev.* **C7**, 466.

CLARKSON, R.G., von BRENTANO, P., and HARNEY, H.L., 1971. *Nucl. Phys.* **A161**, 49.

CLEMENT, D.M. and BARANGER, E., 1968. *Nucl. Phys.* **A120**, 25.

CLOUGH, A.S., BATTY, C.J., BONNER, B.E., TSCHALÄR, C., WILLIAMS, L.E., and FRIEDMAN, E., 1969. *Nucl. Phys.* **A137**, 222.

—— —— —— and WILLIAMS, L.E., 1970. *Nucl. Phys.* **A143**, 385,

—— TURNER, G.K., ALLARDYCE, B.W., BATTY, C.J., BAUGH, D.J., McDONALD, J.D., RIDDLE, R.A.J., WATSON, L.H., CAGE, M.E., PYLE, G.J., and SQUIER, G.T.A., 1973. *Phys. Lett.* **43B**, 476.

CODDINGTON, P., ATKISS, M., BRODBECK, T.J., LOCKE, D.H., MORRIS, J.V., NEWTON, D., and SLOAN, T., 1974. *DL/P 211 preprint*.

COHEN, B.L. and PRICE, R.E., 1961. *Phys. Rev.* **121**, 1441.

COHEN, S., 1960. *Phys. Rev.* **118**, 489.

COLE, R.K. Jr., 1968. *Phys. Rev.* **177**, 164.

COLLARD, H., HOFSTADTER, R., HUGHES, E.B., JOHANSSON, A., and YEARIAN, M.R., 1965. *Phys. Rev.* **138B**, 57.

COLOMBANI, P. JACMART, J.C., POFFÉ, N., RIOU, M., STÉPHAN, C., and TYS, J., 1972. *Phys. Lett.* **42B**, 197.

COMPARAT, J., FRASCARIA, R., MARTY, N., MORLET, M., and WILLIS, A., 1974. *Nucl. Phys.* **A221**, 403.

CONDON, E.U. and GURNEY, R.W., 1928. *Nature (Lond.)* **122**, 439.

—— —— 1929. *Phys. Rev.* **33**, 127.

COOPER, M.D., HORNYAK, W.F., and ROOS, P.G., 1974. *Nucl. Phys.* **A218**, 249.

COOR, T., HILL, D.A., HORNYAK, W.F., SMITH, L., and SNOW, G., 1955. *Phys. Rev.* **98**, 1369.

CORLEY, D.M., 1968. *Ph.D. Thesis* University of Maryland.

—— WALL, N.S., PALEVSKY, H., FRIEDES, J.L., SITTER, R.J., BENNETT, G.W., SIMPSON, W.D., PHILLIPS, G.C., IGO, G.W., and STEARNS, R.L., 1972. *Nucl. Phys.* **A184**, 437.

CRAMER, J.G., DEVRIES, R.M., GOLDBERG, D., ZISMAN, M.S., and MAGUIRE, C.F.,

1976. To be published.

CRAWLEY, G.M., NARASHIMHA, RAO B.V., and POWELL, D.L., 1968. *Nucl. Phys.* **A112**, 223.

CROZON, M., CHAVANON, Ph., COURAU, A., LERAY, Th., NARJOUX, J.L., and TOQUEVILLE, J., 1965. *Nucl. Phys.* **64**, 567.

CURRAN, C.S., DRAKE, T.E., JOHNSTON, A., BRAIN, S.W., GILLESPIE, W.A., LEES, E.W., SINGHAL, R.P., and SLIGHT, A.G., 1972. *J. Phys.* **A5**, L39.

CURTIS, T.H., EISENSTEIN, R.A., MADSEN, D.W., and BOCKELMAN, C.K., 1969. *Phys. Rev.* **184**, 1162.

CUSSON, R.Y., TRIVEDI, H., and KOLB, D., 1972. *Phys. Rev.* **C5**, 2120.

CZYZ, W., 1963. *Phys. Rev.* **131**, 2141.

——— 1971. *Adv. nucl. Phys.* **4**, 61.

DABROWSKI, J., 1958. *Proc. phys. Soc.* **71**, 658.

DAHLL, G. and WARKE, C., 1970. *Nucl. Phys.* **A147**, 94.

DALITZ, R.H., 1951. *Proc. R. Soc.* **A206**, 509.

——— and TUAN, S.F., 1960. *Ann. Phys.* **10**, 307.

DAMGAARD, J., SCOTT, C.K., and OSNES, E., 1970. *Nucl. Phys.* **A154**, 12.

DANOS, M., 1973. *Proc. Asilomar Conf. on Photonuclear Reactions and Applications* ed. B.L. Berman (Springfield, Va.: NTIS), p. 43.

DAR, A. and KOZLOWSKY, B., 1966. *Phys. Lett.* **20**, 311; 314.

DAVIDSON, J.P., 1968. *Collective models of the nucleus* (New York: Academic Press).

DAVIES, H., MUIRHEAD, H., and WOULDS, J.N., 1966. *Nucl. Phys.* **78**, 663.

DAVIES, K.T.R. and BARANGER, M., 1970. *Phys. Rev.* **C1**, 1640.

——— ——— TARBUTTON, R.M., and KUO, T.T.S., 1969. *Phys. Rev.* **177**, 1519.

——— and BECKER, R.L., 1971. *Nucl. Phys.* **A176**, 1.

——— KRIEGER, S.J., and BARANGER, M., 1966. *Nucl. Phys.* **84**, 545.

——— and McCARTHY, R.J., 1971. *Phys. Rev.* **C4**, 81.

——— and SATCHLER, G.R., 1974. *Nucl. Phys.* **A222**, 13.

——— WONG, C.Y., and KRIEGER, S.J., 1972. *Phys. Lett.* **41B**, 455.

DAVIS, D.H., LOVELL, S.P., CSEJTHEY-BARTH, M., SACTON, J., SCHOROCHOFF, G., and O'REILLY, M., 1967. *Nucl. Phys.* **B1**, 434.

DEDONDER, J.P., 1971. *Nucl. Phys.* **A174**, 251.

——— 1972. *Nucl. Phys.* **A180**, 472.

DEDRICK, K.G., 1955. *Phys. Rev.* **100**, 58.

DEFOREST, T., 1967. *Ann. Phys.* **45**, 365.

——— 1969. *Nucl. Phys.* **A132**, 305.

——— and WALECKA, J.D., 1966. *Adv. Phys.* **15**, 1.

DELOFF, A. and LAW, J., 1974*a*. *Phys. Rev.* **C10**, 1688.

——— ——— 1974*b*. *Phys. Rev.* **C10**, 2657.

DEUTCHMAN, P.A. and McCARTHY, I.E., 1968. *Nucl. Phys.* **A112**, 399.

DEVINS, D.W., 1965. *Rev. mod. Phys.* **37**, 396.

DEVONS, S. and DUERDOTH, I., 1969. *Adv. nucl. Phys.* **2**, 295.

DEVRIES, R.M., LILLEY, J.S., and FRANEY, M.A., 1976. *Phys. Rev. Lett.* **37**, 481.

——— PERRENOUD, J.L., and SLAUS, I., 1972. *Nucl. Phys.* **A188**, 449.

DICKENS, J.K., DRISKO, R.M., HALBERT, E., and SATCHLER, G.R., 1965*a*. *Phys. Lett.* **15**, 337.

——— PEREY, F.G., and SATCHLER, G.R., 1965*b*. *Nucl. Phys.* **73**, 529.

DIEPERINK, A.E.L., BRUSSARD, P.J., and CUSSON, R.Y., 1972. *Nucl. Phys.* **A180**, 110.

DIETRICH, K., 1962. *Phys. Lett.* **2**, 139.

DIRAC, P.A.M., 1958. *The principles of quantum mechanics,* 4th edition (Oxford: Clarendon Press).

DIXIT, M.S., CARTER, A.L., HINCKS, E.P., KESSLER, D., WADDEN, J.S., HARGROVE, C.K., McKEE, R.J., MES, H., and ANDERSON, H.L., 1975. *Phys. Rev. Lett.* **35**, 1633.

DONNELLY, T.W., 1970. *Nucl. Phys.* **A150**, 393.

——— 1972. *Proc. Int. Summer School on Dynamic Structure of Nuclear States* eds. D.J. Rowe *et al.* (Toronto: University of Toronto Press), p. 141.

——— and WALECKA, J.D., 1973. *Nucl. Phys.* **A201**, 81.

——— ——— 1975. *A. Rev. nucl. Sci.* **25**, 329.

DORENBUSCH, W.E., BELOTE, T.A., and HANSEN, O., 1966. *Phys. Rev.* **146**, 734.

DOST, M., HERING, W.R., and SMITH, W.R., 1967. *Nucl. Phys.* **A93**, 357.

DOVER, C.B. and VAN GIAI, N., 1971. *Nucl. Phys.* **A177**, 559.

——— ——— 1972. *Nucl. Phys.* **A190**, 373.

DRECHSEL, D., 1968. *Nucl. Phys.* **A133**, 665.

DREHER, B., FRIEDRICH, J., MERLE, K., ROTHAAS, H., and LÜHRS, G., 1974. *Nucl. Phys.* **A235**, 219.

——— LEMB, M., and LENZ, F., 1973. *Proc. Asilomar Conf. on Photonuclear Reactions and Applications* ed. B.L. Berman (Springfield, Va.: NTIS). p. 191, and private communication.

DRELL, S.D. and ZACHARIASEN, F., 1961. *Electromagnetic structure of nucleons* (Oxford: Clarendon Press).

DRISKO, R.M. and RYBICKI, F., 1966. *Phys. Rev. Lett.* **16**, 275.

——— SATCHLER, G.R., and BASSEL, R.M., 1963. *Phys. Lett.* **5**, 347.

DUBLER, T., SCHELLENBERG, L., ENGFER, R., FRICKE, B., SCHNEUWLY, H., VUILLEMEIER, J.L., WALTER, H.K., and ZEHNDER, A., 1973. *Nucl. Phys.* **A219**, 29.

DUGUAY, M., BOCKELMAN, C.K., CURTIS, T.H., and EISENSTEIN, R.A., 1967. *Phys. Rev.* **163**, 1259.

DUHM, H.H., 1968. *Nucl. Phys.* **A118**, 563.

DUMITRESCU, O. and SANDULESCU, A., 1967. *Nucl. Phys.* **A100**, 456.

DUPONT, Y. and CHABRE, M., 1968. *Phys. Lett.* **26B**, 362.

EBERSOLD, P., AAS, B., DEY, W., EICHLER, R., HARTMANN, J., LEISI, H.J., and SAPP, W.W., 1974. *Phys. Lett.* **53B**, 48.

ECKHAUSE, M., KANE, F.R., KANE, J.R., MARTIN, P., MILLER, G.H., SPENCE, C.B., and WELSH, R.E., 1972. *Nucl. Phys.* **B44**, 83.

EDEN, R.J., 1959. *Nuclear reactions* (Amsterdam: North-Holland), p. 1.

EDMONDS, A.R., 1957. *Angular momentum in quantum mechanics* (Princeton, N.J.: Princeton University Press).

EGIDY, T. von and POVEL, H.P., 1974. *Nucl. Phys.* **A232**, 511.

EHLERS, J.W. and MOSZKOWSKI, S.A., 1972. *Phys. Rev.* **C6**, 217.

EHRENBERG, H.F., HOFSTADTER, R., MEYER-BERKHOUT, U., RAVENHALL, D.G., and SOBOTTKA, S.E., 1959. *Phys. Rev.* **113**, 666.

EHRENFEST, P., 1922. *Nature (Lond.)* **109**, 745.

EHRLICH, R.D., 1968. *Phys. Rev.* **173**, 1088.

EISEN, Y., 1971. *Phys. Lett.* **37B**, 33.

EISENBERG, J.M. and GREINER, W., 1972. *Microscopic theory of the nucleus* (Amsterdam: North-Holland).

EISENBERG, Y. and KESSLER, D., 1963. *Phys. Rev.* **130**, 2352.

ELBECK, B. and TJOM, P.O., 1969. *Adv. nucl. Phys.* **3**, 259.

ELLEGAARD, C., KANTELE, J., and VEDELSBY, P., 1967. *Phys. Lett.* **25B**, 212.

ELLIOTT, J.P. and SKYRME, T.H.R., 1955. *Proc. R. Soc.* **A232**, 561.

ELTON, L.R.B., 1950. *Proc. phys. Soc.* **A63**, 1115.

—— 1961*a*. *Nucl. Phys.* **23**, 681.

—— 1961*b*. *Nuclear sizes* (Oxford: Clarendon Press).

—— 1966. *Nucl. Phys.* **89**, 69.

—— 1967*a*. *Proc. Int. Conf. on Electromagnetic Sizes of Nuclei* eds. D.J. Browne, M.K. Sundaresan, and R.D. Barton (Ottawa: Carleton University), p. 267.

—— 1967*b*. *Phys. Rev.* **158**, 970.

—— and SUNDBERG, O., 1972. *Nucl. Phys.* **A187**, 314.

—— and SWIFT, A., 1967. *Nucl. Phys.* **A94**, 52.

—— and WEBB, S.J., 1970. *Phys. Rev. Lett.* **24**. 145.

—— —— and BARRETT, R.C., 1969. *Proc. Conf. on High Energy Physics and Nuclear Structure* (New York: Plenum Press), p. 67.

ENGELBRECHT, C.A. and WEIDENMÜLLER, H., 1972. *Nucl. Phys.* **A184**, 385.

ENGFER, R., SCHNEUWLY, H., VUILLEMEIER, J.L., WALTER, H.K., and ZEHNDER, A., 1974. *Atom. Data Nuc. Data Tables.* **14**, 509.

ENGLER, J., HORN, K., KONIG, J., SCHLUDECKER, P., SCHOPPER, H., SIEVERS, P., ULLRICH, H., and RUNGE, K., 1968. *Phys. Lett.* **26B**, 64.

ENGLER, J., HORN, K., MÖNNIG, F., SCHLUDECKER, P., SCHMIDT-PARZEFALL, W., SCHOPPER, H., SIEVERS, P., ULLRICH, H., HARTUNG, R., RUNGE, K., and GALAKTIONOV, Yu., 1970. *Phys. Lett.* **32B**, 716.

EPP, C.D. and GRIFFY, T.A., 1970. *Phys. Rev.* **C1**, 1633.

EPSTEIN, G.L. and DAVIS, S.P., 1971. *Phys. Rev.* **A4**, 464.

ERB, K.A., HOLDEN, J.E., LEE, I.Y., SALADIN, J.X., and SAYLOR, T.K., 1972. *Phys. Rev. Lett.* **29**, 1010.

ERICSON, M. and ERICSON, T.E.O., 1966. *Ann. Phys.* **36**, 323.

——— ——— and KRELL, M., 1969. *Phys. Rev. Lett.* **22**, 1189.

ERICSON, T.E.O., 1969. Selected topics in pion-nucleus interaction *Rep. No. TH 1093-CERN.*

——— and HÜFNER, J., 1970. *Phys. Lett.* **33B**, 601.

——— and SCHECK, F., 1970. *Nucl. Phys.* **B19**, 450.

FAESSLER, A., GALONSKA, J.E., EHLERS, J.W., and MOSZKOWSKI, S.A., 1972*a*. *Nuovo Cim.* **11A**, 63.

——— ——— and GOEKE, K., 1972*b*. *Z. Phys.* **250**, 436.

——— and WOLTER, H.H., 1969. *Z. Phys.* **223**, 192.

——— KREWALD, S., and WAGNER, G.J., 1975. *Phys. Rev.* **C11**, 2069.

FAI, G. and NEMETH, J., 1973. *Nucl. Phys.* **A208**, 463.

FAGERSTRÖM, B. and KÄLLNE, J., 1973. *Phys. Scripta* **8**, 14.

FAIVRE, J.C., KRIVINE, H., and PAPIAU, A.M., 1968. *Nucl. Phys.* **A108**, 508.

FAJARDO, L.A., FICENEC, J.R., TROWER, W.P., and SICK, I., 1971. *Phys. Lett.* **37B**, 363.

FÄLDT, G. and PILKUHN, H., 1972. *Phys. Lett.* **40B**, 613.

FERNANDEZ, B. and BLAIR, J.S., 1970. *Phys. Rev.* **C1**, 523.

FERNBACH, S., SERBER, R., and TAYLOR, T.B., 1949. *Phys. Rev.* **75**, 1352.

FESHBACH, H., 1958. *Ann. Phys.* **5**, 357.

——— 1962. *Ann. Phys.* **19**, 287.

——— GAL, A., and HÜFNER, J., 1971. *Ann. Phys.* **66**, 20.

——— PORTER, C.E., and WEISSKOPF, V., 1954. *Phys. Rev.* **96**, 448.

FEY, G., FRANK, H., SCHÜTZ, W., and THEISSEN, H., 1973*a*. *Z. Phys.* **265**, 401.

FICENEC, J., FAJARDO, L.A., TROWER, W.P., and SICK, I., 1972. *Phys. Lett.* **42B**, 213.

——— TROWER, W.P., HEISENBERG, J., and SICK, I., 1970. *Phys. Lett.* **32B**, 460.

FINDLAY, D.J.S., GARDINER, S.N., MATTHEWS, J.L., and OWENS, R.O., 1970. *J. Phys.* **A7**, L157.

FINK, M., HEBACH, H., and KÜMMEL, H., 1972. *Nucl. Phys.* **A186**, 353.

FISCHER, W., HARTMANN, M., HÜHNERMANN, H., and VOGG, H., 1974. *Z. Phys.* **267**, 209.

—— HÜHNERMANN, H., KRÖMER, G., and SCHÄFER, H.J., 1971. Private communication referred to by Heilig and Steudel (1974).

FITCH, V.L. and RAINWATER, J., 1953. *Phys. Rev.* **92**, 789.

FIVOZINSKY, S.P., PENNER, S., LIGHTBODY, J.W. Jr., and BLUM, D., 1974. *Phys. Rev.* **C9**, 1533.

FLOCARD, H., 1975. Private communication to Sick *et al.* (1975).

—— QUENTIN, P., KERMAN, A.K., and VAUTHERIN, D., 1973. *Nucl. Phys.* **A203**, 433.

—— —— and VAUTHERIN, D., 1973. *Phys. Lett.* **B46**, 304.

FOETH, H., HOLDER, M., RADEMACHER, E., STAUDE, A., DARRIULAT, P., DEUTSCH, J., KLEINKNECHT, K., RUBBIA, C., TITTEL, K., FERRERO, M.I., and GROSSO, C., 1970. *Phys. Lett.* **31B**, 544.

FOLDY, L.L. and WALECKA, J.D., 1969. *Ann. Phys.* **54**, 447.

FORD, K.W. and RINKER, G.A., 1973. *Phys. Rev.* **C7**, 1206.

—— —— 1974. *Phys. Rev.* **C9**, 2444.

—— and WILLS, J.G., 1969. *Phys. Rev.* **185**, 1429.

FRAHN, W.E., 1966. *Nucl. Phys.* **75**, 577.

—— 1967. *Fundamentals in nuclear theory* (Vienna: IAEA), p. 3.

—— 1971. *Phys. Rev. Lett.* **26**, 568.

—— and VENTER, R., 1963. *Ann. Phys.* **24**, 243.

—— —— 1964. *Ann. Phys.* **27**, 135.

—— and WIECHERS, G., 1966. *Phys. Rev. Lett.* **16**, 810.

FRANCO, V., 1970. *Phys. Rev. Lett.* **24**, 1452.

—— 1972. *Phys. Rev.* **C6**, 748.

FRENCH, J.B., 1964. *Phys. Lett.* **13**, 249.

—— 1965. *Phys. Lett.* **15**, 327.

FRIAR, J.L. and NEGELE, J.W., 1973. *Nucl. Phys.* **A212**, 93.

—— —— 1975*a*. *Nucl. Phys.* **A240**, 301.

—— —— 1975*b*. *Adv. nucl. Phys.* **8**, 219.

—— and ROSEN, M., 1972. *Phys. Lett.* **39B**, 615.

FRICKE, B. and WABER, J.T., 1972. *Phys. Rev.* **B5**, 3445.

FRICKE, M.P., GROSS, E.E., MORTON, B.J., and ZUCKER, A., 1967. *Phys. Rev.* **156**, 1207.

—— —— and ZUCKER, A., 1967. *Phys. Rev.* **163**, 1153.

—— and SATCHLER, G.R., 1965. *Phys. Rev.* **139**, B567.

FRIEDMAN, A.M., SIEMSSEN, R.H., and CUNINGHAME, J.G., 1972. *Phys. Rev.* **C6**, 2219.

FRIEDMAN, E., 1969. *Phys. Lett.* **29B**, 213.

—— 1971*a*. *Nucl. Phys.* **A170**, 214.

—— 1971*b*. *Phys. Lett.* **35B**, 543.

FRIEDMAN, E., JAFFE, A.A., NIR, D., and TUCHMAN, Y., 1972. Unpublished.

——— and MANDELBAUM, B., 1969. *Nucl. Phys.* **A135**, 472.

FRIEDRICH, J., 1972. *Nucl. Phys.* **A191**, 118.

——— and LENZ, F., 1972. *Nucl. Phys.* **A183**, 523.

FRITSCH, W., LIPPERHEIDE, R., and WILLE, U., 1973. *Phys. Lett.* **B45**, 103.

FROSCH, R.F., HOFSTADTER, R., McCARTHY, J.S., NÖLDEKE, G.K., VAN OOSTRUM, K.J., YEARIAN, M.R., CLARK, B.C., HERMAN, R., and RAVENHALL, D.G., 1968. *Phys. Rev.* **174**, 1380.

FUIJISAWA, T., KAMITSUBO, H., WADA, T., and IGARISHI, M., 1969. *J. phys. Soc. Japan* **27**, 278.

FUJII, S., 1963. *Prog. theor. Phys.* **29**, 274.

FULLER, E.G., 1973. *Proc. Asilomar Conf. on Photonuclear Reactions and Applications* ed. B.L. Berman (Springfield, Va.: NTIS), p. 1201.

FULLER, G.H. and COHEN, V.W., 1969. *Nuc. Data* **A5**, 433.

FULLING, S.A. and SATCHLER, G.R., 1968. *Nucl. Phys.* **A111**, 81.

FULMER, C.B., BALL, J.B., SCOTT, A., and WHITTEN, M.L., 1967. *Phys. Lett.* **24B**, 505.

——— and HAFELE, J.C., 1972. *Phys. Rev.* **C5**, 1969.

——— ——— 1973*a*. *Phys. Rev.* **C7**, 631.

——— ——— 1973*b*. *Phys. Rev.* **C8**, 172.

——— ——— 1973*c*. *Phys. Rev.* **C8**, 200.

GAILLARD, M., BOUCHÉ, R., FEUVRAIS, L., GAILLARD, P., GUICHARD, A., GUSAKOW, M., LEONHARDT, J.L., and PIZZI, J.R., 1968. *Nucl. Phys.* **A119**, 161.

——— ——— ——— ——— ——— ——— ——— ——— 1969. *Nucl. Phys.* **A131**, 353.

GAL, A., GRODZINS, L., and HÜFNER, J., 1968. *Phys. Rev. Lett.* **21**, 453.

GAMOW, G., 1928. *Z. Phys.* **51**, 204.

GARDINER, S.N., 1971. *Ph.D. Thesis* University of Glasgow (unpublished).

——— MATTHEWS, J.L., and OWENS, R.Q., 1973. *Proc. Asilomar Conf. on Photonuclear Reactions and Applications* ed. B.L. Berman (Springfield, Va.: NTIS), p. 979; *Phys. Lett.* **46B**, 186.

GARRON, J.P., JACMART, J.C., RIOU, M., RUHLA, C., TEILLAC, J., CAVERZASIO, C., and STRAUCH, K., 1961. *Phys. Rev. Lett.* **7**, 261.

——— ——— ——— ——— ——— and STRAUCH, K., 1962. *Nucl. Phys.* **37**, 126.

GARVEY, J., PATRICK, B.H., RUTHERGLEN, J.G., and SMITH, I.L., 1965. *Nucl. Phys.* **70**, 241.

GAUVIN, H., LE BEYEC, Y., LEFORT, M., and DEPRUN, C., 1972. *Phys. Rev. Lett.* **28**, 697.

GELL-MANN, M., GOLDBERGER, M.L., and THIRRING, W., 1954. *Phys. Rev.* **95**, 1612.

GEOFFRION, B., MARTY, N., MORLET, M., TATISCHEFF, B., and WILLIS, A., 1968. *Nucl. Phys.* **A116**, 209.

GIBSON, B.F. and VAN OOSTRUM, K.J., 1967. *Nucl. Phys.* **A90**, 159.

GIBSON, E.F., RIDLEY, B.W., KRAUSHAAR, J.J., RICKEY, M.E., and BASSEL, R.H., 1967. *Phys. Rev.* **155**, 1194; 1208.

GILLESPIE, W.A., 1973. *Ph.D. Thesis* University of Glasgow.

——— BRAIN, S.W., CURRAN, C.S., JOHNSTON, A., LEES, E.W., SINGHAL, R.P., and SLIGHT, A.G., 1973. *Proc. Int. Conf. on Nuclear Physics Munich* eds. J. de Boer and H.J. Mang (Amsterdam: North-Holland/New York: American Elsevier), p. 626.

GILLET, V., GIRAUD, B., and RHO, M., 1969. *Phys. Rev.* **178**, 1695.

——— and MELKANOFF, M.A., 1964. *Phys. Rev.* **133**, 1190.

——— and SANDERSON, E.A., 1967. *Nucl. Phys.* **A91**, 292.

GILMAN, F.J., 1969. *Rep. SLAC-PUB-289* (Int. Conf. on Particle Interactions at High Energies, Toronto).

GILS, H.J. and REBEL, H., 1976. *Phys. Rev.* **C13**, 2159.

GIRAUD, B., LE TOURNEUX, J., and WONG, S.K.M., 1970. *Phys. Lett.* **32B**, 23.

GLASHAUSER, C. and THIRION, J., 1969. *Adv. nucl. Phys.* **2**, 79.

GLASSGOLD, A.E. and GREIDER, K.R., 1960. *Ann. Phys.* **10**, 100.

GLAUBER, R.J., 1959. *Lectures in theoretical physics* vol. I (New York, London: Interscience), p. 315.

——— 1967. *Proc. Conf. on High Energy Physics and Nuclear Structure* ed. G. Alexander (Amsterdam: North-Holland), p. 311.

GLENDENNING, N.K., 1963. *A. Rev. nucl. Sci.* **13**, 447.

——— 1965. *Phys. Rev.* **137**, B102.

——— 1967. *Lectures Int. School of Physics 'E. Fermi'* Course XL (New York, London: Academic Press).

——— and VENERONI, M., 1966. *Phys. Rev.* 144, 839.

GOBLE, A.T., SILVER, J.D., and STACEY, D.N., 1974. *J. Phys.* **B7**, 26.

GOLDBERG, D.A. and SMITH, S.M., 1972. *Phys. Rev. Lett.* **29**, 500.

——— ——— and BURDZIK, G.F., 1974. *Phys. Rev.* **C10**, 1362.

——— ——— PUGH, H.G., ROOS, P.G., and WALL, N.S., 1973. *Phys. Rev.* **C7**, 1938.

GOLDBERGER, M. and WATSON, K.M., 1964. *Collision theory* (New York: John Wiley).

GOLDFARB, L.J.B., 1965. *Nucl. Phys.* **72**, 537.

GOLDHABER, M. and TELLER, E., 1948. *Phys. Rev.* **74**, 1046.

GOLDRING, G., SAMUEL, M., WATSON, B.A., BERTIN, M.C., and TABOR, S.L., 1970. *Phys. Lett.* **32B**, 465.

GOMES, L.C., 1959. *Phys. Rev.* **116**, 1226.

GOMPELMAN, H., BLAAUW, H.J., and DE JAGER, C.W., 1976, to be published.

GOODING, T.J. and PUGH, H.G., 1960. *Nucl. Phys.* **18**, 46.

GOTTFRIED, K., 1958. *Nucl. Phys.* **5**, 557.

——— and YENNIE, D., 1969. *Phys. Rev.* **182**, 1595.

GRAY, W.S., KENEFICK, R.A., KRAUSHAAR, J.J., and SATCHLER, G.R., 1966. *Phys. Rev.* **142**, 735.

GREEN, A.E.S., 1956. *Phys. Rev.* **102**, 1325.

——— SAWADA, T., and SAXON, D.S., 1968. *The nuclear independent particle model* (New York, London: Academic Press).

GREEN, A.M., 1967. *Phys. Lett.* **24B**, 384.

GREENLEES, G.W., MAKOFSKE, W., and PYLE, G.J., 1970. *Phys. Rev.* **C1**, 1145.

——— and PYLE, G.J., 1966. *Phys. Rev.* **149**, 836.

——— ——— and TANG, Y.C., 1968*a*. *Phys. Rev.* **171**, 1115.

——— ——— ——— 1968*b*. *Phys. Lett.* **26B**, 658.

GREIDER, K., 1970. *Nuclear reactions induced by heavy ions* eds. R. Bock and W.R. Hering (Amsterdam: North-Holland), p. 217.

GRIFFY, T.A., ONLEY, D.S., REYNOLDS, J.T., and BIEDENHARN, L.C., 1962. *Phys. Rev.* **128**, 833.

GRIMM, R.C., McCARTHY, I.E., and STORER, R.G., 1971. *Nucl. Phys.* **A166**, 330.

GROENVALD, K.O., MEYER-SCHÜTZMEISTER, L., RICHTER, A., and STROHBUSCH, U. 1972. *Phys. Rev.* **C6**, 805.

GROSS, D.H.E. and LIPPERHEIDE, R., 1970. *Nucl. Phys.* **A150**, 449.

GROTCH, H. and YENNIE, D.R., 1969. *Rev. mod. Phys.* **41**, 350.

GUSTAFSSON, Ch. AND BERGGREN, T., 1971. *Phys. Lett.* **35B**, 546.

GUTBROD, H., WINN, W.G., and BLANN, M., 1973. *Phys. Rev. Lett.* **30**, 1259.

GUTH, E., 1934. *Anz. Akad. Wiss. Wien* **24**, 299.

HADDOCK, R.P., ZELLER, M., and CROWE, K.M., 1964. *UCLA Rep. No. MPG-64-2.*

HAEBERLI, W. and KNUTSON, L.D., 1973. *Phys. Rev. Lett.* **30**, 986.

HAHN, B., RAVENHALL, D.G., and HOFSTADTER, R., 1956. *Phys. Rev.* **101**, 1131.

HALLOWELL, P.L., BERTOZZI, W., HEISENBERG, J., KOWALSKI, S., MARUYAMA, X., SARGENT, C.P., TURCHINETZ, W., WILLIAMSON, C.F., FIVOZINSKY, S.P., LIGHTBODY, J.W. Jr., and PENNER, S., 1973. *Phys. Rev.* **C7**, 1396.

HAMMERSTEIN, G.R., HOWELL, R.H., and PETROVICH, F., 1973. *Nucl. Phys.* **A213**, 45.

HAND, L.N., MILLER, D.G., and WILSON, R., 1963. *Rev. mod. Phys.* **35**, 335.

HARADA, K., 1961. *Prog. theor. Phys.* **26**, 667.

——— 1962. *Prog. Theor. Phys.* **27**, 430.

——— and RAUSCHER, E.A., 1968. *Phys. Rev.* **169**, 818.

HARNEY, H.L., 1968. *Nucl. Phys.* **A119**, 591.

——— WIEDNER, C.A., and WURM, J.P., 1968. *Phys. Lett.* **26B**, 204.

HARRIS, R.J., SHULER, W.B., ECKHAUSE, M., SIEGAL, R.T., and WELSH, R.E., 1968. *Phys. Rev. Lett.* **20**, 505.

HARTMANN, H., HOFFMANN, H., MECKING, B., and NÖLDECKE, G., 1973. *Proc. Asilomar Conf. on Photonuclear Reactions and Applications* ed. B.L. Berman (Springfield, Va.: NTIS), p. 967.

HARVEY, J.D. and JOHNSON, R.C., 1971. *Phys. Rev.* **C3**, 636.

HASAN, T. and NAQVI, J.H., 1974. *Nucl. Phys.* **A220**, 114.

HAUSER, G., LÖHKEN, R., NOWICKI, G., REBEL, H., SCHATZ, G., SCHWEIMER, G., and SPECHT, J., 1972. *Nucl. Phys.* **A182**, 1.

——— ——— REBEL, H., SCHATZ, G., SCHWEIMER, G.W., and SPECHT, J., 1969. *Nucl. Phys.* **A128**, 81.

HAYBRON, R.M., 1966. *Nucl. Phys.* **79**, 33.

——— JOHNSON, M.B., and METZGER, R.J., 1967. *Phys. Rev.* **156**, 1136.

——— and McMANUS, H., 1965. *Phys. Rev.* **140**, B638.

HEER, E., HIRT, W., MARTIN, M., MICHAELIS, E.G., SERRE, C., SKAREK, P., and WRIGHT, B.T., 1966. *Proc. Williamsburg Conf. on Intermediate Energy Physics* vol. 1 (Williamsburg, Va.: College of William and Mary), p. 277.

HEILIG, K., SCHMITZ, K., and STEUDEL, A., 1963. *Z. Phys.* **176**, 120.

——— and STEUDEL, A., 1974. *Atom. Data Nucl. Data Tables* **14**, 613.

HEIMLICH, F.H., KÖBBERLING, M., MORITZ, J., SCHMIDT, K.H., WEGENER, D., ZELLER, D., BIENLEIN, T.K., BLECKWENN, J., and DINTER, H., 1974. *Nucl. Phys.* **A2231**, 509.

HEISENBERG, J., 1973. Private communication to Friar and Negele (1973).

——— 1974. Private communication to A. Zehnder (1974).

——— HOFSTADTER, R., McCARTHY, J.S., SICK, I., CLARK, B.C., HERMAN, R., and RAVENHALL, D.G., 1969. *Phys. Rev. Lett.* 23, 1402.

——— McCARTHY, J.S., and SICK, I., 1970. *Nucl. Phys.* **A157**, 435.

——— ——— ——— 1971*a*. *Nucl. Phys.* **A164**, 353.

——— ——— ——— and YEARIAN, M.R., 1971*b*. *Nucl. Phys.* **A164**, 340.

——— and SICK, I., 1970. *Phys. Lett.* **32B**, 249.

HELM, R.H., 1956. *Phys. Rev.* **106**, 1466.

HENDRIE, D.L., CHABRE, M., and PUGH, H.G., 1966. *UCRL Rep. No. 16580.*

HERMAN, R. and HOFSTADTER, R., 1960. *High-energy electron scattering tables* (Stanford: University Press).

HERSCOVITZ, V., 1971. *Nucl. Phys.* **A161**, 321.

HETHERINGTON, J.H. and BORYSOWICZ, 1974. *Nucl. Phys.* **A219**, 221.

HINDS, S., MIDDLETON, R., BJERREGAARD, J.H., HANSEN, O., and NATHAN, O., 1966. *Nucl. Phys.* **83**, 17.

HINTERBERGER, F., MAIRLE, G., SCHMIDT-ROHR, Y., WAGNER, G.J., and TUREK, P., 1968. *Nucl. Phys.* **A111**, 265.

HIRAMATSU, H., KAMAE, T., MURUMATSU, H., NAKAMURA, K., IZUTSU, N., and WATASE, Y., 1973. *Phys. Lett.* **44B**, 50.

HIRT, W., 1969. *Nucl. Phys.* **B9**, 447.

HITLIN, D., BERNOW, S., DEVONS, S., DUERDOTH, I., KAST, J.W., MACAGNO, E.R., RAINWATER, J., WU, C.S., and BARRETT, R.C., 1970. *Phys. Rev.* **C1**, 1184.

HNIZDO, V., KARBAN, O., LOWE, J., GREENLEES, G.W., and MAKOFSKE, W., 1971. *Phys. Rev.* **C3**, 1560.

HODGSON, P.E., 1961. *Phys. Rev. Lett.* **6**, 358.

——— 1966. *Adv. Phys.* **15**, 329.

——— 1968. *Adv. Phys.* **17**, 563.

——— 1971. *Nuclear reactions and nuclear structure* (Oxford: Clarendon Press).

HOFSTADTER, R., 1956. *Rev. mod. Phys.* **28**, 214.

——— 1957. *A. Rev. nucl. Sci.* **7**, 231.

HOLM, H., SCHEID, W., and GREINER, W., 1969. *Phys. Lett.* **29B**, 473.

HORIKAWA, Y., TORIZUKA, Y., NAKADA, A., MITSUNOBU, S., KOJIMA, Y., and KIMURA, M., 1971. *Phys. Lett.* **36B**, 9.

HOROWITZ, Y.S., 1972. *Nucl. Phys.* **A193**, 438.

HÜFNER, J., 1964. *Nucl. Phys.* **60**, 427.

——— 1975. *Phys. Lett.* **21C**, 1.

——— and DE SHALIT, A., 1965. *Phys. Lett.* **15**, 52.

HUGHES, D.J. and ECKART, D., 1930. *Phys. Rev.* **36**, 694.

HULTZSCH, H., 1970. *Ph.D. Thesis* University of Mainz.

IACHELLO, F. and LANDE, A., 1971. *Phys. Lett.* **35B**, 205.

IGO, G., 1958. *Phys. Rev. Lett.* **1**, 72.

——— 1959. *Phys. Rev.* **115**, 1665.

——— BARNES, P.D., FLYNN, E.R., and ARMSTRONG, D.D., 1969. *Phys. Rev.* **177**, 1831.

INOPIN, E.V. and BEREZHNOY, Yu.A., 1965. *Nucl. Phys.* **63**, 689.

INOUE, M., 1968. *Nucl. Phys.* **A119**, 449.

IOANIDES, A. and JOHNSON, R.C., 1976. *Phys. Lett.* **61B**, 4.

IRVINE, J.M., 1967. *Nucl. Phys.* **A98**, 161.

——— 1968. *Nucl. Phys.* **A120**, 576.

ISHIMATSU, T., HAYASHIBE, S., KAWAMURA, N., AWAYA, T., OHMURA, H., NAKAJIMA, Y., and MITARAI, S., 1972. *Nucl. Phys.* **A185**, 273.

JACKSON, D.F., 1964. *Phys. Lett.* **14**, 118.

——— 1965. *Rev. mod. Phys.* **37**, 393.

——— 1967*a*. *Nuovo Cim.* **51B**, 49.

——— 1967*b*. *Phys. Rev.* **155**, 1065.

——— 1968. *Adv. Phys.* **17**, 481.

——— 1969*a*. *Nucl. Phys.* **A123**, 273.

——— 1969*b*. *Nuovo Cim.* **63A**, 343.

——— 1970*a*. *Nuclear reactions* (London: Methuen) chap. 8.

——— 1970*b*. *Phys. Lett.* 32B. 232.

——— 1971*a*. *Nucl. Phys.* **A173**, 225.

——— 1971*b*. *Phys. Lett.* **35B**, 99.

——— 1975. *J. Phys. (Paris)* **36**, C5-1.

——— and BERGGREN, T., 1965. *Nucl. Phys.* **62**, 353.

——— and ELTON, L.R.B., 1965. *Proc. phys. Soc.* **85**, 659.

——— HILTON, J.M., and ROBERTS, A.C.M., 1976. Preprint.

——— and JOHNSON, R.C., 1974. *Phys. Lett.* **49B**, 249.

——— and MORGAN, C.C., 1968. *Phys. Rev.* **175**, 1402.

——— and MURUGESU, S., 1970. *Nucl. Phys.* **A149**, 261.

——— and RHOADES-BROWN, M., 1976*a*. *Nucl. Phys.* **A266**, 61.

——— ——— 1976*b*. *Ann. Phys.* to be published.

——— and SHAH, M.B., 1969. *Symp. Nuclear Reaction Mechanisms* (Quebec: Laval University Press), p. 493.

JACKSON, H.E., KOVAR, D.G., MEYER-SCHÜTZMEISTER, L., SEGAL, R.E., SCHIFFER J.P., VIGDOR, S., WANGLER, T.P., BURMAN, R.L., DRAKE, D.M., GRAM, P.A.M., REDWINE, R.P., LIND, V.G., HATCH, E.N., OTTESON, O.H., McADAMS, R.E., COOK, B.C., and CLARK, R.B., 1975. *Phys. Rev. Lett.* **35**, 641.

——— MEYER-SCHÜTZMEISTER, L., WANGLER, T.P., REDWINE, R.P., SEGAL, R.E., TONN, J., and SCHIFFER, J.P., 1973. *Phys. Rev. Lett.* **31**, 1353.

JACOB, G. and MARIS, Th.A.J., 1962. *Nucl. Phys.* **31**, 139.

——— ——— 1973. *Rev. mod. Phys.* **45**, 6.

JACOBSOHN, B.A., 1954. *Phys. Rev.* **96**, 1637.

JACQUOT, C., SAKAMOTO, Y., JUNG, M., BAIXERAS-AIGUABELLA, C., GIRARDIN, L., and BRAUN, H., 1970. *Nucl. Phys.* **A148**, 325.

JAFFE, R.L. and GERACE, W., 1969. *Nucl. Phys.* **A125**, 1.

JAGER, C.W.de, 1973. *Ph.D. Thesis* University of Amsterdam.

——— VRIES, H.de, and VRIES, C.de, 1974. *Atom. Data nucl. Data Tables* **14**, 479.

JAIN, A.K. GROSSIORD, J.Y., CHEVALLIER, M., GAILLARD, P., GUICHARD, A., GUSAKOW, M., and PIZZI, J.R., 1973. *Nucl. Phys.* **A216**, 519.

——— and SARMA, N., 1974. *Nucl. Phys.* **A233**, 145.

——— ——— and BANERJEE, B., 1969. *Nuovo Cim.* **62**, 219.

——— ——— ——— 1970. *Nucl. Phys.* **A142**, 330.

JAIN, A.P., 1964. *Nucl. Phys.* **50**, 157.

JAIN, B.K., 1968. *Nucl. Phys.* **A116**, 256,

JAIN, B.K., 1969. *Nucl. Phys.* **A129**, 145.

—— and JACKSON, D.F., 1967. *Nucl. Phys.* **A99**, 113.

JAIN, M., ROOS, P.G., PUGH, H.G., and HOLMGREN, H.D., 1970. *Nucl. Phys.* **A153**, 49.

JAMES, A.N., ANDREWS, P.T., BUTLER, P., COHEN, N., and LOWE, B.G., 1969*a*. *Nucl. Phys.* **A133**, 89.

—— —— KIRKBY, P., and LOWE, B.G., 1969*b*. *Nucl. Phys.* **A138**, 145.

JÄNECKE, J., 1969. *Isospin in nuclear physics* ed. D.H. Wilkinson (Amsterdam: North-Holland), chap. 8.

JANISZEWSKI, J. and McCARTHY, I.E., 1972*a*. *Nucl. Phys.* **A192**, 85.

—— —— 1972*b*. *Nucl. Phys.* **A181**, 97.

JANSEN, J.A., PEERDEMAN, R.Th. and VRIES, C.de, 1972. *Nucl. Phys.* **A188**, 337.

JARVIS, O.N., HARVEY, B.G., HENDRIE, D.L., and MAHONEY, J., 1967. *Nucl. Phys.* A102, 1967.

—— WHITEHEAD, C., and SHAH, M.B., 1972. *Nucl. Phys.* **A184**, 615.

JEANS, A.F., DARCEY, W., DAVIES, W.G., JONES, K.N., and SMITH, P.K., 1969. *Nucl. Phys.* **A128**, 224.

JENKINS, D.A. and CROWE, K.M., 1966. *Phys. Rev. Lett.* **16**, 637.

—— and KUNSELMANN, R., 1966. *Phys. Rev. Lett.* **17**, 1148.

—— —— SIMMONS, M.K., and YAMAZAKI, T., 1966. *Phys. Rev. Lett.* **17**, 1.

—— POWERS, R.J., MARTIN, P., MILLER, G.H., WELSH, R.E., and KUNSELMAN, A.R., 1970*a*. *Phys. Lett.* **32B**, 267.

—— —— and KUNSELMANN, A.R., 1970*b*. *Phys. Rev.* **C2**, 458.

—— —— MARTIN, P., MILLER, G.H., and WELSH, R.E., 1971. *Nucl. Phys.* **A175**, 73.

JENSEN, A.S. and WONG, C.Y., 1970. *Phys. Rev.* **C1**, 1321.

JEUKENNE, J.-P., LEJEUNE, A., and MAHAUX, C., 1974. *Phys. Rev.* **C10**, 1391.

JOHANSSON, A., SVANBERG, U., and HODGSON, P.E., 1961. *Ark. Fys.* **19**, 541.

JOHNSON, C.S., HINCKS, E.P., and ANDERSON, H.L., 1962. *Phys. Rev.* **125**, 2102.

JOHNSON, M.H. and TELLER, E., 1954. *Phys. Rev.* **93**, 357.

JOHNSON, R.C., 1971. *Polarization phenomena in nuclear reactions* (Madison: University of Wisconsin).

—— and MARTIN, D.C., 1972. *Nucl. Phys.* **A192**, 496; and erratum 1973. *Nucl. Phys.* **A211**, 617.

—— and SANTOS, F.D., 1967. *Phys. Rev. Lett.* **19**, 364.

—— —— 1971. *Particles Nuclei* **2**, 285.

—— and SOPER, P.J.R., 1970. *Phys. Rev.* **C1**, 976.

—— —— 1972. *Nucl. Phys.* **A182**, 619.

JOHNSTON, R.R. and WATSON, K.M., 1961. *Nucl. Phys.* **28**, 383.

JONES, G.D., DURELL, J.L., LILLEY, J.S., and PHILLIPS, W.R., 1974. *Nucl. Phys.* **A230**, 173.

JONES, L.W., AYRE, C.A., GUSTAFSON, H.R., LONGO, M.J., and RAMANA MURTHY, P.W., 1974. *Phys. Rev. Lett.* **33**, 1440.

——— LONGO, M.J., McCORRISTON, T.P., PARKER, E.F., POWELL, S.T., and KREISLER, M.N., 1971. *Phys. Lett.* **36B**, 509.

JONES, P.B., 1958. *Phil. Mag.* **3**, 33.

KADMENSKII, S.G., 1970. *Sov. J. nucl. Phys.* **10**, 422.

——— and KALECHITS, V.E., 1971. *Sov. J. nucl. Phys.* **12**, 37.

KALINSKY, D., CARDMAN, L.S., YEN, R.E., LEGG, J.R., and BOCKELMAN, C.K., 1973. *Nucl. Phys.* **A216**, 312.

KALLIO, A. and KOLTVEIT, K., 1964. *Nucl. Phys.* **53**, 87.

KALLIO, A.J., TOROPAINEN, P., GREEN, A.M., and KOUKI, T., 1974. *Nucl. Phys.* **A231**, 77.

KÄLLNE, J., 1974. *Gustaf Werner Institute Report* GWI-PH 2/74. Revised version to be published in *Phys. Scripta.*

——— and FAGERSTRÖM, B., 1974. *Proc. Fifth Int. Conf. on High Energy Physics and Nuclear Structure* ed. G. Tibell (Amsterdam: North-Holland), p. 369.

KALVIUS, G.M. and SHENOY, G.K., 1974. *Atom. Data Nucl. Data Tables* **14**, 639.

KANNENBERG, S., 1968. *Ph.D. Thesis* Northeastern University, Boston.

KAPLAN, I., 1951. *Phys. Rev.* **81**, 962.

KAST, J.W., 1970. *Ph.D. Thesis* Columbia University, New York.

——— BERNOW, S., CHENG, S.C., HITLIN, D., LEE, W.Y., MACAGNO, E.R., RUSHTON, A.M., and WU, C.S., 1971. *Nucl. Phys.* **A169**, 62.

KEMBHAVI, V.K., 1971. *Ph.D. Thesis* University of Surrey.

——— and JACKSON, D.F., 1969. *Proc. Third Int. Conf. on High Energy Physics and Nuclear Structure* ed. S. Devons (New York: Plenum Press), p. 850.

KENT, J.J., MORGAN, J.F., and SEYLER, R.G., 1972. *Nucl. Phys.* **A197**, 177.

KERMAN, A.K., McMANUS, H., and THALER, R.M., 1959. *Ann. Phys.* **8**, 551.

——— SVENNE, J.P., and VILLARS, F.M.H., 1966. *Phys. Rev.* **147**, 710.

KESSLER, D., 1971. *Proc. Muon Phys. Conf. Fort Collins, Colorado* ed. P. Chand (New York: Dekker).

——— McKEE, R.J., HARGROVE, C.K., HINCKS, E.P., and ANDERSON, H.L., 1970. *Can. J. Phys.* **48**, 3029.

——— MES, H., THOMPSON, A.C., ANDERSON, H.L., DIXIT, M.S., HARGROVE, C.K., and McKEE, R.J., 1975. *Phys. Rev.* **C11**, 1719.

KHVASTUNOV, V.M., AFANAS'EV, N.G., AFANAS'EV, V.D., GUL'KAROV, I.S., OMELAENKO, A.S., SAVITSKII, G.A., KHOMICH, A.A., SHEVCHENKO, N.G., ROMANOV, V.S, and RUSANOVA, N.V., 1970. *Nucl. Phys.* **A146**, 15.

KIDWAI, H.R. and ROOK, J.R., 1971. *Nucl. Phys.* **A169**, 417.

KIENLE, P., 1968. *Hyperfine structure and nuclear radiations* eds. E. Matthias and D.A. Shirley (Amsterdam: North-Holland), p. 27.

——— KALVIUS, G.M., and RUBY, S.L., 1968. *Hyperfine structure and nuclear radiations* eds. E. Matthias and D.A. Shirley (Amsterdam: North-Holland), p. 971.

KIM, J.K., 1965. *Phys. Rev. Lett.* **14**, 29.

——— 1967. *Phys. Rev. Lett.* **19**, 1079.

KING, W.H., 1963. *J. opt. Soc. Am.* **53**, 638.

——— 1971. *J. Phys.* **B4**, 288.

——— STEUDEL, A., and WILSON, M., 1973. *Z. Phys.* **265**, 207.

KISSLINGER, L.S., 1955. *Phys. Rev.* **98**, 761.

——— BURMAN, R.L., KOCH, J.H., and STERNHEIM, M.M., 1972. *Phys. Rev.* **C6**, 469.

KLINE, F.J., CRANNELL, H., O'BRIEN, J.T., McCARTHY, J., and WHITNEY, R.R., 1973. *Nucl. Phys.* **A209**, 381.

KNOLL, J., 1973. *Nucl. Phys.* A201, 289.

KÖBBERLING, M., MORITZ, J., SCHMIDT, K.H., WEGENER, D., ZELLER, D., BIENLEIN, J.K., BLECKWENN, J., DINTER, H., and HEIMLICH, F.H., 1974. *Nucl. Phys.* **A231**, 504.

KOCH, H., 1973. *Proc. Fifth Int. Conf. on High Energy Physics and Nuclear Structure* ed. G. Tibell (Amsterdam: North-Holland), p. 225.

——— KRELL, M., VON DER MALSBURG, CH., POELZ, G., SCHMITT, H., TAUSCHER, L., BACKENSTOSS, G., CHARALAMBUS, S., and DANIEL, H., 1969. *Phys. Lett.* **29B**, 140.

——— POELZ, G., SCHMITT, H., TAUSCHER, L., BACKENSTOSS, G., CHARALAMBUS, S., and DANIEL, H., 1968. *Phys. Lett.* **28B**, 279.

KOFOED-HANSEN, O., 1973. *Nucl. Phys.* **B54**, 42.

——— and MARGOLIS, B., 1969. *Nucl. Phys.* **B11**, 455.

——— and WILKIN, C., 1971. *Ann. Phys.* **63**, 209.

KÖHLER, H.S., 1965. *Phys. Rev.* **138**, B381.

——— 1969. *Nucl. Phys.* **A139**, 353.

——— 1970. *Nucl. Phys.* **A144**, 407.

KÖHLER, H.S. and LIN, Y.C., 1969. *Nucl. Phys.* **A136**, 49.

——— and McCARTHY, R.J., 1967. *Nucl. Phys.* **A106**, 313.

KOHMURA, T., 1971. *Prog. theor. Phys.* **46**, 167.

KÖLBIG, K.S. and MARGOLIS, B., 1968. *Nucl. Phys.* **B6**, 85.

KOLTUN, B.S., 1969. *Adv. nucl. Phys.* **3**, 71.

——— 1972. *Phys. Rev. Lett.* **28**, 182.

——— 1974. *Phys. Rev.* **C9**, 484.

KOOPMANS, T., 1934. *Physica* **1**, 104.

KOPALEISHVILI, T.I. and JIBUTI, R., 1963. *Nucl. Phys.* **44**, 34.

KOREN, M., 1969. *Ph.D. Thesis* Massachusetts Institute of Technology.

KORNER, H.J. and SCHIFFER, J.P., 1971. *Phys. Rev. Lett.* **27**, 1457.

KOSSANYI-DEMAY, P., ROUSSEL, P., FARAGGI, H., and SCHAEFFER, R., 1970.

Nucl. Phys. **A148**, 181.

KRAMER, E., MAIRLE, G., and KASCHL, G., 1971. *Nucl. Phys.* **A165**, 353.

KRELL, M., 1971. *Phys. Rev. Lett.* **26**, 584.

——— and BARMO, S., 1970. *Nucl. Phys.* **B20**, 461.

——— and ERICSON, T.E.O., 1969. *Nucl. Phys.* **B11**, 521.

KRIEGER, S.J., 1970. *Phys. Rev.* **C1**, 76.

——— BARANGER, M., and DAVIES, K.T.R., 1966. *Phys. Lett.* **22**, 607.

——— and MOSZKOWSKI, S.A., 1972. *Phys. Rev.* **C5**, 1440.

KROHN, V.E. and RINGO, G.R., 1973. *Phys. Rev.* **D8**, 1305.

KUDEYAROV, Yu.A., KURDYUMOV, I.V., NEUDATCHIN, V.G., and SMIRNOV, Yu.F., 1969. *Nucl. Phys.* **A216**, 36.

——— ——— ——— ——— 1971. *Nucl. Phys.* **A163**, 316.

——— SMIRNOV, Yu.F., and CHERBOTAREV, M.A., 1967. *Sov. J. nucl. Phys.* **4**, 751.

KUHN, H.G., 1969. *Atomic spectra* 2nd edn. (London: Longmans Green), p. 369.

KUJAWSKI, E., 1970. *Phys. Rev.* **C1**, 1651.

——— and AITKEN, M., 1974. *Nucl. Phys.* **A221**, 60.

KULLANDER, S., LEMEILLEUR, F., RENBERG, P.U., LANDAUD, G., YONNET, J., FAGERSTRÖM, B., JOHANNSON, A., and TIBELL, G., 1971*a*. *Nucl. Phys.* **A173**, 357.

——— RENBERG, P.U., FAGERSTRÖM, B., JOHANSSON, A., TIBELL, G., LANDAUD, G., LEMEILLEUR, F., and YONNET, J., 1971*b*. *Phys. Lett.* **34B**, 197.

KUNSELMAN, R. and GRIN, G. A., 1970. *Phys. Rev. Lett.* **24**, 838.

KUNZ, P.D., 1971. *Phys. Lett.* **35B**, 16.

KUO, T.T.S. and BROWN, G.E., 1966. *Nucl. Phys.* **85**, 40.

LAKIN, W.L., HUGHES, E.B., O'NEILL, L.H., OTIS, J.N., and MANDANSKY, L., 1970. *Phys. Lett.* **31B**, 677.

LAMBERT, E. and FESHBACH, H., 1972. *Phys. Lett.* **38B**, 487.

LANDAU, L., 1944. *J. Phys. U.S.S.R.* **8**, 201.

LANDAU, R.H., PHATAK, S.C., and TABAKIN, F., 1972. *Ann. Phys.* **78**, 299.

LANDAUD, G., YONNET, J., KULLANDER, S., LEMEILLEUR, F., RENBERG, P.U., FAGERSTRÖM, B., JOHANSSON, A., and TIBELL, G., 1971. *Nucl. Phys.* **A173**, 337.

LANDE, A., MOLINARI, A., and BROWN, G.E., 1968. *Nucl. Phys.* **A115**, 241.

——— and SVENNE, J.P., 1971. *Nucl. Phys.* **A164**, 49.

LANE, A.M., 1957. *Rev. mod. Phys.* **29**, 191.

——— 1962. *Phys. Rev. Lett.* **8**, 171.

——— and MEKJIAN, A.K., 1973. *Phys. Lett.* **43B**, 105.

LAPIKÁS, L., DIEPERINK, A.E.L., and BOX, G., 1973. *Nucl. Phys.* **A203**, 609.

LASSEY, K., 1972. *Nucl. Phys.* **A192**, 177.

——— and VOLKOV, A.B., 1971. *Phys. Lett.* **36B**, 4.

LE BEYEC, Y., LEFORT, M., and VIGNY, A., 1971. *Phys. Rev.* **C3**, 1268.

LEE, H.C. and CUSSON, R.Y., 1971. *Nucl. Phys.* **A170**, 439.

LEE, H.K. and McMANUS, H., 1971. *Nucl. Phys.* **A167**, 257.

LEE, J.K.P., MARK, S.K., PORTNER, P.M., and MOORE, R.B., 1967. *Nucl. Phys.* **A106**, 357.

LEE, P.L. and BOEHM, F., 1973. *Phys. Rev.* **C8**, 819.

LEE, W.Y., BERNOW, S., CHEN, M.Y., CHENG, S.C., HITLIN, D., KAST, J.W., MACAGNO, E.R., RUSHTON, A.M., WU, C.S., and BUDICK, B., 1969. *Phys. Rev. Lett.* **23**, 648.

LEES, E.W., 1974. Private communication to de Jager *et al.* (1974).

LEFORT, M., NGO, C., PETER, J., and TAMAIN, B., 1972. *Nucl. Phys.* **A197**, 485.

LEITH, D.W.S., 1969. *Proc. Third Int. Conf. on High Energy Physics and Nuclear Structure* ed. S. Devons (New York: Plenum Press), p. 395.

LEMMER, R.H., 1966. *Rep. Prog. Phys.* **29**, 131.

——— MARIS, Th.A.J., and TANG, Y.C., 1959. *Nucl. Phys.* **12**, 619.

LENZ, F., 1968. *Diplomarbeit* University of Freiburg.

——— 1969. *Z. Phys.* **222**, 491.

LEON, M. and SEKI, R., 1974. *Phys. Lett.* **B48**, 173.

LEONARDI, R., 1972. *Phys. Rev. Lett.* **28**, 836.

——— 1973. *Phys. Lett.* **43B**, 455.

LERNER, G.M., HIEBERT, J.C., RUTLEDGE, L.L., and BERNSTEIN, A.M., 1972. *Phys. Rev.* **C6**, 1254.

——— ——— ——— PAPANICOLAS, C., and BERNSTEIN, A.M., 1975. *Phys. Rev.* **C12**, 778.

——— and MARION, G.B., 1972. *Nucl. Phys.* **A193**, 593.

——— and REDISH, E.F., 1972. *Nucl. Phys.* **A193**, 565.

LESNIAK, L. and WOLEK, H., 1969. *Nucl. Phys.* **A125**, 665.

LEVINGER, J.S., 1951. *Phys. Rev.* **84**, 43.

——— 1960. *Nuclear photo-disintegration* (Oxford: Clarendon Press).

LI, G.C., SICK, I., WALECKA, J.D., and WALKER, G.E., 1970. *Phys. Lett.* **32B**, 317.

——— ——— and YEARIAN, M.R., 1971. *Phys. Lett.* **37B**, 282.

——— YEARIAN, M.R., and SICK, I., 1974. *Phys. Rev.* **C9**, 1861.

LIEBERT, R.B., PURSER, K.H., and BURMAN, R.L., 1973. *Nucl. Phys.* **A216**, 335.

LILLEY, J.S., 1971. *Phys. Rev.* **C3**, 2229.

LIM, K.L. and McCARTHY, I.E., 1964. *Phys. Rev.* **133**, B1006,

——— ——— 1966. *Nucl. Phys.* **88**, 433.

LIN, W.-F., 1972. *Phys. Lett.* **39B**, 447.

LIN, Y.C., 1970. *Nucl. Phys.* **A140**, 359.

LIND, V.G., PLENDL, H.S., FUNSTEN, H.O., KOSSLER, W.J., LIEB, B.J., LANKFORD, W.F., and BUFFA, A.J., 1974. *Phys. Rev. Lett.* **32**, 479.

LINDNER, A., 1968. *Z. Phys.* **211**, 195.

LINGAPPA, N. and GREENLEES, G.W., 1970. *Phys. Rev.* **C2**, 1329.

LITVINENKO, A.S., SHEVCHENKO, N.G., BUKI, A.Yu., SAVITSKIĬ, G.A., KHAVASTUNOV, V.M., and KONDRAT'EV R.L., 1972. *Sov. J. nucl. Phys.* **14**, 23.

LÖBNER, K.E.G, VETTER, M., and HÖNIG, V., 1970. *Nucl. Data Tables* **A7**, 495.

LOCK, W.O. and MEASDAY, D.F., 1970. *Intermediate energy nuclear physics* (London: Methuen).

LOMBARD, R.J., 1970. *Phys. Lett.* **32B**, 652.

——— 1973. *Ann. Phys.* **77**, 380.

——— and WILKIN, M., 1975. *Lett. nuovo Cim.* **13**, 463.

——— AUGER, J.P., and BASILE, R., 1971. *Phys. Lett.* **36B**, 480.

LOMBARDI, J.C., BOYD, R.N., ARKING, R., and ROBBINS, A.B., 1972. *Nucl. Phys.* **A192**, 641.

LOVE, W.G., 1968. *Phys. Lett.* **26B**, 271.

——— 1972. *Particles and Nuclei* **3**, 318.

——— 1974. *Nucl. Phys.* **A226**, 319.

——— and SATCHLER, G.R., 1967. *Nucl. Phys.* **A101**, 424.

——— ——— 1970. *Nucl. Phys.* **A159**, 1.

LYMAN, E.M., HANSON, A.O., and SCOTT, M.B., 1951. *Phys. Rev.* **84**, 626.

MAAS, R. and JAGER, C.W. de, 1974. *Phys. Lett.* **48B**, 212.

MACAGNO, E.R., 1968. *Ph.D. Thesis* Columbia University, New York.

——— BERNOW, S., CHENG, S.C., DEVONS, S., DUERDOTH, I., HITLIN, D., KAST, J.W., LEE, W.Y., RAINWATER, J., WU, C.S., and BARRETT, R.C., 1970. *Phys. Rev.* **C1**, 1202.

McALLEN, G.M., PINKSTON, W.T., and SATCHLER, G.R., 1971. *Particles and Nuclei* **1**, 412.

McCARTHY, I.E., 1959. *Nucl. Phys.* **10**, 583; **11**, 574.

——— 1968. *Introduction to nuclear theory* (New York, London: John Wiley).

——— PAL, D., STORER, R.G., and THOMAS, A.W., 1971. *Rep. No. FUPH-R-47* Flinders University, Australia.

McCARTHY, J.S., SICK, I., WHITNEY, R.R., and YEARIAN, M.R., 1970. *Phys. Rev. Lett.* **25**, 884.

McCARTHY, R.J., 1969. *Nucl. Phys.* **A130**, 305.

McCARTHY, R.J. and DAVIES, K.T.R., 1970. *Phys. Rev.* **C1**, 1644.

McDONALD, F. and HULL, M., 1966. *Phys. Rev.* **143**, 838.

MacFARLANE, M.H. and FRENCH, J.B., 1960. *Rev. mod. Phys.* **32**, 567.

McINTOSH, J.S., PARK, S.C., and RAWITSCHER, G.H., 1964. *Phys. Rev.* **134**, B1010.

McINTYRE, J.A., BAKER, S.D., and WANG, K.H., 1962. *Phys. Rev.* **125**, 548.

——— WANG, K.H., and BECKER, L.C., 1960. *Phys. Rev.* **117**, 1337.

McKINLEY, H.M., 1969. *Phys. Rev.* **183**, 196.

McKINLEY, W.A. and FESHBACH, H., 1948. *Phys. Rev.* **74**, 1759.

MACKINTOSH, R.S., 1972. *Nucl. Phys.* **A198**, 343.

——— 1973. *Nucl. Phys.* **A210**, 245.

——— 1976. *Nucl. Phys.* **A266**, 379.

——— and SWINIARSKI, R. de, 1975. *Phys. Lett.* **57B**, 139.

——— and TASSIE, L.J., 1974. *Nucl. Phys.* **A222**, 187.

McVOY, K.W. and ROMO, W.J., 1969. *Nucl. Phys.* **A126**, 161.

——— and VAN HOVE, L., 1962. *Phys. Rev.* **125**, 1034.

MADSEN, V.A., 1966. *Nucl. Phys.* **80**, 177.

——— and TOBOCMAN, W., 1965. *Phys. Rev.* **139**, B864.

MAGGIORE, C.J., GRUHN, C.R., KUO, T.Y.T., and PREEDOM, B.M., 1970. *Phys. Lett.* **33B**, 571.

MAHALANABIS, J., 1969. *J. Phys. A.* **2**, 66; *Nucl. Phys.* **A134**, 376.

MAILANDT, P., LILLEY, J.S., and GREENLEES, G.W., 1972. *Phys. Rev. Lett.* **28**, 1075.

——— ——— ——— 1973. *Phys. Rev.* **C8**, 2189.

MAIRLE, G. and WAGNER, G.J., 1974. *Phys. Lett.* **B50**, 252.

MALECKI, A. and PICCHI, P., 1973. *Proc. Asilomar Conf. on Photonuclear Reactions and Applications* ed. B.L. Berman (Springfield, Va.: NTIS), p. 987.

MAMASAKHLISOV, V.I. and DZHIBUTI, R.I., 1962. *Sov. Phys.—JETP* **14**, 1066.

MANG, H.J., 1960. *Phys. Rev.* **119**, 1069.

——— 1964. *A. Rev. nucl. Sci.* **14**, 1.

——— and RASMUSSEN, J.O., 1962. *Mat.—Fys. Meddr.* **2**, no. 3.

MANI, G.S., 1971*a*. *Nucl. Phys.* **A165**, 225.

——— 1971*b*. *Nucl. Phys.* **A169**, 194.

——— 1971*c*. *Nucl. Phys.* **A177**, 197.

——— and JACQUES, D., 1971. *Nucl. Phys.* **A177**, 448.

MANUZIO, G., RICCIO, G., SANZONE, M., and FERRARO, L., 1969. *Nucl. Phys.* **A133**, 225.

MARCHESE, C.J., CLARKE, N.M., and GRIFFITHS, R.J., 1972. *Phys. Rev. Lett.* **29**, 660.

MARGOLIS, B., 1968. *Nucl. Phys.* **B4**, 433.

——— and TROUBETSKOY, E.S., 1957. *Phys. Rev.* **106**, 105.

MARTIN, A.D., 1963. *Nuovo Cim.* **27**, 1359.

MARTIN, B.R. and SAKITT, M., 1969. *Phys. Rev.* **183**, 852.

MARTIN, P., 1972. *Ph.D. Thesis* College of William and Mary, Williamsburg.

MASTERSON, K.S. and LOCKETT, A.M., 1963. *Phys. Rev.* **129**, 776.

MATTHEWS, J.L., BERTOZZI, W., KOWALSKI, S., SARGENT, C.P., and TURCHINETZ, W., 1968. *Nucl. Phys.* **A112**, 654.

——— OWENS, R.O., GARDINER, S.N., and FINDLAY, D., 1973. *Proc. Asilomar Conf. on Photonuclear Reactions and Applications* ed. B.L. Berman (Springfield, Va.: NTIS), p. 981.

MATTHIAE, G., 1967. *Nucl. Phys.* **87**, 809.

MAXIMON, L.C., 1969. *Rev. mod. Phys.* **41**, 193.

MELDNER, H., 1969. *Phys. Rev.* **178**, 1815.

——— and PEREZ, J.D., 1971. *Phys. Rev.* **A4**, 1388.

MELKANOFF, M.A., MOSZKOWSKI, S.A., NOVDIK, J.S., and SAXON, D., 1956. *Phys. Rev.* **101**, 507.

——— NOVDIK, J.S., and SAXON, D., 1957. *Phys. Rev.* **106**, 793.

MENET, J.J.H., GROSS, E.E., MALANIFY, J.J., and ZUCKER, A., 1971. *Phys. Rev.* **C4**, 1114.

MERCER, R.L. and RAVENHALL, D.G., 1974. *Phys. Rev.* **C10**, 2002.

MERLE, K., 1973. *Proc. Asilomar Conf. on Photonuclear Reactions and Applications* ed. B.L. Berman (Springfield, Va.: NTIS), pp. 889, 893, and private communication.

MERTON, T.R., 1919. *Proc. R. Soc.* **A96**, 388.

MEYER, W.T., BROWMAN, A., HANSON, K., OSBORNE, A., SILVERMAN, A., TAYLOR, F.E., and HORWITZ, N., 1972. *Phys. Rev. Lett.* **28**, 1344.

MEYER-BERKHOUT, U., FORD, K.W., and GREEN, A.E.S., 1959. *Ann. Phys.* **8**, 119.

MICHEL, F., 1976. *Thesis (Docteur en Sciences Physiques)* University de l'Etat-Mons, Belgium.

MILLENER, D.J. and HODGSON, P.E., 1973. *Nucl. Phys.* **A209**, 59.

MILLER, G.H., ECKHAUSE, M., SAPP, W.W., and WELSH, R.E., 1968. *Phys. Lett.* **27B**, 663.

MILLER, H.G., BUSS, W., and RAWLINS, J.A., 1971. *Nucl. Phys.* **A163**, 637.

MO, L.W., and TSAI, Y.S., 1969. *Rev. mod. Phys.* **41**, 205.

MOLDAUER, P., 1962. *Phys. Rev. Lett.* **9**, 17.

——— 1963. *Nucl. Phys.* **47**, 65.

MONIZ, E.J. and DIXON, G.D., 1969. *Phys. Lett.* **30B**, 393.

MOREIRA, J.R., SINGHAL, R.P., and CAPLAN, H.S., 1971. *Can. J. Phys.* **49**, 1434.

MORGAN, C.G. and JACKSON, D.F., 1969. *Phys. Rev.* **188**, 1758.

MORGENSTERN, J., ALVES, R.N., JULIEN, J., and SAMOUR, C., 1969. *Nucl. Phys.* **A123**, 561.

MOSZKOWSKI, S.A., 1957. *Handb. Phys.* **39**, 411 (Berlin: Springer-Verlag).

——— 1970. *Phys. Rev.* **C2**, 402.

MOTT, N.F., 1929. *Proc. R. Soc.* **A124**, 425.

MOTTELSON, B.R., 1960. *Int. School of Physics 'E. Fermi'* Course XV (New

York: Academic Press).

MOUGEY, J., BERNHEIM, M., BUSSIÈRE, A., GUILLEBERT, A., PHAN XUAN HO, PRIOU, M., ROYER, D., SICK, I., and WAGNER, G.J., 1976. *Nucl. Phys.* **A262**, 461.

MOYER, R.A., COHEN, B.L., and DIEHL, R.C., 1970. *Phys. Rev.* **C2**, 1898.

MEUHLLEHNER, G., POLTORAK, A.S., PARKINSON, W.C., and BASSEL, R.H., 1967. *Phys. Rev.* **159**, 1039.

MURPHY, J.J. II, SHIN, Y.M., and SKOPIK, D.M., 1974. *Phys. Rev.* **C9**, 2125.

MURUGESU, S., 1971. *Ph.D. Thesis* University of Surrey.

MUTHUKRISHNAN, R., 1967. *Nucl. Phys.* **A93**, 417.

MYERS, W.D., 1968. *Ph.D. Thesis* University of California.

——— 1969. *Phys. Lett.* **30B**, 452.

——— 1970. *Nucl. Phys.* **A145**, 387.

——— 1973. *Nucl. Phys.* **A204**, 465.

——— and SWIATECKI, W.J., 1969. *Ann. Phys.* **55**, 395.

NAKADA, A., TORIZUKA, Y., and HORIKAWA, Y., 1971. *Phys. Rev. Lett.* **27**, 745.

NAKAMURA, K., HIRAMATSU, S., KAMAE, T., MURUMATSU, H., IZUTSU, N., and WATASE, Y., 1974. *Phys. Rev. Lett.* **33**, 853.

——— 1976. *Nucl. Phys.* **A268**, 381.

NEGELE, J.W., 1970. *Phys. Rev.* **C1**, 1260.

——— 1971. *Nucl. Phys.* **A165**, 305.

——— and RINKER, G.A. Jr., 1975. To be published.

——— and VAUTHERIN, D., 1972. *Phys. Rev.* **C5**, 1472.

NEMETH, J., 1970. *Nucl. Phys.* **A156**, 183.

——— and BETHE, H.A., 1968. *Nucl. Phys.* **A116**, 241.

——— and GADIOLI ERBA, E., 1971. *Phys. Lett.* **34B**, 117.

——— and RIPKA, G., 1972. *Nucl. Phys.* **A194**, 329.

——— and VAUTHERIN, D., 1970. *Phys. Lett.* **32B**, 561.

NEUDATCHIN, V.G. and SMIRNOV, Yu.F., 1965. *Atom. Energy Rev.* **3**, 157.

——— ——— 1968. *Prog. nucl. Phys.* **10**, 273.

NEUHAUSEN, R., 1975. *Proc. Saclay Conf. on Electron Scatt. at Intermediate Energies* 1975 eds. J.B. Bellicard, M. Bernham, N. de Botton, A. Bussière, J. Duclos, B. Frois, and C. Schunl, p. 315.

——— LIGHTBODY, J.W. Jr., FIVOZINSKY, S.P., and PENNER, S., 1972. *Phys. Rev.* **C5**, 124.

NEWMAN, E., BECKER, L.C., PREEDOM, B.M., and HIEBER, J.C., 1967. *Nucl. Phys.* **A100**, 225.

NEWMEYER, J. and TREFIL, J.S., 1970. *Nucl. Phys.* **B23**, 315.

NOLEN, J.A. and SCHIFFER, J.P., 1969. *A. Rev. nucl. Sci.* **19**, 471.

——— ——— and WILLIAMS, N., 1968. *Phys. Lett.* **27B**, 1.

O'CONNELL, J., 1973. *Proc. Asilomar Conf. on Photonuclear Reactions and Applications* ed. B.L. Berman (Springfield, Va.: NTIS), p. 71.

ONLEY, D.S., 1968. *Nucl. Phys.* **A118**, 436.

——— GRIFFY, T.A., and REYNOLDS, J.T., 1963. *Phys. Rev.* **129**, 1689.

——— REYNOLDS, J.T., and WRIGHT, L.E., 1964. *Phys. Rev.* **134**, B945.

OSTGAARD, E., 1965. *Nucl. Phys.* **64**, 289.

OWEN, L.W. and SATCHLER, G.R., 1963. *Oak Ridge Nat. Lab. Rep. No. ORNL-3525.*

——— ——— 1970. *Phys. Rev. Lett.* **25**, 1720.

PAANS, A.M.J., PUT, L.W., and MALFLIET, R.A.R.L., 1973. *Proc. Munich Int. Conf. on Nuclear Physics* eds. J. de Boer and H.J. Mang (Amsterdam: North-Holland), p. 340.

PAL, M.K. and STAMP, A.P., 1967. *Phys. Rev.* **158**, 924.

PARK, J.Y. and SATCHLER, G.R., 1971. *Particles and Nuclei* **1**, 233.

PARKER, E.F., DOBROWOLSKI, T., GUSTAFSON, H.R., JONES, L.W., LONGO, M.J., RINGIA, F.E., and CORK, B., 1970. *Phys. Lett.* **31B**, 250.

PARKINSON, W.C., HENDRIE, D.L., DUHM, H.H., MAHONEY, J., SAUDINOS, J., and SATCHLER, G.R., 1969. *Phys. Rev.* **178**, 1976.

PARKS, D.R., TABOR, S.L., TRIPLET, B.B., KING, H.T., FISHER, T.R., and WATSON, B.A., 1972. *Phys. Rev. Lett.* **29**, 1264.

PAUL, P., AMANN, J.F., and SNOVER, K.A., 1971. *Phys. Rev. Lett.* **27**, 1013.

PEREY, C.M. and PEREY, F.G., 1963. *Phys. Rev.* **132**, 755.

PEREY, F.G., 1963. *Phys. Rev.* **131**, 745.

——— 1964. *Argonne Nat. Lab. Rep. No. ANL 6848*, p. 114.

——— 1966. *Proc. Second Int. Symp. on Polarization Phenomena of Nucleons* eds. M. Huber and S. Schopper (Karlsruhe: Birkhäuser Verlag), p. 191.

——— and BUCK, B., 1962. *Nucl. Phys.* **32**, 353.

——— and SATCHLER, G.R., 1967. *Nucl. Phys.* **A97**, 515.

——— and SCHIFFER, J.P., 1966. *Phys. Rev. Lett.* **17**, 324.

PETERSON, G.A. and BARBER, W.C., 1962. *Phys. Rev.* **128**, 812.

PETROVICH, F.L., 1970. *Ph.D. Thesis* Michigan State University.

PFEIFFER, H.-J., SPRINGER, K., EGIDY, T. von, and DANIEL, H., 1973. *Proc. Munich Int. Conf. on Nuclear Physics* eds. J. de Boer and H.J. Mang (Amsterdam: North-Holland), p. 313.

PHAN-XUAN, Ho., BELLICARD, J., LECONTE, Ph., and SICK, I., 1973. *Nucl. Phys.* **A210**, 189.

PHILPOTT, R.J., PINKSTON, W.T., and SATCHLER, G.R., 1968. *Nucl. Phys.* **A119**, 241.

PICARD, J., BEER, O., EL BEHAY, A., LOPATO, P., TERRIEN, Y., VALLOIS, G., and SCHAEFFER, R., 1969. *Nucl. Phys.* **A128**, 481.

PINKSTON, W.T. and SATCHLER, G.R., 1965. *Nucl. Phys.* **72**, 641.

PITTEL, S. and AUSTERN, N., 1972. *Phys. Rev. Lett.* **29**, 1403.

——— ——— 1974. *Nucl. Phys.* **A218**, 221.

PIZZI, J.R., GAILLARD, M., GAILLARD, P., GUICHARD, A., GUSAKOW, M., REBOULET, G., and RUHLA, C., 1969. *Nucl. Phys.* **A136**, 496.

PLENDL, H.S., 1973. *Proc. Fifth Int. Conf. on High Energy Physics and Nuclear Structure* ed. G. Tibell (Amsterdam: North-Holland), p. 289.

PLIENINGER, R.D., EICHELBERGER, W., and VELTEN, E., 1969. *Nucl. Phys.* **A137**, 20.

POELZ, G., SCHMITT, H., TAUSCHER, L., BACKENSTOSS, G., CHARALAMBUS, S., DANIEL, H., and KOCH, H., 1968. *Phys. Lett.* **26B**, 331.

POGGENBURG, I.K., 1965. *Ph.D. Thesis Rep. No. UCRL-16187.* University of California, Berkeley.

PORTER, C.E., 1955. *Phys. Rev.* **100**, 935.

POWERS, R.J., 1968. *Phys. Rev.* **169**, 1.

——— BOEHM, F., VOGEL, P., ZEHNDER, A., KING, T., KUNSELMAN, A.R., ROBERTSON, P., MARTIN, P., MILLER, G.H., WELSH, R.E., and JENKINS, D.A., 1975. *Phys. Rev. Lett.* **34**, 492.

PREEDOM, B.M., 1972. *Phys. Rev.* **C5**, 587.

PRESTON, M.A., 1962. *Physics of the nucleus* (Reading, Mass.: Addison-Wesley).

PUGET, J.L., 1970. *CERN Rep. No. Th-1201.*

PUGH, H.G., HENDRIE, D.L., CHABRE, M., and BOSCHITZ, E., 1965. *Phys. Rev. Lett.* **14**, 434.

——— ——— ——— ——— and McCARTHY, I.E., 1967. *Phys. Rev.* **155**, 1054.

——— and RILEY, K.R., 1961. *Proc. Rutherford Jubilee Int. Conf.* (Manchester: Heywood), p. 195.

——— WATSON, J.W., GOLDBERG, D.A., ROOS, P.G., BONBRIGHT, D.I., and RIDDLE, R.A.J., 1969. *Phys. Rev. Lett.* **22**, 408.

PYLE, G.J. and GREENLEES, G.W., 1969. *Phys. Rev.* **181**, 1444.

QUITMANN, D., 1967. *Z. Phys.* **206**, 113.

——— ENGFER, R., HEGEL, U., BRIX, P., BACKENSTOSS, G., GOEBEL, K., and STADLER, B., 1964. *Nucl. Phys.* **51**, 609.

RADHAKANT, S., 1972*a*. *Nucl. Phys.* **A188**, 353.

——— 1972*b*. *Phys. Lett.* **40B**, 70.

RAFELSKI, J., MÜLLER, B., SOFF, G., and GREINER, W., 1974. *Ann. Phys.* **88**, 419.

RAND, R.E., FROSCH, R.F., and YEARIAN, M.R., 1966. *Phys. Rev.* **144**, 859; erratum *Phys. Rev.* **148**, 1246.

RAPAPORT, J. and KERMAN, A.K., 1968. *Nucl. Phys.* **A119**, 641.

——— SPERDUTO, A., and SALOMA, M., 1972. *Nucl. Phys.* **A197**, 337.

RASMUSSEN, J.O., 1963. *Nucl. Phys.* **44**, 93.

RAUSCHER, E.A., RASMUSSEN, J.O., and HARADA, K., 1967. *Nucl. Phys.* **A94**, 33.

RAVENHALL, D.G. and YENNIE, D.R., 1954. *Phys. Rev.* **96**, 239.

——— 1957. *Proc. phys. Soc.* **A70**, 857.

RAWITSCHER, G.H., 1966*a*. *Nucl. Phys.* **83**, 259.

——— 1966*b*. *Phys. Rev.* **151**, 846.

——— 1967. *Medium energy nuclear physics with electron linear accelerators* eds. W. Bertozzi and S. Kowalski (Cambridge, Mass.: MIT), p. 167.

——— 1970. *Phys. Lett.* **33B**, 445.

——— 1972. *Phys. Rev.* **C6**, 1212.

——— and SPICUZZA, R.A., 1971. *Phys. Lett.* **37B**, 221.

RAWLINS, J.A., GLAVINA, C., KU, S.H., and SHIN, Y.M., 1968. *Nucl. Phys.* **A122**, 128.

REBEL, H., LÖHKEN, R., SCHWEIMER, G.W., SCHATZ, G., and HAUSER, G., 1972*a*. *Z. Phys.* **256**, 258.

——— and SCHWEIMER, G.W., 1973. *Z. Phys.* **262**, 59.

——— ——— SCHATZ, G., SPECHT, J., LÖHKEN, R., HAUSER, G., HABS, D., and KLEWE-NEBENIUS, H., 1972*b*. *Nucl. Phys.* **A182**, 145.

REDISH, E.F., STEPHENSON, G.J., and LERNER, G.M., 1970. *Phys. Rev.* **C2**, 1665.

REEHAL, B.S. and SORENSEN, R.A., 1971. *Nucl. Phys.* **A161**, 385.

REITAN, A., 1962. *Nucl. Phys.* **36**, 56.

RENBERG, P.U., MEASDAY, D.F., PEPIN, M., SCHWALLER, P., FAVIER, B., and RICHARD-SERRE, C., 1972. *Nucl. Phys.* **A183**, 81.

RINKER, G.A. Jr., 1971. *Phys. Rev.* **C4**, 2150.

——— 1976. *Phys. Rev.* **A14**, 18.

——— and WILETS, L., 1973. *Phys. Rev.* **D7**, 2629.

——— ——— 1975. *Phys. Rev.* **A12**, 748.

RIPKA, G., 1968. *Adv. nucl. Phys.* **1**, 183.

——— 1970. *Fast neutrons and the study of nuclear structure* (New York: Gordon and Breach).

——— and GILLESPIE, J.G., 1970. *Phys. Rev. Lett.* **25**, 1624.

RISKALLA, R., 1971. *Ph.D. Thesis* University of Paris *(Orsay Rep. No. LAL 1243)*.

ROBERT-TISSOT, B., 1975. *Ph.D. Thesis* University of Fribourg.

ROBSON, D., 1963. *Nucl. Phys.* **42**, 592.

ROLLAND, C., GEOFFRION, B., MARTY, N., MORLET, M., TATISCHEFF, B., and WILLIS, A., 1966. *Nucl. Phys.* **80**, 625.

ROOK, J.R., 1962. *Nucl. Phys.* **39**, 479.

——— 1963. *Nucl. Phys.* **43**, 363.

——— 1965. *Nucl. Phys.* **61**, 219.

ROOK, J.R., 1968*a*. *Nucl. Phys.* **B6**, 543.

——— 1968*b*. *Nucl. Phys.* **B9**, 441.

——— 1970. *Nucl. Phys.* **B20**, 14.

——— 1971. RHEL *meeting on stopping kaons and nuclear structure, Rep. No. RHEL/M/H9* Rutherford High Energy Laboratory, Chilton, Oxon.

——— and MITRA, D.; 1964. *Nucl. Phys.* **51**, 96.

——— and WYCECH, S., 1972. *Phys. Lett.* **39B**, 469.

ROOS, P.G., KIM, H., JAIN, M., and HOLMGREN, H.D., 1969. *Phys. Rev. Lett.* **22**, 242.

——— PUGH, H.G., JAIN, M., HOLMGREN, H.D., EPSTEIN, M., and LINDEMANN, C.A., 1968. *Phys. Rev.* **176**, 1246.

——— and WALL, N.S., 1965. *Phys. Rev.* **140B**, 1237.

ROSENBLUTH, M.N., 1950. *Phys. Rev.* **79**, 165.

ROSENTHAL, J.E. and BREIT, G., 1932. *Phys. Rev.* **41**, 459.

ROSS, A., MARK, H., and LAWSON, R.D., 1956. *Phys. Rev.* **102**, 1613; **104**, 401.

ROSS, D.K., 1968. *Phys. Rev.* **173**, 1965.

ROST, E., 1967. *Phys. Rev.* **154**, 994.

——— 1968. *Phys. Lett.* **26B**, 184.

——— and EDWARDS, G.W., 1971. *Phys. Lett.* **37B**, 247.

ROWLEY, N., 1974. *Nucl. Phys.* **A219**, 93.

ROYER, D., ARDITI, M., BIMBOT, L., DOUBRE, H., FRASCARIA, N., GARRON, J.P., and RIOU, M., 1970. *Nucl. Phys.* **A158**, 516.

ROYNETTE, J.C., ARDITI, M., JACMART, J.C., MAZLOUM, F., RIOU, M., and RUHLA, C., 1967. *Nucl. Phys.* **A95**, 545.

RUHLA, C., ARDITI, M., DOUBRE, H., JACMART, J.C., LIU, M., RICCI, R.A., RIOU, M., and ROYNETTE, J.C., 1967. *Nucl. Phys.* **A95**, 526.

——— RIOU, M., GUSAKOW, M., JACMART, J.C., LIU, M., and VALENTIN, L., 1963. *Phys. Lett.* **6**, 282.

——— ——— RICCI, R.A., ARDITI, M., DOUBRE, H., JACMART, J.C., LIU, M., and VALENTIN, L., 1964. *Phys. Lett.* **10**, 326.

RUTHERFORD, E., 1929. *Proc. R. Soc.* **A123**, 323.

SAKAI, M. and KUBO, K.I., 1972. *Nucl. Phys.* **A185**, 217.

SAKAMOTO, Y., 1964. *Phys. Rev.* **134**, B1211.

SAKURAI, J.J., 1967. *Advanced quantum mechanics* (Reading, Mass.: Addison-Wesley).

SANDERSON, E.A., 1961. *Nucl. Phys.* **26**, 420.

——— 1962. *Nucl. Phys.* **35**, 557.

SANDULESCU, A., 1962. *Nucl. Phys.* **37**, 332.

——— 1963. *Nucl. Phys.* **48**, 345.

——— and DUMITRESCU, O., 1965. *Phys. Lett.* **19**, 404.

SANZONE, M., RICCIO, G., COSTA, S., and FERRARO, L., 1970. *Nucl. Phys.* **A153**, 401.

SAPERSTEIN, A.M. 1973. *Phys. Rev. Lett.* **30**, 1257.

SAPLAKOGLU, A., BOLLINGER, Z.M., and COTÉ, R., 1958. *Phys. Rev.* **109**, 1258.

SAPP, W.W., ECKHAUSE, M., MILLER G.H., and WELSH, R.E., 1972. *Phys. Rev.* **C5**, 690.

SATCHLER, G.R., 1960. *Nucl. Phys.* **21**, 116.

——— 1965. Unpublished results quoted by Roos and Wall (1965).

——— 1966. *Nucl. Phys.* **A77**, 481.

——— 1967*a*. *Nucl. Phys.* **A95**, 1.

——— 1967*b*. *Nucl. Phys.* **A100**, 481.

——— 1967*c*. *Nucl. Phys.* **A92**, 273.

——— 1969. *Isospin in nuclear physics*. ed. D.H. Wilkinson (Amsterdam: North-Holland), chap. 9.

——— 1970. *Phys. Lett.* **33B**, 385.

——— 1971*a*. *Phys. Lett.* **35B**, 279.

——— 1971*b*. *Particles and Nuclei* **2**, 265.

——— 1971*c*. *Particles and Nuclei* **1**, 397.

——— 1971*d*. *Phys. Lett.* **36B**, 169.

——— 1972*a*. *Nucl. Phys.* **A195**, 1.

——— 1972*b*. *J. math. Phys.* **13**, 1118.

——— 1972*c*. *Phys. Lett.* **39B**, 495.

——— 1973. *Z. Phys.* **260**, 209.

——— 1975. *Phys. Lett.* **B55**, 167.

——— and HAYBRON, R.M., 1964. *Phys. Lett.* **11**, 313.

——— OWEN, L.W., ELWYN, A.J., MORGAN, G.L., and WALTER, R.L., 1968. *Nucl. Phys.* **A112**, 1.

SAVITSKIĬ, G.A., AFANAS'EV, N.G., ANDREEVA, I.V., GUL'KAROV, I.S., KRUGOVAYA, L.M., KHVASTUNOV, V.M., KHOMIC, A.A., and SHEVCHENKO, N.G., 1969. *Bull. Acad. Sci. USSR Phys. Ser.* **33**, 50.

SCHAEFFER, R., 1969*a*. *Nucl. Phys.* **A132**, 186.

——— 1969*b*. *Nucl. Phys.* **A135**, 231.

——— 1970. *Nucl. Phys.* **A158**, 321.

SCHAFER, R.E., 1967. *Phys. Rev.* **163**, 1451.

——— 1973. *Phys. Rev.* **D8**, 2313.

SCHECK, F., HÜFNER, J., and WU, C.S., 1976. *Muon. physics* vol. I, eds. V.W. Hughes and C.S. Wu (New York: Academic Press).

SCHEERBAUM, R.R., 1969. *Ph.D. Thesis* Cornell University, Ithaca, New York.

SCHIER, H. and SCHOCH, B., 1974. *Nucl. Phys.* **A229**, 93.

SCHIFF, L.I., 1956. *Phys. Rev.* **103**, 443.

SCHIFF, L.I., 1968. *Phys. Rev.* **176**, 1390.

SCHIFFER, J.P., 1968. *Suppl. J. phys. Soc. Japan* **24**.

——— and KORNER, H.J., 1973. *Phys. Rev.* **C8**, 841.

SCHIMMERLING, W., DEVLIN, T.J., JOHNSON, W., VOSBURGH, K.G., and MISCHKE, R.E., 1971. *Phys. Lett.* **37B**, 177.

SCHMIT, C., 1970. *Lett. Nuovo Cim.* **4**, 454.

SCHMITT, H., 1973. Private communication to de Jager *et al.* (1974).

——— TAUSCHER, L., BACKENSTOSS, G., CHARALAMBUS, S., DANIEL, M., KOCH, H., and POELZ, G., 1968. *Phys. Lett.* **27B**, 530.

SCHUCAN, T., 1965. *Nucl. Phys.* **61**, 417.

SCHÜTZ, W., 1973. *Ph.D. Thesis* T.H. Darmstadt.

SCHWALLER, P., FAVIER, B., MEASDAY, D.F., PEPIN, M., RENBERG, P.U., and SERRE, C., 1972. *CERN Rep. No. 72-12.*

SCHWINGER, J., 1949. *Phys. Rev.* **75**, 651.

SELTZER, E.C., 1969. *Phys. Rev.* **188**, 1916.

SENGUPTA, S., 1960. *Nucl. Phys.* **21**, 542.

SERBER, R., 1947. *Phys. Rev.* **72**, 1114.

SETH, K., 1966. *Nucl. Data* **A2**, 299.

SETH, K.K., 1969. *Nucl. Phys.* **A138**, 61.

SHAFER, R.E., 1967. *Phys. Rev.* **163**, 1451.

——— 1973. *Phys. Rev.* **D8**, 2313.

SHAH, M.B., 1971. *Ph.D. Thesis* University of Surrey.

SHAKIN, C.M., WAGHMARE, Y.R., and HULL, M.H., 1967. *Phys. Rev.* **161**, 1006.

SHALIT, A. de and FESHBACH, H., 1974. *Theoretical nuclear physics* vol. I *Nuclear structure* (New York: John Wiley), chap. 6.

SHANTA, R., 1973. *Nucl. Phys.* **A199**, 624.

——— and JAIN, B.K., 1971. *Nucl. Phys.* **A175**, 417.

SHAPIRO, I.S., KOLYBASOV, V.M., and AUGST, J.P., 1965. *Nucl. Phys.* **61**, 353.

SHAW, G.L., 1959. *Ann. Phys.* **8**, 509.

SHAW, R.R., SWIFT, A., and ELTON, L.R.B., 1965. *Proc. phys. Soc.* **86**, 513.

SHERIF, H., 1969. *Nucl. Phys.* **A131**, 532.

——— and BLAIR, J.S., 1968. *Phys. Lett.* **26B**, 489.

——— ——— 1970. *Nucl. Phys.* **A140**, 33.

SHERIF, H.S. and PODMORE, B.S., 1972. *Proc. Conf. on Few Nucleon Problems in the Nuclear Interaction* eds. I. Slaus, S.A. Moskowski, R.P. Haddock, and W.H.T. van Oers (Amsterdam: North-Holland/New York: American Elsevier), p. 691.

SHEVCHENKO, N.G., AFANAS'EV, N.G., SAVITSKIĬ, G.A., KHVASTUNOV, V.M., KOVALEV, V.D., OMELAENKO, A.S., and GUL'KAROV, I.S., 1967*a*. *Sov. J. nucl. Phys.* **5**, 676.

——— ——— ——— ——— GUL'KAROV, I.S., OMELAENKO, A.S., and KOVALEV, V.D.,

1976*b*. *Nucl. Phys.* **A101**, 187.

SHKLYARESKII, G.M., 1959, *JETP-Sov. Phys.* **9**, 1057.

——— 1962. *JETP-Sov. Phys.* **14**, 170.

SHLOMO, S., 1972. *Phys. Lett.* **42B**, 146.

SICK, I., 1973*a*. *Phys. Lett.* **44B**, 62,

——— 1973*b*. *Nucl. Phys.* **A208**, 557.

——— 1974. *Nucl. Phys.* **A218**, 509.

——— 1975. *Proc. 6th Int. Conf. on High Energy Physics and Nuclear Structure, Santa Fe.*

——— BELLICARD, J.B., BERNHEIM, M., BUSSIERE DE NERCY, A., FROIS, B., HUET, M., LECONTE, Ph. MOUGEY, J., PHAN XUAN HO, ROYER, D., and TURCK, S., 1975. *Phys. Rev. Lett.* **35**, 910.

——— and McCARTHY, J.S., 1970. *Nucl. Phys.* **A150**, 631.

——— ——— and WHITNEY, R.R., 1976. *Phys. Lett.* **64**, B33.

SIEMENS, P.J., 1970. *Phys. Rev.* **C1**, 98.

SIEMSSEN, R.H., 1972. *Nuclear spectroscopy* vol. II (New York, London: Academic Press) chap. 4.

SILBAR, R.R. and STERNHEIM, M.M., 1972. *Phys. Rev.* **C6**, 764.

SILVER, J.D. and STACEY, D.N., 1973. *Proc. R. Soc.* **A332**, 129, 139.

SIMBEL, M.H., 1974. *Phys. Rev.* **C10**, 1083.

——— and ABUL-MAGD, A.Y., 1971. *Nucl. Phys.* **A177**, 322.

SIMENOG, A. and SITENKO, N., 1966. *Nucl. Phys.* **80**, 643.

SINGH, P.P., MALMIN, R.E., HIGH, M., and DEVINS, D.W., 1969. *Phys. Rev. Lett.* **23**, 1124.

SINGHAL, R.P., MOREIRA, J.R., and CAPLAN, H.S., 1970. *Phys. Rev. Lett.* **24**, 73.

SINHA, B.B.P., PETERSON, G.A., SICK, I., and McCARTHY, J.S., 1971. *Phys. Lett.* **35B**, 217.

——— ——— WHITNEY, R.R., SICK, I., and McCARTHY, J.S., 1973. *Phys. Rev.* **C7**, 1930.

SKARDHAMAR, H.F., 1970. *Nucl. Phys.* **A151**, 154.

SKYRME, T.H.R., 1959. *Nucl. Phys.* **9**, 615.

SLANINA, D. and McMANUS, H., 1968. *Nucl. Phys.* **A116**, 271.

SMILANSKY, U., 1970. *Nuclear reactions induced by heavy ions* eds. R. Bock and W.R.Hering (Amsterdam: North-Holland), p. 392.

SMITH, S.M., TIBELL, G., COWLEY, A.A., GOLDBERG, D.A., PUGH, H.G., RIECHART, W., and WALL, N.S., 1973. *Nucl. Phys.* **A207**, 273.

SOOD, P.C. and GREEN, A.E.S., 1957. *Nucl. Phys.* **5**, 274.

SPECHT, J., REBEL, H., SCHATZ, G., SCHWEIMER, G.W., HAUSER, G., and LÖHKEN, R., 1970. *Nucl. Phys.* **A143**, 373.

——— SCHWEIMER, G.W., REBEL, H., SCHATZ, G., LÖHKEN, R., and HAUSER, G., 1971. *Nucl. Phys.* **A171**, 65.

SPICER, B.M., 1969. *Adv. nucl. Phys.* 2, 1.

SPRINGER, A. and HARVEY, B.G., 1965. *Phys. Lett.* **14**, 116; *Phys. Rev. Lett.* **14**, 316.

SPRUNG, D.W. and BANERJEE, P.K., 1971. *Nucl. Phys.* **A168**, 273.

——— MARTORELL, J., and CAMPI, X., 1976. *Nucl. Phys.* in the press.

SRIVASTAVA, D.K., GANGULY, N.K., and HODGSON, P.E., 1974. *Phys. Lett.* **51B**, 439.

STACEY, D.N., 1964. *Proc. R. Soc.* **A280**, 439.

——— 1966. *Rep. Prog. Phys.* **29**, 171.

STANFIELD, K.C., CANIZARES, C.R., FAISSLER, W.L., and PIPKIN, F.M., 1971. *Phys. Rev.* **C3**, 1448.

STARODUBSKY, V.E. and DOMCHENKOV, O.A., 1972. *Phys. Lett.* **42B**, 319.

STEARNS, M.B., 1957. *Prog. nucl. Phys.* **6**, 108.

STEINWEDEL, H., JENSEN, J.H.D., and JENSEN, P., 1950. *Phys. Rev.* **79**, 1019.

STELSON, P.H. and GRODZINS, 1965. *Nucl. Data* **A1**, 21.

STEPHENSON, G.J., REDISH, E.F., LERNER, G.M., and HAFTEL, M.I., 1972. *Phys. Rev.* **C6**, 1559.

STERNHEIM, M.M. and AUERBACH, E.H., 1970. *Phys. Rev. Lett.* **25**, 1500.

STOCKER, W., 1971. *Nucl. Phys.* **A166**, 205.

STODOLSKY, L., 1967. *Phys. Rev. Lett.* **18**, 135.

STOVALL, T., GOLDEMBERG, J., and ISABELLE, D.B., 1966. *Nucl. Phys.* **86**, 225.

SUMBAEV, O.I., MEZENTSEV, A.F., MARUSHENKO, V.I., RYLNIKOV, A.S., and IVANOV, G.A., 1969. *Sov. J. nucl. Phys.* **9**, 529.

SUMNER, W.Q., 1974. *Ph.D. Thesis* University of Washington.

SUNDBERG, O. and KÄLLNE, J., 1969. *Ark. Fys.* **39**, 323.

——— and TIBELL, G., 1969. *Ark. Fys.* **39**, 397.

SÜSSMANN, G., 1970. *UCRL Rep. No. 19960.* University of California, Berkeley.

SUZUKI, A., 1967. *Phys. Rev. Lett.* **19**, 1005.

SWIATECKI, W.J., 1951. *Phys. Rev.* **83**, 178.

SWIFT, A. and ELTON, L.R.B., 1966. *Phys. Rev. Lett.* **17**, 484.

TAMAIN, B., NGÔ, C., PÉTER, J., and HANAPPE, F., 1975. *Nucl. Phys.* **A252**, 187.

TANG, Y.C., WILDERMUTH, K., and PEARLSTEIN, L.D., 1962. *Nucl. Phys.* **32**, 504.

TARBUTTON, R.M. and DAVIES, K.T.R., 1968. *Nucl. Phys.* **A120**, 1.

TASSIE, L.J., 1956. *Aust. J. Phys.* **9**, 407.

——— 1957. *Nuovo Cim.* **5**, 1497.

TATISCHEFF, B., 1967. *Nucl. Phys.* **A98**, 384.

——— and BRISSAUD, I., 1970. *Nucl. Phys.* **A155**, 89.

——— ——— and BIMBOT, L., 1972. *Phys. Rev.* **C5**, 234.

TAUSCHER, L., 1973. Unpublished.

——— BACKENSTOSS, G., FRANSSON, K., KOCH, H., NILSSON, A., and RAEDT, J. de, 1975. *Phys. Rev. Lett.* **35**, 410.

TERRIEN, Y., 1973. *Nucl. Phys.* **A215**, 29.

THEISSEN, H., 1972. *Habilitationsschrift* T.H. Darmstadt; Springer Tracts on Modern Physics ed. G. Höhler (Berlin: Springer-Verlag) **65**, 1.

——— PETERSON, R.J., ALSTON, W.J. III, and STEWART, J.R., 1969. *Phys. Rev.* **186**, 1119.

THIRION, J., 1973. *Proc. Fifth Int. Conf. on High Energy Physics and Nuclear Structure* ed. G. Tibell (Amsterdam: North-Holland), p. 168.

THOMAS, G.L. and BURGE, E.J., 1969. *Nucl. Phys.* **A128**, 545.

——— and SINHA, B.C., 1971. *Phys. Rev. Lett.* **26**, 325.

——— ——— and DUGGAN, F., 1973. *Nucl. Phys.* **A203**, 305.

THOMPSON, A.C., 1969. *Ph.D. Thesis* Carnegie-Mellon University, Pittsburgh.

THOMPSON, D.R. and TANG, Y.C., 1971. *Phys. Rev.* **C4**, 306.

THOMPSON, G.E., EPSTEIN, M.B., and SAWADA, T., 1970. *Nucl. Phys.* **A142**, 571.

THOMPSON, W.J., CRAWFORD, G.E., and DAVIS, R.H., 1967. *Nucl. Phys.* **A98**, 228.

TIBELL, G., 1969. *Phys. Lett.* **28B**, 638.

——— SUNDBERG, O., and MIKLAVZIC, 1962. *Phys. Lett.* **2**, 100.

——— ——— and RENBERG, P.U., 1963. *Ark. Fys.* **25**, 433.

TOEPFFER, C. and DRECHSEL, D., 1970. *Phys. Rev. Lett.* **24**, 1131.

TOMLINSON, W.J. and STROKE, H.H., 1964. *Nucl. Phys.* **60**, 614.

TOWNER, I.S., 1967. *Nucl. Phys.* **A93**, 145.

——— 1969. *Nucl. Phys.* **A126**, 97.

——— 1973. *Nucl. Phys.* **A216**, 589.

——— and HARDY, J.C., 1969. *Adv. Phys.* **18**, 401.

TRIPATHI, R.K., FAESSLER, A., and MacKELLAR, A.D., 1973. *Phys. Rev.* **C8**, 129.

TUCKER, T.C., ROBERTS, L.D., NESTOR, C.W., CARLSON, T.A., and MALIK, F.B., 1969. *Phys. Rev.* **178**, 998.

TYREN, H., KULLANDER, S., SUNDBERG, O., RAMACHANDRAN, R., ISACSSON, P., and BERGGREN, T., 1966. *Nucl. Phys.* **79**, 321.

——— and MARIS, Th.A.J., 1957. *Nucl. Phys.* **3**, 52; **4**, 637, 632.

ÜBERALL, H., 1971. *Electron scattering from complex nuclei* parts A and B (New York: Academic Press).

UEHLING, E.A., 1935. *Phys. Rev.* **48**, 55.

ULLRICH, H., BOSCHITZ, E.T., ENGELHARDT, H.D., and LEWIS, C.W., 1974. *Phys. Rev. Lett.* **33**, 433.

URETSKY, J.L., 1967. *Proc. Second Int. Conf. on High Energy Physics and Nuclear Structure* ed. G. Alexander (Amsterdam: North-Holland), p. 395.

URONE, P.P., PUT, L.W., CHANG, H.H., and RIDLEY, B.W., 1971*a* *Nucl. Phys.* **A163**, 225.

——— ——— and RIDLEY, B.W., 1972. *Nucl. Phys.* **A186**, 344.

——— ——— ——— and JONES, G.D., 1971*b*. *Nucl. Phys.* **A167**, 383.

VAN DER MERWE, J.J. and HEYMANN, G., 1969. *Z. Phys.* **220**, 130.

VAN DER WERF, S.Y., KOOISTRA, B.R., FRYSZCZYN, B., HESSELINK, W.H.A., IACHELLO, F., PUT, L.W., and SIEMSSEN, R.H., 1974. *Proc. Int. Conf. on Nuclear Structure and Spectroscopy.* eds. H.P. Blok and A.E.L. Dieperink (Amsterdam: Scholars Press), pp. 175, 176.

VAN GIAI, N., VAUTHERIN, D., VENERONI, M., and BRINK, D.M., 1971. *Phys. Lett.* **35B**, 135.

VAN NIFTRIK, G.J.C., 1969. *Nucl. Phys.* **A131**, 574.

VAN OERS, W.H.T., 1971. *Phys. Rev.* **C3**, 1550.

VAN OERS, W.T.H. and HAW, H., 1973. *Phys. Lett.* **45B**, 227.

VAUTHERIN, D., 1969. *Thesis* University of Paris.

——— 1973. *Phys. Rev.* **C7**, 296.

——— and BRINK, D.M., 1970. *Phys. Lett.* **32B**, 149.

——— ——— 1972. *Phys. Rev.* **C5**, 626.

——— and VENERONI, M., 1967. *Phys. Lett.* **25B**, 175.

——— ——— 1968. *Phys. Lett.* **26B**, 552.

——— ——— 1969. *Phys. Lett.* **29B**, 203.

VENTER, R.H. and FRAHN, W.E., 1964*a*. *Ann. Phys.* **27**, 385.

——— ——— 1964*b*. *Ann. Phys.* **27**, 491.

VIOLLIER, R. and ALDER, K., 1971. *Helv. Phys. Acta* **44**, 77.

VOGEL, P., 1973*a*. *Phys. Rev.* **A7**, 63.

——— 1973*b*. *Phys. Rev.* **A8**, 2292.

——— 1974. *Atom. Data nucl. Data Tables* **14**, 599.

VOGT, E., 1962. *Phys. Lett.* **1**, 84; *Rev. mod. Phys.* **34**, 723.

——— 1968. *Adv. nucl. Phys.* **1**, 261.

WAGNER, G.J., 1969. *Princeton Univ. Rep. No. PUC-937 534.*

——— 1973. *Proc. Minerva Symp. on Nuclear Structure Physics* eds. U. Smilansky, I. Talmi, and H. Weidenmüller (Berlin: Springer-Verlag), p. 16.

WALL, N.S., 1964. *Argonne Nat. Lab. Rep. No. ANL-6848*, p. 108.

——— and ROOS, P.G., 1966. *Phys. Rev.* **150**, 811.

WATANABE, S., 1958. *Nucl. Phys.* **8**, 484.

WATSON, J.W., 1962. *Nucl. Phys.* **A198**, 129.

——— PUGH, H.G., ROOS, P.G., GOLDBERG, D.A., RIDDLE, R.A.J., and BONBRIGHT, D.I., 1971. *Nucl. Phys.* A172, 513.

WATSON, K.M., 1953. *Phys. Rev.* **89**, 575.

——— 1957. *Phys. Rev.* **105**, 388.

——— 1958. *Rev. mod. Phys.* **30**, 565.

WATSON, P.J.S. and SUNDARESAN, M.K., 1974. *Can. J. Phys.* **52**, 207.

WEINBERG, S., 1967. *Phys. Rev. Lett.* **19**, 1264.

——— 1971. *Phys. Rev. Lett.* **27**, 1688.

WEISE, W., 1973*a*. *Phys. Rev. Lett.* **31**, 773.

——— 1973*b*. *Proc. Asilomar Conf. on Photonuclear Reactions and Applications* ed. B.L. Berman (Springfield, Va.: NTIS), p. 95.

——— and HUBER, M.G., 1971. *Nucl. Phys.* **A162**, 330.

WEISSER, D.C., LILLEY, J.S., HOBBIE, R.K., and GREENLEES, G.W., 1970. *Phys. Rev.* **C2**, 544.

WEISSKOPF, V., 1951. *Science* **113**, 101.

WEIZSÄCKER, C.F., 1935. *Z. Phys.* **96**, 431.

WENDLING, R.D. and WALTHER, V.H., 1974. *Nucl. Phys.* A219, 450.

WENG, W.T., KUO, T.T.S, and RATCLIFF, K.F., 1974. *Phys. Lett.* **52B**, 5.

WESOLOWSKI, J.J., SCHWARCZ, E.H., ROOS, P.G., and LUDEMANN, 1968. *Phys. Rev.* **169**, 878.

WEST, D., 1958. *Rep. Prog. Phys.* **21**, 271.

WHEELER, J.A., 1947. *Phys. Rev.* **71**, 320.

WHITEHEAD, C., McMURRAY, W.R., AITKEN, M.J., MIDDLEMAS, N., and COLLIE, C.M., 1958. *Phys. Rev.* **110**, 941.

WHITEN, M.L., SCOTT, A., and SATCHLER, G.R., 1972. *Nucl. Phys.* **A181**, 417.

WHITNEY, R.R., 1971. *Ph.D. Thesis* Stanford University.

WICHMANN, E.H. and KROLL, N.M., 1956. *Phys. Rev.* **101**, 843.

WIEGAND, C.E., 1969. *Phys. Rev. Lett.* **22**, 1235.

——— GALLUP, J.M., and GODFREY, G.L., 1972. *Phys. Rev. Lett.* **28**, 621.

——— and MACK, D.A., 1967. *Phys. Rev. Lett.* **18**, 685.

WILDENTHAL, B.H., PREEDOM, B.M., NEWMAN, E., and CATES, M.R., 1967. *Phys. Rev. Lett.* **19**, 960.

WILDERMUTH, K., 1962. *Nucl. Phys.* **31**, 478.

——— and McCLURE, W., 1966. *Cluster representation of nuclei* (Berlin: Springer-Verlag).

WILETS, L., 1954. *Mat.-Fys. Meddr.* 29, no. 3.

——— 1956. *Phys. Rev.* **101**, 1805.

——— 1958. *Handb. Phys.* **38/1**, 96 (Berlin: Springer-Verlag).

——— 1972. *Gordon Conf. on Nuclear Chemistry* unpublished.

WILETS, L. and CHINN, D. 1963. Unpublished, quoted by Wu and Wilets (1969).

——— and RINKER, G.A., 1975. *Phys. Rev. Lett.* **34**, 339.

WILKIN, C., 1970. *Lett. nuovo Cim.* **4**, 491.

WILKINSON, D.H., 1961. *Proc. Rutherford Jubilee Conf.* ed. J.R. Berks (Manchester: Heywood), p. 339.

——— 1968. *Proc. Symp. on the Use of Nimrod for Nuclear Structure Physics, Rutherford High Energy Lab. Rep. No. RHEL/R166.*

——— 1970. *Phys. Lett.* **31B**, 447.

——— 1971. *Phys. Rev. Lett.* **27**, 1018.

——— 1973. *Nucl. Phys.* **A205**, 363.

——— and HAY, W.D., 1966. *Phys. Lett.* **21**, 80.

——— and MAFETHE, M.E., 1966. *Nucl. Phys.* **85**, 97.

WILLE, U., GROSS, D.H.E., and LIPPERHEIDE, R., 1971. *Phys. Rev.* **C4**, 1070.

——— and LIPPERHEIDE, R., 1972. *Nucl. Phys.* **A189**, 113.

WILLIAMS, R.W., 1955. *Phys. Rev.* **98**, 1387.

WILLIS, A., GEOFFRION, B., MARTY, N., MORLET, M., ROLLAND, C., and TATISCHEFF, B., 1968. *Nucl. Phys.* **A112**, 417.

WILLIS, N., BRISSAUD, I., LE BORNEC, Y., TATISCHEFF, B., and DUHAMEL, G., 1973. *Nucl. Phys.* **A204**, 454.

WINSBORROW, L.A., THOMAS, G.L., COLEMAN, C.F., and CONLON, T.W., 1972. *Nucl. Phys.* **A180**, 19.

WOHLFAHRT, H.D., SCHWENTKER, O., FRICKE, G., and ANDRESEN, H.G., 1973. *Proc. Munich Int. Conf. Nucl. Phys.* eds. J. de Boer and H.J. Mang (Amsterdam: North-Holland), p. 623.

WOLTER, H.H., FAESSLER, A., and SAUER, P.U., 1971. *Nucl. Phys.* **A167**, 108.

WONG, C., ANDERSON, J.D., McCLURE, J., POHL, B., MADSEN, V.A., and SCHMITTROTH, F., 1967. *Phys. Rev.* **160**, 769.

WONG, C.W., 1967. *Nucl. Phys.* **A104**, 417.

——— 1970. *Nucl. Phys.* **A151**, 323.

WONG, C.Y., 1972*a*. *Phys. Lett.* **42B**, 186.

——— 1972*b*. *Phys. Lett.* **41B**, 451.

——— 1973. *Ann. Phys.* **77**, 239.

WOODS, R.D. and SAXON, D.S., 1954. *Phys. Rev.* **95**, 577.

WOODS, T.J., WHITTEN, C.A., and IGO, C.J., 1972. *Nucl. Phys.* **A198**, 542.

WOOLLAM, P.B., GRIFFITHS, R.J., GRACE, J.F., and LEWIS, V.E., 1970. *Nucl. Phys.* **A154**, 513.

WRIGHT, L.E., 1969. *Nucl. Phys.* **A135**, 139.

WU, C.S. and WILETS, L., 1969. *A. Rev. nucl. Sci.* **19**, 527.

WYCECH, S., 1967. *Acta Phys. Polon.* **32**, 161.

——— 1971. *Nucl. Phys.* **B28**, 541.

——— 1973. *Proc. Fifth Int. Conf. on High Energy Physics and Nuclear*

Structure ed. G. Tibell (Amsterdam: North-Holland), p. 239.

YANG, G.C., SINGH, P.P., DRENTJE, A.G., and PAANS, A.M.J., 1973. *Proc. Munich Int. Conf. on Nuclear Physics* eds. J. de Boer and H.J. Mang (Amsterdam: North-Holland), p. 338.

YEBOAH-AMANKWAH, D., GRODZINS, L. and FRANKEL, R.B., 1967. *Phys. Rev. Lett.* **18**, 791.

YENNIE, D.R., RAVENHALL, D.G., and WILSON, R.R., 1954. *Phys. Rev.* **95**, 500.

YNTEMA, J.L. and SATCHLER, G.R., 1967. *Phys. Rev.* **161**, 1137.

YUASA, T. and HOURANY, E., 1967. *Proc. Gatlinburg Nuclear Physics Conf.* (New York, London: Academic Press), p. 72.

ZAIDER, M., ALSTER, J., ASHERY, D., AUERBACH, N., COCHAVI, S., MOINESTER, M.A., WARSZAWSKI, J., AND YAVIN, A.I., 1973. *Proc. 5th Int. Conf. on High Energy Physics and Nuclear Structure* ed. G. Tibell (Amsterdam: North-Holland), p. 219.

ZAIDI, S.A.A. and DARMODJO, S., 1967. *Phys. Rev. Lett.* **19**, 1446.

ZEHNDER, A., 1974. *Diplomarbeit* ETH Zurich.

ZIEGLER, J.F., 1967. *Yale Univ. Rep. No. 2726E-49.*

——— and PETERSON, G.A., 1968. *Phys. Rev.* **165**, 1337.

ZOFKA, J. and RIPKA, G., 1971. *Nucl. Phys.* **A168**, 65.

INDEX